ENVIRONMENTAL
BIOLOGY
(Principles of Ecology)

[For B.Sc., B.Sc. (Hons.) and M.Sc. Students of all Indian Universities]

Dr. P.S. Verma
M.Sc., Ph.D., FESI, FAZ
Former Reader
Department of Zoology
Meerut College, Meerut

V K Agarwal
M.Sc., Ph.D.
Former Head
Department of Zoology
Meerut College, Meerut

S. CHAND
PUBLISHING

S Chand And Company Limited
(ISO 9001 Certified Company)

S.CHAND PUBLISHING

S Chand And Company Limited

(ISO 9001 Certified Company)

Head Office: D-92, Sector–2, Noida – 201301, U.P. (India), Ph. 91-120-4682700

Registered Office: A-27, 2nd Floor, Mohan Co-operative Industrial Estate, New Delhi – 110 044, Phone: 011-49731800

www.**schandpublishing.com**; e-mail: **info@schandpublishing.com**

Marketing Offices:

Chennai	:	Ph: 23632120; chennai@schandpublishing.com
Guwahati	:	Ph: 2738811, 2735640; guwahati@schandpublishing.com
Hyderabad	:	Ph: 40186018; hyderabad@schandpublishing.com
Jalandhar	:	Ph: 4645630; jalandhar@schandpublishing.com
Kolkata	:	Ph: 23357458, 23353914; kolkata@schandpublishing.com
Lucknow	:	Ph: 4003633; lucknow@schandpublishing.com
Mumbai	:	Ph: 25000297; mumbai@schandpublishing.com
Patna	:	Ph: 2260011; patna@schandpublishing.com

First Edition 1983
Second Edition 1985; Reprints 1986, 89, 92, 93, 95, 96, 98, 2000, 2001, 2006, 2007, 2008, 2009, 2010, 2012 (Twice), 2013, 2014, 2015, 2016 (Twice) , 2017, 2018, 2019 Reprint 2020 (Twice), 2021, 2022

Reprint 2023 (Twice)

ISBN: 978-81-219-0859-7 **Product Code:** H6ENB68LFSC10ENAB85O

PRINTED IN INDIA

By Vikas Publishing House Private Limited, Plot 20/4, Site-IV, Industrial Area Sahibabad, Ghaziabad–201010 and Published by S Chand And Company Limited, A-27, 2nd Floor, Mohan Co-operative Industrial Estate, New Delhi – 110 044.

Dedicated to

Dr. S. Z. Qasim and Shri Sunder Lal Bahuguna
for Their Great Contribution in the
Field of Conservation
Ecology.

Authors

PREFACE

In recent years, man, first time in his history, has confronted with certain entirely new kinds of tragic problems like speedily recurring floods and droughts ; grave pollution of air, water and land ; ever shrinking natural landscapes ; fastly vanishing wild life and depleting natural resources ; human population explosion and resultant continuously increasing greed, competition, anxiety, tension, feeling of insecurity, fear, confusion, violence, wars, poverty in people, to name a few. All of these problems have cropped up only due to thoughtless exploitation of various natural resources by man and also due to his absolutely egoist attitude towards nature. Science of ecology has come forward to human rescue to save him from this state of self-destruction, dilemma and paradox. This encyclopaedic biological discipline has provided answers for many biological riddles and in fact, it has opened new vistas in divergent fields of human interest like agriculture, forestry, fisheries, city-planning, sewage and garbage disposal, conservation, medicine, human behaviour, etc.

Timely recognition of the significance of the subject of ecology has resulted in its inclusion in the syllabi of various Indian Universities. Visualising the current ecological awareness, the authors have prepared this text book "**Principles of Ecology**" for the students of B.Sc., B.Sc. (Hons.), and M. Sc. of different Universities in India. This book will serve the purpose of the students of botany, zoology, agriculture, horticulture, forestry, fishery, limnology and oceanography. Authors have attempted to present the subject matter in most simple, lucid but exhaustive way, so that it becomes easily understandable. Effusive Indian examples are cited in the various chapters for the benefit of the readers. Readers may find various significant references at the end of the book for the further pursuance of the subject. Glossary, index and revision questions have been included at appropriate places. The authors hope that this book will serve the purpose of readers more adequately.

The authors express their thanks and gratitudes to various Indian Ecologists for generously supplying the reprints of their research papers. Thanks are also due to Shri Shyam Lal Gupta, "Padma **Shri**" and T. N. Goel of M/s S. Chand and Co. Ltd. New Delhi, for their whole hearted co-operation in the publication of this book.

Suggestions for the **further** improvement of the book shall be highly appreciated and would be incorporated in future editions.

Department of Zoology,
Meerut College,
Meerut.

Authors

CONTENTS

Chapters *Pages*

Part 1—Preliminary of Ecology

1. **Ecology : definition, preview, branches and scope** **3—12**
Origin ; definition ; preview ; objectives , subdivisions ; scope.

2. **Historical background of ecology** **13—22**
Ecology in Indian classics ; ecology in ancient Greek and Roman literature ; ecology from 12th centnry to 17th century , ecology from 18th century to 20th century.

3. **Growth of ecology in India** **23—33**
Growth of plant ecology in India ; growth of animal ecology in India ; growth of oceanography in India ; growth of limnology in India.

4. **Ecological tools and techniques** **34—60**
Sampling techniques of populations ; methods of estimating primary production ; methods of estimating consumer production ; methods of measurements of environmental factors ; mapping the study area ; remote sensing.

5. **Statistical applications in ecology** **61—91**
Definitions ; development ; need of biometry in ecology ; data in biology ; descriptive statistics ; application of Boolean algebra in depiction of life patterns of animals and environmental effects.

6. **Concept of levels of organisation and biosphere** **92—99**
Hierarchy of levels of organisation ; deduction of certain basic principles from the levels of organisation; biological importance of increasing complexity and organisation ; the biosphere ; from biosphere to ecological niche.

7. **The ecosystem** **100—108**
Systems ; structure of an ecosystem ; example of ecosystem ; function of an ecosystem ; types of ecosystem.

Part 2—Environment and Abiotic Components of Ecosystem

8. **Atmosphere and air** **111—121**
Physico-chemical structure of atmosphere ; air as an ecological factor ; air as a medium for living organisms ; wind as an ecological factor.

9. **Lithosphere and soil** **122—154**
The lithosphere ; the soil ; origin of soil (pedogenesis) ; the soil profile ; soil texture ; structure of soil ; soil colour ; physical

properties of soil ; chemical properties of soil ; classification of soils ; major soil types of India ; soil as a habitat for animals ; soil fauna and soil flora.

10. Hydrosphere and water 155—192

Structure of water ; physical properties of water ; chemical properties of water ; hard and fresh-waters ; hydrologic cycle; effect of factor of aquatic environment on organisms ; ecological groups of organisms and ecological adaptations.

11. Abiotic environmental factors 193—247

Nature of response of organisms to abiotic environmental factors ; essential elements and limiting factors : Liebig-Blackman Law of limiting factors, threshold and rate, Shelford's law of tolerance ; temperature : nature of temperature—radiant energy exchange, electromagnetic spectrum, solar radiation and organisms, temperature fluctuations in different environments, range of temperature tolerance, classification of organisms according to temperature tolerance and regulation, factors affecting organismal response to temperature, effect of temperature on plants and animals, thermal adaptations of plants and animals ; light : light variations in different environments, effects of light on plants, effect of light on animals ; pressure ; geomagnetism and gravity ; tide ; fire ; chemical abiotic factors; microclimate : causes of fluctuation of environmental factors in microclimates, topography and microclimate, microclimates of valleys, microclimates of the city.

12. Autecology of species 248—252

Comparison of **autecology** and synecology ; outline for study of ecological life history of the plant ; outline for study of ecological life history of the animal.

Part 3—Biotic Components of Ecosystems

13. Biotic community and ecological niche 255—291

Definition and types ; size and structure of biotic community ; ecotone and edge effect ; ecological niche ; community stability; correlation of environmental gradients and community gradients ; nature of community ; classification of community.

14. Ecological succession and community evolution 292—312

Ecological succession : types, general process of succession, concept of climax, examples of succession, heterotrophic succession, trends of succession, community retrogression, influence on succession ; community evolution.

15. Population 313—328

Population density ; natality of population : fecundity ; mortality of population, life tables ; age distribution of population : age pyramids, sex ratio, biotic potential and environmental resistance ; **growth** form of population ; growth rate of population , population dispersion : emigration, immigration, migration ; **regulation** of population size.

16. **Intraspecific and interspecific interactions** 329—361

Intraspecific interactions : types of association, colonisation, aggregations, social organisation and behaviour, psychological factors ; interspecific interactions : neutralism, commensalism, protocooperation, mutualism, amensalism, parasitism, predation, competition.

17. **Ecological genetics** 362—370

Population genetics and natural selection ; adaptations and ecotypes ; speciation ; ecological indicators.

Part 4—Dynamics of Ecosystem

18. **Energy flow in ecosystems** 373—385

Energy in an ecosystem ; patterns of flow of energy through the ecosystems : grazing food chain, detritus food chain ; ecological pyramids ; ecological efficiencies.

19. **Biogeochemical cycles** 386—403

Organic and abiotic phases of geochemical cycles ; types of biogeochemical cycles ; water cycle, gaseous cycles—the carbon cycle, the nitrogen cycle ; sedimentary cycles—sulphur cycle, phosphorus cycle.

Part 5—Diversity of Ecosystems

20. **Aquatic ecosystems : fresh-water communities** 407—426

Subdivisions of aquatic ecosystems ; fresh-water ecosystems ; physico-chemical nature of fresh-water ; lentic ecosystems : lakes and ponds, physico-chemical properties of lakes and ponds, kinds of lakes, origin of lakes and ponds ; biotic communities of lakes and ponds ; lotic ecosystems : characteristic of lotic environment, rapidly flowing water, slowly flowing water.

21. **Aquatic ecosystems : estuaries and marine communities** 427—449

Estuaries : physico-chemical aspects of estuaries, biotic communities of estuaries ; marine ecosystems : physico-chemical aspects of marine environment ; marine communities ; biotic communities of oceanic region, biotic communities of continental shelf ; coral reef as a specialized oceanic ecosystem, biotic communities of coral reef.

22. **Terrestrial ecosystems** 450—464

Physico-chemical nature of terrestrial ecosystems and their comparison with aquatic ecosystems ; biogeographic realms ; biomes : tundra biome, high altitude or the alpine biome, forest biomes, tropical savanna biomes, grassland biomes, desert biome.

23. **Biogeography** 465—482

Patterns of distribution of biota : continuous distribution, discontinuous distribution, endemic ; descriptive phytogeography : major plant communities (biomes) of the world, chief vegetational belts of the earth, phytogeography of India, vegetation of India : moist-tropical forests, dry tropical forests, montane sub-tropical forests, montane temperate forests, alpine forests ; descriptive zoogeography : zoogeographical regions, palaerctic region, nearctic region, neotropical region, ethiopian region, oriental region, Australian region ; life zones, biotic provinces and biomes ; dynamic biogeography : dispersal dynamics, dispersal pathways.

Part 6—Applied Ecology

24. **Ecology of human population** 485—492

Demography ; history of human population growth : primitive man, agricultural man, industrial man ; factors regulating human population.

25. **Pollution** 493—517

Origin of pollution ; pollutants : the creators of pollution ; types of pollution ; air pollution : kinds and sources of air pollutants, methods of detection and measurement of air pollution, ecology of air pollution, control of air pollution ; water pollution : types of water pollution, kinds and sources of water pollutants, ecology of water pollution—sewage pollution, industrial pollution, thermal pollution, silt pollution, estuarine and oceanic pollution, control of water pollution ; land pollution ; pesticide and herbicide contamination ; radioactive pollution ; noise pollution, effects of noise pollution.

26. **Ecology and human welfare (Conservation of natural resources)** 518—537

Conservation of natural resources ; minerals and their conservation : terrestrial mineral resources, marine mineral resources, conservation of terrestrial mineral resources, ecological aspects of mining ; energy and its conservation ; food, agriculture and aquaculture : shifting cultivation, sedentary cultivation, new sources of food ; forestry ; range management (grassland management) ; wild-life management ; water conservation ; space utilization on earth ; land use ; soil conservation,

27. **Wild life management in India** 538—550

Wild life of India ; reasons for depeletion of wild life ; necessity for wild life conservation ; modes of wild life conservation : protection by law, protected species of Indian wild life, establishment of sanctuaries and national parks, other conservation measures.

Glossary 551—561

Bibliography 562—580

Index 581—591

Part I
PRELIMINARY OF ECOLOGY

As the discovery of genetic properties of DNA in the field of molecular genetics has aroused immense curiousity among the biologists, academicians, and common man regarding the mode of inheritance in living organisms, similarly, the recent detection of environmental pollution in the field of ecology, has attracted the attention of the persons of every walk of life. In the last two decades, most industrially and technologically advanced countries of west and east alike, felt an urgent need to combat against the environmental pollution with their all might and force, and during their war-like preparations against this 'invisible blood-thirsty ghost' which seemingly had threatened the very existence of human-life on the planet earth, much has been studied and investigated about the environment, its relation with man and other organisms, its limitations, its exploitation in early years for short-sighted gains, and consequent problems such as exhaustion of natural resources, extermination of wild-life, soil erosion, increased epidemics and floods, population explosion and decreased terrestrial space for human dwellings, etc. In consequence, within a few years, ecology progressed from a rather quiet and obscure branch of biology to a subject of national and international concern. . Education, business, politics, law, agriculture, engineering, medicine, public health and even international affairs are all affected by the sudden upsurge of ecological and environmental concern. Ecology has occupied the consciousness of modern man in such a terrible manner that it has become a most common word of his day-to-day vocabulary and much is spoken about it in every sphere of human life, be a news paper, magzine, radio, TV, cinema, theatre, fiction, novel, poetry, painting, group discussion, scientific gathering, or a political lecture.

1

Ecology : Definition, Preview, Branches and Scope

1·1. INTRODUCTION

Life does not occur in a vacuum. Every living organism is surrounded by materials and forces which constitutes its environment and from which it must derive its needs. Thus, for its survival, a plant, an animal, or a microbe cannot live completely sealed in an impervious skin or shell but requires from its environment—a supply of energy, a supply of materials, and a removal of waste products. For these basic requirements each living organism has to depend and also to interact with different non-living or abiotic and living or biotic components of the environment. The abiotic environmental components include basic inorganic elements and compounds such as water and carbon dioxide, calcium and oxygen, carbonates and phosphates and an array of organic compounds, the by-products of organism's activity or death. They also include such physical factors as soil, rainfall, temperature, moisture, winds, currents, and solar radiation with its concomitants of light and heat. The biotic environmental factors comprise plants, animals, and microbes, all of which interact in a fundamentally energy-dependent fashion. The scientific study of the interactions of organisms with their physical environment and with each other, is called **ecology** (**Helena Curtis, 1975**). It mainly concerns with the directive influences of abiotic and biotic environmental factors over the growth, distribution, behaviour and survival of organisms (**Herreid II,** 1977).

1·2. ORIGIN OF THE TERM ECOLOGY

The first author of the term 'ecology' is uncertain, however, many biologists grant credit to the German zoologist **Ernst Haeckel,** who used the term as *oekologie* in 1866 to refer to the inter-relationships of living organisms and their environment (see **Southwick,** 1976). In fact, a few years earlier, in 1857, **Henry David Thoreau** had also spoken of ecology among other fields of biology and natural history, but he apparently did not provide a definition of the term (see **Kormondy,** 1969).

Paradoxically, seven years earlier to **Haeckel,** two more terms were coined for ecology—the **ethology** by the French zoologist

I.G.S. Hilaire and **hexicology** by English naturalist **S.G.J. Mivart. Hilaire** defined ethology as "the study of the relations of the organisms within the family and society in the aggregate and in the community." **Mivart** defined hexicology as the study of the relations which exist between the organisms and their environment as regards the nature of the locality they frequent, the temperatures and amounts of light which suit them, and their relations to other organisms as enemies, rivals or accidental and involuntary benefactors." However, both terms remained quite vague and unpopular than the term ecology.

The word ecology comes from two Greek words '*oikos*', meaning "household" or "home" or "place to live" or "habitation", and '*logos*' meaning "discourse" or "study". Thus literally ecology deals with the organism and its place to live. Basically, organism's place to live is its environment, so ecology is also called **environmental biology (Smith**, 1977).

1·3. DEFINITION OF THE TERM ECOLOGY

Conventionally, ecology has been defined variously by different classical and modern ecologists with different viewpoints. Quite oddly, no universally accepted definition of ecology has been formulated by any ecologist so far, and for the better understanding of scope, limitations, purpose, and mode of study of different ecological phenomena, one has to go through following chronologically arranged definitions of ecology :

Ernst Haeckel (1866) defined ecology "*as the body of knowledge concerning the economy of nature—the investigation of the total relations of animal to its inorganic and organic environment.*" An American plant ecologist, **Frederick Clements** (1916) considered ecology to be "*the science of community.*" British ecologist **Charles Elton** (1927) defined ecology as "*the scientific natural history concerned with the sociology and economics of animals.*" **Taylor** (1936) preferred to define ecology as "*the science of the relations of all organisms to all their environments.*" A much broader definition of ecology of **Allee** *et al.*, (1949), considered ecology as "*the science of inter-relation between living organisms and their environment, including both the physical and biotic environments, and emphasizing inter-species as well as intra-species relations.*" **G. L. Clarke** (1954) defined ecology as "*the study of inter-relations of plants and animals with their environment which may include the influences of other plants and animals present as well as those of the physical features.*" **Woodbury** (1955) regarded ecology as "*the science which investigates organisms in relation to their environment : a philosophy in which the world of life is interpreted in terms of natural processes.*" British ecologist, **A. Macfadyen** (1957) defined ecology as "*a science which concerns itself with the inter-relationships of living organisms, plants and animals, and their environments.*" Likewise, **S. C. Kendeigh** (1961, 1974) defined ecology as "*the study of animals and plants in their relation to each other and to their environment.*"

Further, **Andrewartha** (1961) defined ecology as *"the scientific study of the distribution, and abundance of organisms."* **T. Lewis** and **L. R. Taylor** (1967) have defined ecology as *"the study of the way in which individual organisms, populations of some species and communities of populations respond to these changes."* **G. A. Petrides** (1968) defined ecology as *"the study of environmental interactions which control the welfare of living things, regulating their* **distribution**, *abundance, production and evolution."* American ecologist **Eugene Odum** (1971) preferred to define ecology as *"the study of the structure and function of ecosystems or broadly of nature."* **C. J. Krebs** (1972) considered Odum's definition of ecology as quite ambiguous and instead of it, he provided a modified version of Andrewartha's definition of ecology, which defined ecology as *"the scientific study of interactions that determine the distribution and abundance of organisms."*

Recently, certain modern ecologists have provided somewhat more broader definitions of ecology. Thus, **M. E. Clark** (1973) considers ecology as *"a study of ecosystems or the totality of the reciprocal interactions between living organisms and their physical surroundings."* **Pinaka** (1973) defined ecology as *"the study of relations between organisms and the totality of the biological and physical factors affecting them or influenced by them."* **Southwick** (1976) defined ecology as *"the scientific study of the relationships of living organisms with each other and with their environments."* He further elaborates his definition of ecology by saying that *"it is the science of biological interactions among individuals, populations, and communities ; and it is also the science of ecosystems—the inter-relations of biotic communities with their non-living environments."* **R. L. Smith** (1977), prefers to consider ecology as *"a multidisciplinary science which deals with the organism and its place to live and which focuses on the ecosystem."*

1·4. A PREVIEW OF ECOLOGY

Ecology is a vast and encyclopaedic biological subject. To a beginner, at the outset, ecology may appear to be quite confusing and discouraging because it seems so diffuse and incoherent. After all, it must take into account the life habits of over two million different kinds of animals and plants and it considers all manner of influence and interactions among them. Fortunately, however, the innumerable facts of ecology can be distilled into following some relatively basic and simple principles :

The ecology is the study of ecosystems or the totality of the reciprocal interactions between living organisms and their environment. The term **organism** refers to an individual unit constituted to carry on the activities of life. It is a dynamic biological unit which is greatly influenced by an enveloping and fluctuating environment. For a given organism, the **environment** includes all the surrounding physical and biological factors, with which it interacts. The **factor** is any external force, substance, or condition that affects organisms in any way. Thus, environment is the sum total of everything that directly influences the animal's chances of survival and reproduction

(**Maelzer**, 1965). The intimately local and immediate surrounding of the organism is called **microenvironment**, while, the **macroenvironment** is the sum total of the physical and the biotic conditions existing external to the organism and its microenvironment.

The total life-containing and life-supporting environment of the world is restricted to a very thin and irregular veil or film around the globe. This thin veil of living material of earth is called **ecosphere** or **biosphere.** Thus, the biosphere is that part of the earth in which life exists (**Hutchinson**, 1970). The biosphere too is not entirely hospitable to life, in one region called **parabiosphere,** severe environmental conditions permit inhabitation by some organisms only in the resistant stages of their life-cycles. The parabiosphere includes such broad areas as the higher altitudes, the polar regions, the deepest ocean troughs, the most extreme deserts ; and such localized regions as volcanoes, geysers, and grossly-polluted areas of land and water. The remaining portion of biosphere is called **eubiosphere,** in which active metabolic processes of organisms are possible. Besides the biosphere, the three other main compartments of the world are **air (atmosphere)**, **water (hydrosphere)**, and **earth** or **land (lithosphere).** The entire world can be considered as a single vast ecosystem of the universe as its biotic or living part is build by the biosphere (organisms) and its abiotic or non-living portion is formed by atmosphere, hydrosphere, and lithosphere.

The ecosystem is any spatial or organizational unit which includes living organisms and non-living substances interacting to produce an exchange of materials between the living and non-living parts. The ecosystem can be studied from either structural and functional aspects. The structural aspects include a description of the arrangement, types and numbers of species and their life histories, along with a description of the physical features of the environment. The functional aspects of the ecosystem include the flow of energy and the cycling of nutrients. The non-living part of the ecosystem includes different kinds of habitats such as air, water and land, and a variety of abiotic factors. **Habitat** is the natural abode or locality of an animal, plant or person. It includes all features of the environment in a given locality. For instance, water is used as habitat by aquatic organisms and it comprises three major categories—**marine, brackish** and **freshwater habitats.** Each of these may be subdivided into smaller units, such as freshwater habitat may exist as a large lake, a pond, a puddle, a river or a stream. Similarly, the land is used as a habitat for numerous terrestrial organisms. It includes many major categories of landmasses which are called **biomes**. Biomes are distinct large areas of earth with relatively homogeneous climatic factors and flora and fauna, *e.g.*, deserts, prairie, tropical forest, etc. Soil is also used as a habitat by a variety of microbes, plants and animals.

Among the main abiotic factors of the ecosystem are such **climatic factors** as solar radiation, temperature, wind, water currents, rainfall, and such **physical factors** as light, fire, pressure, geomagnetism, and such **chemical factors** as acidity, salinity and the availability of inorganic nutrients needed by plants. The biological (biotic) factors of ecosystem include all the living organisms—plants, animals, bacteria and viruses. Each kind of living organism found in an ecosystem is called a **species.** A species includes individuals which are genetically alike and which are capable of freely inter-breeding and producing fertile offsprings. In an ecosystem, there exist various relationships between species. Two species may have a negative effect upon one another (**competition**), a neutral effect (**neutralism**) or a beneficial effect (**protocooperation** and **mutualism**). Other relationships occur in which one species benefits and the other is harmed (**predator-prey, herbivore-plant, parasite-host**), or where one benefits but does not affect the other (**commensalism**), or where one species is harmed by any other species that derives no benefit (**amensalism**). Likewise, species may aggregate, or separate, or show a random relationship to one another.

The distribution and growth of plants and animals in an ecosystem are controlled by both abiotic and biotic features of the environment. Any factor acting to alter the growth and survival of a population is called a **limiting factor.** A **population** is a group of **inter-acting individuals,** usually of the same species, in a definable **space.** Thus, one can speak of population of deer on an island, the population of locusts in a crop-field, and the population of fishes in a pond. The size of a population of any given species is determined by a balance between its **reproductive potential** and the **environmental resistance.** Thus, population size is determined by the relative number of organisms added to or removed from the group. Recruitment into the population is a function of **birth rate** and **immigration rate.** Loss from the population is a function of **death rate** and **emigration.** Population regulation is achieved by following factors—1. physical attributes of the environment (*e.g.*, climate), 2. food (quantity and quality), 3. disease (host-parasite relationships), 4. predation, and 5. competition (inter-specific and intra-specific).

An ecosystem usually contains numerous populations of different species of plants, animals and microbes, all interacting with one another as a community and with the physical environment as well. A **community** or **biotic community**, thus, consists of the populations of plants and animals living together in a given place. For example, one refers to the community of an oak forest, a grassland, a coral reef, a desert or a pond. Further, each organism in a community has a distinct **ecological niche**; different species tend to live in different habitats, eat different foods and live with different life-styles. The ecological niche, thus, is a more inclusive term that includes not only the physical space occupied by an organism, but also its functional role in the community (*e.g.*, its trophic position) and its position in environmental gradients of temperature, moisture, pH, soil, and other conditions of existence. Further, all biological communities exhibit some form of layering or **stratification**, which largely reflects the life forms of the plants and which influences the nature and distribution of animal life in the community.

From the energetic viewpoint, the ecosystem may be divided into three types of organisms : **producers, consumers**, and **reducers.** Photosynthetic algae, plants and bacteria are the producers of the ecosystem ; all other organisms depend upon them directly or indirectly for food. Consumers are **herbivorous, carnivorous,** and **omnivorous** animals ; they eat the organic matter produced by other organisms. Reducers are heterotrophic organisms like animals ; they are fungi and bacteria that decompose dead organic matter. Thus the ecosystem is composed of organisms eaten by, or eating other organisms. Species are related by their feeding behaviour in **food chains** or **food webs.** There are two basic types of food chains : the consumer food chain includes the sequence of energy flow from producer → herbivore → carnivore → reducer ; the detritus food chain bypasses the consumers, going from producer → reducer.

Further, it is found that approximately 30 natural elements are essential to organisms ; these are cycled through the abiotic compartments of world, *i.e.*, atmosphere, hydrosphere, and lithosphere as well as the producers, consumers, and decomposers of the biological community (biosphere). Organisms take in inorganic nutrients (*e.g.*, carbon, oxygen, hydrogen, sodium, nitrogen, phosphorus, calcium, sodium, sulphur, potassium, chlorine, etc.) from the soil, water, and air ; they eliminate their wastes, and their bodies are rendered into inorganic molecules once again. Thus, water and minerals shuttle among the air, land, and water with an occasional pause in an organism. Such cycling of nutrients in circular paths in between biotic and abiotic components of the ecosystem are called **biogeochemical cycles.**

Moreover, changes occur within ecosystems. The most dramatic are those associated with **succession.** Ecological succession is a sequential change of organisms as an area progresses from a condition in which it is relatively poor in species composition and organic content, until it becomes a diverse and rich biological community. **Primary succession** begins on rock, sand, or mud where little or no organic material has been previously present. **Secondary succession** begins where organic matter has been present, such as a burned out forest region, or a bulldozered landscape. The final stage in succession is called **climax.** It is predictable for a given area and depends upon rainfall, temperature and soil conditions. Even within stable ecosystems, species and their activities vary in time and space. Recurrent **daily** and **seasonal rhythms** exist in the lives of all organisms. These are driven by environmental changes or by internal **"biological clocks".** A biological clock is a hypothetical internal mechanism by which an organism can keep track of time and govern its activities.

1·5. OBJECTIVES OF ECOLOGY

Ecology is a distinct science because it is a body of knowledge not similarly organized in any other division of biology, because it uses a special set of techniques and procedures ; and because it has a unique point of view (see **Kendeigh,** 1974). The essence of ecology lies in the comprehensive understanding of the meaning of following phenomena :

1. The local and geographical distribution and abundance of organisms (habitat, niche, community, biogeography).

2. Temporal changes in the occurrence, abundance and activities of organisms (seasonal, annual, successional, geological).

3. The inter-relations between organisms in populations and communities (population ecology).

4. The structural adaptations and functional adjustments of organisms to their physical environment (physiological ecology).

5. The behaviour of organisms under natural conditions (ethology).

6. The evolutionary development of all these inter-relations (evolutionary ecology).

7. The biological productivity of nature and how this may best serve mankind (ecosystem ecology).

8. The development of mathematical models to relate interaction of parameters and predict effects (systems analysis).

9. The conservation and management of natural resources and pollution (applied ecology).

1·6. SUBDIVISIONS OF ECOLOGY

Early ecologists have recognized two major subdivisions of ecology in particular reference to animals or to plants, hence **animal ecology** and **plant ecology.** But when it was found that in the ecosystems plants and animals are very closely associated and interrelated, then, both of these major ecological subdivisions became vague. However, when animals and plants are given equal emphasis, the term **bioecology** is used.

Further, ecology is often broadly divided into **autecology** and **synecology.** Autecology deals with the ecological study of one species of organism. Thus, an **autecologist** may study the life history, population dynamics, behaviour, home range and so on, of a single species, such as the Mexican free-tailed bat, Indian bull frog, or maize-borer, *Chilo partellus*. Synecology deals with the ecological studies of communities or entire ecosystems. Thus, a **synecologist** might study deserts, or caves or tropical forests. He is interested in describing the overall energy and material flow through the system rather than in concentrating on the finer details of a particular organism. In the words of **Herreid** II (1977) "the two types of study, autecology and synecology, inter-relate, the synecologist painting with a broad brush the outline of the picture and autecologist stroking in the finer details."

Besides these major ecological subdivisions, there are following specialized branches of ecology :

1. Habitat ecology. It deals with ecological study of different habitats on planet earth and their effects on the organisms living there. According to the kind of habitat, ecology is subdivided into **marine ecology (oceanography), estuarine ecology, fresh water ecology (limnology)**, and **terrestrial ecology**. The terrestrial ecology in its turn is classified into **forest ecology, cropland ecology, grassland ecology, desert ecology**, etc., according to the kinds of study of its different habitats.

2. Community ecology. It deals with the study of the local distribution of animals in various habitats, the recognition and composition of community units, and succession.

3. Population ecology (Demecology). It deals with the study of the manner of growth, structure and regulation of population of organisms.

4. Evolutionary ecology. It deals with the problems of niche segregation and speciation.

5. Taxonomic ecology. It is concerned with the ecology of different taxonomic groups of living organisms and eventually includes following divisions of ecology : **microbial ecology, mammalian ecology, avian ecology, insect ecology, parasitology, human ecology** and so on.

6. Human ecology. It involves population ecology of man and man's relation to the environment, especially man's effects on the biosphere and the implication of these effects for man.

7. Applied ecology. It deals with the application of ecological concepts to human needs and thus, it includes following applications of ecology : **wild-life management, range management, forestry, conservation, insect control, epidemiology, animal husbandry, aquaculture, agriculture, horticulture, land use** and **pollution ecology**.

8. Ecosystem dynamics. It deals with the ecological study of the processes of soil formation, nutrient cycling, energy flow, and productivity.

9. Production ecology. It deals with the gross and net production of different ecosystems like fresh water, sea water, agriculture, horticulture, etc., and tries to do proper management of these ecosystems so that maximum yield can be get from them.

10. Ecological energetics. It deals with energy conservation and its flow in the organisms within the ecosystem. In it thermodynamics has its significant contribution.

11. Physiological ecology (ecophysiology). The factors of environment have a direct bearing on the functional aspects of organisms. The ecophysiology deals with the survival of populations as a result of functional adjustments of organisms with different ecological conditions.

12. Chemical ecology. It concerns with the adaptations of animals of preferences of particular organisms like insects to particular chemical substances.

13. Ecological genetics (genecology). An ecologist recognised a kind of genetic plasticity in the case of every organism. In any environment only those organisms that are favoured by the environment can survive. Thus, genecology deals with the study of variations of species based upon their genetic potentialities.

14. Palaeoecology. It is the study of environmental conditions, and life of the past ages, to which palynology, palaeontology, and radioactive dating methods have made significant contribution.

15. Geographic ecology (ecogeography). It concentrates on the study of geographical distribution of animals **(zoogeography)** and plants **(phytogeography)**, and also of palaeoecology and biomes.

16. Space ecology. It is a modern subdivision of ecology which is concerned with the development of partially or completely regenerating ecosystems for supporting life of man during long space flights or during extended exploration of extra-terrestrial environments.

17. Pedology. It is a branch of terrestrial ecology and it deals with the study of soils, in particular their acidity, alkalinity, humus contents, mineral contents, soil-types, etc., and their influence on the organisms.

18. Radiation ecology. It deals with the study of gross effects of radiations and radioactive substances over the environment and living organisms.

19. Ethology. It is the interpretation of animal behaviour under natural conditions. In it, often, detailed life history studies of particular species are amassed.

20. Sociology. It is the study of ecology and ethology of mankind.

21. Systems ecology. It is the modern branch of ecology which is particularly concerned with the analysis and understanding of the function and structure of ecosystem by the use of applied mathematics, such as advanced statistical techniques, mathematical models, characteristics of computer sciences.

1·7. SCOPE OF ECOLOGY

Ecology is a multidisciplinary science and it includes not only the life sciences but chemistry, physics, geology, geography, meteorology, climatology, hydrology, palaeontology, archeology, anthropology, sociology and mathematics and statistics as well. For explaining the behaviour of an organism or biotic community in a given environment, an ecologist has to integrate the data which is obtained from many sources—morphology, taxonomy, genetics, physiology, soil science, climatology, geology, physics and chemistry.

The scope of ecology is quite vast. The study of ecological principles provides a background for understanding the fundamental relationships of the natural community and also the sciences dealing with particular environment such as forest, soil, ocean and inland waters. Many practical applications of this subject are found in agriculture, **horticulture**, forestery, limnology, oceanography, **fishery** biology, biological survey, game management, pest control, **public** health, toxicology, pollution control, conservation, etc. Ecological knowledge helps in discovering new sources of food (algae, krill, etc.), new unpolluting sources of energy (*i.e.*, solar energy), and new methods of pest control such as biological control which causes no environmental pollution. By applying certain ecological techniques, some ecologists have been successful in diagnosing the cause of desertness of certain Australian deserts and they investigated that these deserts lack certain trace elements like zinc, copper and molybdenum, so they called them **trace element deserts** (**Anderson** and **Underwood**, 1959). Now they have cured their ecological diseases and have converted them into new rich agricultural lands.

1·8. REVISION QUESTIONS

1. Trace the origin of the term 'Ecology', in history ?
2. How will you define ecology ?
3. What is the scope and objectives of ecology ?
4. Describe different subdivisions of ecology.
5. Describe, why ecology is significant for human being ?

2

Historical Background of Ecology

2·1. INTRODUCTION

The early growth of ecological knowledge is intimately related with evolutionary and cultural development of man. Early man (*i.e.*, preagricultural, hunter and food-gatherer man) was a successful practicing ecologist. He survived in a rich and competitive biotic community, and his relationship to this community was continually intimate. In comparison to other members of his biotic community, he had several advantages. For instance, he had a close effective social organization, he had an unusual manipulative ability and he had, of course, an emerging intelligence. He developed tools and fire in early paleolithic time (50,000 to 1,000,000 years ago) and he accumulated knowledge at a faster rate than other primates. Much of this knowledge was ecological in nature. It was the knowledge of his environment and the most effective use of it. It was detailed knowledge of food and water resources. Thus, the prehistoric man used environmental information in hunting, fishing, traping animals, locating and gathering of edible vegetation, avoiding enemies and finding shelter to survive hardship imposed by nature. An increased knowledge of the importance of environmental conditions led quite naturally to religious rituals, myths, worship of weather gods and rain dances, all of which are still quite prevalent in different social groups and tribes of Indian, African, Australian, Red Indian, and other ancient civilizations.

The establishment of agricultural civilization increased the need to learn about the practical ecology of domestication of plants and animals for greater productivity and control over the means of subsistence. Consequently, the development of pastoral and agricultural life in the Neolithic period of 10,000 B.C. to 6,000 B.C. altered the entire pattern of human existence.

2·2. ECOLOGY IN INDIAN CLASSICS

Early Indian philosophers and thinkers had some vague knowledge of ecological nature, as have been revealed by the Indian classical writings, in the Vedic, Brahamanic, Upanishadic, Pauranic, and Epic literature. **Charaka** considered the important factors of *Vayu* (air and gases), *Jala* (water), *Desha* (topography), and *Kal* (time), in regulating the life of plants (see **R. Misra**, 1962). Our two great

Epics—*Ramayana* (**Valmeki**) and *Mahabharata* (**Vyas**) have displayed their ecological awareness at several occasions. The classical poet **Kalidasa** has exhibited his ecological faculty in his classics such as '*Ritu Sanhar*', '*Meghdoot*' and '*Abhigyan Shakuntlam*'. Similarly, one may come across ecological instances in the Sanskrit classics of many other poets and thinkers such as **Baa Bhutt's** *Kadumbry*, **Sudrak's** *Mrichkaticum* and **Kautilya's** *Arthashastra*.

2·3. ECOLOGY IN ANCIENT GREEK AND ROMAN LITERATURE

The beginnings of ecological knowledge for modern man lay in the development of biology and medicine as scientific disciplines. The fourth century Greek, **Hippocrates** (460–377 ? B.C.), often called the "Father of Medicine", emphasised certain ecological factors in the genesis of health and disease. His work, "On *Airs, Waters and Places*" was environmental in emphasis—"*whoever wishes to investigate medicine properly, should proceed thus : in the first place to consider the seasons of year, and what effects each af them produces......... Then the winds, the hot and cold, ...the qualities of waters, for as they differ from one another in taste and weight.......*"

Another Greek philosopher **Aristotle** (384–322 B.C.), sometimes referred to as the "Father of Biology", classified animals according to their habits and habitats, *e.g.*, gregarious vs. solitary, carnivorous vs. herbivorous, resident vs. migratory, and so forth. Although he was a good naturalist he could not be considered an ecologist, for he did not really consider the inter-relations of plants, animals, and their environments other than in the most superficial terms. **Theophrastus** (372–287 ? B.C.), a student of Aristotle was regarded by some scholars such as **Ramaley** (1940) as the 'first ecologist', though the word ecology was not used until much later. **Theophrastus** discussed plants and their environments in a moderately systematic way. He studied plant types and forms in relation to attitude, moisture, and light exposure.

About the time of **Christ**, the Hebrews had some ecological notions, but of a very general and rather obvious sort. An example is the Parable of Sower (**Luke**, 8 : 4–8) which illustrates the dependence of seed germination on the soil.

Pliny the Elder (23–79 A.D.) was a Roman naturalist with a system of classification somewhat like that of **Aristotle** ; that is, he classified plants and animals according to their habits and habitats. He was a great compiler and cataloguer of facts, but most of this was basic natural history and could not be considered truly ecological in orientation.

2·4. ECOLOGY FROM 12th CENTURY TO 17th CENTURY

After a considerable long gap of many centuries, in 12th century certain naturalists began their scientific studies in natural history which accelerated rapidly in the 16th century to flourish as a scholarly endeavour. The description and catalouging of animals became a favourite scientific activity. **Albertus Magnus** (1193–1280) was a leading naturalist of his time who wrote on plants and their environment. In the 16th century, **Konrad Gesner** (1516–1565), **Ulisse Aldrovandi** (1522–1605), **Pierre Belon** (1517–1565), **Cordus** (1515–1544) and **Andrea Caesalpino** (1519–1603) all started to relate biology and natural history to geography and chemistry.

In 17th century, **Robert Boyle** (1627–1691), sometimes called the first modern chemist, related the origin of chemistry to vital processes in plants and animals. He did experiments on the effects of law of air pressure on a variety of animals from frogs to chickens and mice. In one series of experiments he showed the ability of ducks to withstand a shortage of air longer than chickens, and he related this to their aquatic environment and habit of feeding under water. Other important naturalists of 17th century were **Gaspard Bauhin** (1550–1624), **Graunt** (1664), **Rudolph Camerarius** (1665–1721) and **John Ray** (1627–1705).

2·5. ECOLOGY FROM 18th CENTURY TO 20th CENTURY

From the late seventeenth century through the nineteenth century, following five major fields of study laid the foundation for the development of ecology as a distinct scientific discipline in the twentieth century : (1) natural history and faunal exploration, (2) environmental physiology and the study of responses to environmental changes, (3) evolution and theories of natural selection, (4) population studies and demography, and (5) ecological geography and conservation (**Southwick**, 1976).

2·5·1. Natural History and Faunal Exploration

Natural history has been a primary developmental source of ecology since the time of Aristotle. Some of the important naturalists of late 17th, 18th and 19th centuries who contributed significantly for early growth of modern ecology are following :

Rene Reaumur (1683–1757), a noted French naturalist and physicist wrote a six-volume work entitled *Memoires pour servir a l'histoire des insectes*, in which he discussed the life habits and community patterns of social insects. He also investigated the natural history of crustaceans and molluscs. The first great modern systematist, **Carolus Linnaeus** (1707–1778) a Swedish professor, physician and naturalist, in his classical work, entitled **Systemmae Naturae**, established the binomial system of nomenclature of organisms, furthered the sciences of plant and animal geography, and studied plant associations in different regions. **Georges Louis Leclerc Buffon** (1707–1788), an eminent French naturalist, wrote the forty-four volume *Histoire naturelle, generale et particuliere*, a vast compendium of most of the natural history knowledge of the day. He considered that animals and plants developed adaptations which enabled them to favourably respond to their environment and termed this **environmental induction.** He regarded climate as a variable compelling modification of plants and animals. Buffon explained how the great fertility of every species was counterbalanced by innumerable agents of destruction. In some ways his writings anticipated the evolutionary ideas of Darwin, but were not fully expressed or developed. He definitely opposed Linnaeus' ideas of the fixity of species, and he amassed considerable evidence demonstrating variation and mutability in living organisms.

Gilbert White (1720–1793), an English naturalist, described the natural history of Southern England in his book "*The Natural History of Selbourne.*" An American naturalist **James J. Audubon** (1785–1851) compiled the volume of "*Birds of America.*" **Louis Agassiz** (1807–1873), the Swiss American naturalist, compiled four volumes of *Bibliographia Zoologica*, along with other books on fossil fishes, glacial activity, and naturalistic expeditions to Brazil. He initiated the researches in marine ecology (oceanography) by establishing many laboratories for marine ecological researches. German naturalist **Alexander Von Humboldt** (1769–1859), after his extensive travels of tropical and temperate South America, in 1807 discussed the Geographical distribution of plants and animals in relation to climate.

Charles Darwin (1809–1882), a great British naturalist and evolutionist, wrote more than a dozen outstanding books and monographs on a variety of biological topics from 1842, when he published '*Coral reefs*', to 1881, when he published '*Action of earthworms on the processes of soil formation*'. His most famous works, of course, were the '*Origin of Species*' (1859), and the '*Descent of Man*' (1871), in which he presented his great interpretive theory of natural selection and evolution. From ecological point of view, he increased our knowledge of island or insular life with their characteristically different types of environmental conditions. **Edward Forbes** (1815–1854), an English marine biologist, studied the dynamic relations existing between marine organisms and their environment. **Alfred Russell Wallace** (1823–1913), a leading British naturalist of nineteenth century, wrote '*Island life*' and several other leading books on the distribution of plants and animals in the Malay Archipelago. **Henri Fabre** (1823–1915), a French naturalist, published ten volumes on the '*Life of Insects.*' **H. W. Bates** (1825–1892), an English naturalist, wrote '*Naturalist on the Amazon*' in 1863 and **A. E. Brehm** (1829–1884), a German naturalist, published his thirteen volume compilation '*Tierleben*' (Animal life) in 1911–1918. In 1838, a German botanist, **August Grisebach,** studied the plants and animals in the form of integrated communities. **Stephen A. Forbes** (1844–1930), an American ecologist, published **a series** of papers on the concept of ecological communities. In one of his classical paper entitled, '*The Lake as a Microcosm*'. Forbes suggested that the species assemblage in a lake was an organic complex and that by affecting one species one exerted some influence on the whole assemblage. He concluded that each species **maintained** a 'community of interest' with the other species and one should not limit his studies to a single species. **C. Hart Merriam** (1855–1942) devised a classification of major life zones–a system of habitat classification, still in wide use. **Mobius** (1877), a German naturalist, investigated oyster banks and introduced the term "**biocoenosis**" to designate a group of organisms as an ecological unit. **J.E.B. Warming**, a Danish botanist, emphasized the unity of plant communities in 1895.

Animal ecologist **C.C. Adams** (1873–1955) in 1906 and 1909 recognized and described many animal communities in his ecological surveys of northern Michigan **and** Isle Royale in Lake superior. In 1913, an American animal ecologist **V.E. Shelford** (1877–1968) presented a classical study of animal communities in temperate America. In 1927, **Charles Elton** published an outstanding analysis of community dynamics. **F.E. Clements** and **V.E. Shelford** (1939) emphasized the significance of early investigation of **J.G. Cooper** (1859) that whole community is one biotic unit, rather than one unit of plants and another of animals. In recent years many workers such as **D. Ramon Margalef**, **Robert MacArthur**, and others have analyzed the structure of these biotic communities in respect to such **phenomena** as species diversity, niches and how they came **about through** evolution (**Kendeigh**, 1974).

2-5·2. Environmental Physiology and the Study of Responses to Environmental Changes

Environmental physiology or physiological ecology had its historical beginnings in the correlation of biological phenomena with variations in temperature stimulated by **Galileo's** invention of a hermetically sealed thermometer about 1612 A.D. **Rene Reaumur** and **Georges Buffon** were primarily naturalists of the eighteenth century, but they were also interested in the responses of organisms to temperature, light and altitude. **Reaumur** summed the mean daily temperatures for April, May, and June in 1734 and again in 1735, and correlated the earlier maturing of fruit and grain during the first year with the greater accumulation of heat. A discovery of parallel significance was of oxygen in 1774 by the English clergyman, **Priestley**, and the finding by **Lavoisier**, a Frenchman, in 1777 that oxygen was an essential part of air. The observations of **Priestley** and **Scheele** that plants produce oxygen led to an understanding of the inter-relationship of plants and animals. The discovery that green plants use carbon dioxide and water to form organic matter for the use of animals and result in the release of oxygen and water as wastes, provided the concept of food-chains.

Justus von Liebig (1803–1873) initiated the idea of carbon and nitrogen cycles in nature. He also developed his analysis of limiting-factors (1840). He was interested primarily in physical factors which limit the distribution and abundance of plant and animal life. We recognize his basic contribution today in **Liebig's Law of Minimum**, which states that an organism will be **limited** by that factor most closely approaching its minimum life supporting requirements. **Jacques Quetelet** (1796–1874) studied dormancy and hibernation in plants and animals in relation to temperature (1846), and shortly there after **Alphonse de Candolle** (1806–1893) studied plant germination in relation to temperature and moisture (1865). **Claude Bernard,** a French physiologist, enunciated the principle of homeostasis in 1876. This concept originally referred to regulatory mechanisms which maintained the "internal environment" of body constant in the phase of changing external conditions. Later the concept came to be applied also to maintenance of community inter-relations. **Vant Hoff**, a Dutch scientist, contributed to physiological ecology in 1884 in describing how the speed of chemical reactions increased two or three fold with each rise of 10°C.

Two men who gave environmental physiology its greatest advance in the nineteenth century were **K.G. Semper** (1832–1893) and **Charles Davenport** (1866–1944). **Semper** wrote an important ecological treatise in 1881, entitled *Animal Life as Affected by the Natural conditions of Existence*, in which he established the 'principle of food chains' and what we now call a 'pyramid of numbers'. He was interested in predator-prey interactions, protective colouration, and the spatial relations of animals. He showed that pond snails and fresh-water isopods would be stunted if crowded in a small volume of water, and he postulated the existence of chemical factors passing between animals in the environment and influencing growth and development. We now recognize these as **"ectocrines"** or **environmental hormones**, and they represent a major area of current research. **Davenport** wrote two volumes in 1897 and 1899, entitled *Experimental Morphology*. He studied animal behaviour and development in a variety of artificial environments and media, and he analyzed a wide range of environmental responses to chemicals, light, moisture, temperature and gravity,

More recent summaries of knowledge and methods in physiological ecology have been made by **V.E. Shelford** in *Laboratory and Field Ecology*, **Samuel Brody** in *Bioenergetics and Growth* (1945) and by **F.J. Vernberg** and **W.B.Vernberg** in *The animal and the Environment* (1970). **J. L. Cloudseley-Thompson** (1956, 1975) worked extensively on the ecology of terrestrial arthropods such as centipede, scorpions, beetles, etc., of arid environments (deserts) and he specifically studied **effect** of high temperature on them.

2-5·3. Evolution and Theories of Natural Selection

The growth of evolutionary theories of 18th and 19th centuries paved the way to the early development of ecology. **Georges Buffon** had developed the ideas of evolutionary change in his *Histoire naturelle* in 1749 which stated '*if we again consider each species in different climates we shall find obvious varieties both as regards size and form, all are influenced more or less strongly by the climate. These changes only take place slowly and imperceptibly; the great workman of nature is time : he walks always with even strides, uniform and regular, he does nothing by leaps ; but by degrees,* *by gradations, by succession, he does everything; and these changes at first imperceptible, little by little become evident......*"

Jean Baptiste de Lamarck (1744—1829) was environmental evolutionist of the 18th and early 19th century. Like **Buffon** he believed that plants and animals were changed by environmental influence. In his two major books *Philosophie Zoologique* in 1809, and *Histoire Naturelle Des Animaux Sans Vertebres* in 1815 **Lamarck** presented his evolutionary ideas—"*the use and disuse theory*" and "*inheritance of acquired characteristics*". He believed, for example, that the giraffe's neck had become long because he continually stretched higher for leaves to feed on, and thus each generation was born with a slightly longer neck as a result of stretching. Similarly, he believed that blind cave fish lost their eyes because they failed to use them. Although we can recognize elements of truth in these theories, we now know that Lamarck's understanding of the mechanism was incorrect. It is well established now that environmental changes affecting only somatic cells of the body are not transmitted to subsequent generations. However, some environmental changes which affect germ cells, such as radiation or certain chemicals (mutagens) can produce genetic changes that are passed on to later generations. So in this sense, environmental factors can influence subsequent generations.

The evolutionary theories of **Charles Darwin** and **Alfred R. Wallace**, which emerged about 40 years after **Lamarck** were similarly ecological in emphasis, but radically different in their explanation of how evolution occurred. They developed their views independently and almost concurrently, placing emphasis on population pressure and struggle for existence as the driving forces of evolutionary change. They envisaged survival of the fittest and natural selection as the mechanism of evolution, working on the foundation of inherited variations in plants and animals.

2-5·4. Population Studies and Demography

From the sixteenth century onward, a few scholars were worried about population growth and pressures. **Machiavelli** (1469–1527), **Botero** (1540–1617), **Buffon**, and **Franklin** (1707–1790) all partially anticipated **Malthus** by expressing concern about population growth, starvation, disease and war. However, **Malthus** was primarily responsible for focusing controversial attention on these topics.

Thomas Robert Malthus, a professor of history and political economy at Haileybury College, Reverend, published his controversial book, *An essay on population* (1798), in which he observed that human populations tend to increase geometrically (1, 2, 4, 8, 16, 32...) and thus exceed the means of subsistence, which increase arithmatically (1, 2, 3, 4, 5, 6, 7...). He felt that unless populations were controlled by voluntary constraint they surpassed their level of support and were then decimated by starvation, disease, war, and poverty.

Malthusian theory produced a storm of protest and has remained controversial to modern times. Many scholars believed that Malthus failed to give any cognizance to the great potential of modern science to overcome famine and disease. Some felt it presented too dismal a view of human potential. **Donbleday** (1841) contradicted Malthusian idea by stating that whenever a species was threatened, nature made a corresponding effort to preserve it by increasing the fertility of its members. Human populations that were undernourished had the highest fertility ; those that were well fed had the lowest fertility. He explained these effects by the over supply of mineral nutrients in well-fed populations. Recently, however, Malthusian doctrine has come back into favour, and it is being taken more seriously now than in the 1930's (**Hardin**, 1964). Nowa-days, even many writers who believe in the potential of science and technology to solve the problems of population pressure admit that the Malthusian fears of starvation, war and disease still haunt the earth.

Incidentally, it is an odd quirk of scientific history that **Darwin** was favourably influenced by **Malthus**, and relied heavily on his doctrine to establish his theory of natural selection. Subsequently, Darwin's theories were accepted in scientific circles long before those of Malthus.

Population studies resumed great importance in the twentieth century with the work of **Raymond Pearl** (1925), **Lotka** (1925), **Volterra** (1931), **Thomas Park** (1933), **Charles Elton (1933) Nicholson** (1933), **Gause** (1935), **Andrewartha** and **Birch** (1954) **and many** others.**Pearl** analyzed mathematically the characteristics of population growth. **Lotka** and **Voltera** developed theoretical mathematical equations to show the manner inwhich populations of different species interact. **Nicholson** (1933) studied the factors that stablize populations at particular levels. **Gause** (1935) studied interacting populations of predators and prey. **Andrewartha** and **Birch** emphasized the importance of climate and other factors in determining the size of populations.

2-5·5. Ecological Geography and Conservation

Modern geographic ecology dates back to the generalzations on the worldwide distribution of animals made by **Buffon** and the explorations of **Humboldt**. There was lively interest and many important contributions in this general field during the nineteenth century; notably, the life-zone concept of **C. Hart Merriam** (1890–1910) needs special mention. During the present century the concept of biotic provinces is identified with **L.R. Dice** (1943) and the biome concept with **F.E. Clements** and **V.E. Shelford** (1939). The broad survey of ecological animal geography made by **R. Hesse** in 1924 exerted considerable influence.

In 1864, **George Perkins Marsh** of North America, published a book *Man and Nature*, or *Physical Geography as Modified by Human Action* and laid the foundations for ecological geography and conservation. **Marsh**'s basic theme in *Man and Nature* was that man's economic progress has often disrupted the balance of nature to his own detriment. He clarified the reciprocal interaction of man and environment. He traced the history of civilizations, particularly those in the Mediterranean area, in terms of physical geography, natural resources, and land changes induced by man. He first called attention to the striking effects of deforestation and land scarring on erosion, agriculture, and water resources. He thus forced his readers to look upon man in a different perspective–not as a conqueror but as a despoiler.

Another American, **John Muir** (1838–1914) developed a philosophy of conservation similar to Marsh. He extensively travelled throughout the world, and wrote poetically of the values and beauties of natural wilderness. Through articles, books, and lectures, he worked effectively for the conservation of lands and forests and was ultimately influential in the designation of 148,000,000 acres, as reserve forests in the United States. Through his painstaking efforts many national parks such as **Yosemite** and **Sequoia National Parks**, **Mt. Rainier**, **King's Canyon**, and **Grand Canyon National Parks** were established in U.S.A. in 1890. Later on, certain staunch supporters of **Marsh** and **Muir** but political man like **Gifford Pinchot** (1865–1947) and President **Theodore Roosevelt** (1858–1919) spearheaded the environmental conservation movement in U.S.A.

In regards to other divisions of ecology, the crystallization of studies in oceanography may be credited to **Edward Forbes** (1843), **Maury** (1855), **Alexander Agassiz** (1888), **Peterson** and his colleagues (1911), and **Murray** and **Hjort** (1912); limnology to **Forel** (1869), **Birge** (1893), **Juday** (1896), **Ward** and **Whipple** (1918), **Thienemann** (1913, 1935) and **Naumann** (1918–32) and wildlife management to **Aldo Leopold** (1933). Attention to ecological succession though began in 1685, but, in modern context it started only in 1899 when the American plant ecologist **Henry Cowels** (1869–1938) studied succession of plants on the sand dunes at the South end of Lake Michigan. In 1916, another American plant ecologist, **Frederis E. Clements** (1874–1945) worked on plant succession. **C. C. Adams** (1906,1909) and **V.E. Shelford** (1913) were among the first to apply the concept of ecological succession to animals. In the 1930's, attention to biological productivity began in connection with practical pond-fish culturing and the limnological studies of **Thienemann** in Europe and of **Birge** and **Juday** at the University of Wisconsin, but the modern crystallization of the subject came with the freshwater and marine investigations of **Lindeman**, **Hutchinson**, **Riley** at Yale University (**Ivlev**, 1945) and of **Eugene** and **Howard Odum**. An early study of energy relation within terrestrial communities is that of **Stanchinsky** (1931).

Further development of ecology during 20th century has been greatly stimulated by organization of ecologists. The *British Ecological Society* was founded in 1913 and had a membership of about 2000 in 1974. Following the British, the *Ecological Society*

America was founded in 1915 and in 1974 had a membership of 4700. The Society in America has given birth to several offsprings; the *Wildlife Society*, *Society of Limnologists and Oceanographers*, the *Nature Conservancy*, the *Animal Behaviour Society*. In 1971, the *Inter-American Institute of Ecology* was established in U.S.A.

The Ecological Society of America publishes three periodicals–*Bulletin*, *Ecology*, and *Ecological monographs*. The British Ecological Society also publishes three periodicals : *Journal of Ecology*, *Journal of Animal Ecology*, and *Journal of Applied Ecology*. *Oikos* began publication in 1949 to represent ecologists in Denmark, Finland, Iceland, Norway and Sweden. The *Polish Academy of Science* has been publishing *Ekologia Polska* since 1953. *Oecologia*, a German periodical derived from *Zeitschrift fur Morphologie und Okologie der Tiere*, began in 1968. The *U.S. S.R. Academy of Sciences* started the publication of the periodical *Ekologiya* (*a Soviet Journal of Ecology*) in 1970. The *International Society for Tropical Ecology*, including ecologists of India and adjacent countries, was founded in 1960 and publishes *Tropical Ecology*. The *New Zealand Ecological Society* came into existence in 1952 and the *Ecological Society of Australia* in 1960, and both Societies publish their annual proceedings. The *Japanese Journal of Ecology*, begun in 1954, is the official publication of *The Ecological Society of Japan*. Researches on population Ecology, started in 1959, are published by the *Japanese Society of Population Ecology*. In 1968, an *International Association for Ecology* was formed within the International Union of Biological Sciences to coordinate ecological work in various countries. It started publishing the *Intecol Bulletin* in 1969.

Research papers of Ecological nature usually also appearing in the following biological journals : *Science*, *Nature*, *Evolution*, *Bioscience*, *Endeavour*, *Journal of Soil and Water Conservation*, *Advance Ecological Research*, *Archivio di Oceanographia Limnologia*, *Biological Bulletin*, *Journal of Experimental Botany*, *Journal of Experimental Zoology*, *Biological Review*, *Journal of Wildlife Management*, *Geographical Review*, *Marine Biology*, *Journal of Parasitology*, *Scientific American*, *New Biology*, *Journal of Theoretical Biology*, *Turtox News*, *Canadian Journal of Research*, *Annals of Entomological Society of America*, *Journal of Morphology*, etc.

2-6. REVISION QUESTIONS

1. Describe, how did ecology grow before 17th century ?
2. Write about historical growth of ecology after 18th century.
3. Where will you find the literature concerning ecology ? Enumerate certain national and international journals.
4. Write an essay on "Historical background of ecology"

3
Growth of Ecology in India

3-1. INTRODUCTION

India exhibits a panorma of the ecological conditions of rest of the world within her limited geographical boundaries. Almost each and every kind of environmental condition and ecological habitat of rest of the world is reflected by some geographical area of Indian subcontinent. For instance, the cold deserts of Bolvia-Arctic can be compared with the cold desert of Ladakh and likewise, Sahara-Arabia-Gobi hot desert-complex can be compared with the South-Rajasthan hot desert-complex of India. Further, in India there occur so much variation in ecological conditions from one place to another that one can study simultaneously different ecological phenomena within her geographical boundaries. Due to wide climatic and seasonal fluctuations, the ecological conditions of a given area do not remain static for a long time and the flora and fauna of the Indian subcontinent have evolved wide range of adaptations to cope with them.

Indians have a traditional and natural deep involvement in different ecological and environmental issues. Even a common man is well aware about the inter-relationships of vegetation, animals, habitat and environment. Many villagers have the unique capacity of forecasting about forthcoming natural calamities merely by observing changes in plants, animals and climate in their surrounding. Throughout India numerous laboratories have been engaged and still are engaged in investigating some specific ecological aspects such as soil ecology, plant ecology, animal ecology, limnology, oceanography, pollution biology, management and conservation of natural resources like land space, soil, water, wildlife, minerals, etc. During last two or three decades much ecological insight has been achieved about Indian subcontinent but despite these efforts, a coherent and complete picture of Indian ecology has not been emerged so far. However, history or growth of ecology in India can be studied separately for various ecological branches as follows :

3-2. GROWTH OF PLANT ECOLOGY IN INDIA

In India, the earliest contributions to the modern plant ecology were mainly from some forest officers who presented purely descriptive accounts of forests. First forest conservation efforts were made in 1864 by **Sir Dietrich Brandis**, the **Inspector** General of forests and first Indian Forest Act came into existence in 1885. In 1893, **Voelcker**, a German agricultural scientist, submitted a report to Indian Government regarding the role of forests vis-a-vis agriculture and he stressed the **need** for formulating a forest policy with a definite bias for serving agricultural interests more directly than before. ccordingly, Indian Government issued **a resolution** dated 19th october, 1894 regarding her policy about the management of Indian forests on scientific lines **(Sagreiya, 1967).**

Early work of plant ecology was of **synecological nature. Troup** (1921) worte three volumes on the *Silviculture of Indian Trees.* **Dudgeon** (1921) contributed a comprehensive ecological account of the upper gangetic plains and he employed the concept of succession in his investigations. His concept of succession was **further** elaborated by **Saxton** (1922) but later on contradicted by one of the great Indian plant ecologist **R. Misra** (1946, 1958, 1959). **Champion** (1929, 1936, 1938) studied the ecology of different forest types of India. **Kashyap** (1932) worked out the ecology of alpine vegetation of Himalaya and Tibet. **Bor** (1942) investigated synecology of grasses of U.P. forests. **Champion** and **Griffith** (1948) published their report '*Manual of General Silviculture in India*'. **G. S. Puri** (1950) studied the distribution of conifers in Kulu Himalayas and also investigated the geology of this area. In 1953, he studied the litter production in Dehradun forest. In 1954, he studied seasonal variations in forest flora, different forest types of India and soil climate of some Indian forests. He wrote the book—'*Indian Manual of Plant Ecology*' in the coauthorship of **R. Misra** in 1954. **Raji** (1955, 1956) made observations on the phytogeography of the Mysore hill tops. **Arora** (1960) studied the ecology of Coorg forests. **Ahuja** (1961) studied ecology, phytosociology and phytogeography of the vegetation of the humid tropics. **Shindhu** (1961) worked on the ecology of the mangroves. **Gupta** (1961) studied the vegetation of the catchment areas of river Ganga and Yamuna with reference to their soil and water conservation problems. **Sharma** (1961) worked out the ecology of the vegetation of arid zones. **Homji** (1962) studied the bioclimates of India in relation to vegetational criteria.

R. Misra (1944–48) initiated comprehensive autecological studies on herbaceous plants of different habitats such as aquatic, ravines, eroded river banks and low lying areas. **Misra and Rao** (1948) studied autecology of *Lindenbergia polyantha.* **Pandeya** (1950, 1951) studied autecology of *Bothriochloa pertusa* and *Dichanthium annulatum.* In 1953-1967, he studied different ecological aspects of grasslands of Sagar, Madhya Pradesh. **Whyte** *et al.*, (1955) published the account of different grasslands of India. **Bhatia** (1954) worked on the ecology of teak (*Tectona grandis*) forest of Madhya Pradesh. **Trivedi** (1955) worked out the ecology of *Sesbania bispinosa.* **Mall** and coworkers made autecological studies of *Chrozophora rottleri* and *Achyrathus aspera* in 1956 and of *Cassia tora*, *Cassia obtusifolia* and *Tridax procumbens* in 1957. **Ramakrishnan** (1958) studied ecotypic differentiation in some plants of Varanasi. **Joshi** (1958) investigated phytosociology and autecology of *Anogeissus latifolia* of Madhya Pradesh. **Jain** (1958) made phytosociological and autecological studies on sal (*Shorea robusta*) in Madhya Pradesh. **Ramam** (1961) studied the soil root relationships in grassland communities of Varanasi and in 1968, he studied certain wood land ecosystems.

A. N. Puri (1925) made a critical study of the hygroscopic coefficient of soil. **Raychoudhuri** (1941) studied soil profile of red soils of India. In 1966, he published his book '*India–Land and Soil*'. **Upadhyaya** (1955) investigated soil formation in relation to plant cover. **Jenny** and **Raychoudhuri** (1960) studied the effects of climate and cultivation on nitrogen and organic matter reserves in Indian soils. **R. N. Singh** (1961) wrote a monograph on the '*Role of Blue-Green Algae in Nitrogen Economy of Indian Agriculture*'. **Rao** (1962) published his book '*Soil Conservation in India*'. **Seth**, **Kaul** and **Gupta** (1963) made some observations on nutrition cycle and return of nutrients in plantations of New Forest. **J. S. Singh** (1962)) studied the effect of teak plantation on soil fertility in U. P. (gangetic alluvium). **Champion** and **Seth** (1968) published their report '*A Revised Survey of the Forest Types of India*'.

Saksena (1955) worked out ecological factors governing the distribution of soil microfungi in some forest soils of Sagar. The events of heterotrophic succession have been investigated by various workers like **Khanna** (1964), **Yadav** (1966) and **Sharma** (1971–1974). **Satyanarayana** *et al.*, (1977) studied seasonal variations in the mycoflora of nests of the birds like crow, house sparrow, etc.

J. S. Singh (1967, 1968) estimated above ground productivity in the grasslands at Varanasi. **K. P. Singh** (1968) estimated litter production and nutrient turnover in deciduous forests of Varanasi. **R. Misra** (1968) studied the mode of energy transfer along the terrestrial food chains and in 1969, he estimated primary productivity of terrestrial communities at Varanasi. In 1970, he made estimation of the primary production of Chakia Forest and also made IBP/PT study of organic poductivity and nutrient cycling in the monsoon forests, grassland and cropland. **Pandeya** and **Kuruvilla** (1968) studied natural forest communities, their net biomass and status of underlying biogenic salts in Dang Forests, Gujrat. **Rao**, **Dabral** and **Pandey** (1972) estimated litter production in forest plantations of chir (*Pinus roxiburghii*), Teak (*Tectona grandis*), and sal (*Shorea robusta*) at New Forest Dehradun. **Gopal** (1973) made a survey of Indian studies on ecology and production of wetland and shallow water communities.

3-3. GROWTH OF ANIMAL ECOLOGY IN INDIA

Animal ecology in India is still in the state of infancy. Except some progress in the understanding of insecian ecology and fish ecology little advance has been made in this field in India. Some of the notable contributions of certain animal ecologists can be listed in a chronological order which are as follows.

Sterndale (1884) studied the natural history of Indian mammals. **Baker** (1921, 1935) **made certain ecological observations** on brids of India, Burma and Ceylon. **Dharmakumar-Sinhji** (1937-1975) studied extensively the biology and ecology (migration, etc.) of water fowls and other birds of Gujrat **Ayyar** and **Margabandhu** (1941) made ecological observations on the cotton stem weevil. *Pempharulus affinis* under controlled physical conditions. **Trehan** (1945) studied the ecology of the soil fauna of cotton-fields of Lyallpur (Pakistan). **Mani** (1953–1974) worked on the ecology of certain insects specially ladybird beetle of high altitudes (Himalaya). **Atwal** (1955) **recorded** the effects of temperature, photoperiod and food on the **rate** of development, longevity, fecundity, etc., of the diamond black moth, *Plutella*. **Prakash** (1958, 1961, 1963) studied the ecology of a variety of vertebrates including reptiles, bats and other mammals of Rajasthan deserts. In the year 1975, in the coauthorship of **Gupta**, **Prakash** published a book on "*Wildlife Ecology and Conservation*', in which the wildlife ecology of Rajasthan deserts has been discussed. **Purohit** (1968) studied ecology of mammals in the Great Indian Desert.

Ramalingam (1961) studied ecology of larval forms of the trematode *Echinochasmus bugulei*. **Southwick** *et al.*, (1961) carried out a population survey of rhesus monkeys (*Macaca mulatta*) in villages, towns and temples of northern India. **Beeson** (1961) published the book '*The Ecology and Control of Forest Insects of India and Neighbouring Countries*'. **Ananthasubramaniam** and **Ananthakrishnan** (1963) studied ecology of microarthropods such as *Symphylella* of pasture soil. **Bhalla** and **Venkataraman** (1963) studied the ecology of parasitism of a hymenopteran *Vipio deisae* on stalk borer . *Chilo zonellus*. **Nayar** and **Prabhu** (1965) worked on the soil fauna of tea-growing tracts of Kerala. **Oppenheimer** (1973) studied the effects of environmental factors on the activity of village dwelling langurs (primates) in West Bengal. In 1975, **Singh** *et al.*, and **Bains** and **Shukla** made autecological observations on the effect of temperature on population dynamics, fecundity, and development of *Chilo partellus*. **Davidar** (1976) studied population ecology of thar in Nilgiris, Tamil Nadu.

3-4. GROWTH OF OCEANOGRAPHY IN INDIA

The ecology of Indian seas and oceans is comparatively most extensively worked out than any other ecological discipline. The knowledge of the oceanography of Indian seas has been mainly acquired from the observations of expeditions by the following ships:

Challenger (1872–1876), *Gazelle* (1874–1876), *Investigator* (1884–1925), *Vitiaz* (1886–1909), *Egeria* (1894), *Valdivia* (1898–1899), *Gauss* (1901–1903), *Discovery I* (1901–1904), *Sea Lark* (1905), *Planet* (1906–1907), *Aurora* (1911–1912), *Dana I* (1921–1922), *Dana II* (1935), *Snellius* (1929–1930), *Discovery II* (1929–1931), *Mahabis* (1933–1938), *Albatross* (1947–1948), *Vema* (1953), etc. During International Geographical year (IGY) programme following research ships have added to our ecological knowledge : *Vitiaz* (1957-1960), *Michael Lomonosov* (1957–1960), *Severyanka* (1958), *Chain* (1958–1960), *Argo* (1959–1960), *Woelkof* (1959), *Shokalski*, (1960), etc. During years 1960–1964, an International Indian Ocean Expedition (IIOE) was arranged to study the Indian oceans. In this expedition 20 countries including India participated with 38 research vessels and a big team of scientists. India participated in this field programme of expedition from September 1962 to December 1965. The vessels *INS Kistna*, *R. V. Varuna*, *R. V. Conch*, and *F. V. Bangada* were used for making detailed observations on several aspects of physical, chemical and biological oceanography as well as the fisheries, meterology, marine geology and geophysics.

On the conclusion of the Indian programme of IIOE, the *National Institute of Oceanography* (NIO) was set up in January, 1966 as one of the national laboratories under the control of Council of Scientific and Industrial Research (CSIR). NIO took over all the activities and centres of the Indian Ocean Expedition in India and it organized them into different divisions of oceanography. After being temporarily located at New Delhi, NIO shifited its head quarter to Dona Paula (Goa) on July 15, 1969.

NIO and its regional centres at Bombay and Cochin consist of following seven divisions : (i) The Indian Ocean Biological Centre at Cochin; (ii) Physical Oceanography; (iii) Chemical Oceanography ; (iv) Geological Oceanography ; (v) Biological Oceanography ; (vi) Data and Documentation ; and (vii) Oceanographic Instrumentation. Two more regional centres of NIO are going to be established at Waltair and Calcutta. Different centres of oceanographical research of India are tabulated as follows ;

Table 3·1. Different centres of oceanographical research of India.

Institute/Organization	Activities
1. The Indian Meteorological Department, New Delhi.	Collection of marine meteorological and sea surface temperature data.
2. Regional Telecommunication Station, New Delhi (a part of the Global Telecommunication System of world weather watch programme).	Exchange of meteorological and some oceanographic data.

Institute/Organization	Activities
3. The Survey of India, Dehra-Dun.	Operation and maintenance of tide gauges, determination of various tidal datum plane and evaluation of monthly and annual sea level variations at different minor and major ports.
4. The Naval Hydrographic Office, Dehra Dun.	Preparation of bathymetric charts of the seas around India and the approaches to various harbours.
5. Indian Institute of Tropical Meteorology, Poona.	Provides facilities for marine meteorological research.
6. The Bhaba Atomic Resarch Centre, Bombay : (a) The Health Physics Division. (b) Radioactive Waste Treatment Division.	Conducts circulation studies in the nearshore waters off Bombay.
7. The Tata Institute of Fundamental Research, Bombay.	Carries out geochemical studies pertaining to heavy metals and radioisotopes in the seas and sediments.
8. The Institute of Sciences, Bombay.	Carries out research on the biological oceanography and marine resources of the Arabian seas.
9. Taraporevala aquarium and Marine Biological Research Station, Bombay.	—do—
10. The Naval Chemical and Metallurgical Laboratory, Bombay.	Carries oceanographic research under Research and Development Organization of the Ministry of Defence.
11. The Naval Physical and Oceanographic Laboratory, Cochin.	—do—
12. The Physical Research Laboratory, Ahmedabad.	Carries research in oceanography from space in collaboration with National Institute of oceanography and other Institutes.
13. Directorate of Fisheries, Government of Gujrat, Ahmedabad.	Carries out research on the biological oceanography and marine resources of the Arabian sea.
14. The Agricultural University, Bangalore (Mangalore Unit).	—do—
15· University of Kerala, Department of Marine Biology and Fisheries, Trivandrum with branch laboratories.	—do—
16. Department of Marine Sciences, Cochin University, Erna-Kulam, Cochin.	—do—

Institute/Organization	Acitivities
17. Central Marine Fisheries Research Institute, Cochin and Mandapam (with 15 substations and units).	Carries out investigations on marine fisheries, general marine biological surveys along the entire Indian sea board, Indian marine fisheries statistical survey and acts as the centre for fisheries data.
18. Central Institute of Fisheries Technology, Cochin.	Conducts research on fish processing and craft and gear technology.
19. Government of India Deep Sea Fishing Station, Bombay (with substations at Cochin, Veraval, Tuticorin and Visakhapatnam).	Carries out exploratory fishing and offers training in offshore fishing.
20. Indo-Norwegian Project, Ernakulum.	Exploratory fishing.
21. The Central Salt and Marine Chemical Research Institute, Bhavnagar.	Studies recovery of important chemicals from the sea water. conversion of sea water into fresh water using solar evaporation methods and research on marine algae and agar production.
22. Andhra Universty, Waltaire.	Carries out research on the biological oceanography and marine resources of Bay of Bengal.
23. Marine Biological Station, Annamalia University, Porto Novo.	—do—
24. University of Madras, Madras.	—do—
25. Madras State Department of fisheries (with the Marine Biological Station at West Hill, Ennur and Krusadi).	—do—

Growth of different aspects of Indian oceanograpy can be studied separately as follows :

1. **Bal** *et al.*, (1946) accounted preliminary physical and chemical factors of water of Bombay Harbour during 1944–1945. **Chidambaram** and **Rajendran** (1951) collected hydrobiological data on Wedge Bank early in 1949. **Jayaraman** (1952, 1954, 1965) studied chemical oceanography of sea of Bay of Bengal, Gulf of Mannar and Palk Bay near Mandapam. **Lafond** (1952) worked on the marine meterology and its relation to organic production in South-East Asian waters. Later on, **Lafond** (1954, 1957) made certain oceanographic studies in the Bay of Bengal. **Ramamurthy** (1953) made hydrobiological studies of the Madras coastal waters. **Ramasastry** and **Balramamurthy** (1957) discovered thermal field and oceanic circulation along the east coast of India. **Jayaraman**

and **Gogate** (1957) studied variation in salinity and temperature in the surface waters of Arabian sea off the Bombay and Saurashtra coast. **Ramasastry** and **Myrland** (1959) recorded the distribution of temperature, salinity and density in the Arabian sea along the south Malabar coast. **Jayaraman** *et al.*, (1960) studied hydrography of the Laccadivies offshore water. **Remamrithan** and **Jayaraman** (1960) worked on the hydrographic features of the continental shelf waters off Cochin during the years 1958 and 1959. **Panikkar** and **Jayaraman** (1966) recorded the biological and oceanographic differences between the Arabian sea and the Bay of Bengal. **Mahadevan** and **Nayar** (1968) studied general topography and ecology of the rocky bottom in the Gulf of **Mannar** off Tuticorin.

2. Deep sea fauna (benthos) of Indian water is mainly known from the expedition-work of Investigator (1884–1925). **Alcock** (1902) have described the bottom fauna of Indian seas. **Hornell** (1902–1922) worked extensively for the establishment and development of pearl-fisheries in the gulf of Mannar and in Kathiawar, Gujrat. **Graveley** (1927) studied the littoral fauna of Krusadi islands in the Gulf of Mannar. **Bapat** (1951, 1970) made different ecological observations on Bombay duck, *Harpodon nehereus*. **Panikkar** and **Prasad** (1952) investigated the phenomenon of commnsalism in ophiuroid fish, crab, and jelly-fish, *Rhopalima*. **Daniel** (1954) studied seasonal variations and the succession of fouling communities in the Madras Harbour. **Jones** (1960) studied ecological relationship between scyphomedusa *Acromitus flagellatus* and young *Seleroides leptolepis*. **Sukumaran** and **Krishnaswamy** (1961) studied the ecological effect of salinity changes on the gastropd mollusc *Sellana radiata*. **Sokolova** and **Pasternak** (1962) investigated the benthic fauna in the tropical part of Indian ocean to be richer than that of tropical pacific ocean. **Sita** (1963) reported helminth parasites from Madras coast. **Krishnamoorthy** (1963) recorded the distribution of certain species of polychaetes in Adyar estuary, Madras. **Nayar** and **Mahadevan** (1965) recorded the interspecific population interactions between a fish and a sea snake and sea anemones and fishes in the Gulf of Mannar. **Jones** and **Mahadevan** (1965) studied ecological relationship in between pea-crab *Pinnotherus deccaneusis* living inside respiratory tree of sea cucumber *Holothuria scabar*. **Govindankutty** and **Nair** (1966) made preliminary observations on the interstitial fauna of south west coast of India. **Reddiah** (1970) studied ecology of reefs of Appa island in the Gulf of Mannar.

3. The ecology of marine phytoplanktons was worked out by many workers such as **Hornell** and **Nayudu** (1924), **Menon** (1931), **Chidambaram** and **Menon** (1945), **Gonzalves** (1947), **Chacko** (1950), **Ganapati** *et al.*, (1953, 1955), **George** (1953), **Ramamurthy** (1953), **Prasad** (1956), **Subrahmanyan** and **Sharma** (1960), etc.

Nair and **Subrahmanyan** (1955) observed a close relationship between abundance of oil sardine, *Sardinella longiceps* and the diatom *Fragilaria oceanica* which occurs in the Arabian sea, during the south-west monsoon. The blooming of this diatom is thus found to act as ecological indicator for the abundance of oil sardines. Primary production of marine environment have been estimated by various workers, such as **Ryther** and **Menzel** (1965), **Ryther** *et al*., (1966), **Qasim** and coworkers (1958–1973), **Ramachandran** (1970), **Nair** (1970), **Dehadrai** and **Bhargava** (1972), **Gopinathan** (1974, 1976) and so on.

Seshappa and **Jayaraman** (1956, 1957) studied interactions between sea bottom muds and nutrient cycles (phosphorus cycle) in tropical seas and their influence on the primary production. **Reddy** and **Sankaranarayanam** (1968) studied the distribution of phosphates and silicates in the Central, Western and North Indian oceans. **Rao** and **Jayaraman** (1969) investigated the occurence of oxygen maxima and minima in North Western Indian ocean.

4. **Ganapathy** and associates (1954–1962) worked on the ecology of marine planktons, planktonic copepods and chaetognatha of Visakhapatnam coast Certain workers such as **Bhimachar** and **George** (1950), **Prasad** (1953), **Prasad** and **Jayaraman** (1954), **Subrahmanyan** (1954, 1959), and **Panikkar** (1967) reported population explosion of certain zooplanktons such as *Noctiluca miliaris* on one occasion and green *Noctiluca*, *Hornellia marina*, *Trichodesmia*, etc., at other occasions. They observed that outburst of *Noctiluca* was correlated with scarcity of other living organisms in the Indian seas. **Subrahmanyan** (1959) made detailed studies of Zooplanktons off Calicut coast and also studied inter-relationship between phytoplanktons and **zooplanktons. Ramamurthy** *et al*., (1963) studied ecology of planktons at Kandla in the Gulf of Kutch. In 1965, **Ramamurthy** made observations on the ecology of planktons of North Canara Coast in relation to the pelagic fishery. **Prasad** (1968, 1969) published maps of the total zooplankton biomass of the Arabian sea and Indian ocean, in the *International Indian Ocean Expedition Plankton Atlas*. He also studied the distribution pattern and productivity of zooplanktons of different zones. **Pillai** (1968) made observations on the planktons off Bombay coast and also remarked on the hydrographic conditions and fisheries at this coast. **Qasim** (1970) observed the ecological reasons of *Trichodesmium* bloom in the Laccadives and deep sea. **Qasim** and coworkers (1972) worked extensively on the feeding ecology of many marine fishes and determined their status in marine food chains.

3-5. GROWTH OF LIMNOLOGY IN INDIA

Hora (1930) studied ecology, binomics and evolution of the torrential fish fauna with special reference to the organs of attachment. In the year 1947, **Hora** recorded the zoogeographical distribution of torrential fish. **Pruthi** (1933) reported seasonal changes in the physical and chemical conditions of the water in the tank in Indian Museum compound. **Ray** and **Chakravarthy** (1934) studied lunar periodicity in the conjugation of *Conchophtherius lamellidiens*. **Hutchinson** (1937) made certain limnological studies in Tibet. **Philipose** (1940) studied ecology and seasonal succession in a permanent pool at Madras city. In the year 1959, **Philipose** made observations on the ecology of fresh water phytoplanktons of Inland fisheries. **Ganapati** (1941, 1943, 1947) made divergent limnological investigations on the freshwater ponds of Madras City. **Banerjea** and coworkers (1954–1970) studied limnology of various fish ponds of India. **Chacko** and **Krishanamoorthy** (1954) studied ecology of planktons of three freshwater ponds of Madras City. **Biswas** and **Calder** (1955) made floral study on certain conmmon water and marsh plants of India and Burma. **Mitra** (1955, 1966) studied autecology of certain aquatic plants such as *Limnanthemum cristatum*, *L. indicum* and *Hydrilla verticillata*. **Das** and **Shrivastava** (1955–1959) studied the limnology of various freshwater ponds of Lucknow. **Das** and **Moitra** (1955) worked on the feeding ecology of some comnon fishes of U.P. and classified them into surface feeders, mid feeders, and bottom feeders. **Mitra and coworkers** (1956, 1965) worked on feeding ecology of the Indian major carp fry. **Alikunhi** (1956) made certain ecological observations such as fecundity on *Labeo bata*. **Singh** (1956) studied limnological relations of Indian inland waters with special reference to algal blooms. **Das** (1959–1967) studied ecological effects of micronutrients, vitamin B_{12}, density, etc., on the Indian freshwater carps.

George (1961–1966) made ecological observations on the physico-chemical nature af water and zooplanktons and rotifers of certain shallow ponds of Delhi. **Subhramanyam** (1962) published a monograph on certain common aquatic angiosperms of India. **Saigal** *et al.*, (1962) studied ecology of *Mystus seenghala* of Ganga river system. **Bhatt** *et al.*, (1963) reported mineral contents of freshwater and freshwater organisms of Kalyan, Kolhapur and Jaduguda areas. **Michael** (1964) studied limnology of planktons and macrofauna of a freshwater pond of West Bengal. **Sreenivasan** (1964) studied hydrology of tropical impoundment, Bhavanisagar Reservoir, Madras state for the years 1956–1961. In 1968, **Sreenivasan** undertook the studies on the limnology of two upland impoundments of Nilgiris of Madras state. **Banerjee** and **Raychoudhury** (1966) worked out the physico-chemical nature of Chilka lake of Orissa. **Krishnamurthy** (1966) made preliminary studies on the macrofauna of the Tungabhadra Reservoir. **Verma** (1967) studied diurnal variation in a fish pond in Seoni. **Vasisht** (1968) made limnological studies of Sukhana Lake, Chandigarh. **Jha** (1968) worked out ecology of a pond ecosystem of Banaras. **Moitra** and **Bhowmik** (1961) studied seasonal cycles of rotifers in a freshwater fish pond in Kalyani, West Bengal. **David** *et al.*, (1969) studied limnology and fishery of the Tungabhadra reservoir. **Nayar** (1970) studied ecology of rotifer populations of two ponds at Pilani, Rajasthan. **Khan** and **Siddiqui** (1970–1974) investigated the diurnal and seasonal variations in the limnological features of a perennial fish moat of Aligarh, Uttar Pradesh. **Kaul** (1971) studied the production and ecology of some macrophytes of Kashmir lakes. **Patnaik** (1971) worked on the seasonal abundance and distribution of benthos of Chilka lake, Orissa. **Ganapati** (1972) compared the organic production of various types of aquatic ecosystems of India. **Kant** and **Kachroo** (1975) recorded the diurnal changes in the temperature and pH of water and diurnal movement of planktons in Dal lake of Shrinagar. **Pillai** and **Sreenivasan** (1975) estimated the carbon and nitrogen status of some lakes, reserviors and ponds of Tamil Nadu. **Gupta** (1976) studied the limnology of macrobenthic fauna of Loni reservoir of Rewa of Madhya Pradesh. **Sreenivasan** (1977) studied the limnology and fisheries of Tirmoorthy reservoir, Tamil Nadu. **Rao** (1977) investigated the ecology of certain phytoplanktons such as diatoms, Euglenineae and Myxophyceae of three fresh water ponds of Hyderabad.

3-6. REVISION QUESTIONS

1. Describe the historical growth of plant ecology in India.
2. Write an essay on 'growth of animal ecology in India'.
3. Trace the historical growth of oceanography in India.
4. Describe the history of limnology in India.

4

Ecological Tools and Techniques

4-1. INTRODUCTION

Ecologists have to work with living systems and various environmental parameters. Like the naturalist, an ecologist has to begin with observations of plants and animals in natural surroundings. A great deal of careful observation is needed to identify the various organisms and to begin to note the major food chains. After the food chains have been identified, the ecologist can make more detailed measurements. Populations of various kinds of organisms can be counted. The amount of various materials and of energy passing through each part of the ecosystem can be calculated. Various environmental parameters like physico-chemical nature of three basic habitats–soil, water and air; temperature, light, humidity, pH, wind velocity, etc., can be measured. Chemicals such as phosphates and nitrogen can be traced in their movements through the food web. Relationships and dependency between different kinds of organisms can be worked out. In this manner, little by little understanding of how a particular ecosystem is put together and functions can be developed. This knowledge throw light as to how the earth's biosphere functions and what its needs are.

To acquire ecological knowledge about any kind of ecosystem of the biosphere, the ecologists have to use a variety of tools and techniques of other sciences like biology, chemistry, physics, mathematics statistics, climatology, sociology and economics. Some of the most commonly used and simple ecological tools and techniques can be studied as follows :

4-2. SAMPLING TECHNIQUES OF POPULATIONS

One of the major problems in ecology is the determination of population distribution, size and changes in abundance. For different kinds of populations following different methods of measurement (or sampling) of populations are adopted by ecologists :

4-2·1. Sampling Plant Populations

The vegetation or plant community (*i.e.*, plant populations) of a given site can be analyzed by the following methods :

4-2·1A. Methods of Study of Qualitative Characters of Vegetation

(i) Floristic composition. It involves the recording of different species of plants including vascular plants, bryophytes, pteridophytes and lichens, in a plant community. In order to prepare a complete list, inspection and collection in different seasons are necessary. Each species has its own range of ecological amplitude.

The actual recording of the plant community is done by the use of small **sampling units.** These units may be in the form of **area, line,** or **point,** as has been employed in the **quadrate, transect,** and **point** sampling methods, respectively as follows :

(a) Quadrate method. It is a technique which is used when only a part of a large area is sampled. On the basis of this information, the total population of the area is estimated. For example, if we want to know the number of pine trees in a forest, we can make a total count, but this may be too much time consuming, difficult and expensive. Instead, if we count the trees in several blocks or squares **(quadrates)** of the forest and by extrapolating these results for the whole forest area, we can make an estimate of the total number of trees. The quadrate method is used to measure the population density of organisms such as plants, planktons, earthworms, insects and also blood cells in the blood.

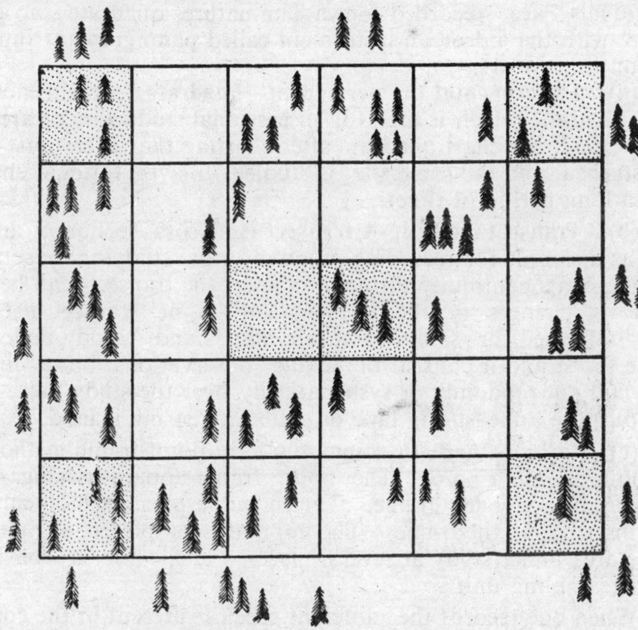

Fig 4·1. Quadrate sampling method for population estimation. The pine forest has been marked off into 30 squares. Thirteen trees are counted in four corner squares and two central squares. Thus, if 13 trees are in 6 squares, we would estimate 65 pine trees to be present in 30 squares (after Herreid II, 1977)

Shape and size of quadrates. A quadrate is a sample unit or plot which is an area of a definite size. In shape, it may be rectangular, square or circular. The size of the quadrate is determined according to the characteristics of the community. The richer the flora, the larger or more numerous the quadrates must be. To sample forest trees the fifth-acre plot is a popular size, but it may be too large if trees are numerous or if many species are involved. Smaller quadrates are used to study shrubs and understory trees. For grass and herbaceous plants, 1 m^2 is the usual size.

Kinds of quadrates. Quadrates are often labelled according to the uses of or data derived from them.

(1) List quadrate. In list quadrate, the organisms found in the area are listed by name.

(2) List-count quadrate. In list-count quadrate, in addition to listing out the species, numerical counts of individuals of each species are also made for estimating the abundance and density. This method is widely used in forest survey work.

(3) Chart quadrate. It is a quadrate that is mapped to scale to show the location of individual plants. In this method, the individuals are recorded on a miniature quadrate on graph paper with the aid of an instrument called **pantograph** at intervals of months or years.

(4) Experimental or **permanent quadrate.** It is a modified chart quadrate which is used for successional studies in an area. In this method, the chart quadrate after charting the vegetation is left undisturbed, and the same site is studied for vegetational changes over a long period of time.

(b) Transect method. A **transect** is a cross section of an area used as a sample for recording, mapping, or studying vegetation. Because of its continuity through an area, the transect can be used to relate changes in the environment. **Line transect method** is commonly used for sampling shrub stands and woody understory of the forest and it consists of taking observations on a line or lines laid out randomly or systematically over the study area. For this purpose a measuring tape or calibrated string is used.

(c) Point method. In commonly used point-frame method, the sampling unit is a point. The point frame apparatus (Fig. 4·2) is made up of a wooden frame. Ten movable pins, each 50 cm long, are inserted in the frame. The apparatus is placed at randomly in the area under study at several places. Each pin is considered as one sampling unit.

When one records the different species present in the community or vegetation of a given area by above mentioned three common methods, then analytical parameters like frequency, density, abundance, etc., of each species are then calculated by quantitative methods to be discussed later.

(ii) Stratification. The way in which plants of different species are arranged in vertical strata in order to make full use of available ecological requirements like light intensity, temperature, moisture contents, organic contents of soil, etc., is known as **layering** or **stratification**. The data on stratification are obtained by the use of bisect, where by constructing the vertical projection one may plot the stature of plant species, in order to find out the distribution pattern of stem and roots of different species.

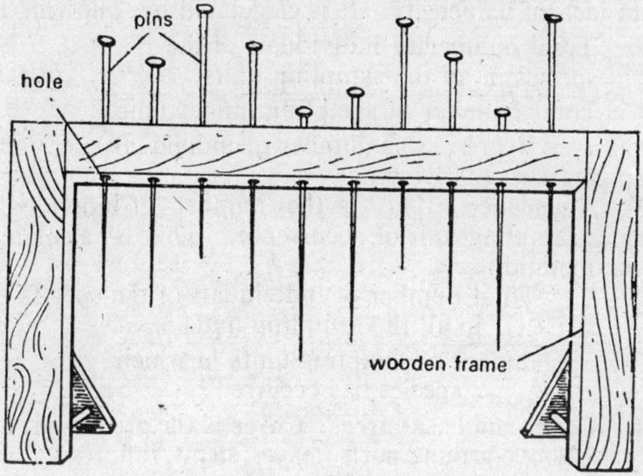

Fig. 4·2. Point-frame apparatus.

(iii) **Phaenology**. Phaenology is the study of organisms as effected by climate, especially dates of seasonal phenomena, as seed germination, flowering, fruiting, leaf-fall, seed and fruit dispersal. Data on phenology of each species in the community are recorded. Phenology of different species may differ and this is influenced by the environment.

(iv) **Vitality**. Vitality means the condition of a plant and its capacity to complete the life cycle. Some species are weak and their seeds fail to germinate. The health of a species is determined by the weight of the plants, stem height, root length, leaf area, leaf number, number and weight of flower, fruit and seed, etc.

(v) **Life-form**. A life-form is the sum of the adaptation of the plant to climate. **Raunkaier** (1934) considered that the way in which different species overcome the adverse environmental conditions, determines their limit of distribution. Thus, the plant's climate can be expressed by the statistical distribution of lifeforms in the flora of a particular region.

4-2·1B. Methods of Calculation of Quantitative Data

Quantitative data for individual species is calculated from qualitative data by following methods :

(1) **Frequency**. It is the number of sampling units (quadrate, transect and point) in percentage, in which a given species occurs. It is determined by following formula :

$$\% \text{ frequency} = \frac{\text{Number of sampling units in which the spcies occurred}}{\text{Total number of sampling units studied}} \times 100$$

(2) **Density**. Density represents the numerical strength of the species in the community. As plants grow at varying distance in respect with each other, so, the number of plants in an unit area gives an idea of its density. It is calculated by following formula :

$$\text{Density} = \frac{\text{Total number of individuals of the species in all the sampling units}}{\text{Total Number of sampling units studied}}$$

Density is expressed as number of individuals per unit area, as for example, per square metre.

(3) **Abundance**. This is the number of individuals of the species per sampling unit of occurrence. This is calculated by the following method :

$$\text{Abundance} = \frac{\text{Total number of individuals of the species in all the sampling units}}{\text{Number of sampling units in which species has occurred}}$$

(4) **Cover and basal area**. **Cover** is the area of ground occupied by the above-ground parts (leaves, stems, inflorescence) as viewed from above. **Basal area** refers to the ground actually penetrated by stems. In forests, cover is more than 100%. Basal area of a plant is determined in terms of its stem girth.

4-2·2. Sampling Phytoplankton and Periphyton

In aquatic communities algae are the dominant vegetation. There are two kinds of growth involved : the **plankton** suspended on the water and the **periphyton** growing attached to some substrate of aquatic habitat.

4-2·2 A. Sampling technique for phytoplankton. Phytoplanktons of a water body can be sampled by following two methods :

(i) In one method, phytoplanktons are obtained by drawing water samples from several depths. Cell counts of algae present in each sample, either normal or concentrated, are made with a **Sedgewich Rafter counting chamber** and a **Whipple ocular**. According to the need, sometime the samples may be concentrated by centrifugation in **forest plankton centrifuge.** The centrifuged samples are then diluted to a suitable volume (100 to 200 ml) in a volumetric flask.

During the counting of algal cells, a separate tally should be kept for each species. This will permit an analysis of community structure at each station. For single-celled forms, the number of cells is recorded; while for colonial forms, the number of colonies is recorded.

(ii) In another method, the phytoplanktons of the sample are first fixed by aqueous solution of formaldehyde, then filtered and ultimately counted by quadrate method.

4-2. 2B. Sampling technique for periphyton. Epiphyton, the periphyton growing on living plants and animals, can be observed in place on the organism if the substrate is thin or transparent enough to allow the transmission of light. The epiphyton growth on one side of the leaf only is scraped away for study. Small leaves are examined over the whole area. Large leaves are sampled in strips marked by grids on a slide or by an ocular micrometer. With large aquatic plants a square is cut from the leaf or stem. If the leaf is too thick to handle under the microscope, the periphyton is scraped off quantitatively and is mounted in a counting cell for examination. The results are then related back to the total surface area.

Another method is used for algae growing on such aquatic animals as turtles, molluscs, etc., and on stones. One easy method employs a simple hollow square instrument with a sharpened edge, which is pressed closely or driven into the substrate. This separates out a small area of given size around which the periphyton is washed away. The instrument is then raised and the periphyton remaining in the sample square is scraped into a collecting bottle.

The periphyton can be counted in a Sedgewich-Rafter cell recording a predetermined number, usually 100 to 1,000, as they appear in the field of view.

4-2· 2C. Sampling stream-bottom organisms. Samples of stream-bottom organisms are collected by a modified **Surber bottom-fauna sampler** (Fig. 4·3). This sampler consists of a brass frame with

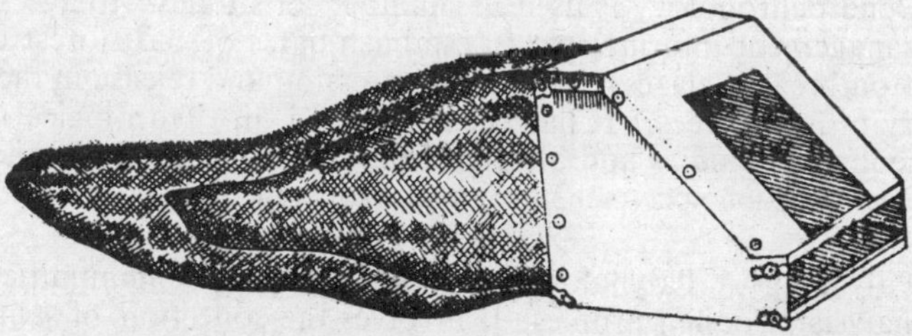

Fig. 4·3. A modified Surber bottom-fauna sampler (after Smith, 1974).

stainless-steel side pieces and a current baffler. To this is attached on a removable brass frame a fine net of 74 meshes/linear inch and a coarse net with 19 meshes/linear inch. The latter is fitted in the sampler in front of the fine net to produce a cone or cone effect. Flanges on the insert prevent its being forced into the fine net in the rear.

This sampler picks up many small organisms along with fine detritus. At a time, it collects two subsamples with respect to size.

This sampler encloses a specified area of stream (500 cm^2), which is the sample unit. Organisms, detritus and trash are scrubbed free from the substrate, and the current washes it into net.

The contents are transferred to a container and are taken back to the laboratory for examination and sorting.

Dendrochronology. Dendrochronology is the science that calculates the accurate dating of past events through the aging of trees. It is based on the variation of growth rings. Growth rings, despite popular belief, are not regular, nor are they all necessarily laid down annually. Because of the failure of cambium to form a sheath of xylem the entire length of the bole, rings may be omitted, especially near the base. This may be caused by the lack of food manufacture in the crown, by drought, fire, extreme cold, insect outbreaks, and so on. At the other extreme are multiple rings produced by multiple waves of cambial activity during the growing season. These are caused by temporary interruptions in the normal growth, such as late spring frost, or by regrowth taking place after normal seasonal growth has ended. Thus, the growth rings reflect the interaction of woody plants and their environment.

The fundamental principle of dendrochronology is **crossdating**, the correlation of distinctive patterns of growth between trees for a given sequence of years. Because no two plants have exactly the same growing environmental conditions and life history, the similarities are relative rather than quantitative. The relative widths of corresponding rings are the same in relation to adjoining rings. By lining up these similarities, the investigator can establish the relative identity of any ring in sequence and aberrant rings in the individual specimen.

Dendrochronology is used in number of studies—to age trees for management information, to establish dates of past forest fires, insect outbreaks, glaze, periods of suppression and release in the life history of forest trees. It has been involved in hydrological and archaeological studies and even in legal cases involving boundary disputes in which specimens are taken from fence posts and witness trees.

Palynology. Palynology is the study of plant communities by the analysis of pollen profiles. It involves the collection of samples of peat in bog by peat borer from different depths and microscopic analysis in laboratory of the sample. Carbon–14 dating method can be used for correct assessment of the age of pollen grains in the sample.

4-2·3. Sampling Animal Populations

The ecological study of animal populations in the ecosystem involves considerably more problems than the study of plants. Animals are harder to see and most are not stationary—they are here one minute and gone the next. The following methods furnish valuable informations regarding animal populations of different kinds :

4-2·3 A. Trapping and collecting animals. The sampling of a population involves collecting animals, either alive for marking and release, or dead. For different types of animals different techniques are adopted which are as follows :

(i) **Trapping and collecting flying insects.** Aerial nets or heavy duty sweep nets which are designed to withstand the hard-wear encountered when put through grass and woody vegetation, are used for the collection of diurnal insects. For nocutrnal insects, traps **containing** ultra-violet light or a mercury-vapour light are used (Fig. 4·4). Insects then picked off the sheet. For killlng the insects, killing jars containing a layer of plaster of paris and potassium cyanide (KCN) on the bottom are used.

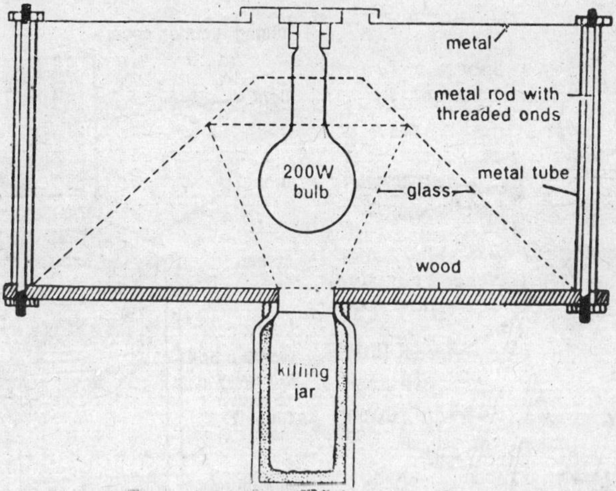

Fig. 4·4. "Rothamsted" light trap for nocturnal flying insects (after Lewis and Taylor, 1967).

(ii) **Trapping and collecting aquatic organisms.** For collecting aquatic organisms are used dip nets for organisms in the water, bottom nets for scraping along the bottom of ponds, wire-baskets scraper nets, and plankton towing nets. For aquatic collecting from the shore, aquatic throw nets are useful. For collecting bottom organisms in deep water, is used a bottom dredge (bag net) lowered from a boat. Fish, tadpoles, and large crustaceans can be collected with seines (large vertical fishing-nets).

(iii) **Trapping and collecting soil organisms.** Different animals of soil are collected by following methods : Soil arthropods are extracted from the soil by means of a Tullgren funnel, an improv-ed version of the Berlese funnel (Fig. 4·5). It consists of a heat source and a smooth funnel, preferably glass, and a shelf of hard-ware cloth on which to place the sample. The heat from the lamp and then desiccation derives the arthropods downward, until they fall into the collecting bottle beneath the funnel.

For collecting soil arthropods, a sample of litter or soil is placed on the hardware cloth shelf, so **fitted** in the funnel that air space is present between the wire and the wall of the funnel. To begin extraction, the lid of the funnel is opened to 90° and the 100-watt bulb is turned on. After about 16 hours, depending upon the sample size and moisture content, the lid is shut and bulb is changed by 15- watt bulb. There occur two periods of arthropod exodus, the first wave due to heat and the second wave due to desiccation. The collecting bottle beneath the funnel may contain alcohol, formalin, or water. The animals are then sorted and identified under the microscope.

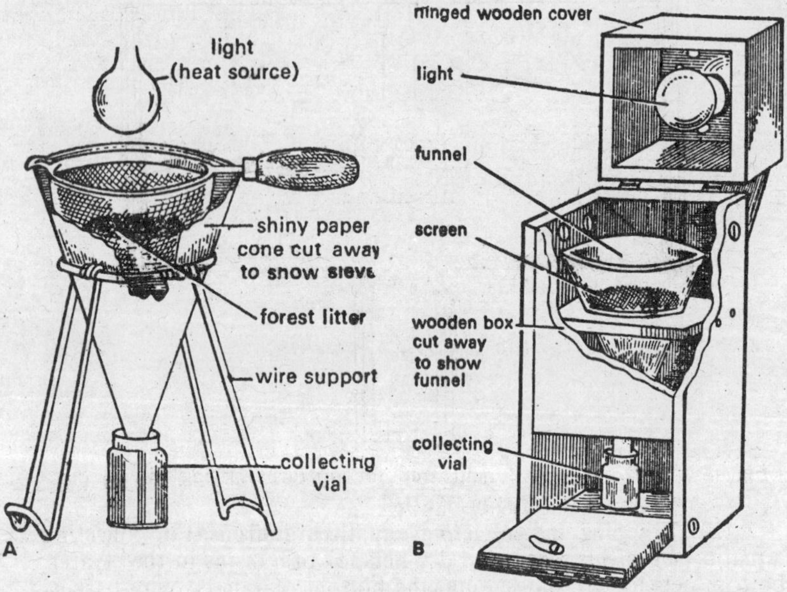

Fig. 4·5. Tullgren funnel for collecting soil arthropods. A—Simple funnel ; B—A much improved funnel (after Smith, 1974).

Larger soil animals like spiders and beetles, can be taken in traps made from funnels and cans set in the soil to ground level. Boards placed on the ground may attract millipedes, centipedes and slugs. Meat bait in small wire traps will attract scavenger insects.

For the collection of earthworms, a dilute solution of formalin (25 ml of 40 per cent formalin to 4·541 litres) is applied to a quadrate 61 cm². Within a few minutes worms will come to the surface. After earthworm movement to the surface stops, a second application of dilute formalin solution is done. When worms cease to come to the surface the second time, the quadrate is dig out as deep as necessary. The soil is hand-sorted for maximum recovery of earthworms. Earthworm cocoons can be extracted by the **floatation** method.

Soil nematodes can be collected by Tullgren funnel method and also by floatation method.

(iv) **Trapping and collecting small animals in vegetation. Sweep nets** with stout frames to withstand sweeps close to the ground and in woody growth are useful for collecting many types of insects and even some arboreal amphibians and reptiles. Drag nets consisting of light tubular frames, to which are attached canvas bags, are useful on flat ground. Overhead vegetation can be sampled by beating the limbs with sticks to dislodge the animals, which should fall into canvas collecting trays beneath.

(v) **Trapping and collecting birds and mammals.** Birds can be trapped for banding in specially constructed traps, cannon nets for larger game birds, and mist nets. For mammals, live traps of wood or wire and snap traps are used. These traps can be baited with natural foods, dripping water, etc.

4-2·3 B. Marking animals. Marking individuals in an animal population is necessary if one wishes to distinguish certain members of a population at some future date, to recognize individuals from their neighbours, to study movements or to estimate **number** of animals in a population.

Arthropods and snails are best marked with a quick-drying cellulose paint. Aquatic insects and molluscs are marked by ship-fouling paint.

Fishes are usually marked by **tagging** in several ways. Strap tags of monel metal (a nickle-base alloy) may be attached to the jaw, the preopercle, or the operculum. Streamer or pennant (a long narrow flag) tags are sometimes attached to various parts of body, usually at the base of the dorsal fin. A plastic tag can be inserted into the body cavity of fish by performing a minor operation. Fish can also be marked by clipping the fins.

Frogs, toads, salamanders and most lizards can be marked by some system of **toe clipping** which involves the removal of the distal part of one or more toes. Snakes and lizards can be marked by removing scales or patches of scales in certain combinations. Birds are usually marked either by serially numbered aluminium bands, by cellulose and aluminium coloured bands or by dyeing plumage by conspicuous or contrasting colour.

Small mammals may be marked by toe-clipping or by notching the ear. Bear, deer, elk, moose, rabbits and hares can be marked with strap tags or plastic discs attached to the ear. Aluminium bands similar to those used on birds can be attached to the forearm of bats.

Radioactive tracers. The use of radioactive tracers in marking of animals, is a particularly useful method for studying animals that are secretive in habits, live in dense cover, spend part or all of their lives underground, or that have radically different phases in their life cycle, such as moths and butterflies.

It is found that if animals are fed small traces of gamma-emitting radioactive meterial along with food, then the radio-active materials are metabolically incorporated into the tissues. The tracer becomes a part of the animal and is passed along to egg or offspring. This technique is useful for studying dispersal, for the identification of specific broods or litters, for obtaining data on population dynamics and natural selection.

4-2·3 C. Aging techniques. The determination of age structure of wild populations is a difficult task. Aging techniques are known only for few animals such as fishes, birds and mammals.

(a) Fishes. Fish aging techniques began first when **Hoffbaur** (1898) published his studies on the scale markings of known-age carp. Since then, this technique become more developed and refined. It is based on the fact that a fish scale starts as a tiny plate and grows as the fish grows. A number of microscopic ridges, the **circuli** (Fig. 4·6), are laid down about the centre of the scale each

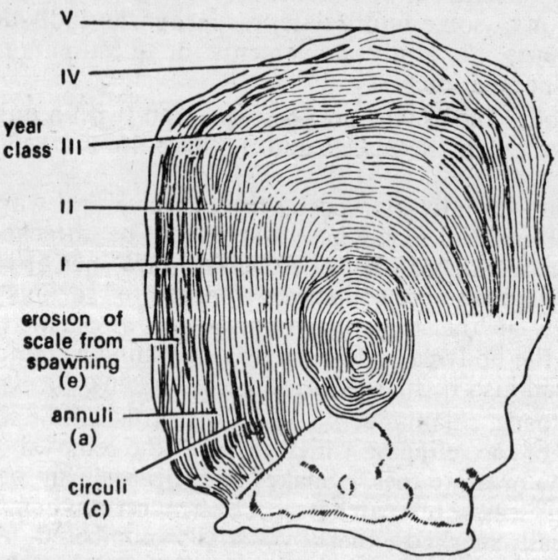

Fig. 4·6. The age of fish can be determined by the growth lines (rings) on the scale (after Smith, 1974).

year when the fish growing well in summer, the ridges are spaced wide apart. During winter, when growth slows down, the ridges are close together. This annual check on growth enables the biologists to determine the age of a fish by counting the number of areas of closed rings, the **annuli**. Because the growth of a scale continues throughout the life of a fish, it also provides information **on the growth rate.** This is obtained by measuring the total radius of the scale, the radius to each year's growth ring, and the total body length of the fish. Then by simple proportion, the yearly growth rate can be determined.

Other techniques in aging fish include the **length-frequency distribution**, **vertebral development**, and rings or **growth layer** in the otolith or ear stone.

Further, due to identification of the large number of **year classes** (animals born in a population during a particular year), one can determine dominant year classes, learn the age when fish reach sexual maturity, estimate production, mortality and the effects of fish harvest.

(b) Birds. The age of birds can be determined by studying the plumage development. In a technique, until moulted, the tail feathers of juvenile waterfowl are notched at tip, in contrast with the normally contoured feathers of the winter plumage. The shape of the primary wing feathers separates adults from young among many gallinaceous game birds. When the whole bird is available, the presence or depth of the **bursa of Fabricus**, a blind pouch lying dorsal to the caecum and opening into the cloaca, indicates juvenile birds.

(c) Mammals. Among mammals the examination of reproductive organs is useful because the majority do not breed until the second year. The presence of the epiphysal cartilage in rabbits and squirrels identifies juveniles up to 6 or 7 months. Likewise, black bars on the pelage of the underside of the tail of juvenile grey squirrels separate the young from the adults. Skull measurements are useful in beavers and musk rats. Annual growth rings on the roots of canine teeth indicate age for the first few years of life in the fur seal. Growth rings also show up in the horns of mountain sheep. The wear and **replacement** of teeth in deer and elk permit the determination of different age classes in these mammals.

The age of some animals like cottontail rabbit, the racoon, the black bear and fur seal and birds can be determined by **lens-weight technique.** Because the lens of the eye of most mammals and birds grows continuously throughout life, and because there is only slight variation between individuals in lens size and growth, the measurement of the lens is a practicable method for aging in these animals. The technique involves the weighing of the dry lens and comparing its weight against a chart of lens weights of known-age individuals.

4-2·3 D. Sex determination. The sex of most mammals can be determined by examining external genitalia and the sex of birds by plumage differences. For example, peacock-peahen, cock and hen, male and female Koel (*Eudynamys*), etc., all have plumage differences in both sexes.

4-2·3 E. Determination of home range and territory. The **home range** of wild animals can be determined by following two simple methods : 1. The area in which all the observations of the movements of individuals have been made, is measured and outlined on a map. If the observations are obtained by trapping, then one must assume that animal could have gone half-way toward an adjacent trap, especially if the traps were set in a regular grid. 2. In another method, the fact that small mammals come readily to bait has been exploited in a manner that does not involve trapping. Bait boxes are made from quart-size paper milk containers and are set out in a 50-ft grid over the study area. These boxes serve as containers for the bait and as receptacles for the faeces. If food at certain stations is stained with a dye, then the visitation by animals that fed on the dyed bait to other bait boxes can be traced through coloured droppings. The distance moved by a particular animal from a station with dyed bait can be determined when coloured scats are recovered at various stations in the grid.

Because territorial boundaries are rather rigidly maintained by birds during the breeding season, **territorial boundaries** can be mapped by observing the movements of the birds during the day, by plotting singing perches, by observing location of territorial disputes, and on occasion by chasing the bird. When the bird arrives at the boundary of its territory, it generally will double back.

4-2·3 F. Estimation of number of animals in population. The numbers of animals in wild populations can be estimated by following three methods : 1. **True census**, a count of all individuals in a given area ; 2. **sampling estimates,** derived from counts on sample plots ; and 3. **indices,** in which the trends of populations from year to year or from area to area are obtained through roadside counts, animal signs, and the like.

(a) True census. A true census implies a direct count of all individuals in a given area. Direct counts can be made only in large and conspicuous animals and in areas of their maximum concentration. Elephants, rhinoceros, deer, etc., in open country, **herds** of elk and caribou, waterfowl on wintering grounds, rookeries, roosts, breeding colonies of birds **and** mammals permit direct counting usually either from the air or from aerial photographs.

(b) Estimates from sampling. This involves following methods :

(i) sample plots. Relatively immobile forms such as barnacles, molluscs and cicada emergence holes, can be estimated by the **quadrate method,** similar to that used for plants. Foliage arthropods may be sampled by a number of strokes with a standard sweep net over a $10 = m^2$ area. Estimates of zooplankton, obtained by pulling a plankton net through a given distance of water at several depths, can be made by filtering a known volume of sample through a funnel using a filter pump. The filter paper is marked off in equal squares and by the help of a hand lens or a binocular microscope, the organisms in each square are counted. The numbers are then related back to the total volume of water sampled.

Very small-sized zooplanktons are counted by **Rafter plankton-counting cell.** This consists of a microscope slide base plate ruled into ten 1 cm squares.

(ii) **Mark-recapture method**. In this method, a group of animals in the population are captured and marked with a band or tag for some time. Then they are released back into the population, where they are distributed among unmarked individuals. Later another group of animals is captured, some of which will be marked and some of which are not. The ratio of marked animals to the number in the sample is assumed to be similar to the ratio of the total animals initially marked to the total number of animals in the entire population.

(iii) **Removal method**. This method involves capturing animals and removing them from the study area. If this technique is used over a short time, the number of animals removed per unit effort should get smaller and smaller. By totalling the accumulative catch through time, it is possible to estimate the number in the area, even though we only remove some of the animals. This method is useful where one desires a relative measure or index figure for small-mammal populations in order to compare one habitat with another.

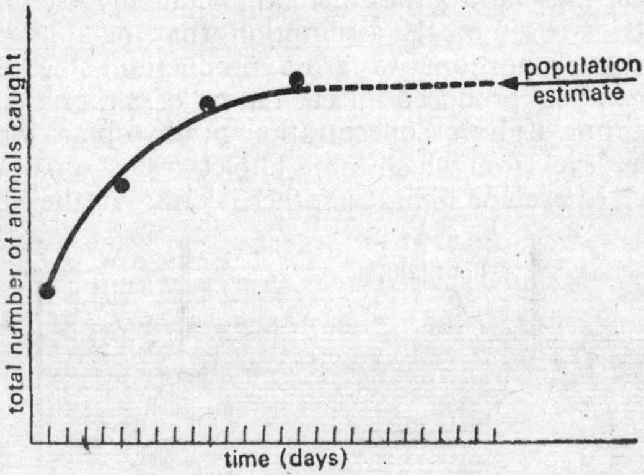

Fig. 4·7. Removal method for population estimation. Animals are captured and removed from an area. The total accumulative catch is followed, allowing an estimate of the total number of catchable animals in the area. (after Herreid II, 1977).

(c) **Indices**. **Indices** are estimates of animal populations derived from counts of animal signs, calls, road-side counts, and so on. The results do not give estimates of absolute populations, but they do indicate trends of populations from year to year and from habitat to habitat. **Call counts** are used for game birds like woodcock, pea cock, quail, etc. **Road side counts** involve counting of animal tracks, browse, signs, active dens and lodges, etc. **Counting pellet** or faecal groups is widely used to estimate big-game populations.

4-2·3 G. Measuring mortality. The measurement of mortality in natural populations is most earnestly required for various ecological aspects like construction of life tables, life equations, survivorship curves, growth curves, etc. The mortality in a wild population can be measured by following methods : (i) If a number of marked animals are found dead a known time after marking, the percentage dying can be plotted against time and a curve drawn through the points. The probability of dying can be read directly from the chart, or it may be calculated by dividing the number of individuals that die during a period of time by the initial population. (ii) If the animals can be readily aged, as is true with deer, rabbits, mountain sheep, and others, and if sufficient lower jaws, eye lenses, or horns can be obtained, then mortality and the probability of dying can be expressed on an age basis.

4-3. METHODS OF ESTIMATING PRIMARY PRODUCTION

The primary production of an ecosystem can be measured by following methods :

4-3·1. O_2 Production or Light and Dark Bottle Method

The light-and-dark-bottle method, commonly used in aquatic environments, is based on the assumption that the amount of oxygen produced is proportional to gross production, since one molecule of oxygen is produced for each atom of carbon fixed. Two bottles containing a given concentration of phytoplankton are suspended at the level from which the samples were obtained. One bottle is black to exclude light; the other is clear. In the light bottle

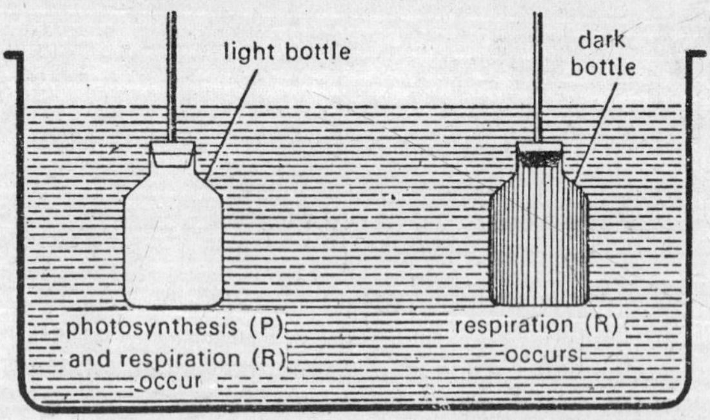

Fig. 4·8. Light and dark bottle method for estimating gross primary production of ecosystem (after Herreid II, 1977).

a quantity of oxygen proportional to the total organic matter fixed (gross production) is produced by photosynthesis. At same time the phytoplankton is using some of the oxygen for respiration. Thus the amount of oxygen left is proportional to the amount of fixed organic matter remaining after respiration or net production. In the dark bottle, oxygen is being utilized but is not being produced.

Thus the quantity of oxygen utilized, obtained by subtracting the amount of oxygen left at the end of the run (usually 24 hours) from the quantity at the start, gives a measure of respiration. The amount of oxygen in the light bottle added to the amount used in the dark bottle provides an estimate of total photosynthesis or gross production.

A modification of this method involves the whole aquatic ecosystem, which becomes the light bottle in day-time and dark bottle in night-time. The oxygen content of the water is taken every 2 to 3 hours during a 24-hour period. The rise and fall of the oxygen during the day and night is plotted as a diurnal curve. For obtaining a correct estimate for the oxygen production of plants, the oxygen exchanged between air and water and between the water and bottom must be estimated and deducted.

4-3·2. Carbon Dioxide Assimilation

This method is suitable for estimation of production of terrestrial ecosystems. It involves the measurement of the uptake of carbon dioxide and its release in respiration. In this method a sample of community, which may be twig and its leaves, a segment of the tree stem, the ground cover and soil surface, or even a portion of total community, is enclosed in a plastic tent. Air is drawn through the enclosure, and the carbon dioxide concentration of the incoming and outgoing air is measured with an infrared gas analyzer. The assumption is that any carbon dioxide removed from the incoming air has been incorporated into organic matter. A similar sample is enclosed within a dark bag. The amount of carbon dioxide produced in the dark bag is a measure of respiration. In the light bag, the quantity of carbon dioxide would be equivalent to photosynthesis minus respiration. The two results added together indicate gross production.

4-3·3. Chlorophyll Method

An estimate of the production of some ecosystems (*e.g.*, aquatic ecosystems) can be obtained from chlorophyll and light data. This method involves the determination of chlorophyll-a content of the plant per gram or per square metre, which under reasonably favourable conditions is the same. Because the quantity of chlorophyll in aquatic (and terrestrial) communities tends to increase or decrease with the amount of photosythesis (which varies at different light intensities), the chlorophyll per square metre indicates the food manufacturing potential at the time. Chlorophyll content in plants is accounted with the help of **colorimeter** (Fig. 4·9).

4-3·4. Harvest Method

This method is widely used in terrestrial ecosystems for estimating production. It is most useful for ecosystems for estimating production of cultivated land and range, where production starts from zero at seeding or planting time, becomes maximum at harvest, and is subject to minimal utilization.

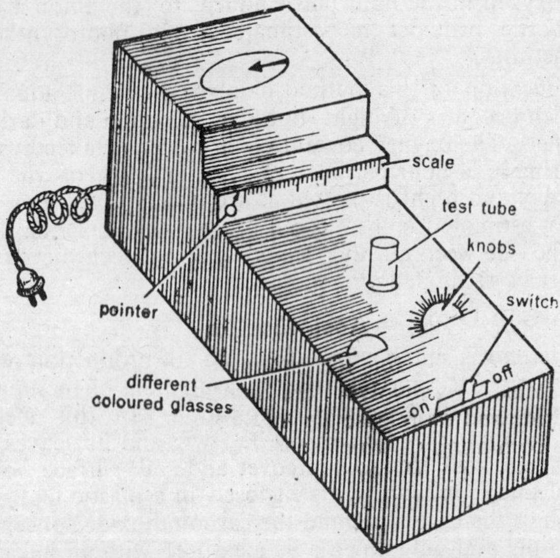

Fig. 4·9. A colorimeter to determine the chlorophyll content.

In brief, harvest method involves the clipping or removal of vegetation at periodic intervals, drying to a constant weight, and expressing that weight as biomass in grams per square metre per year. The caloric value of the material can also be determined by **bomb calorimeter**. Then the biomass is converted to calories and the harvested material expressed as kilocalories per square metre per year. To be accurate, plant material must be sampled throughout the growing season and the contribution of each individual species determined.

Because plants of different age, size, and species make up the forest and shrub ecosystems, a modified harvest method known as **dimension analysis** is used. This involves the measurement of height, the diameter at breast height (DBH), and the diameter growth rate of the trees in a sample plot. A set of sample trees is **cut**, weighed, and measured, usually at the end of growing season. The height to the top of the tree, DBH, depth and diameter of crown and other parameters are taken. Total weight, both fresh and dry, of the leaves and branches are determined, as is the weight of the trunk and limbs. Roots are excavated and weighed. By various calculations the net annual production of wood, bark, leaves, twigs, roots, flowers, and fruits is obtained. From this information the biomass and production of the trees in the sample unit are estimated and then summed for the whole forest.

4·3·5. Radioactive Tracers Method

The most recent method of determining production involves the measurement of the rate of up take of radioactive carbon (^{14}C) by plants. It is a most sensitive method to measure net photosynthesis under field conditions.

This method involves the addition of a quantity of radioactive carbon as a carbonate ($^{14}CO_3$) to a sample of water containing its natural phytoplankton population. After a short period of time, to allow photosynthesis to take place, the plankton material is strained from the water, washed, and dried. Then radioactive counts are taken, and from them calculations are made to estimate the amount of carbon dioxide fixed in photosynthesis.

In a similar way, radioactive phosphorus (^{32}P), has been used for determining the direction of energy flow in an ecosystem.

4-4. METHODS OF ESTIMATING CONSUMER PRODUCTION

Methods for estimating consumer production of an ecosystem, involve the **determination of food consumption, energy assimilation, heat production, maintenance requirements,** and **growth**.

The first step involves some estimation of food consumption. This can be determined in the laboratory or estimated in the field. Laboratory determinations involve feeding the animal a known quantity of its natural foods, allowing it to eat over a period of time, usually 24 hours, then removing the food and weighing the remains. The amount of food consumed equals the amount fed minus the amount removed. The caloric value of the food consumed can be determined by burning a sample in the bomb calorimeter. If the activity periods of the animal and the weight of the food in stomach with hold are known, then consumption can be accurately determined by multiplying the activity periods by the mean weight of observed stomach contents from a sample of animals from the population. Activity periods are used because most animal activity is usually concerned with feeding.

Once consumed, the food must be assimilated. Assimilation can be determined by subtracting the energy voided in faeces from energy consumed.

The energy assimilated is used for maintenance and growth. Energy used for maintenance is lost. The cost of maintenance can be determined by confining the animal to a calorimeter and measuring the heat production directly, or the energy used in maintenance can be determined indirectly by placing the animal in a respirometer and measuring the oxygen consumed or the carbon dioxide produced. These results are then converted to calories of heat. But to do this one must know the respiratory quotient, the ratio of volume of carbon dioxide produced to oxygen consumed. The respiratory quotient varies with the type of food utilized in the body. To estimate accurately the heat production of a population from laboratory determinations, one must also know the daily activity periods, the weight distribution of the population, and the environmental temperature.

Production or storage of energy is estimated by weighing individuals fed on a natural diet in the laboratory or by weighing animals each successive time they are caught in the field.

An indirect and usually more useful method is based on the age distribution of a population, the growth curve for the species, and the caloric value of the animal tissue. Growth curves are obtained for each population under investigation and for each season under study. Once a growth curve is available and the age distribution of the population is known, the weight of the tissue produced in a given period can be estimated for each age category. The weight gain is then converted to caloric equivalents (For details see S. Z. **Qasim**, 1973).

4·5. METHODS OF MEASUREMENTS OF ENVIRONMENTAL FACTORS

Different environmental factors of two basic habitats, soil and water can be measured by following methods :

4-5·1. Measurement of Soil Characteristics

Different parameters of soil can be determined by following methods :

4-5·1 A. Soil atmosphere. The soil gases are collected by an apparatus devised by **Haley** and **Brierley** (1953) and are tested in the laboratory for oxygen and carbon dioxide by **Haldane-Gutherie gas analysis apparatus**.

4-5·1 B. Soil organic matter. A reasonable estimate of soil organic matter can be obtained by loss of weight on ignition. About 5g of oven dried soil is taken, weighed carefully, and then heated red hot in a crucible for about half an hour. It is cooled in a desiccator, is reweighed and heated again. This procedure is continued until weight of soil becomes constant. The loss in weight, expressed as a percentage, represents the amount of oxidizable organic matter present.

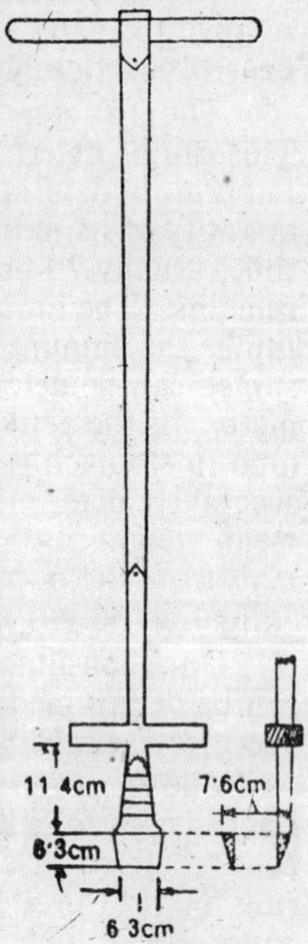

Fig. 4·10. Soil sampling tool. The graduations indicate the depth of the soil core taken (after Lewis and Taylor, 1967).

4-5·1 C. Soil texture and composition. For studying the texture and composition of soil following two methods are used :

(A) Samples of soils are collected from various areas–forests, fields and gardens by the help of a **soil augur** (Fig. 4·10) which enables one to extract soil from known depths and horizons. Each type of soil sample is seen under hand-lens and small amount of it, in dry as well as wet or moist state, is squeezed between thumb and fingers. The texture of sample soil can be identified by following characteristics :

Examination of soil and as felt between thumb and fingers	Texture
Individual grains may be seen or felt (gritty), from a cast when moist soil is squeezed	Sandy
Gritty, form a cast but fall apart when dry soil is squeezed, moist soil forms cast that does not break	Sandy loam
Mellow with a gritty feel, when dry soil is squeezed forms a cast requiring careful handling, moist one forms cast that does not break	Loam
Dry appearing cloddy, with soft feeling, forming cast (dry as well as moist) that does not break, moist soil does not form ribbons	Silt Loam
Breaking into clods or lumps, hard when dry, if moist soil is pinched forming thin ribbon which breaks readily.	Clay Loam
Forms very hard clods or lumps when dry, sticky when wet, if moist soil squeezed, it forms a long, flexible ribbon	Clay

(B) In another standard method, the soil is passed through a series of sieves with different size (diameter) of their meshes (pores), *i.e.*, soil is passed through sieves with decreasing aperture sizes which separate different components of soil, as shown in following table :

Diameter (mm) of the particle	Texture
below 0·002	Clay
0·002—0·02	Silt
0·02 —0·20	Fine sand
0·20 —2·00	Coarse sand

4-5·1 D. Soil colour. The soil sample is spread uniformly over a cardboard sheet. The particles of soil are now matched with chips of a number of colours in the **Munsell's soil colour chart** and colours of each is noted on the chart.

4-5·1 E. Soil temperature. Soil temperature is measured by a special type of thermometer–**soil thermometer**, which has a vertical arm with bulb at one end, and a dial with deflection needle on the other end. The bulb of thermometer is buried at different depths, *i.e.*, 1″, 6″, 12″, 18″ of the soil. The temperature may be read on the dial in °C values. (**Fig. 4·11**).

Highly accurate temperature measurements of microlocations like soil, leaf litter, tree wood, bark, leaf, etc., are taken with resistance thermometers or **thermistors** (Fig. 4·12) and microprobes that are buried in the material or object. Thermistor is based on the principle of **thermocouple**. The microprobe of a thermistor contains thermocouple in glass globule which is lowered at different depths in soil or object. The leads (cable) of a thermistor are projecting above the surface. To take a reading, the wires are attached to the terminals of a potentiometer, on which temperature is read directly.

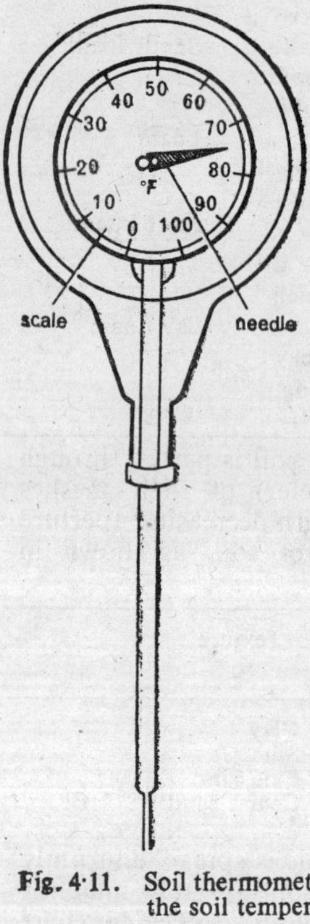

Fig. 4·11. Soil thermometer to measure the soil temperature.

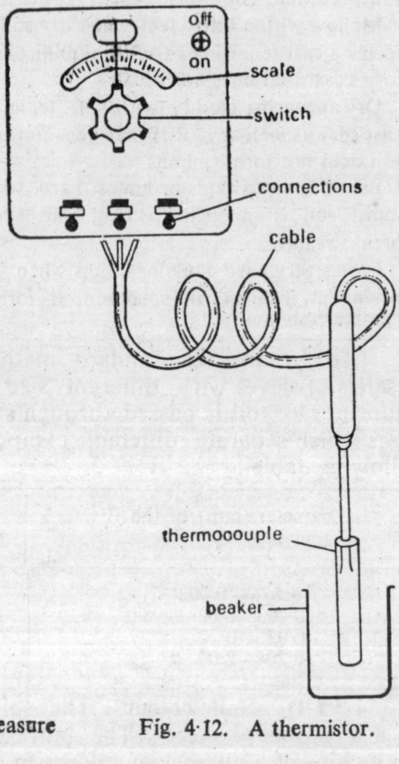

Fig. 4·12. A thermistor.

4-5·1 F. Soil moisture. Soil moisture content is determined in percentage by comparing the weight of the soil samples at the time of collection with their weight in oven-dry condition. In this method, a soil sample of known weight is kept in a hot-air oven at a temperature of 105°—110°C until the weight of sample becomes constant (*i.e.*, usually 24 hours). Then, the soil sample is cooled in the desiccator and weighed. The loss in weight from fresh to dry represents the moisture content or actual amount of water lost. The loss in weight expressed as a percentage is the percent moisture content. It is found as follows :

Fresh weight – Oven dry weight (O.D.) = Moisture content (M.C.)

Percent moisture content (P.M.C.) $= \frac{\text{M.C.} \times 100}{\text{O.D.}}$

4-5·1 G. Carbonate content. For measuring the carbonate contents in different soil samples, a little amount of each type of soil is taken in different test tubes and to each tube few drops of hydrochloric acid (HCl) is added and the degree of effervescence is noted in each test tube. The degree of effervescence produced in all cases may be divided into different groups as +, ++, +++, ++++, or as A, B, C, D. The sample showing maximum degree of effervescence (*i.e.*, D) indicates the maximum amount of carbonate content present.

4-5·1 H. Nitrate content. For measuring the nitrate contents in different soil samples, a clear solution of each type of soil in water (soil 1 part : water 5 parts) is prepared. A small amount of each solution is taken on white spot plate and to each is added 2–3 drops of diphenylamine and 2 drops of conc. H_2So_4 (0·2%). The appearance of blue colour indicates the presence of nitrate in soil sample. The degree of intensity of the blue colour is noted and divided into four arbitrary groups, as done for carbonate content.

4-5·1 I. pH value. For field test. pH of soil is recorded with the help of **universal indicator.** For pH determination, small amount of each type of soil sample is taken in different test tubes and to each equal amount of barium sulphate and 15 ml of water are added. The solution is kept for some time till a clear supernatant fluid is obtained. Now few drops of this fluid is taken on a porcelain tile and an equal amount of universal indicator is added to it. The colour produced is matched with the colour chart for different values pasted on the indicator bottle and pH value of each is noted. Besides such a colorimeteric determination of pH, an electronic **pH-meter** is also used for this purpose.

4-5·2. Measurement of Aquatic Characteristics

Different characteristics of aquatic environment (water) can be measured as follows :

4-5·2 A. Collection of water sample. The collection of samples of water from various depths of lakes, ponds, rivers, and estuaries is a difficult task and is done by **Meyer water sampler.** This apparatus consists of a heavy bottle fitted with a weight heavy enough to sink it when full of air and attached to a wire harness. A stopper with an eyebolt attached is loosely inserted. A cord is attached both to the eyebolt and to the wire harness. When the bottle is lowered to the desired depth (indicated by marked line), the cord is jerked to remove the cork. Sufficient time is allowed for the bottle to fill and it is then pulled to the surface.

4-5·2 B. Dissolved oxygen. The dissolved oxygen of water is determined by rapid **Winkler's method** which involves titration. This method has been described in detail by **Smith** (1974).

4-5·2 C. Free carbon dioxide concertration. The concentration of free CO_2 in water is determined by titration by NaOH solution and phenolpthalein solution (for details see **Smith**, 1974).

4-5·2 D. Hardness and salinity. Temporary hardness of water is caused by carbonate of calcium and magnesium (or by combined CO_2) and is determined by titration method by HCl solution.

Permanent hardness of water is caused by chlorides and sulphates of calcium and magnesium and is accurately measured by a **colorimetric titration method**. This method is based on the ability of sodium versenate (sodium diethylenediamine tetracetate) to form un-ionized complexes with calcium and magnesium. If eriochrome black T, a dark blue dye, is added to this solution containing Ca^{2+} and Mg^{2+} ions, a complex, pink in colour, is formed. By adding sodium versenate solution to this complex, the solution can be turned back to blue again by removing Ca^{2+} and Mg^{2+} from dye complex to the versenate complex again. The end point, when the pink changes back to blue, is used as a measure of total hardness.

Salinity of estuarine water is tested by titration with silver nitrate solution.

4-5·2 E. Acidity or **pH value**. For the determination of pH value of water of different habitats, samples of water are collected during different months of the season, and from different places and depths. From each sample few drops of water are taken and are tested for pH value either by universal indicator, narrow-range pH papers or pH meter.

4-5·2 F. Turbidity. Turbidity of pond water or river water is recorded with the help of **Secchi disc** which is a round disc with different strips of two contrast colours, white and black. The degree of indistinction between two colours in the water determines the index of turbidity of water. For measuring the turbidity index, the disc is lowered slowly and slowly in water until black and white colours become indistinguishable and the depth is noted down. This depth indicates the turbidity index.

4-5·2 G. Temperature. The temperature of water of different habitats is measured at different depths at different times by **thermistor** or by **maximum** and **minimum thermometer**. The latter consists of two thermometers, each with metallic float that in the one thermometer lodges at the highest temperature, in the other at the lowest. For new readings the floats are reset at the top of the mercury **columns**, preferably with a small magnet. For recording the water temperature at different depths of pond, lake or river, to one end of thermometer is attached a measured line and to its other end a weight is attached. The reading is taken vertically from the surface to the bottom at 0·5 or 1·0 m intervals.

4-5·2 H. Light intensity. Light intensity of aquatic environment is measured by the use of **photometer** which consists of a photoelectric cell and an amperemeter. The photocell is exposed to light, and the value is read on the amperemeter, with deflection needle. The light intensities are measured in terms of **lux units.**

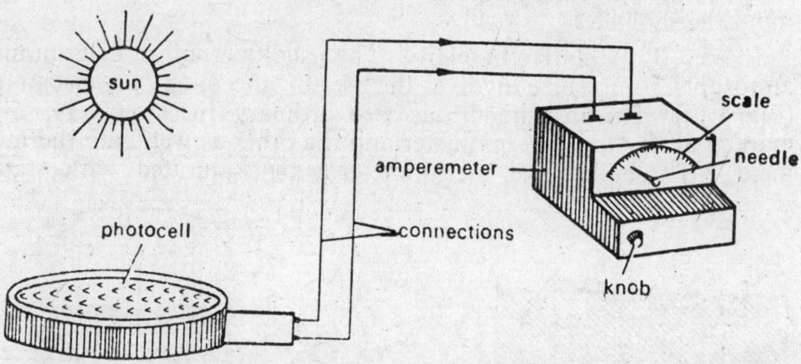

Fig. **4·13.** A photomete r (Luxmeter).

For the use of photometer in water, the photocell is sealed in a transparent waterproof case. The photocell is placed at different sites and depths of pond water to find out variations in light intensity values.

4-5·2 I. Current. The estimates of stream flow or current is made either by **current meters** or by a quick method of **Robin** and **Crawford** (1954). In latter method, a cross section of the stream is chosen where the current and depth are most uniform, and its width is measured. The width is divided into three equal segments by pushing sticks into the bottom or colouring the surface of stones. Then, the depth is recorded at the midpoint of each segment and the velocity of the surface current is determined by dropping a fisherman's float (without projecting arms) attached to 5 ft of limp, monofilament nylon fishing line (0·005–0·01 inch in diameter). The time required for the float to travel the 5 ft depth is recorded by a stopwatch. This is done several times and the average figure recorded and converted into feet per second.

The volume of flow R can be determined for each segment of the cross section by the following formula :

$$R = WDaV$$

where, a = a bottom factor constant (0·8 for rocks and coarse gravel ; 0·9 for mud, sand, hardpan, bedrock)

W = width of the segment

D = depth at the midpoint of the segment

V = surface current velocity taken at the midpoint of the segment

Total flow is determined by adding the flows for the three segments.

4-5·3. Measurements of Characteristics of Aerial Environment

Different aspects of aerial environment or atmosphere are studied by **following methods** :

4-5·3 A. Tempereture. The temperature of air is measured by a standard mercury thermometer or especially by maximum minimum thermometer.

4-5·3 B. Relative humidity. The usual method of determining atmospheric moisture involves the use of the **sling psychrometer** (Fig. 4·14). This instrument has two ordinary thermometers, one marked as dry-bulb thermometer and the other as wet-bulb thermometer. The bulb of wet-thermometer is kept saturated with water

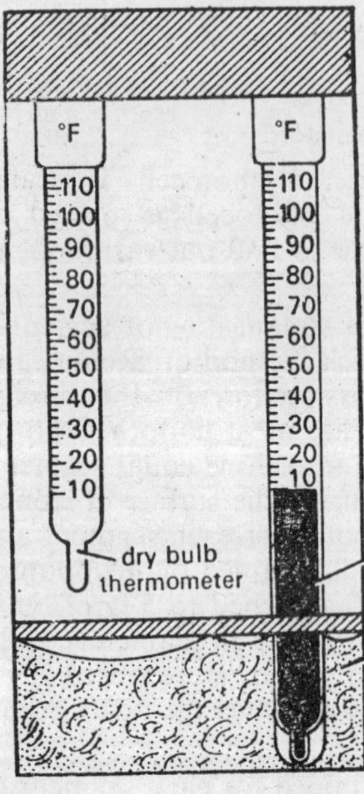

Fig. 4·14. A sling psychrometer.

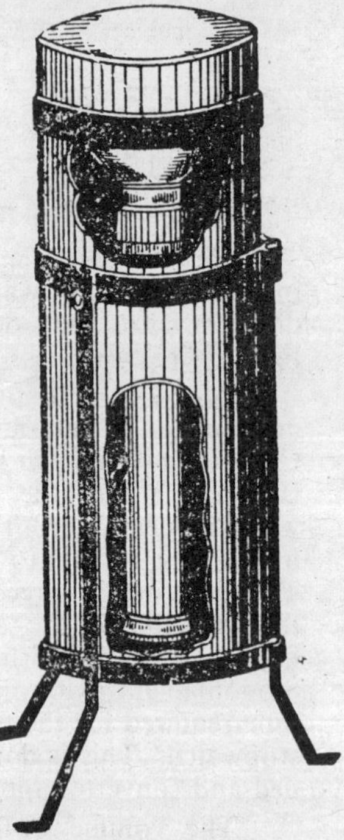

Fig. 4·15. Rain gauge.

by covering it with a piece of moist cotton. When readings are to be taken, the fabric over the wet bulb is soaked in distilled water and instrument is rapidly swung so that the evaporation cools the wet-bulb. When the readings of two bulbs, wet and dry, are constant, the temperatures on both are recorded. Relative humidity is obtained from the dry-bulb reading and the difference between the dry-bulb reading and wet-bulb reading. To convert these readings to relative humidity, a conversion table prepared for this purpose is consulted. If there is no difference in the values of dry bulb and wet bulb thermometers it means relative humidity is 100 per cent.

4-5·3 C. Light intensity. Light intensities are recorded by photometer under different conditions, *i.e.*, shade, open sun, etc., and variations are noted.

4-5·3 D. Rain fall. Precipitation or rain fall of any **region** is measured by an apparatus called **rain gauge** (Fig. 4·15).

4-5·3 E. Wind velocity. The velocity of wind is measured by an instrument called **anemometer** (Fig. 4·16).

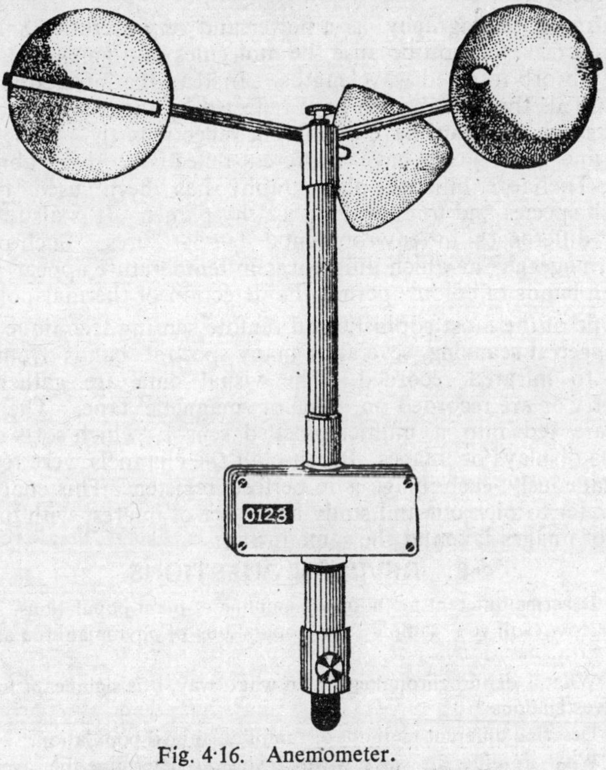

Fig. 4·16. Anemometer.

4-6. MAPPING THE STUDY AREA

In some ecological studies an accurate map is desirable and often necessary. It may be needed to show the location of sample plots, animal concentrations and movements, vegetative cover and the like. The area is mapped by tape and compass according to certain standard land-surveying techniques (**Smith**, 1974). Maps are prepared for land areas, lakes, ponds and a section of streams.

4-7. REMOTE SENSING

Remote sensing is the technique of learning about ecosystems and the environment without direct contact. These methods involve the use of electromagnetic and sound waves, radiant energy and ionizing radiation.

One method of remote sensing uses aerial photography. Black and white **stereoscopic photos** are used to determine the volume of timber stands, to count wild animals, to record vegetational and man made changes in the landscape and to prepare topographic maps. More recently **colour photography** has been widely employed, for its visual contrasts make the identification of tree species, insect damage, plant disease and details of aquatic environment much easier.

Infrared photography is a newer and very useful technique in the study of vegetation because the molecules of pigments in plants do not absorb infrared wavelengths. Instead the infrared is either transmitted through the leaf or is reflected by the cell walls. Cells of one species of plant have a different reflectivity than those of another, and normal cells have a different reflectivity than abnormal ones. Therefore, infrared photography has been used to distinguish species and to detect unhealthy plants. It is also used to detect differences in environmental temperatures. Such pictures or **thermographs** in which differences in temperature appear as contrasting bands of colour, permit the detection of thermal pollution.

One of the most sophisticated remote sensing technique, called **multispectral scanning**, several or many spectral bands from ultraviolet to infrared recorded. The visual data are gathered by telemetry or are recorded on film or magnetic tape. The image data are fed into a multichannelled sensor, which sorts out the various displays or images. Because all the channels were recorded simultaneously, each image is in perfect register. This enables the interpreter to pick out and study an object of interest with different types of images taken at the same instant.

4-8. REVISION QUESTIONS

1. Describe different methods of sampling of plant populations.
2. How will you sample the populations of phytoplankton and periphyton ?
3. What is dendrochronology ? In what way, it is siginficant for ecological investigations ?
4. Describe different methods of sampling animal population.
5. What are different ways by which you can determine the age of wlid animals ?
6. Describe different methods of estimation of primary production of various ecosystems.
7. How will you determine consumer production of an ecosystem ?
8. How will you measure the soil temperature and soil pH ?
9. What is remote sensing ? Describe its different aspects in brief.

5
Statistical Applications in Ecology

5-1. DEFINITIONS

Biostatistics can be defined as *the application of statistical methods to the solution of biological problems* (**Sokal** and **Rohlf,** 1973). It is also called **biological statistics** or **biometry**. Ecology being a biological science requires a wide application of statistical methods. The above definition of biostatistics leaves us somewhat up in the air—"statistics" has not been defined as yet.

Statistics can be defined as *the scientific study of numerical data based on natural phenomena.* **All** parts of this definition **are** significant and deserve emphasis.

Scientific study. In the science of statistics, we are concerned with the commonly accepted criteria of validity of scientific evidence. Thus, objectivity in presentation and in evaluation of data and the general ethical code of scientific methodology must constantly be in evidence.

Data. Statistics generally deals with populations or groups of individuals ; hence it deals with **quantities** of information, not with a single **datum.** Thus, the measurement of a single animal or the response from a single biochemical test will generally not be of interest.

Numerical. Unless the data of a study can be quantified in one way or the other, they would not be liable to statistical analysis. Numerical data can be measurements—the length or width of a structure, or counts—the number of species of microbes, plants and animals in a pond, lake or any habitat.

Natural phenomena. We use this term in a wide sense, including all those events that happen in animate and inanimate nature not under the control of man, plus those evoked by the scientist and partly under his control, as in an experiment.

5-2. DEVELOPMENT OF BIOSTATISTICS

The origin of modern statistics can be traced back into the 17th century, in which it derived from two sources. The first of these related to political science and developed as a quantitative description of the various aspects of the affairs of a government or state (hence the term statistics). This subject also become known as political arithmetic. Taxes and insurance caused people to become interested in problems of census, longevity, and mortality. Such considerations assumed increasing importance, especially in England, as the country prospered during the development of its empire. **John Graunt** (1620–1674) and **William Petty** (1623–1687) were early students of vital statistics and others followed in their footsteps.

Second root of modern statistics existed in the mathematical theory of probability which roused by the interest in games of chance among the leisure classes of the time. Important contributions to this theory were made by **Blaise Pascal** (1623–1662) and **Pierre de Fermat** (1601–1665), both Frenchmen. **Jacques Bernoulli** (1654–1705), a Swiss, laid the foundation of modern probability theory in *Ars conjectandi*, published posthumously. **Abraham de Moivre** (1667–1754), a Frenchman, was the first to combine the statistics of his day with probability theory in working out annuity values. De Moivre also was the first to approximate the improtant normal distribution through the expansion of the binomial.

A later stimulus for the development of statistics came from the science of astronomy, in which many individual observations had to be digested into a coherent theory. Many of the famous astronomers and mathematicians of the 18th century such as **Pierre Simon Laplace** (1749–1827) in France and **Karl Friedrich Gauss** (1777–1855) in Germany were among the leaders in this field. Perhaps the earliest important figure in biostatistic thought was **Adolphe Quetelet** (1796–1874), a Belgian astronomer and mathematician, who in his work combined the theory and practical methods of statistics and applied them to problems of biology medicine, and sociology. **Francis Galton** (1822–1911), a cousin of Charles Darwin, has been called the father of biostatistics and eugenics, two subjects that he studied interrelatedly. Galton's major contribution to biology is his application of statistical methodology to the analysis of biological variation, such as the analysis of variability and his study of regression and correlation in biological measurements. **Karl Pearson** (1857–1936), at University college, London, became intersted in **application** of statistical methods to biology, particularly in the demonstration of natural selection, through the influence of **W.F.R. Weldon** (1860–1906), a zoologist at the same institution. **Weldon**, incidently, is credited with coining the term biometry for the type of studies pursued by him. **Pearson** continued in the tradition of Galton and laid the foundation for much of descriptive and correlations statistics. **Student** (1908) contributed the 't' distribution. The dominant figure in statistics and biometry in this century has been **Ronald A. Fisher** (1890–1962). Statistics today is a broad and extremely active field, whose applications touch almost every science and even the humanities.

5-3. NEED OF BIOMETRY IN ECOLOGY

The basic understanding of biometry is necessary for the students of biology, agriculture and ecology. Firstly, the study of biometry will help the potential ecologist to understand the nature of variability. Secondly, biometry helps us in driving general laws from small samples. It is because the biologist has generally to deal with small samples (the yield of a wheat strain grown on six or eight plots ; the weight gains of four or five pigs fed on a particular diet). The process of **inductive inference** reasoning from the particular to the general–has had considerable influence in statistical science.

Thus, various biometrical principles such as normal distribution, random sampling, tests for significance (X_2—square test, t—test, Z—test, probability, etc.) have a wide application in ecology. Therefore, an understanding of certain basic laws and methods of biometry will be quite useful for the students of ecology for appreciating certain aspects of ecology and in drawing general conclusions for a population from small samples.

5-4. DATA IN BIOLOGY

We shall now define a number of important terms necessary for an understanding of biological data.

5-4·1. Samples and Populations

The data in biostatistics are generally based on **individual observations**. They are observations or measurements taken on the smallest **sampling unit**. These smallest sampling units frequently, but not necessarily, are also individuals in the ordinary biological sense. If we measure weight in 100 rats, then the weight of each rat is an individual observation ; the hundred rats weight together represent the **sample of observations**, which is defined as *a collection of individual observations selected by a specified procedure*. In this example, one individual observation is based on one individual in a biological sense–that is in one rat. However, if we had studied weight in a single rat over a period of time, the sample of individual observations would be the weights recorded on one rat at successive times. Likewise, if we wish to measure temperature in a study of ant colonies, where each colony is a basic sampling unit, each temperature reading for one colony is an individual observation, and the sample of observations is the temperatures for all the colonies considered.

The term "individual observation" and 'sample of observations' as used above defines only the structure but not the nature of the data in a study. The actual property measured by the individual observations is the **character** or **variable**. The more common term employed in general statistics is **variable**. However, in biology the word **character** is frequently used synonymously. More than one character can be measured on each smallest sampling unit. Thus, in a group of 25 mice we might measure the blood pH and the erythrocyte count. Each of the 25 mice (a biological individual) is the smallest sampling unit ; blood pH and red blood cell count would be two characters studied ; the pH readings and cell counts are individual observations, and two samples of 25 observations, on pH and erythrocyte count would result. Or we may speak of a **bivariate sample** of 25 observations, each referring to a pH reading paired with an erythrocyte count.

Population, biologically refers to all individuals of a given species (perhaps of a given life-history stage or sex) found in a circumscribed area at a given time. In statistics, population always means *the totality of individual observations about which inferences are to be made, existing anywhere in the world or at least within a definitely specified sampling area limited in space and time.* For example, if one takes five man and studies the number of leucocytes in their peripheral blood and he is prepared to draw conclusions about all men from this sample of five, then the population from which sample has been drawn represents the leucocyte counts of all extant males of the species *Homo sapiens.* If, on the other hand, one restricts one self to a more narrowly specified sample, such as five male Indians, aged 30, and one is restricting his conclusions to this particular group, then the population from which one is sampling will be leucocyte numbers of all Indian males of age 30. The population **in this statistical sense** is sometimes referred to as the **universe**.

A population may refer to variables of a concrete collection of objects or creatures, such as the tail lengths of all white mice in the world, the leucocyte counts of all the male Indians in the world of age 30, etc. ; or it may refer to the outcomes of experiments, such as all the heart beat frequencies produced in guinea pigs by injections of adrenalin. In the first cases the population is generally finite. Although in practice it would be impossible to collect, count, and examine all white mice in the world or male Indians of age 30, these populations are in fact finite. Certain smaller populations, such as all the rhinoceroses in India or all the pocket gophers in a given colony, may well lie within reach of a total census. However, though populations are generally so theoretically finite in most applications in biology, they are generally so much larger than samples drawn from them, that they can be considered *de facto* infinitely sized populations.

5-4·2. Determinations

The records of the observations are known as **determinations.** For different variables, following two types of determinations are used—**measurements** and **counts**. The variables of continuous nature are often measured. For example, the yield of milk is measured in terms of weight or volume and the size of the wing of *Drosophila* is measured by scale. On the other hand, the variables of discrete or discontinuous nature are determined by counting. For example, the number of hair follicles on animals is determined by counting and the chaeta number and egg- production of *Drosophila* are determined by counting. The counting yield **integers** (whole numbers) and a discrete variable cannot take up the innumerable possible fractional values between the integers.

In order to develop a general theory, it is necessary to use symbols. Instead of writing variable each time, let X denote the variable being studied. Thus X may on one occasion be the grain yield per plot and on another occasion, the number of diseased wheat plants per plot.

Further, let X_1 be the yield of the first plot, X_2 the yield of the second plot, X_3 the yield of the third plot and so on. In genral let X_i (read as X sub *i*) denote the yield on the *i*th plot.

Let X_i=yield of grain from the *i*th plot, $i=1, 2, \ldots\ldots\ldots 20$.

A sign of operation used continually in biometry is the sign of summation, usually written as Σ which is the Greek capital letter *Sigma.*

The sum $X_1+X_2+X_3+X_4+X_5$ is written as $\sum_{i=1}^{5} X_i$ and is read the sum of all the X's from 1 to 5' (or summation X sub *i*, *i* running from 1 to 5'). The subscript is usually a letter taken from the middle of the alphabet. The letters *h, i, j, k, l* are those most commonly used as subscripts. The first value is written below the summation sign and the last above it. These are called the **limit of summation**. When it is obvious what these limits are, they are often omitted.

5-4·3. Variables in Biology

Each biological discipline has its own set of variables, which may include conventional morphological measurements, concentrations of chemicals in body fluids, rates of certain biological processes, frequencies of certain events as in genetics and radiation biology and many more. A **variable** can be defined as *a property with respect to which individuals in a sample differ in some as certainable way*. If the property does not differ within a sample at hand or at least among the samples being studied, it cannot be of statistical interest. Length, height, weight, number of teeth, vitamin C contents and genotypes are examples of variables in ordinary, genetically and phenotypically diverse, groups of organisms. Warm-bloodedness in a group of mammals is not, since they are all alike in this regard, although body temperature of individual mammals would, of course, be a variable.

5-4·3A. Types of biological variables. Biological variables has been classified into following types :

1. **Measurement variables.** *Measurement variables are all those whose differing states can be expressed in a numerically ordered fashion.* They are divisible into two kinds. The first of these are **continuous variables,** which at least theoretically can assume an infinite number of values between any two fixed points. For example, between two length measurements 1·5 and 1·6 cm there is an

infinite number of lengths that could be measured if one were so inclined and had a precise enough method of calibration to obtain such measurements. Any given reading of a continuous variable, such as a lenght of 1·57 mm is, therefore, an approximation to the exact reading, which in practice is uncommon. Some common examples of biological continuous variables are lengths, areas, volumes, weights, angles, temperatures, periods of time, percentages, and rates.

Contrasted with continuous variables are the **discontinuous variables**, also known as **meristic** or **discrete variables**. These are variables that have only certain fixed numerical values, with no intermediate values possible in between. Thus, the number of segments in a certain insect appendage may be 4 or 5 or 6, but **never** 5½ or 4·3. Examples of discontinuous variables are numbers of certain structure (such as segments, bristles, teeth or glands), the numbers of offspring, the numbers of colonies of micro-organisms or animals, or the numbers of plants in a given **quadrate.**

2. **Ranked variables**. Some variables cannot be measured but at least can be ordered or ranked by their magnitude. Thus, in an experiment one might record the rank order of emergence of ten pupae without specifying the exact time at which each pupa emerged. In such cases the data is coded as a **ranked variable**, the order of emergence. Thus, by expressing a variable as a series of ranks, such as 1, 2, 3, 4, 5, we do not imply that the difference in magnitude between, say, ranks 1 and 2 is identical to or even proportional to the difference between 2 and 3.

3. **Attributes**. Variables that cannot be measured but must be expressed qualitatively are called **attributes**. These are all properties, such as black or white, pregnant or not-pregnant, dead or alive, male or female. When such attributes are combined with frequencies, they can be treated statistically. For example, of 80 mice, we may state that four were black, two agouti, and the rest gray. When attributes are combined with frequencies into tables suitable for statistical analysis, they are referred as **enumeration data**. Thus, the enumeration data on colour in mice just mentioned would be arranged as follows :

Colour	Frequency
Black	4
Agouti	2
Gray	74
Total number of mice=80	

5-4·3 B. Variate. Variate is a single reading, score or observation of a given variable. Thus, if we have measurements of the length of the tails of five mice, tail length will be a continuous variable, and each of the five readings of length will be a variate.

5-4·3C. Derived variables. There is an other important class of variables in biological research that we may call the **derived** or **computed variables**, which are generally based on two or more independently measured variables whose relations are expressed in a certain way. We are referring to ratios, percentages, indices, rates, and the like.

A **ratio** expresses as a single value the relation that two variables have one to the other. In its simplest form it is expressed as in 64 : 24, which may represent the number of wild type versus mutant individuals or the number of males versus females, or the proportion of parasitized individuals versus those not parasitized and so on. The above example implied ratios based on counts ; a ratio based on a continuous variable might be similarly expressed as 1·2 : 1·8, which may represent the ratio of width to length in a sclerite of an insect or the ratio between the concentrations of two minerals contained in water or soil. Ratios may also be expressed as fractions ; thus the two ratios above could be expressed as $\frac{64}{24}$ and $\frac{1\cdot2}{1\cdot8}$. However, for computational purposes it is most useful to express the ratio as a quotient. The two ratios cited above would, therefore be 2·666... and 0·666..., respectively. These are pure numbers, not expressed in measurement units of any kind. **Percentages** are also a type of ratio. Ratios and percentages are basic quantities in biological research, and are widely used and generally familiar.

An **index** is the ratio of one anatomic variable divided by a larger, so called standard one. An index could be the average of two measurements—either simply, such as 1/2 (length of A+length of B), or in weighted fashion, such as 1/3 [(2×length of A)+length B]. Another example of an index is the cephalic index in physical anthropology.

Rates will be important in many experimental fields of biology. The amount of a substance liberated per unit weight or volume of biological material, weight gain per unit time, reproductive rates per unit population size and time (birth rates), and death rates would fall in this category.

5-4·4. Frequency and Frequency Distribution

The term **frequency** means the number of occurrence of a observation. The frequency of milking means the number of times the animal is milked. The frequency of a particular observation is the number of times the same observation occurs. **Total frequency** is the term generally used for the sum of the frequencies for the individual items. If the frequency of an observation X_1 is f_1, that of X_2 is f_2, etc., then the total frequency is $f_1+f_2+...+f_k=N$, if there are k items and the total number of observations including the repititions is N.

Further, after data have been obtained in a given study they have to be arranged in a form suitable for computation and interpretation. We **may** assume that variates are randomly ordered initially or are in the order in which the measurements had been taken. A simple arrangement would be an **array** of the data by order of magnitude. Thus, for example, the variates 7, 6, 5, 7, 8, 9, 6, 7, 4, 6, 7 could be arrayed in order of decreasing magnitude as follows : 9, 8, 7, 7, 7, 7, 6, 6, 6, 5, 4, where there are some variates of the same value, such as the 6's and the 7's in this fictitious example, a time-saving device might immediately have occurred to us—namely, to list a frequency for each of the recurring variates, thus : 9, 8, 7 (4 ×), 6 (3 ×), 5. 4. Such a shorthand notation is one way to represent a **frequency distribution**, which is simply *an arrangement of the classes of variates with the frequencies of each class indicated.* Conventionally, a frequency distribution is stated in tabular form as follows for the above example :

Variable Y	Frequency f
9	1
8	1
7	4
6	3
5	1
4	1

The above is an example of a **quantitative frequency distribution**, since Y is clearly a measurement variable. A quantitative frequency distribution based on meristic variates is shown in Table 5·1. This is an example from plant ecology : the number of plants per quadrate sampled are listed at the left in the variable column ; the observed frequency is shown at the right.

Table 5·1. A meristic frequency distribution : number of plants of the sedge *Carex flacca* found in the 500 quadrates (Archibald, 1950).

No. of plants per quadrate (Y)	Observed frequency (f)
0	181
1	118
2	97
3	54
4	32
5	9
6	5
7	3
8	1
Total	500

However, arrays and frequency distributions need not be limited to quantitative variables. We can also make frequency distributions of attributes, called **qualitative frequency distributions**. In these, the various classes are listed in some logical or arbitrary order. For example, in genetics we might have a qualitative frequency distribution as follows :

	f
$A-$	86
aa	32

This exhibits that there are two classes of individuals, those identified by the A—phenotype, of which 86 were found, and the homozygote recessive aa, of which 32 were seen in the sample. In ecology, it is quite common to have a species list of inhabitants of a sampled ecological area. The arrangement of such tables is usually alphabetical. An example of an ecological list constituting a qualitative frequency distribution of families is shown in Table 5·2. In this example the families of insects referred to are listed alphabetically ; in other cases the sequence may be by convention, as for the families of flowering plants in botany.

Table 5·2. A qualitative frequency distribution : number of individuals of Heteroptera tabulated by family in total samples from a summer foliage insect community (**Whittaker**, 1952).

Family	f
Alydidae	2
Anthocoridae	37
Coreidae	2
Lygaeidae	318
Miridae	373
Nabide	3
Neididae	5
Pentatomidae	25
Piesmidae	1
Reduviidae	3
Rhopalidae	2
Sadidae	1
Thyreocoridae	10
Tingidae	62
Total Heteroptera	=851

5-4·5. Apparent and True Class Intervals and Class Limits

If a plant breeder measures the heights of 100 wheat plants and each height is measured to the nearest centimetre, the classes might be 'labelled' (in cm) 50–59, 60–69, 70–79, 80–89, etc. The plant whose height is recorded as 74 cm will be put into the class 'labelled' 70 to 79 cm. The interval 70 cm to 79 cm is called the **apparent class interval**. A interval between 69·5 cm to 79·5 cm is called the **true class interval**. The limits of the true class interval are mid-way between two recorded measurements—the lower limit (69·5 cm) of the true class interval being mid-way between the upper limit (69) of the preceding apparent class interval (60–69) and the lower limit (70) of the apparent class interval under consideration (70–79).

5-4·6. Graphical Representation of Data

Graphs are more intelligible than the numerical statement as far as a layman is concerned. Moreover graphs provide quick conclusions if the observer is more interested in the trend than in the actual magnitude of changes.

The commonly used graphs to represent frequency tables are the histogram, frequency polygon, frequency curve and bar diagram.

(i) Histogram. In the histogram, which is used for continuous variables, the true class intervals are represented on the horizontal axis, which need not begin at zero. Above these intervals, rectangles are erected whose areas are proportional to the frequencies of the various classes. For a distribution whose class ranges are all the same, these rectangles are obtained by marking on the vertical axis (which must begin at zero) either the frequency per interval or the frequency per unit range. The frequency per interval gives a quick visual appreciation of the number in any particular class.

When drawing a histogram for a frequency distribution with unequal class ranges, since the areas have to be proportional to the frequencies, the vertical scale must show the frequency per unit range (instead of the frequency per interval). The frequency per unit range is the frequency per interval divided by the range of that particular interval. Irrespective of the scale chosen for the vertical axis, it is important that the histogram be self-explanatory.

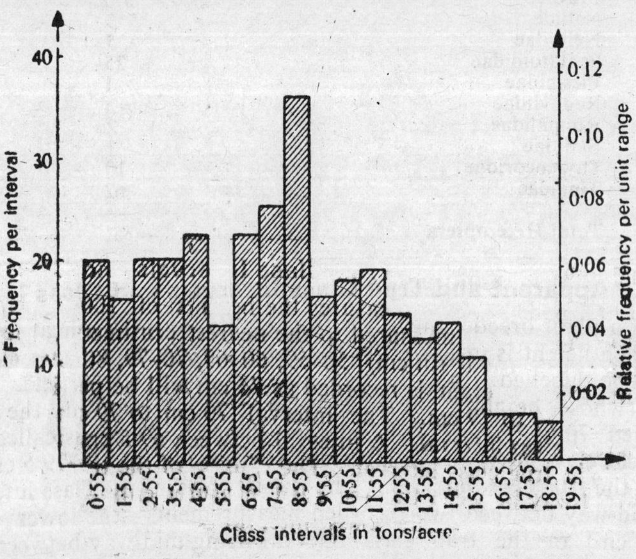

Fig. 5·1. Histogram of peach yields.

Thus the scale on each axis should be marked and each axis clearly labelled.

(ii) Frequency polygon. The figure obtained by joining by the straight lines of the bars is known as the **frequency polygon.** Thus, in a frequency polygon, instead of a rectangle of the appropriate height being erected over each class interval, a dot is put at the same height over the midpoint of the class interval. These dots are then joined by straight lines to form the frequency polygon. Vertical lines can be drawn from the horizontal axis to the dots above the midpoints of the first and last class intervals. While frequency polygons are more quickly constructed than histograms, they do not give as accurate picture as that given by histograms, since the areas above the various intervals are not exactly proportional to the frequencies.

Another way to present data contained in a frequency distribution is by means of a **cumulative frequency curve or ogive.** The cumulative frequency to a particular value X_q, of the viriable is the sum of the frequencies corresponding to all values of the variable less than or equal to X_q, *i.e.*, cumulative frequency to $X_q = \Sigma f_i$

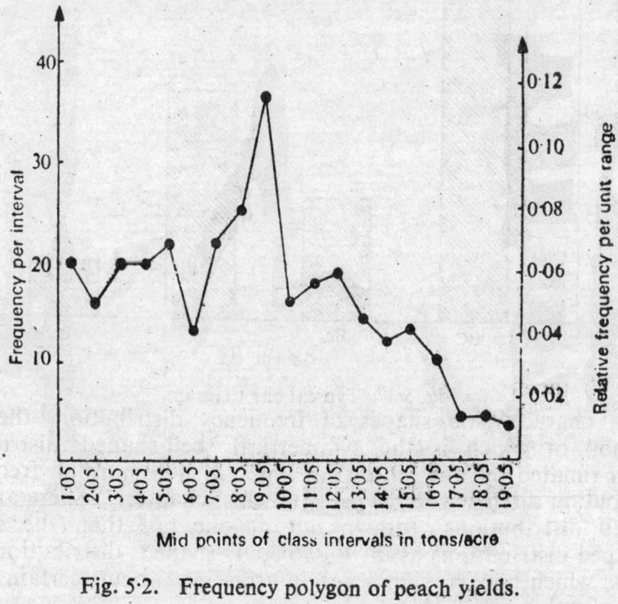

Fig. 5·2. Frequency polygon of peach yields.

where the summation is over those value of i such that $X_i \leqslant X_q$. The ogive is useful for determining the value of median.

(iii) Bar diagram. The bar diagram is often used with discrete qualitative variables. It consists of equally spaced vertical rectangles of equal width, placed on a common horizontal base line. The heights of the rectangles are proportional to the frequencies.

Lastly, the distribution of a variable is of considerable biological interest. If we find that the distribution is asymmetrical and drawn out in one direction, it tells us that there is, perhaps, selection for or against organisms falling in one of the tails of the distribution, or possibly that the scale of measurement chosen is such as to bring about a distortion of the distribution. If, in a sample of immature insects, we discover that the measurements are bimodally distributed (two peaks), this would indicate that the population is dimorphic; different species or races may have become intermingled in our sample, or the dimorphism could have arisen from the presence of both sexes or of different instars. There are

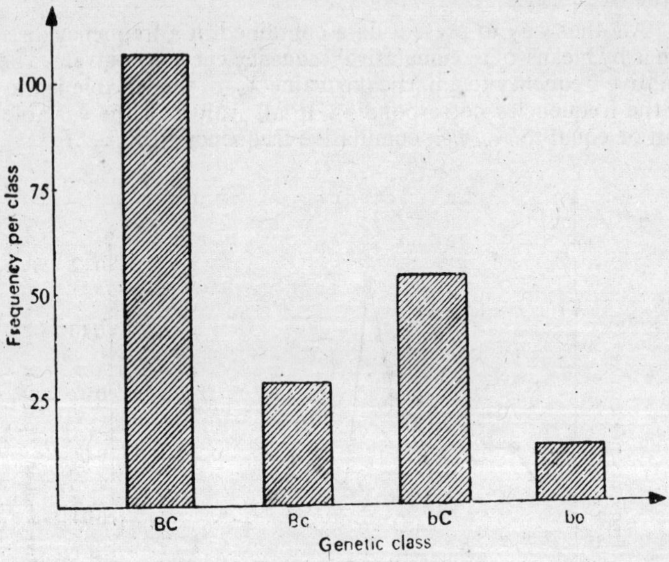

Fig. 5·3. Typical bar diagram.

several characteristic shapes of frequency distribution, the most common of which is the symmetrical bell-shaped distribution (approximated by the graph in Fig. 5·2), the normal frequency distribution discussed elsewhere in this chapter. There are also skewed distributions (drawn out at one tail than the other), L-shaped distributions as in Fig. 5·3, U-shaped distribution, and others, which impart significant information about certain types of relationships.

5-4·7. The Handling of Data

The skillful and expeditious handling of data is essential to the successful practice of statistics. Although a desk calculator is well on its way to becoming as indispensable a research tool as a microscope, through one circumstance or another, one may find him without such a computational aid. In such instances, statistical computations must be carried out by so-called pencil and paper method.

5-5. DESCRIPTIVE STATISTICS

In a sizeable sample it would obviously be tedious to obtain knowledge of the material by contemplating all the individual observations. We need some form of summary to permit us to deal with the data in manageable form, as well as to be able to share our findings with others in scientific talks and publications. A histogram or bar diagram of the frequency distribution would be one type of summary. However, for most purposes, a numerical summary is needed to describe concisely, yet accurately, tne properties of the observed frequency distribution. Quantities providing such a summary may be called **descriptive statistics.** They may be of following two kinds : **statistics of location** describe the position of a sample along a given dimension representing a variable. Thus, when measuring the length of certain animals, we would like to know whether the sample measurements lie in the vicinity of 2 cm or 20 cm. A statistics of location, therefore, must yield a representative value for the sample of observations. However, such a statistics (sometimes also known as a measure of **central tendency)** will not describe the shape of a frequency distribution. This may be long or very narrow ; it may be humped or U-shaped ; it may contain two humps, or it may be markedly asymmetrical. Quantitative measures of such aspects of frequency distributions are required. To this end we need to define and study the **statistics of dispersion.** Thus, statistics of location include such aspects as arithmetic mean, geometric mean, harmonic mean, the median and the mode. The statistics of dispersion include such aspects as range, standard deviation and coefficient of variation.

5-5·1. The Arithmetic Mean

The **arithmetic mean, mean** or **average** is calculated by summing all the individual observations or items of a sample and dividing this sum by the number of items in the sample. For example, as the result of a gas analysis in a respirometer an investigator obtains the following four readings of oxygen peraentages :

$$\begin{array}{r} 14\cdot9 \\ 10\cdot8 \\ 12\cdot3 \\ 23\cdot3 \\ \hline \text{Sum}=61\cdot3 \end{array}$$

He calculates the mean oxygen percentage as the sum of the four items divided by the number of items—here, by four. Thus the average oxygen percentage is

$$\text{Mean}=\frac{61\cdot3}{4}=15\cdot325\%$$

Calculating a mean presents us with the opportunity for learning statistical symbolism. An individual observation is symbolized by Y_i, which stands for the ith observation in the sample. Four observations could be written symbolically as Y_1, Y_2, Y_3, Y_4.

We shall define n, the **sample size,** as the number of items in a sample. In this particular instance, the sample size n is 4. Thus, in a large sample, we can symbolize the array from the first to the nth item as follows : $Y_1, Y_2..., Y_n$. When we wish to sum items, we use the following notation :

$$\sum_{i=1}^{i=n} Y_i = Y_1 + Y_2 + ... + Y_n$$

The capital greek sigma, Σ, simply means the sum of items indicated. The $i=1$ means that the items should be summed, starting with the first one, and ending with the nth one as indicated by the $i=n$ above the Σ. The subscript and superscript are necessary to indicate how many items should be summed. Below are seen increasing simplifications of the complete notation shown at the extreme left :

$$\sum_{i=1}^{i=n} Y_i = \sum_{i=1}^{n} Y_i = \sum_{i} Y_i = \sum^{n} Y = \Sigma Y$$

5-5·2. The Geometric Mean and Harmonic Mean

Variables are sometimes transformed into their logarithms or reciprocals. If we calculate the means of such transformed variables and then change them back into the original scale, these mean will not be the same as if we had computed the arithmetic means of the original variables. The resulting means have received special names in statistics. The back-transformed mean of the logarithmically transformed variable is called the **geometric mean.** It is computed as

$$\text{G.M. } Y = \text{antilog } \frac{1}{n} \log Y$$

which indicates that the geometric mean G.M. Y is the antilogarithm of the mean of the logarithms of variable Y.

The reciprocal of the arithmetic mean of reciprocals is called the **harmonic mean**. If we symbolize it by H_Y, the formula for the harmonic mean can be written in concise form as :

$$\frac{1}{H_Y} = \frac{1}{n} \Sigma \frac{1}{Y}$$

For example, the arithmetic mean 15·325% of the four oxygen percentages will be modified into 14·65% and 14·09% by the geometric and harmonic means, respectively.

5-5·3. The Median

The **median** (M) is a statistics of location occasionally useful in biological research. It is defined as that value of the variable (in an ordered array) that has an equal number of items on either side of it. Thus, the median divides a frequency distribution into two halves. In the following sample of five measurements,

14, 15, 16, 19, 23

M = 16, since the third observation has an equal number of observations on both sides of it. We can visualize the median easily if we think of an array from largest to smallest—for example, a row of men lined up by their heights. The median individual will then be that person having an equal number of men on his right and left sides. His height will be the median height of the sample considered. This quantity is easily evaluated from a sample array with an odd number of individuals. When the number in the sample is even, the median is conventionally calculated as the midpoint between the $(n/2)$th and the $[(n/2)+1]$th variate. Thus, for the sample of four measurements,

14, 15, 16, 19

the median would be the midpoint between the second and third items, or 15·5.

Median has wide application in behaviour biology, pollution (*i.e.*, the median lethal or effective dose, LD_{50} or ED_{50}) and mutation (the median time for a mutation to appear in a number of lines of species).

5-5·4. The Mode

The mode refers to the most "fashionable" value of the variable in a frequency distribution, or the value represented by the greatest number of individuals. For example, the number of tillers per plant in a sample of ten plants were—10, 9, 10, 11, 9, 12, 10, 10, 9, 12 and the arithmetic mean ($\bar{x}$, read as x bar) and mode of the numbers of tillers per plant can be calculated as follows :

$$\bar{x} = \Sigma x_i/10,$$
$$= (10+9+10+11+9+12+10+10+9+12)/10$$
$$= 10 \cdot 2$$

In this sample 9 occurs three times, 10 four times, 11 once and 12 twice. Thus 10 is the mode.

When seen on a frequency distribution, mode is the value of the variable at which the curve peaks. In biology mode does not have many applications.

Distributions having two peaks (equal or unequal in height) are called **bimodal** ; those with more than two peaks are **multimodal.** In U-shaped distribution, the low point at the middle of the distribution as an **antimode.**

5-5·5. The Range

One simple measure of dispersion is the **range.** It is the difference between the largest and the smallest items in a sample. Thus, the range of the four oxygen percentages listed earlier is

$$\text{Range} = 23 \cdot 3 - 10 \cdot 8 = 12 \cdot 5\%$$

5-5·6. Standard Deviation

While the arithmetic mean is the most commonly used measure of central tendency, the **variance,** or its square root, the **standard deviation**, is the most commonly used measure of spread in biological statistics.

It may be thought that an indication of spread or distribution could be obtained by summing deviations from the mean, by the following formula :

$$\sum_{i=1}^{N} (X_i - \mu) \quad \text{or} \quad \sum_{i=1}^{n} (X_i - \bar{x}).$$

However, both these are zero since in each case the sum of the positive deviations cancel with the sum of the negative deviations. To avoid this cancelling, absolute deviations from the mean might be considered, and an infrequently used measure of dispersion is the arithmetic mean of the absolute deviations $\Sigma \mid X_i - \mu$ /N for a finite population of size N, or $\Sigma \mid X_i \quad \bar{x} \mid /n$ for a sample of size n. This quantity is called the **mean deviation.**

For a finite population of size N, the **variance** is denoted *b* σ^2 (σ=the small Greek letter sigma and this is read as sigma squared) and is defined to be

$$\sigma^2 = \sum_{i=1}^{N} (X_i - \mu)^2/(N-1)$$

The positive square root of the population variance is called the **standard deviation.**

For a sample of size n, the **sample variance** is defined to be

$$S^2 = \sum_{i=1}^{n} (X_i - \bar{x})^2/(n-1)$$

and the positive square root of the sample variance is called the **sample standard deviation.**

The importance to be attached to the size of the standard deviation clearly depends on the values of the variables. Thus a standard deviation (s. d.) of one foot for a **population** of heights of trees is of less significance than an equal s. d. for a population of heights of wheat plants.

The ratio of the standard deviation to the population mean, or the sample standard deviation to the sample mean, expressed as a percentage is called the **coefficient of variation.** It is an absolute measure of dispersion in the sense that it is **independent** of the unit employed. Using this coefficient, a comparison of the variabilities of populations or samples having different means may be made. For a biological data, coefficient of variation of the order of 10% to 15% are common. For very homogeneous material this figure may be reduced to 5%, while a coefficient of variation of 25% would indicate very considerable variability.

5-5·7. Sample Statistics and Parameters

A random sample from a given population is a sample chosen in such a manner that each possible sample has an 'equal chance' of being drawn. In sampling at random we are attempting to permit the frequencies of various events occurring in nature to be reproduced unalteredly in our records ; that is, we hope that on the average the frequencies of these events in our sample will be the same as they are in the natural situation. Quantities which characterize populations are known as **parameters** while those which characterize samples are called **statistics**. **A parameter** is a fixed quantity, not subject to variation, whereas a **statistics** is a **variate**, since in general different samples from the same population give different values of the statistics. Conventionally, parameters are designated by letters from the Greek alphabet, while statistics are represented by letters from the English or Roman alphabet.

5-5·8. Standard Error

A random sample, precludes any personal bias but it need not necessarily be quite representative of the population. The mean calculated from the observations in the sample need not be exactly the same as the parameter of the population. This can be easily seen if a number of random samples of the same population have been studied. The difference between the estimate from a sample and the actual population parameter is arising out of sampling and as such it is called sampling error. As the average of the means of several samples is an estimate of the population average if k samples are studied we can expect k different sample means and then the standard deviation of these sample means is one estimate of the sampling error ; this standard deviation of the means is termed the **'standard error'** (S. E.) of the mean, which is the mean of the means. The magnitude of the standard error tells how precise the mean of the sample is.

Usually, an experimenter has to be satisfied with a single sample. In such cases the mean of the single sample studied is that statistics and its standard error is calculated by dividing the standard deviation of the sample—observations with the square root of the number of observations in the sample.

i.e., $$\text{S. E. }(\bar{x})=\frac{s}{\sqrt{(n)}}$$

because $$\text{S. E. }(\bar{x})=\sqrt{(\text{var.})}\ \bar{x}=\sqrt{\left[\text{var.}\left\{\frac{(\Sigma X)}{n}\right\}\right]}=\sqrt{\left[\frac{\text{var. }(\Sigma X)}{n^2}\right]}$$
$$=\sqrt{\left[\frac{n\text{ var. }(X)}{n^2}\right]}=\sqrt{\left(\frac{s^2}{n}\right)}$$

5-5·9. Probability

Probability is a measure of the relative chance of the occurrence of an event from among a set of alternatives. If there are N equally likely ways in which an event can happen and if n is the number of ways in which the desired can occur, then the probability of the occurrence of the desired event is given by n/N and is generally denoted by p. Since p is a measure of the relative chance to the totality of equally likely events, its value does not exceed 1. All the values of p lie between 0 and 1, both limits inclusive.

The rolling of an ordinary, cubical die may result in any one of the six different faces facing upward. The group of six possible results is said to be **exhaustive**. The possibility of die standing on its edge is not considered ! The six possible results are said to be **mutually exclusive** since the occurrence of one result precludes the happening of the other. The rolling of the die is called **a trial.**

Suppose that the die is a perfect cube made of homogeneous material and that there is no reason to expect that any particular face will come upper most frequently than any other. The six possible results are said to be **equally likely**, or **equally probable.**

If a trial may result in any one of n exhaustive, mutually exclusive and equally likely outcomes and if m of these outcomes entail the occurrence of an **event** E, then the probabilities that E will happen as the result of the trial is given by $P(E)=p=m/n$.

Since $0 \leqslant m \leqslant n$, p is a positive number less than, or equal to unity. Since the number of outcomes involving the failure of E is $n-m$, the probability of the failure of E, denoted $\bar{E}$, is given by

$$p(\bar{E})=q=(n-m)/n=1-m/n=1-p$$

further,

$$p+q=1.$$

Example. Sex determination in bisexual organism is a chance event. For a normal **fertilization** of an ovum, there are only two equally likely ways. The zygote may develop into a male or female depending on the type of gamete the male parent provided in animals (man, *Drosophila*, etc.) but in poultry it depends on the type of gamete the egg contains because in poultry female is the heterogametic sex. The gamete produced by the heterogametic sex may or may not contain the X-chromosome and these are equally likely too. So the chance that a fertilized egg will develop into particular sex is 1/2.

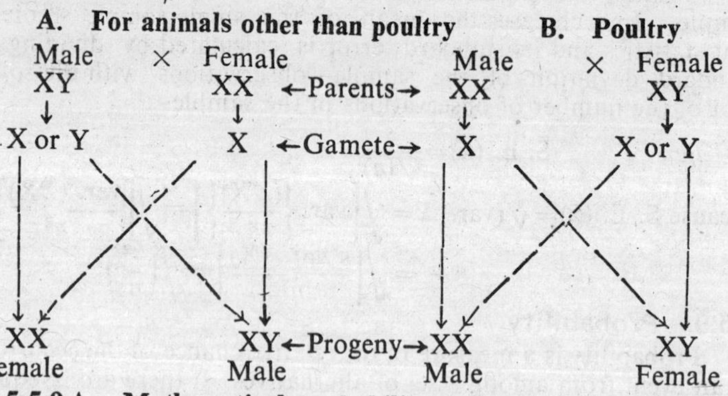

5-5·9 A. Mathematical probability and statistical probability. In the case of above example of sex determination, the probabilities have been calculated on deductive reasoning even before any trial or experiment is conducted. So these probabilities are known as **mathematical** or **apriori probabilities.**

But, in practice, the actual probability in trials conducted may not be coinciding with the apriori probability. For instance, suppose a coin is tossed and it falls with the face E up. Though the mathematical probability of its occurrence is only 1/2, in this case $P(E)=1$ and $P(E')=0$.

But if the coin is tossed 10 times, the number of times E appears can be 0, or 1 or 2......, or 10, the extreme cases being very rare with an unbiased coin. Suppose E was shown up in 4 out of 10 trials. Considering the occurrence of E as the favourable events the 4 occurrences out of the 10 equally likely cases gives the relative frequency 4/10 for the occurrence of E. (The apriori probability is 1/2. But if the number of trials is increased from 10 to 20 it is likely that the ratio of number of times E comes up out of 20 trials becomes more close to 1/2.

In general, if there are n actual occurrences of the favourable event, say, E out of N trials of equally likely ways as far as E is concerned, then the relative frequency of the event is n/N. The limit of this relative frequency as N becomes indefinitely large is known as the **statistical probability**:

$$i.e., P(E) = \underset{N \to \infty}{Lt} \frac{n}{N}.$$

5-5·9 B. Mutually exclusive events and additional theorem. Events are said to be mutually exclusive if the occurrence of one excludes the possibility of the occurrence of the other events of the set. For example, the two faces of a coin are equally likely and they are mutually exclusive. Thus such events are equally likely and mutually exclusive events.

Additional theorem states that the probability of occurrence of one or the other of a set of equally likely and mutually exclusive events in the sum of the separate probabilities of the occurrences of the separate events of the set. That is,

$$P(E_1 \text{ or } E_2 \text{ or} \ldots\ldots E_n) = P(E_1) + P(E_2) + \ldots\ldots + P(E_n).$$

For example, when a coin is tossed, the probability of the occurrence of one or the other face is probability of the occurrence of one face plus probability of the occurrence of the other face.

Symbolically,

$$\begin{aligned} P(E \text{ or } E') &= P(E) + P(E') \\ &= 1/2 + 1/2 = 1, \end{aligned}$$

a certainty which it should be as one of the two faces has to be on the top.

Problem 1. A herd has 4 *Haryana* calves and 4 *Tharparkar* calves, all of the same sex and almost of the same age. Find the probability that (i) one calf selected at random is a Haryana (2) two calves selected at random are both of the same breed ?

Solution. (1) As all the 8 calves have equal chances of selection and as there are 4 Haryana calves out of 8 the probability of selection of a Haryana calf in a random choice is $4/8 = 1/2$.

(2) The two calves selected can be both Haryana or both Tharparkar. So the required probability = probability of selecting two Haryana calves + probability of selection of two Tharparkar calves because these are two equally likely and mutually exclusive cases.

The number of ways in which two calves of the same breed can be selected from calves is 6 because if calves are numbered H_1, H_2, H_3, H_4, T_1, T_2, T_3 and T_4, the number of ways are H_1H_2,

H_1H_3, H_1H_4, H_2H_3, H_2H_4 and H_3H_4 of Haryana and T_1T_2, T_1T_3, T_1T_4, T_2T_3, T_2T_4 and T_3T_4, of Tharparkar. The total number of ways in which calves can be selected is 28, considering the non-repeated combinations of H_1, H_2...H_4, T_1,...T_2 two at a time

(**Note** : The number of combinations of 2 from 8 different things is $8C_2=\frac{8\times7}{1\times2}=28$. **The number of combinations of 2 from** 4 different things is $4C_2=\frac{4\times3}{1\times2}=6$. **In general, the number of** combinations of 2 from different things is $^nC_2=\frac{n\times(n-1)}{1\times2}$, **and the number of combinations of 3 from different things is** $^nC_3=\frac{n\times(n-1)\times(n-2)}{1\times2\times3}$ and so on). **So probability of getting at** random two calves of Haryana breed=6/28 and the probability of getting two calves of Tharparkaı breed **at random** is 6/28. Then the required probability of selecting at random two Haryana calves or two Tharparkar calves is $\frac{6}{28}+\frac{6}{28}=\frac{3}{7}$.

5-5·9 C. Independent events and multiple theorem. A set of events are said to be independent if the occurrence of any one event does not affect the chance of the occurrence of any other event of the set. Thus, the two faces of two coins falling up when they tossed together are independent of each other. Likewise, the sex of a calf to be born is independent of the sex of the calf in the previous calving by the same dam.

The **multiple theorem** states that the probability of the occurrence of a set of independent events simultaneously is the product of the separate probabilities of the several independent events.

If E_1, E_2,......E_n are n independent events then

$$P(E_1, E_2......E_n)=P(E_1)\times P(E_2)\times......\times P(E_n).$$

For example, if two coins are tossed simultaneously, probability that both show the face E up is $P(E_1E_2)=P(E_1)\times P(E_2)=\frac{1}{2}\times\frac{1}{2}=\frac{1}{4}$.

Problem : In a herd 16 calves were born in a month and their male to female ratio is 5:3. If three calves are drawn from these at a time find the probability (1) they are all male, (2) they are all female, (3) they are of one sex only.

Solution. Since the sex ratio is 5 : 3 for male to female we know there are 10 male and 6 females. The number of ways in which 3 calves can be drawn from 16 is $^{16}C_3=\frac{16\times15\times14}{1\times2\times3}=560$.

(1) The number of ways in which 3 male calves can be selected is $^{10}C_3=\frac{10\times9\times8}{1\times2\times3}=120$.

So the probability of getting 3 male calves in a draw is

$$\frac{^{10}C_3}{^{16}C_3}=\frac{120}{560}=\frac{3}{14}$$

(2) The number of ways in which 3 female calves are selected in a draw is ${}^6C_3=\frac{6\times5\times4}{1\times2\times3}=20$

So the probability that 3 female calves are got when three calves drawn at a time is $\frac{{}^6C_3}{{}^{16}C_3}=\frac{20}{560}=\frac{1}{28}$.

(3) In a draw of 3 calves at a time, the probability of getting all of them as one sex is the probablity that the three calves are either all male or three calves selected are all female. Since these are considered exclusive cases.

Probability that the three calves are of the same sex=probability that they are all male+probability that they are all famale.

5-5·9 D. Law of probability and binomial expression. The law of the probability of coincident independent events states that the *chance or probability of the simultaneous occurrence of two or more independent events equal to the product of the probability that each will occur separately.* Thus, when two independent events occur with the probability p **and** q respectively then the probability of of their joint occurrence is pq.

For example, if 10 coins are tossed together the most frequent results will be 5 heads and 5 tails, 4 heads and 6 tails, 3 heads and 7 tails, etc., occur less often and either 10 heads or 10 tails is rare. If we draw a graph of the results of the heads and tails in the tossess of coin then we would have a symmetrical graph (curve). This **symmetrical** or **normal probability curve** is usually derived by plotting on expansion of expression $(a+b)^n$ or $(p+q)^n$, which is known as **binomial expression**. Here, a or p stands for the probability of the head, while b or q stands for the probability of the tail and n stands for number of coins. According to the different values of n, following expansion of $(p+q)^n$ can be received :

$$(p+q)^1=p+q$$
$$(p+q)^2=p^2+2pq+q^2$$
$$(p+q)^3=p^3+3p^2q+3pq^2+q^3$$
$$(p+q)^4=p^4+4p^3q+6p^2q^2+4pq^3+q^4$$
$$(p+q)^5=p^5+5p^4q+10p^3q^2+10p^2q^3+5pq^4+q^5.$$

Here, the coefficients (the numbers before the powers of p and q) express the number of ways a particular outcome is obtained. An easy method for evaluating the coefficients of the expanded terms of the binomial expression is through the use of **Pascal's triangle**, which is shown below :

k or n	
1	1 1
2	1 2 1
3	1 3 3 1
4	1 4 6 4 1
5	1 5 10 10 5 1
...	....

Thus, Pascal's triangle provides the coefficients of the binomial expression—that is, the number of possible outcomes of various combinations of events.

Binomial expression has its wide use in population genetics and ecology. For example, suppose we have a population of insects, exactly 40% of which are infected with a given virus X. If we take samples of $k=5$ insects each and examine each insect separately for presence of virus, what distribution of samples could we expect if the probability of infection of each insect in a sample were independent from that of other insects in the sample ? In this case $p=0{\cdot}4$, the proportion infected, and $q=0{\cdot}6$, the proportion not infected. It is assumed that the population is large and that the question of whether sampling is with or without replacement is irrelevent for practical purposes. The expected proportions would be the expansion of the binomial :

$$(p+q)^k=(0{\cdot}4+0{\cdot}6)^5$$

with the aid of Pascal's triangle this expansion is the probability space

$$\{p^5+5p^4q+10p^3q^2+10p^2q^3+5pq^4+q^5\}$$

or $(0{\cdot}4)^5+5\,(0{\cdot}4)^4\,(0{\cdot}6)+10\,(0{\cdot}4)^3\,(0{\cdot}6)^2+10\,(0{\cdot}4)^2\,(0{\cdot}6)^3$
$+5\,(0{\cdot}4)\,(0{\cdot}6)^4+(0{\cdot}6)^5$

representing the expected proportion of samples of five infected insects, four infected and one non-infected insects, three infected and two non-infected insects, and so on. Thus, in "five-infected insects" there is a probability of occurrence—in this case $(0{\cdot}4)^5=0{\cdot}01024$.

5-5·10. Probability Distribution

In statistical analysis, one has to deal also with twin problems —"Estimation" and "Testing of Hypothesis". **Estimation** is the method of estimating the characteristic value of any character (variable) which has a population of values. In previous sections, we have observed that as the size of the sample (from a population) increases the distribution of the character is close to a symmetrical **unimodal** curve. This shape of the curve is important as it is found that in nature the distribution of many of the characters follow this trend. So, a general functional form of this curve has been derived. This form should naturally be based on characteristic values of the population. These characteristic values are the parameters of the algebraic equation to the curve. The algebraic equation for the curve has been derived by considering the nature of the variable, *viz.*, discrete or continuous and the probability of occurrence of this random variable. These considerations gave the way to different probability distribution, the fundamental three probability distributions known as 'Normal', 'Binomial' and 'Poisson'.

5-5·10 A. Binomial distribution. This probability distribution has been formulated by **James Bernoulli**. This is the distribution of a discrete **random** variable. A random variable X is defined as follows :

X is said to be a random variable if takes values $x_1, x_2......x_n$ with probabilities $p_1, p_2, ... p_n$ respectively such that $p_1+p_2+...$ $...+p_n=1$.

As the name of the distribution itself indicates, this distribution is for the variable or event which can be in two alternative forms. These can be considered as success and failure of the event.

For instance, in tossing a coin we can consider the coin showing up the face E_1 as success and the showing up of the face E_2 as failure. Here the probability of success, $P(E_1)$ and the probability of failure, $P(E_2)$ are both equal to 1/2. But if we consider the throwing of a die which is unbiassed, considering the appearance of ace (the face numbered one) at the top and the appearance of any other face at the top as failure, the probability of success $P(F_1)=1/6$, and the probability of failure, $P(f)=P(F_2)+P(F_3)$ $+P(F_4)+P(F_5)+P(F_6)=5/6$. Similarly, if the male birth is considered success, $P(M)=1/2$, then the female birth is the alternative form (failure) with the probability $P(F)=1/2$.

5-5·10 B. Normal distribution. This is the most important distribution in biometry because, most of the characters in agriculture, biology and genetics follow this distribution. The algebraic form of this distribution was first derived in 1733 by **De Moivre** (1667–1754) but this was later rediscovered and developed by **Gauss** (1809) and by **Leplace** (1812). For these reasons the normal distribution is also called the **De Moivre, Gaussian** or **Leplace distribution**. Further, it is also known as **Gaussian law of error** because **Gauss** found that the errors of observations on any character follow the normal distribution. The term 'normal' is not used to mean the normal trend of any character nor it stands for the norm of the character whose distribution is under consideration. In fact, the normal distribution is the distribution of random variable which varies continuously. The shape of the curve of the normal distribution is unimodal, symmetrical and the ends of the curve tail off asymptotically to the base. Graphically, the total arc under the curve is unity.

The algebraic equation to the curve is

$$P(X)=\frac{1}{\sigma\sqrt{2\pi}}\,e-\frac{(x-\mu)^2}{2\sigma^2}\,; \quad \begin{array}{l}\mu=\text{Mean}\\ \sigma=\text{S. D.}\end{array}$$

When the total number of observations (total frequency) on a character is N, if the distribution follows normal, then, the frequency with which Xi of the character X occurs is,

$$F(Xi)=\frac{N}{\sigma\sqrt{2\pi}}-e-\frac{(Xi-\mu)^2}{2\sigma^2}$$

here μ and σ are called parameters.

5-5·10 C. Poisson distribution. This is the distribution of a discrete (discontinuous) random variable whose probability of occurrence is very small but the mean and variance of the distribution are equal and finite. Thus, the Poisson distribution is the distribution of a rare event with a finite mean and variance. The equality of mean and variance is an important property of Poisson distribution. The Poisson distribution is named after its inventor French mathematician **S. D. Poisson**, who described it in (1837).

If X follows a Poisson distribution, the probability that X takes the value X_i is

$$P(X_i)=\frac{mX_ie^{-m}}{X_i}.$$

Here, m is the only parameter that is estimated by the mean of the distribution.

5-5·11. Tests of Hypotheses

To test the agreement between hypothesis and observation, two biometrical tests called **chi-square test** (X^2-test) and **Student's *t*-test,** are used.

5-5·11A. Chi-square test (or x^2-test) of goodness of fit. In calculating expected results of coin tosses, we do not expect to achieve results exactly equal to our calculations but we expect certain deviations due to chance from our expectations. Mendel in his crosses get hundreds or thousands of individuals, but his results always were not equal to his expectations. To record such deviations of results from the expectations, chi-square test (x^2-test) has been employed. The chi-square test can be understood well by following example :

If, we toss six coins—then generally we expect the bionomial ratio of 1 : 1 from the test. Instead of 1 : 1 ratio, the ratio of 4 : 2 easily can be tolerated, but, if in a test 600 coins are tossed and they give the ratio of 400 : 200, then such gross deviation from the expected results cannot be neglected. Similarly, if we test 40 coins and find a deviation of 10 in each class, this deviation seems serious :

Observed (O)	30	10
Expected (E)	20	20
Deviation (d)	10	10

But, if we test 200 coins, the same numerical deviation of 10 reasonably enough explains as purely a chance effect—

Observed (O)	90	110
Expected (E)	100	100
Deviation (d)	10	10

Such deviations of these statistical tests are squared and each squared deviation is then divided by the expected number in its class. The resulting quotients are then all added together to give a single value, called the chi-square (χ^2), χ is a Greek letter which is read as chi for the distribution. Thus

$$\chi^2=\Sigma\frac{(O-E)^2}{E}$$

$$=\Sigma \frac{d^2}{E}$$

here O stands for observed number, E stands for expected number and d stands for normal deviation.

Now, we can calculate the value of χ^2 for this example to show this value relates the magnitude of the deviation to the size of the sample :

Table 5·1. Calculation of chi-square value.

	Toss of 40 coins		Toss of 200 coins	
	Head	Tail	Head	Tail
Observed number (O)	30	10	90	110
Expected number (E)	20	20	100	100
Deviation (d)	10	10	10	10
d^2	$10^2=100$	$10^2=100$	$10^2=100$	$10^2=100$
$\frac{d^2}{E}$	$\frac{100}{20}=5$	$\frac{100}{20}=5$	$\frac{100}{100}=1$	$\frac{100}{100}=1$
$\chi^2=\Sigma\left(\frac{d^2}{E}\right)$	$\chi^2=10$		$\chi^2=2$	

From this table we can note that the value χ^2 is much larger for the less number of tosses than large number of tosses, however, the degree of deviations in both are numerically same. Thus, deviations are related to the number of tosses or trials. In genetics, chi-square test is used for observing the numerical frequencies of deviations in F_2 offsprings of monohybrid, dihybrid or polyhybrid crosses from the expected ratios. Precisely the same kind of test can be applied more generally, whenever in fact we want to examine a hypothesis which specifies the frequency with which the observations should fall into certain classificatory groups. This is usually referred to as a **goodness of fit test.**

Test of significance. The first step in a test of significance is the formulation of a **null hypothesis** which often states that the treatment (The **treatment** is used to indicate any kind of factor whose different states are under investigation, *e.g.*, breeds of animals or strains of a cereal) has had no effect or that the treatments are having the same effect. On the basis of this assumption the probability of obtaining, due to chance, results as extreme as or more extreme than those observed is calculated. If this probability is small (less than 5% or 1%), the null hypothesis is rejected and the observed result is said to be **statistically significant.**

A test of significance consists of following four steps :

(i) Formulation of the null hypothesis.

(ii) Choosing an appropriate level of significance (α), (*i.e.*, 0·05 or 0·01 level).

(iii) On the assumption that the null hypothesis is true and by using statistical theory, calculation of the probability of the observed results (and those less favourable to the null hypothesis) being due to chance or random variation.

(iv) If the probability, p, found in (iii) is less that in (ii), the null hypothesis is rejected. If not, judgment is suspended. If $p \leqslant 0{\cdot}01$, the result is said to be **highly significant**: if $0{\cdot}01 < p \leqslant 0{\cdot}05$, it is said to be **significant.**

5-5·11B. T-test or T-distribution. When a biologist has to interpret and draw the decision from the very small samples of data, he often made use of so called t-test for it. Sometimes size of a sample remains so small that one has to depend on it for drawing conclusions. For example, the number of animals with some rare disease that are available for experimentation may well be quite small, and are obliged to make the best of it. The t-test can have following distribution :

$$t = \frac{\bar{X} - \mu}{S/\sqrt{(n)}}$$

Where sample of n observations has mean $\bar{x}$ and standard deviation s. We want to compare $\bar{x}$ with a hypothetical mean μ, but the true standard deviation σ is unknown. The t-distribution is used in significance testing and interval estimation of a small sample. It was first found by **W. S. Cosset** (1876–1937) who wrote under the *nom de plume* of "student". For this reason the distribution has come to be known as **student's t-distribution.**

5-5·12. Simple Linear Regression

In biological and agricultural research the problem frequently arises of relating variation in one variable with the variation occurring in one or more other variables. For example, it may be desired to study the way in which wheat yields vary with change in plant density, or the manner in which animal weights vary with protein level of feed, or how the number of eggs laid by a given species of mosquito varies with the saline content of the water in which it is living.

In many such problems, it may be possible to distinguish between the variable of primary interest and one or more secondary variables which are of interest only in so far as they affect the main one. In dealing with these it is conventional to refer to the main one as the **dependent variable** and to denote it by Y, and to refer to the secondary variables as the **independent variables** and to denote them by $X_1, X_2, ..., X_k$, or if there is only one independent variable, by X. Thus, in studying the way in which yield of wheat vary in relation to change of amount of fertilizer applied yield would constitute the dependent variable Y, and the fertilizer levels the independent variable X.

In general, regression problem can be divided into two categories—**simple** and **multiple**, depending on whether problems concerning two (simple) or more than two (multiple) variables are being studied. Simple regression problem may in turn be classified into **linear** and **non-linear** types according to whether or not a mathematical function representing a straight line can be used to describe the relationship between the two variables. In other words, a linear regression problem is one in which the same change in the dependent variable Y can be expected for a change in the independent variable X, irrespective of the value of X.

Example : As an illustration of a simple linear regression problem, consider the regression of the following weights of 7 calves (expressed in tens of pounds) on their ages (expressed in months).

Weight (Y)	**Age (X)**
4	0
7	2
14	3
10	4
20	6
28	8
36	12
$\Sigma Y=119$	$\Sigma X=35$
$\bar{y}=17$	$\bar{x}=5$

As a starting point in this regression problem, the relationship between the 'dependent variable (weight) and the independent variable (age) may be illustrated in the form of a **scatter diagram** in which the weights are plotted against the ages (fig. 5·4)

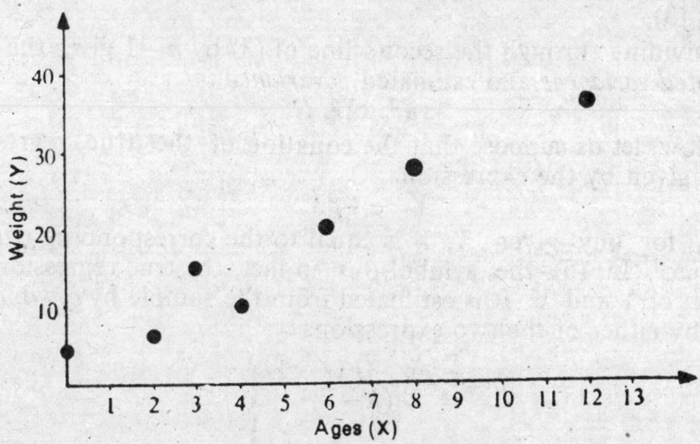

Fig. 5·4. Scatter diagram.

From the scatter diagram, it becomes evident that there is a linear relationship between weight and age, that weight is affected by age or speaking in statistical terms, that there is a *linear regression* of *weight on age.*

Calculation of regression coefficient. The basic observations are n pairs of associated observations represented by (X, Y) in which variations in one variable, Y are related to variation in other, variable X. For the calculation of regression coefficient, first the basic quantities are calculated from the data as follows :

$$\left.\begin{array}{l} n \\ \Sigma X \; \Sigma Y \\ \Sigma X^2 \; \Sigma Y^2 \; \Sigma XY \end{array}\right\} \qquad ...(1)$$

Now, the sums of squares and product about the means can be obtained as follows :

$$\left.\begin{array}{r} \Sigma (X-\bar{x})^2 = \Sigma X^2 - \frac{1}{n} (\Sigma X)^2, \\ \Sigma (Y-\bar{y})^2 = \Sigma Y^2 - \frac{1}{n} (\Sigma Y)^2, \\ \Sigma (X-\bar{x})(Y-\bar{y}) = \Sigma XY - \frac{1}{n} (\Sigma X)(\Sigma Y) \end{array}\right\} \qquad ...(2)$$

The last equation in (2) is a new form, but its affinity with the other two is evident. We can now extend the array in (1) by adding first a row of "correction factors". Substracting each of these from the number above then gives the required expressions on the left of (2). Thus we have the additional rows

$$\left.\begin{array}{l} \frac{1}{n} (\Sigma X)^2, \; \frac{1}{n} (\Sigma Y)^2, \; \frac{1}{n} (\Sigma X) \, \Sigma Y) \\ \Sigma (X-\bar{x})^2, \; \Sigma (Y-\bar{y})^2, \; \Sigma (X-\bar{x})(Y-\bar{y}) \end{array}\right\} \qquad ...(3)$$

where the quantities in the first line of (3) are substracted from the corresponding quantities in the last line of (1) to give the second line of (3),

Dividing through the second line of (3) by $n-1$ gives the two *estimated variances* and estimated *covariance* :

$$sX^2, \; sX^2, \; C. \qquad ...(4)$$

Now, let us suppose that the equation of the true regression line is given by the expression

$$Y = \alpha + \beta X \qquad ...(5)$$

where, for any given X, Y is equal to the corresponding true Y-mean μx. In (5) the symbol β is, in fact, the true regression coefficient of Y and X. It is estimated from the sample by b, which is given by either of the two expressions :

$$\left.\begin{array}{l} b = \frac{\Sigma (X-\bar{x})(Y-\bar{y})}{\Sigma (X-\bar{x})^2} \\ \quad = \frac{c}{sx^3}. \end{array}\right\} \qquad ...(6)$$

The constant α is now estimated by

$$a=\bar{y}-b\bar{x}, \qquad \ldots(7)$$

The fitted regression line is thus

$$Y=a+bx \qquad \ldots(8)$$

where a and b are given by (7) and (8).

Further, variance σ_2 of the deviations of Y from the regression line is estimated by

$$s^2=\frac{1}{n-2}\left\{\Sigma\,(Y-\bar{y})^2-\frac{[\Sigma\,(X-\bar{x})\,(Y-\bar{y})]^2}{\Sigma\,(X-\bar{x})^2}\right. \qquad \ldots(9)$$

5-5·13. Correlation of Measurements

In biological research, sometimes many types of paired measurements (X, Y) have bivariate normal distribution or distribution approximating this. A popular index of the degree of association between continuous variables is the **correlation coefficient** (ρ, small Greek rho)

$$\Phi \times Y=\frac{1}{2\pi\sigma_X\sigma_Y\,(1-\rho^2)^{1/2}}\exp T$$

where
$$T=-\frac{1}{2\,(1-\rho^2)}\left\{\frac{(X-\mu_X)^2}{\sigma_X}-2\rho\left(\frac{X-\mu_X}{\sigma_X}\right)\left(\frac{Y-\mu_Y}{\sigma_Y}\right)+\left(\frac{Y-\mu_Y}{\sigma Y}\right)^2\right\}$$

where μ_X, μ_Y, σ_X, σ_Y, ρ are parameters; σ_{X_1} $\sigma_Y > 0$, $-1 > \rho > +1$. The parameters μ_X, μ_Y are the means, and σ_X, σ_Y are the standard deviations, of the X and Y populations respectively ; ρ is called the correlation coefficient.

5-6. APPLICATION OF BOOLEAN ALGEBRA IN THE DEPICTION OF LIFE PATTERN OF ANIMALS AND ENVIRONMENTAL EFFECTS

In writing the life patterns of animals and various arthropods much information needs to be brought together, relating not only to the morphological states but also including statements about the speed of development, limiting conditions, steady states and alternative possibilities. Such relationships can only be stated concisely if the printed sentence is replaced by a symbol. This can be done by the help of Boolean Algebra as follows (**Hohn**, 1960 ; **Clarke** 1973).

First of all one must have symbols to represent the major morphological changes in the development of the animal : *z* can stand for zygote, meaning the fertilized egg of the animal ; *e* for embryonic development ; *l* for larva, meaning the post-embryonic to pre-adult stage of the life cycle ; *a* for adult, meaning the body form the animal has during the reproductive cycles of the life history. The life pattern is terminated by death, indicated by *d* written at the end of the symbols. Thus the stages in the life pattern of an arthropod or any animal with indirect development could be indicated thus :

$$z\,e\,l\,a\,d \qquad \text{...(i)}$$

the symbols written in sequence without punctuation indicating that the events they represent follow one another in the order stated. Where it is necessary to recognize more stages in the life history extra terms can be introduced, as for example, in endopterygote insects where, if *p* stands for pupa, the symbolic representation would be

$$z\,e\,l\,p\,a\,d. \qquad \text{...(ii)}$$

If more than one larval stage needs to be recognized terms can be introduced thus :

$$z\,e\,n\,m\,c\,f\,o\,a\,d \qquad \text{...(iii)}$$

for the *Euphausiacea* (Crustacea, Malacostraca) where n=nauplius, m=metanauplius, c=calyptopsis larva, f=furcilia larva, o=cyrtopia larva and a=adult.

Where names have not been given to the larval forms a sequence of larval stages could be designated :

$$l_1\, l_2\, l_3\, l_4\, l_5\, l_6\, l_7\, l_8, \text{ etc.}$$

Each of the stages so named has properties which must be taken into account : these can be indicated by a symbol which can be written as subscript to the appropriate stage. Since all these properties must be present they are enclosed in a bracket and as they are all of equal importance the symbols are separated by a full stop. This system of notation says all must occur ; all are of equal value to the animal. These properties and their symbols are : *g* for growth, *df* for differentiation, *r* for reproduction, and *dp* for diapause.

Thus a larval stage may exhibit growth and differentiation and would be symbolized thus :

$$l\,(g.df)$$

or growth only

$$lg$$

or be a resting diapause stage

$$l\,dp$$

A sequence of larval stages showing these events could be symbolized thus :

$$l_1\,(g.df)\; l_2\,(g)\; l_3\,(dp)\; p\; a\; d$$

An entire life history of an endopterygote could be written thus :

$$z\; e\,(df)\; l\,(g.df)\; p\,(df\; dp\; df)\; a\,(g.r)\; d$$

where $l\,(g.df)$ indicates that in larval development growth and differentiation occur together, but in the pupal stage the absence of stops in the term of $(df\; dp\; df)$ indicates that these events, differentiation, diapause and differentiation must occur in the sequence.

Further, the rates of development and tolerances of physical conditions for each stage differ ; they may be symbolically represented thus : T=temperature in degrees centigrade, $R.H$=relative humidity and P=photoperiod, the number of hours of light and dark in a day to which the stage is subjected. The limits of temperature are indicated by an integral sign with the figures written the upper limit at the top, the lower at the bottom of the sign, and the photoperiod by the number of hours light/number of hours dark. Thus a larval period may be written

$$l\,(g.df)\left(T\int_{10}^{30} R.H.\int_{20}^{90} P\frac{16}{8}\right)$$

indicating that larval growth and differentiation normally take place within a temperature range 10–30°C, a relative humidity 20–90%, a photoperiod of 16 hours daylight and 8 hours darkness. If necessary the shape of the developmental curves for temperature can be written following the integral. For a pupa with diapause, the diapause conditions can be written following the symbol in subscript thus :

$$p\left(df\,T\int_{10}^{20} dp\,T\int_{-4}^{6} df\,T\int_{10}^{20}\right).$$

The formula for a complete life history of the tropical bug *Dysdercus intermedius*, a serious pest on cotton in Africa, can be written thus :

$$z\,e\left(T\int_{17\cdot5}^{32\cdot5} D\int_{16\cdot2}^{4\cdot1}\right) l_1\,(g)\left(T\int_{17\cdot5}^{35} D\int_{9.4}^{2\cdot9}\right) l_2$$

$$(g)\left(T\int_{17\cdot5}^{35} D\int_{18\cdot8}^{3\cdot2}\right) l_3\,(g)\left(T\int_{17\cdot5}^{35} D\int_{12\cdot7}^{4\cdot5}\right) l_4\,(g)$$

$$\left(T\int_{17\cdot5}^{35} D\int_{13\cdot5}^{5\cdot6}\right) l_5\,(g)\left(T\int_{20}^{32\cdot5} D\int_{16}^{6\cdot8}\right) a\,(g.r)$$

$$\left(T\int_{17\cdot5}^{32\cdot5} D\int_{19\cdot6}^{7\cdot5} Pr\int_{8\cdot2}^{2\cdot1} Pro\int_{19\cdot9}^{7\cdot4} I.O.\int_{8}^{5\cdot5}\right) d$$

D=duration of instar in days,

Pr=pre-copulatory period in days,

Pro=pre-oviposition period in days,

$I.O$=inter-oviposition period in days.

5-7. REVISION QUESTIONS

1. Why knowledge of biometry (biostatistics) is must for ecology ?
2. Describe 'the data in biology'.
3. Describe arithmetic mean and geometric mean.
4. What are median and mode ?
5. What is histogram ?
6. Describe standard deviation.
7. Explain probability.
8. What is binomial expression ?
9. Describe simple linear regression.

6

Concept of Levels of Organisation and Biosphere

6-1. INTRODUCTION

An evident unit of life in nature is the individual organism. Linguistically as well as biologically, the root of the word '*organism*' is organisation : a characteristic **structural order**. Each kind of living organism possesses a certain degree of organisation which enables them to perform different life activities. In fact, close examination of the any aspect of the living world reveals that everything has a number of interrelated component parts. For example, a human being possesses in his body various organ systems such as digestive system, circulatory system, respiratory system, etc., for sustaining the life of an individual. Each organ system, in its turn, contains numerous different organs. For example, the digestive system is built up of such organs as mouth, buccal cavity, oesophagus, stomach, intestine, rectum, anus, liver, pancreas, etc. Each organ too is made up of different kinds of tissues.)Thus, stomach possesses epithelial tissues, muscles, glandular tissues and so on. Even each kind of tissue contains numerous cells, as muscles of stomach possess muscle cells.

Further, each cell contains different cellular organelles such as mitochondria, endoplasmic reticulum, Golgi bodies, lysosomes, nucleus and so on. Each organelle is built up of different kinds of macromolecules such as polypeptides, polysaccharides, lipids and nucleic acids (DNA and RNA). Each kind of organic macromolecule is built up by several simple molecules (monomers), such as polypeptides contain many molecules of amino acids. Each molecule contains atoms of different elements, such as a molecule of amino acid consists of atoms of many elements like carbon, nitrogen, hydrogen. Lastly, each atom has its own structural components in the form of subatomic particles such as electrons, protons and neutrons. Thus, in the living world, there exist different separate levels of organisation. In non-living world too we may find different levels of structures of matter.

6-2. HIERARCHY OF LEVELS OF ORGANISATION

The smallest structural units of matter (living as well as non-living) are **subatomic particles.** The next larger units are **atoms,** each of which consists of subatomic particles. Atoms in turn form still more complex combinations called **chemical compounds** (molecules). Chemical compounds are variously joined together as even more elaborate units or **complexes of compounds** (macromolecules). These units are representing successively **higher levels of organisation of matter.** They form a pyramid or hierarchy, in which any given level contains all lower levels as components and is itself a component of all higher levels. For example, atoms contain subatomic particles as components and atoms are themselves components of molecules.

All structural levels upto and including those of macromolecules are encountered both in the non-living and the living world. For example, water and table salt are two chemical compounds which are found in non-living and living matter. Examples of complexes of compounds in the non-living world are rocks which are composed of several types of compounds. In living world, complexes of compounds often occur as microscopic and submicroscopic bodies called **organelles.** But even in their most elaborated and complicated forms, complexes of compounds (macromolecules like proteins, lipids, carbohydrates, nucleic acids, etc.) or organelles (i.e., mitochondria, endoplasmic reticulum, chloroplasts, etc.) cannot qualify as living units. To reach the level of life, we must go to the next higher structural level, that of cells.

A cell is a specific combination of organelles. It is usually a microscopic bit of matter organised just complexly enough to contain all the necessary apparatus for the performance of metabolism and self-perpetuation. A cell in effect represents the least-elaborate known structure that can be fully alive. It follows that a living organism must consists of atleast one cell. In fact, **unicellular** (single celled) organisms like Monera (Bacteria and blue-green algae) and Protista (unicellular eukaryotic organisms like Protozoa and algae) constitute the majority of living creatures on earth. All other organisms are **multicellular**, each composed of up to hundred of trillions of joined cells.

Further, within a multicellular organism, several distinct levels of organisations can be distinguished. The simplest multicellular types contain comparatively small number of cells. If all such cells are more or less alike, the organism is often referred to as a **cellular colony**. If two or more different groups of cells are present each such group usually forms a **tissue**. Thus, there occur cellular colony in bacteria, two-tissue colony in slime-molds and three-tissue colony in *Volvox*. In structurally more complex organisms, there are not only several tissues, but some of the tissues also joined-further as one or more units called **organs**. The most complex organisms contains not only many tissues and organs, but also groups of organs united as one or more **organ systems.** Thus, living organisms exhibit atleast five levels of structural complexity: the single-celled form, the colonial form, the organisms with tissues, the type with organs and the type with organ systems.

Thus, whether of one cell or of trillions cells, the individual is the relevant unit as regards physiological organisation, total metabolism, responsiveness, reproduction, growth and development, and certain other essential phenomena of living things. **The individual is the separate and objectively concrete biological unit.** However, several still higher forms of life can be distinguished beyond the level of the individual organism.

An individual cannot live in isolaton in nature. For its survival and for perpetuation of its race, an individual often associates with other individuals of its own kind. All the similar individuals of a given locality which potentially form a single sexually interbreeding group, form a next higher level of organisation called **population**. Since mendelian laws of heredity apply to the transmission of genes among the individuals of a population, so, it is also termed as **mendelian population** by the geneticists. The size of a population may vary, but it is usually considered to be a local group which is called **local population** or **deme**. Each individual member of a deme has an equal chance of mating with any other member of the opposite sex. For example, all individuals of *Paramecium* in a fresh-water pond form a deme or local population or population.

A group of similar demes or of actually interbreeding natural populations is called **species**. Thus, a species is a group of like individuals that live together and interbreed to produce fertile offspring. Thus all corn plants on earth, all bull frogs, all the human beings, each represent a species.

Many groups of populations belonging to different species form the next higher level of organisation, the **biotic community**. For example, different species of micro-organisms, plants and animals inhabiting a fresh water pond along the population of *Paramecium* form a biotic community. A biotic community together with the physical space in which it exists represents an ecological unit, the **ecosystem**. For example, a fresh water pond with its biotic community represents an ecosystem. The sum of all ecosystems existing on earth form the biggest ecological organisation called **biosphere**. Biosphere is a huge ecosystem which contains living components in the form of biological world (living matter) and non-living or abiotic components in the form of earth or soil (**lithosphere**), water (**hydrosphere**) and air (**atmosphere**).

6-3. DEDUCTION OF CERTAIN BASIC PRINCIPLES FROM THE LEVELS OF ORGANISATION

From the hierarchy of levels in the organisation of non-living and living matter, following generalizations can be made :

1. **Each level of organisation has fewer units than the preceding lower level. Thus, there are fewer communities than species and fewer cells than organelles.**

2. Each level of organisation is structurally more complex than lower ones. It includes the complexity of all the lower levels but has some additional special features of its own. This fact supports the **principle of emergent properties** by showing the emergence of new properties at each level of organisation when parts are organised to form a living system.

3. In any hierarchy of levels, a jump from one level to the next can often be achieved only at the expense of energy. It takes energy to build atoms into chemical compounds, and it takes energy to create cells out of chemical complexes. Indeed once a higher living level has been created, energy expenditures must usually continue thereafter to maintain that level. For example, if the energy supply to the cell, the organ, or the organism is stopped, death and decomposition soon follow and reversion to lower levels occur.

6-4. BIOLOGICAL IMPORTANCE OF INCREASING COMPLEXITY AND ORGANISATION

The trend of increasing complexity and organisation from lower level toward higher one produces following new properties : **cooperation, division of labour** and **specialization.**

One property which arises at each new higher level of organisation is united, integrated function. Disunited structure means independent function and by extension, **competition** ; while united structure means joint function and by extension, **cooperation**. For example, the cells can remain struturally independent and can compete for space and raw materials. But if they form a multicellular unit, they surrender their independence and become a cooperative, integrated system.

The fundamental advantage of cooperation is **operational efficiency** : the cooperating whole is more efficient in performing the functions of life than its lower-level components separately and competitively. For example, separate cells must expend more energy and materials for survival than if that same number of cells were integrated as a tissue.

One underlying reason for this difference is that, in the integrated unit, duplication of effort can be avoided. In a set of separate cells, for example, every cell is exposed to the environment on all sides and must, therefore, expend energy and materials on all sides to cope with the impact of the environment. However, if the same cells are grouped together as a compact tissue only the outermost cells are in direct contact with the environment, and inner cells then need not to channel their resources into protective activities.

Besides avoiding duplication of efforts, cooperative groupings make possible continuity of efforts. For example, an unicellular form must necessarily carry out all survival functions in its one cell. Sometimes, the performance of even one of these functions requires most or all of the capacities of all the cell. Thus, the entire cell surface must often serve as gateway for entering nutrients and departing wastes. And all parts of the cell may have to participate directly in locomotion or in feeding, for example.

By contrast, in multicellular organisms continuity of a given effort becomes possible through **divison of labour**. The total task of survival can be divided up into several subtasks, and each of these can become the continuous responsibility of particular cells only. Some cells might function in feeding, continuously so, and others in locomotion, again continuously. The division of labour becomes so pronounced that many or most cells are permanently limited in functional capacity : they can perform only certain jobs and no others. For example, mature nerve cells can conduct nerve impulses only and are quite unable to reproduce or move. Thus, different cells of multicellular organisms exhibit **specialization** for particular set of functions.

6-5 THE BIOSPHERE

The entire world basically can be broadly classified into **abiotic world** and **biotic world**. The abiotic world can be further divided into three packages : the **atmosphere** (air), the **lithosphere** (earth) and the **hydrosphere** (water). To these three ecological components we can add the biological world or biotic world, the **biosphere**. All four subdivisions thus represent four major global components of the world ecosystem. These four spheres continuously exchange matter with one another, but obviously there are differences in the relative and total amounts of the natural elements in each organisms take in inorganic nutrients from the soil, water and air ; they eliminate their wastes, and their bodies are rendered into inorganic molecules once again. Water and minerals shuttle among the air, land and water with an occasional pause in an organism.

Thus, the biosphere is that part of the earth in which life exists (**Hutchinson**, 1970). More specifically, the sum of those portions of the hydrosphere, lithosphere and atmosphere into which life penetrates is the biosphere. However, together with the geological, chemical features of the totality of our habitats, these (air, water and earth, and organisms) are sometimes grouped under the term **ecosphere** (**Cole**, 1958). Biosphere or ecosphere may be thought of as a biochemical system capable of capturing, converting, storing, and utilizing the energy of the sun. Approximately three hundred thousand species of green plants and micro-organisms are recognised as **primary producers** which utilize inorganic elements and compounds to synthesize the organic materials of life. Their productivity is consumed by more than a million other species of organisms which convert this organic storehouse into animal form, adding to the beauty, and value of the biosphere as well as its complexity. Still other species, primarily bacteria and fungi, accomplish the recycling process by returning plant and animal wastes and residues to inorganic form so the process may be renewed. In this cyclic process of life, many elements are shared from a common global pool and are converted from inorganic to organic form and back again. Examples of such elements which commonly shuttle among the air, water, earth and organism are carbon, oxygen, hydrogen, nitrogen, phosphorus, sulphur, sodium, potassium, calcium, magnesium, iron, manganese, cobalt, copper, and zinc.

The oldest known photosynthetic organisms are at least two billion years old, and throughout this vast period of time, the entire biosphere has matured and evolved as a self-regulatory system with internal checks and balances. It has developed a remarkable degree of homeostasis and ecological control within its various components. In forthcoming chapters ; we will consider in more detail the process of photosynthesis or primary production, the cyclic passage of elements, the flow of energy through living organisms, the growth and development of populations, and the diversity and maturation of communities and ecosystems.

In physical terms, the biosphere is a relatively **thin and incomplete** envelope covering most of the world. It represents a mosaic of different biotic communities from simple to complex, aquatic to terrestrial, and tropical to polar. It does not exist in the extremities of the polar regions, the highest mountains, the deepest ocean troughs, the most extreme deserts, or the most highly polluted areas of land and water. Its total thickness, including all portions of the earth where living organisms can exist, is less than 26 kilometres. Its zone of active biological production, in terms of photosynthesis, is much narrower, and varies from a few centimetres to over 100 metres. This zone would, for instance, be only a few centimetres in muddy or turbid water, whereas in very clear ocean water, it could be more than 100 metres in thickness. On land, the zone of biological production might be only a few millimetres in a desert or rock environment, whereas it might again be more than 100 metres in a sequoia or tropical rain forest. Living organisms can exist, of course, beyond the range of active biological production ; some insects or birds may be airborne to altitudes above 20,000 feet, and viable spores, seeds of plants, and micro-organisms may be found in the atmosphere and mountain tops above 25,000 feet. In the ocean depths, many animals can exist well below a thousand feet—one hydra-like animal has been photographed at a depth of 15,900 feet in the south Atlantic. However, in both extreme altitude and extreme depth, the organisms depend upon the much thinner zone of active biological production, that portion of the system which converts the energy of sunlight into the chemical and physical energy of living organisms.

6-6. FROM BIOSPHERE TO ECOLOGICAL NICHE

The biosphere can be divided into a series of **biogeographical regions**, each inhabited by distinctive species of animals and plants. These organisms are able to move freely from place to place within each region, but not from one region to another, this being prevented by various natural barriers like river, lake, sea, mountain, desert, etc. Such biogeographical regions include Eurasia, South America, Africa and Australia.

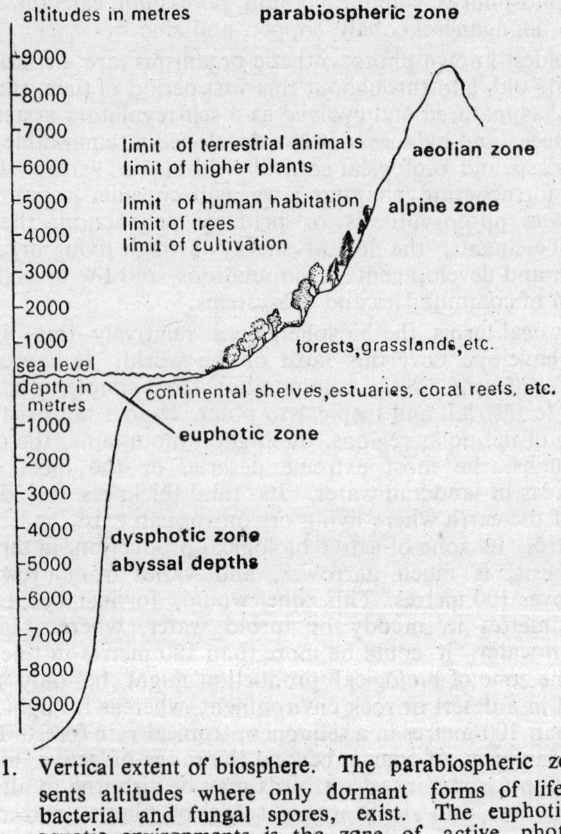

Fig. 6·1. Vertical extent of biosphere. The parabiospheric zone represents altitudes where only dormant forms of life, such as bacterial and fungal spores, exist. The euphotic zone in aquatic environments is the zone of active photosynthesis (after Southwick, 1976).

Although they may possess a different fauna and flora, two such geographical regions may be virtually identical in their climatic conditions, and this is reflected in similar adaptations shown by their respective inhabitants. These equivalent areas, cutting right across different continents, are lumped together as **biomes**. Each biome has its unique set of conditions, and each supports a particular type of flora and fauna. Such areas include the tundra of the Arctic and Antarctic regions ; grasslands, deserts, and the different kinds of forests (coniferous, temperate and tropical). The oceans can be regarded as a single biome (the marine biome), though coral reefs are generally regarded as a separate biome. Within each biome only those organisms with the necessary adaptations for surviving the physical conditions are found. Thus coniferous forest and tundra biomes only have organisms capable of withstanding long periods of extreme cold. Desert organisms, on the other hand, must be able to cope with intense heat and drought. Marine organisms must be able to thrive in salt water, and so on.

Biomes are subdivided into small units called **zones, each with** its own particular set of physical conditions. For example, a forest biome can be divided into **ground zone** and **canopy zone**; the desert into **surface** and **subterranean zones** and the marine biome into **surface, abyssal** and **intertidal (littoral) zones**. Individual organisms may be specially adapted to live in one of these different zones, in some cases providing them with a means of avoiding the extremes of climatic conditions met with in the biome as a whole. For example, many desert animals avoid the heat of the mid-day sun by burrowing into the subterranean zone.

Within each biome are numerous **habitats**, specific localities each with a particular set of conditions and an appropriately adapted community of organisms. Typical habitats include freshwater ponds, slow-flowing streams, rock pools, and hedegrows.

Within a particular habitat an organism is generally confined to a restricted situation. For example, in a slow-flowing stream, crayfish live in burrows in the river-bank, flatworms and shrimps under stones, fishes in the open water, and so on. This idea is embodied in the term **ecological niche** which means more than just the precise place where it lives. The term ecological niche embraces the organism's entire way of life, its 'profession', so to speak. The ecological niche of amphibians includes both aquatic and terrestrial habitats, while that of migratory birds span several biomes.

6-7 REVISION QUESTIONS

1. Explain the concept of 'level of organisation with suitable example.
2. What is the biological significance of increasing complexity and organisation in living beings ?
3. Describe the biosphere.
4. Describe different divisions of the biosphere.

7

The Ecosystem

7-1. INTRODUCTION

Modern ecology focuses on the basic functional ecological unit, the ecosystem. An ecosystem is any spatial or organisational unit which includes a community of living organisms and non-living substances of environment interacting to produce an exchange of materials between the living and non-living parts. It may be as small as a puddle or as large as the entire earth (biosphere or ecosphere). Further, an ecosystem may be **natural** as a pond, a lake, a river, an estuary, an ocean, a forest, etc., or it may be **man-made** or **artificial** like an aquarium, a dam, a cropland, a garden, an orchard, a city and so on. The term 'ecosystem' was coined by **A. G. Tansley** (1935)—its '*eco*' part means environment and the '*system*' part implies, a complex of coordinated units. The latter requires some more explanation which is as follows.

7-2. SYSTEMS

As discussed in previous chapter of 'levels of organisation and biosphere', a system is a collection of interdependent parts or events that make up a whole. For example, a radio consists of various transistors, transductors, wires, a speaker, and control knobs, among other things. Each part has a specific function, yet the expression of their role depends on the proper functioning of all the other parts. The whole system fails to function unless there is some kind of **input** from the outside which the system can act to produce some **output**. For the radio the outside input is electrical energy, on which the system acts to pick up certain radio waves, which are transmitted as an output—sound. Thus the radio with all its parts functions as a total system. The ecosystem functions in a similar manner. Energy from the sun, the input, is fixed by plants and transferred to animal components. Nutrients are withdrawn from the substrate; deposited in tissues of plants and animals; cycled from one feeding group to another, released by decomposition to the soil, water, and air; and then recycled. The deserts, forests, grasslands, and seas are not independent of each other. The energy and nutrients in one find their way to another, so that ultimately all parts of the earth are interrelated, each comprising a part of the total system that keeps the biosphere functioning.

7·2·1. Types of Systems

There are two types of systems that are of ecological interest —open and cybernetic. **Open systems** depend on an outside environment to provide inputs and accept outputs. The radio is an example of an open system. It receives an input of electrical energy from the outside, acts upon it, and transmits sound waves to the outside. The ecosystem. too, is an open system, for it receives energy from an outside source, the sun ; fixes and utilizes it ; and ultimately dissipates heat to space. If electrical energy is shut off to the radio, it will not function. If the sun's energy were to be cut off, the ecosystem (and life on earth) would cease to function. An open system can consist of a number of components or parts, called **subsystems** ; and each subsystem can consist of a number of **elements.**

A **cybernetic system**, which can also be an open system, has some sort of feedback system to make it self-regulating. It maintains itself by some forms of information feedback into the system. To function in such a manner the cybernetic system has an ideal state or **set point** about which it operates. In a purely mechanical system that set point can be fixed specifically. For example, a dehumidifier which is set for a humidity level of 50 per cent, represents a cybernetic system. When the humidity of the air in a room exceeds 50 per cent, the switch on the dehumidifier turns on and the fans start to pull air over the refrigerated coils on which the water condenses, to be carried away by a hose. When sufficient water has been wrung out of the air, the dehumidifier shuts off. The feedback of information on humidity that causes the dehumidifier to shut off or turn on is called **negative feedback,** because it halts or reverses a movement away from a set point.

Most living systems possess cybernetic systems that can function at various levels, from the cell to the community, but always functioning through organisms—the difference being that in living systems the set point is not fixed. Rather, organisms have a limited range of tolerances, called **homeostatic plateaus**, within which conditions must be maintained. If environmental conditions exceeds the operating limits of the system, it goes out of control. Instead of negative feedback governing the system, **positive feedback** takes over. Positive feedback is a continual movement away from a system's set point or homeostatic plateau that ultimately can destroy the system. A **feedback loop** can be defined as the relationship in which a change in some original state of being 'feedback" onto that original state to alter the rate or direction of further change.

An example of cybernetic control in a living system is temperature regulation in homeothermal animals, such as humans

(Fig. 7·1). The normal temperature for humans is 98·6 °F. If the temperature of the environment rises, sensory mechanisms detect

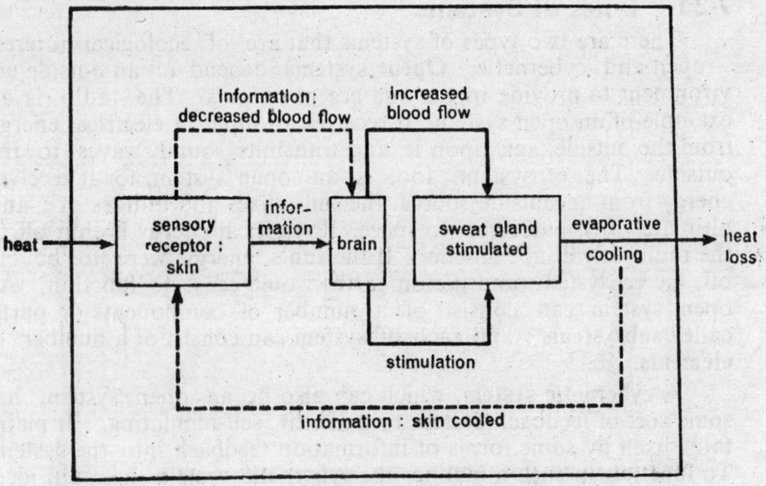

Fig. 7·1. An example of a cybernetic system in living organisms : temperature regulation in the human species (after Smith, 1977).

it, send the message to the brain which acts (involuntarily) on the information and relays the message to the effector mechanims that increase bloodflow to the skin and induce sweating. Water excreted through the skin evaporates, cooling the body. If the environmental temperature falls below a certain point, a similar action in the system takes place, this time reducing blood flow and causing shivering, an involuntary muscular exercise producing more heat. If environmental temperature becomes extreme, the cybernetic system breaks down. If the temperature becomes too hot, the body is unable to lose heat fast enough to hold the temperature at normal. Body metabolism speeds up, further increasing body temperature and eventually ending in heat stroke or death. If the environmental temperature drops too low, metabolic processes slow down, further decreasing body temperature and eventually resulting in death by freezing.

Cybernetic systems produce **stable systems.** A stable system is one that responds to stimuli by developing forces to restore it to its original condition. In natural ecosystems this stability may mean that the system can survive many changes but still preserve a similar structure. Stability also means persistence, *i.e.*, the system remains much the way it was.

Fig. 7·2 shows an ecological example of positive and negative feedbacks In both, we start with a given number of animals. In the positive feedback loop, production of a large number of offspring leads to a larger population than the original. This then becomes the new reproducing population, which can produce even more offsprings. The result is a self-sustaining population increase,

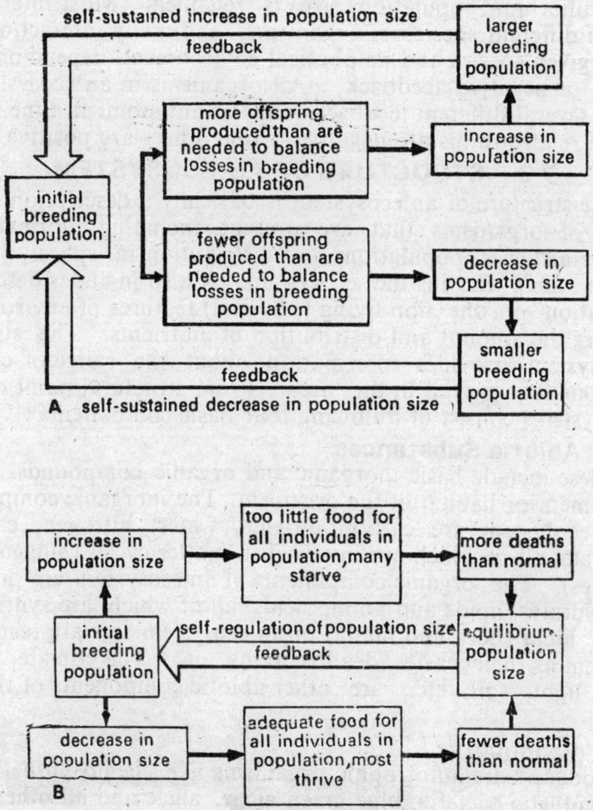

Fig. 7·2. Generalized diagram of the operation of feedback mechanisms in natural ecosystems. A and B figures refer to the operation of these mechanisms in the regulation of population size in a hypothetical population of animals : positive feedback leading to self-sustained change (A), and negative feedback leading to self-regulated homeostasis (B) (after, Clapham, Jr., 1973).

driven by the offspring feeding back into the breeding population. In the same way, if more breeders die than are replaced each year (which happens in some populations), this loss will be reflected in a smaller breeding population and will result in self-sustaining decline in population number by the same type of the positive-feedback loop. In the negative feedback-loop, the population is controlled by external factors, such as food supply. If the population increase in size, there will not be sufficient food to go around, and more individuals than usual will die from starvation ; conversely, if the population decrease in size, there will be more than enough food to go around, and the animals will be well-nourished, leading to a smaller than normal die off. In either case, the equilibrium population size is regained. Most interactions between different species of organisms, and many interactions between a given species and its physical environment, depend on either positive or negative feedback. All organisms in an ecosystem are part of several different feedback loops at any point in time. Some of these relationships are negative, while others are positive.

7-3. STRUCTURE OF AN ECOSYSTEM

The structure of an ecosystem is basically a description of the species of organisms that are present, including information on their life histories, populations and distribution in space. It is a guide to who's who in the ecosystem. It also includes descriptive information on the non-living (physical) features of environment, including the amount and distribution of nutrients. The structure of ecosystem provides information about the range of climatic conditions that prevail in the area. From structural point of view all ecosystems consist of following four basic **components** :

7-3·1. Abiotic Substances

These include basic inorganic and organic compounds of the environment or habitat of the organism. The inorganic components of an ecosystem are carbon dioxide, water, nitrogen, calcium, phosphate, all of which are involved in matter cycles (**biogeochemical cycles**). The organic components of an ecosystem are proteins, carbohydrates, lipids and amino acids, all of which are synthesized by the **biota** (flora and fauna) of an ecosystem and are reached to ecosystem as their wastes, dead remains, etc. The climate, temperature, light, soil, etc., are other abiotic components of the ecosystem.

7-3·2. Producers

Producers are autotrophic organisms like chemosynthetic and photosynthetic bacteria, blue green algae, algae and all other green plants. They are called ecosystem producers because they capture energy from non-organic sources, especially light, and store some of the energy in the form of chemical bonds, for the later use. Algae of various types are the most important producers of aquatic ecosystems, although in estuaries and marshes, grasses may be important as producers. Terrestrial ecosystems have trees, shrubs, herbs, grasses, and mosses that contribute with varying importance to the production of the ecosystem.

Since heterotrophic organisms depend on plants and other autotrophic organisms like bacteria and algae for their nutrition, the amount of energy that the producers capture, sets the limit on the availability of energy for the ecosystem. Thus, when a green plant captures a certain amount of energy from sunlight, it is said to "**produce**" the energy for the ecosystem.

7-3·3. Consumers

They are heterotrophic organisms in the ecosystem which eat other living creatures. There are **herbivores**, which eat plants, and **carnivores**, which eat other animals. They are also called **phagotrophs** or **macroconsumers.** Sometimes herbivores are called **primary macroconsumers** and carnivores are called **secondary macroconsumers.**.

7-3·4. Reducers or Decomposers

Reducers, decomposers, saprotrophs or microconsumers are heterotrophic organisms that breakdown dead and waste matter. Fungi and certain bacteria are the prime representatives of this category. Enzymes are secreted by their cells into or onto dead plant and animal debris. These chemicals digest the dead organism into smaller bits or molecules, which can be absorbed by the fungi or bacteria (saprotrophs). The decomposers take the energy and matter that they harvest during this feeding process for their own metabolism. Heat is liberated in each chemical conversion along the metabolic pathway.

No ecosystem could function long without decomposers. Dead organisms would pile up without rotting, as would waste products. **It would not be long before an essential element, phosphorus, for example, would be first in short supply and then gone altogether,** because the dead corpses littering the landscape would be hoarding the entire supply. Thus, the importance of the decomposers to the ecosystem is that they tear apart organisms and in their metabolic processes release to the environment atoms and molecules that can be reused again by autotrophic organisms. They are not important to the ecosystem from the energy point of view but from the material (nutrient) point of view. Energy cannot be recycled, but matter can be. Energy must be fed into ecosystem to keep up with the dissipation of heat or the increase in entropy. Matter must be recycled again and again by an ecological process called **biogeochemical cycle**, which will be discussed later on in a separate chapter.

7-4. EXAMPLE OF ECOSYSTEM

A classic example of an ecosystem is a small lake or pond (Fig. 7·3). The abiotic or non-living parts of a freshwater pond include the water, dissolved oxygen, carbon dioxide, inorganic salts such as phosphates, nitrates and chlorides of sodium, potassium, and calcium, and a multitude of organic compounds such as amino acids, humic acids, etc. The living part of the freshwater pond can be subdivided according to the functions of the organisms, *i.e.*, what they contribute toward keeping the ecosystem operating as a stable, interacting whole. In a freshwater pond there are two types of producers, the larger plants growing along the shore or floating in shallow water, and the microscopic floating plants, most of which are algae, that are distributed throughout the water as deep as light will penetrate. These tiny plants, collectively referred to as **phytoplankton**, are usually not visible unless they are present in great abundance and give the water a greenish tinge. Phytoplanktons have more significance as food producers for the freshwater pond ecosystem than are the more readily visible plants.

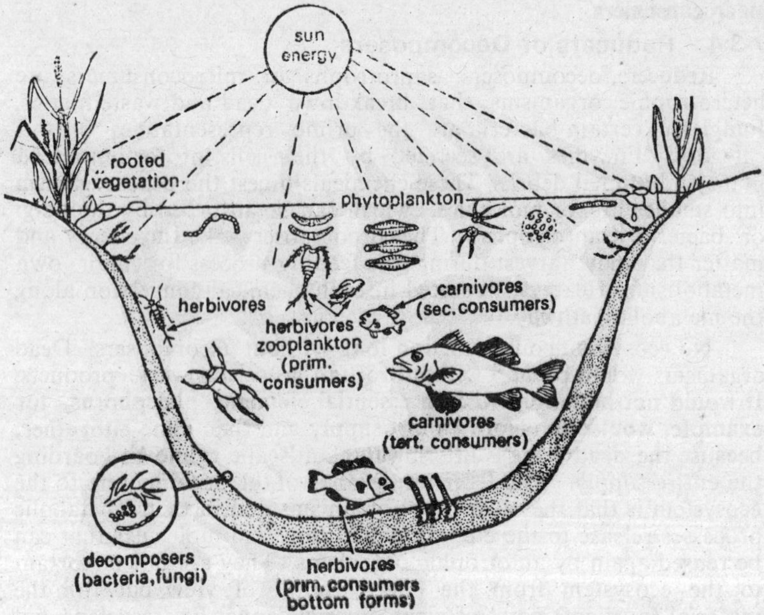

Fig. 7·3. A small freshwater pond as an example of an ecosystem. The component part—producer, consumer and decomposer or reducer organisms, plus the non-living or abiotic components are indicated (after Odum, 1971).

The macroconsumers or phagotrophs of pond ecosystems include insects and insect larvae, Crustacea, fish and perhaps some freshwater clams. Primary consumers such as **zooplankton** (animal plankton) found near the surface of water and **benthos** (bottom forms) are the plant eaters (herbivores) and secondary consumers are the carnivores that eat the primary consumers. There might be some tertiary consumers that eat the carnivores (secondary consumers).

The ecosystem is completed by saprotrophs or decomposer organisms such as bacteria, flagellate protozoans and fungi, which break down the organic compounds of cells from dead producer and consumer organisms either into small organic molecules, which they utilize themselves, or into inorganic substances that can be used as raw materials by green plants.

7-5. FUNCTION OF AN ECOSYSTEM

When we consider the function of an ecosystem, we must describe the flow of energy and the cycling of nutrients. That is, we are interested in things like how much sunlight is trapped by plants in a year, how much plant material is eaten by herbivores, and how many herbivores are eaten by carnivores. The flow of energy in an ecosystem will be discussed in detail in forthcoming chapter, but a brief idea about food chains **and** food webs can be given here.

7-5·1. Food Chains and Food Webs

Organisms of an ecosystem are linked together in **food chains.** A sheep may eat some grass, and in turn it may be eaten by a person. A mouse may eat some grain, and in turn it may be eaten by an owl, hawk, or fox. The algae of a lake will be eaten by many zooplankton, such as crustaceans and insect larvae. The latter are eaten by small fish, such as bluegills, and they in turn are eaten by bass, which may be caught by a bear. It becomes clear from all these cases that plants form the only link between the biotic and abiotic components of ecosystem. They draw water and minerals from the soil and combine them with sunlight and carbon dioxide from the air (photosynthesis) to make carbohydrates, fats, proteins, vitamins and usable minerals. Small herbivorous organisms, such as caterpillars, field mice, and copepods, consume this nutrient-rich vegetable matter and convert it to animal material. They may then serve as food to support meat-eating animals, who may, in turn, be eaten by larger cornivores ; this sequence of eating and being eaten, with the resultant transfer of energy, is known as **food chain.** All of these examples of food chains involve several links or steps and ofcourse, there is energy dissipation along the way because of the inefficiency of the processes. Each step is known as **trophic level** and the study of energy flow through these steps is called **trophic ecology.**

Further, food chains are not isolated from one another. A mouse in a field may eat several types of seeds and be preyed upon by several different carnivores. Each of the carnivore may consume more than one type of prey and so on. Consequently, the food chains interlock with one another, and this inter-connecting network of species is spoken of as a **food web.**

7-6. TYPES OF ECOSYSTEM

Ellenberg (1973) has classified the world into a hierarchy of ecosystems. Biosphere is the largest and all-encompassing ecosystem. Next lower level is **mega-ecosystems** such as **marine ecosystems** (*i.e.*, ecosystems of saline water of sea, ocean and lake); **limnic ecosystems** (*i.e.*, ecosystems of fresh-water) ; **semi-terrestrial ecosystems** (*i.e.*. ecosystems of wet soil and air) ; **terrestrial ecosystems** (*i.e.*, ecosystems of aerated soil and air) ; and **urban-industrial ecosystems** (man-made ecosystems like cropland, city, etc.). The next-lower level is represented by **macro-ecosystems** (forests, etc.)

within each mega-ecosystem. **Meso-ecosystems** (*e.g.*, a cold deciduous broad leaf forest with its fauna) are considered as the basic units of this scheme. Meso-ecosystems are sub-divided into **micro-ecosystems** which depart with respect to a certain component (a low land, mountain, or sub-alpine cold deciduous broad leaf forest with its fauna). **Nano-ecosystems** are further small ecosystems which are spatially contained within larger ecosystems and that show a certain individuality of their own.

7·6·1. Natural and Artificial Ecosystems

Different types of ecosystems like the pond, lake, river, stream, spring, sea, ocean, estuary, forest, grassland, desert, coral reef, cropland, city, etc., differ from each other not only in productivity but also in their species composition. All of them have a similar fundamental plan of their gross structure and function as have been already stated earlier. However, some ecosystems like pond, lake, river, stream, spring, sea, ocean, estuary, forest, grassland, desert, coral reef, etc., are self-regulating systems without much direct human interference and manipulations, and are called **natural ecosystems.** On the other hand, the city and cropland ecosystems are man-made (or man-engineered)-ecosystems, so are called **artificial ecosystems.** In these ecosystems man plans manipulations (*e.g.*, addition of fertilizers to increase the yield, use of chemicals to control diseases, proper irrigation, etc.) in the physico-chemical environment of the cropland ecosystem to get more food, cloth, timber, medicines and other useful plant products.

7·6·2. Incomplete Ecosystems

There are certain ecosystems which do not contain all the four basic components of ecosystem (*i.e.*, abiotic substances, producers, consumers and decomposers) and they may lack one or more basic components of an ecosystem. Such ecosystems are called **incomplete ecosystems** (**Southwick**, 1976). For example, certain incomplete ecosystems such as abyssal depths of the sea and caves lack producers but contain only consumers and decomposers. As in the realm of complete darkness green plants cannot survive.

7·7. REVISION QUESTIONS

1. What is system ? Describe different kinds of systems.
2. Describe the structure of an ecosystem by taking an example of pond ecosystem.
3. Describe the function of ecosystem.
4. Into how many types the ecosystems can be classified ?

Part 2

ENVIRONMENT AND ABIOTIC COMPONENTS OF ECOSYSTEM

Each and every living organism has its specific surrounding, medium or environment to which it continuously interacts and remains fully adapted. The **environment** is a collective term embracing all the conditions in which an organism lives, for example, light, temperature, water and other organisms. In other words, the environment is the sum total of physical (abiotic) and biotic conditions influencing the responses of the organism (**S. C. Kendeigh**, 1974).

An organism's environment, thus, may be subdivided into the physical (abiotic) and biotic environments. In broader terms, abiotic or physical environment of all organisms existing on earth includes three main media which also act like factors, *viz.*, **atmosphere (air)**, **lithosphere (soil)** and **hydrosphere (water)**, and a variety of other physical factors like **temperature**, **light**, **pH**, etc. The study of abiotic factors and substances has been described as **environmental ecology** by A. S. **Boughey** (1971). When environmental ecology concerns agricultural ecosystems, it is studied by agronomists. If it involves effects on populations of domestic animals, it is investigated by workers in animal husbandry ; if it relates to cultivated plants, it is examined by crop physiologists.

8

Atmosphere and Air

8-1. INTRODUCTION

The multilayered gaseous envelope surrounding the planet earth is called **atmosphere**. It is an ocean of air which blends into outer space some 1,000 km or so above the earth's surface. The atmosphere maintains its contacts with all the major types of environments of earth, interacting with them and greatly affecting their ability to support life. The atmosphere is a reservoir of several elements essential to life and it serves many functions, including the filtering of radiant energy from the sun, insulation from heat loss at the earth's surface, and stabilization of weather and climate owing to the heat capacity of the air. Several cycles are also present in it that relate to the movement of matter between an organism and its environment. These biogeochemical cycles of matter include the water cycle, the carbon cycle, the nitrogen cycle, the phosphorus cycle and others.

8-2. PHYSICO-CHEMICAL STRUCTURE OF ATMOSPHERE

There are five concentric layers within the atmosphere, which can be differentiated on the basis of temperature. These include the troposphere, the stratosphere, the mesosphere, the heterosphere and the exosphere (Fig. 8·1), each of which has following physico-chemical structure.

8-2·1. Troposphere

The lowest layer of atmosphere in which man along with other living organisms live, is called **troposphere**. The troposphere is roughly 10 km thick, being somewhat thicker in the equatorial region than it is at the poles. It is characterized by a steady decrease in temperature and is a mixture of several different gases, most of which are fairly constant in their abundance. The composition of troposphere excepting water vapours and air-borne particles (dust), is shown in the table 8·1. Both water vapour and dust occur in the troposphere in extremely variable concentrations. The concentration of water vapour in the troposphere ranges from virtually 0 per cent to more than 4 per cent. As a rule, its abundance is a function of altitude and temperature. It is most common at the lowest levels of the atmosphere, being almost entirely absent above about 8–10 km, and it is more abundant in warm air than it is in cold air. Dust is even more variable component of troposphere. It is even more limited to the lower levels of the atmosphere than is water vapour, and it has no specific relationship with any other feature of the atmosphere.

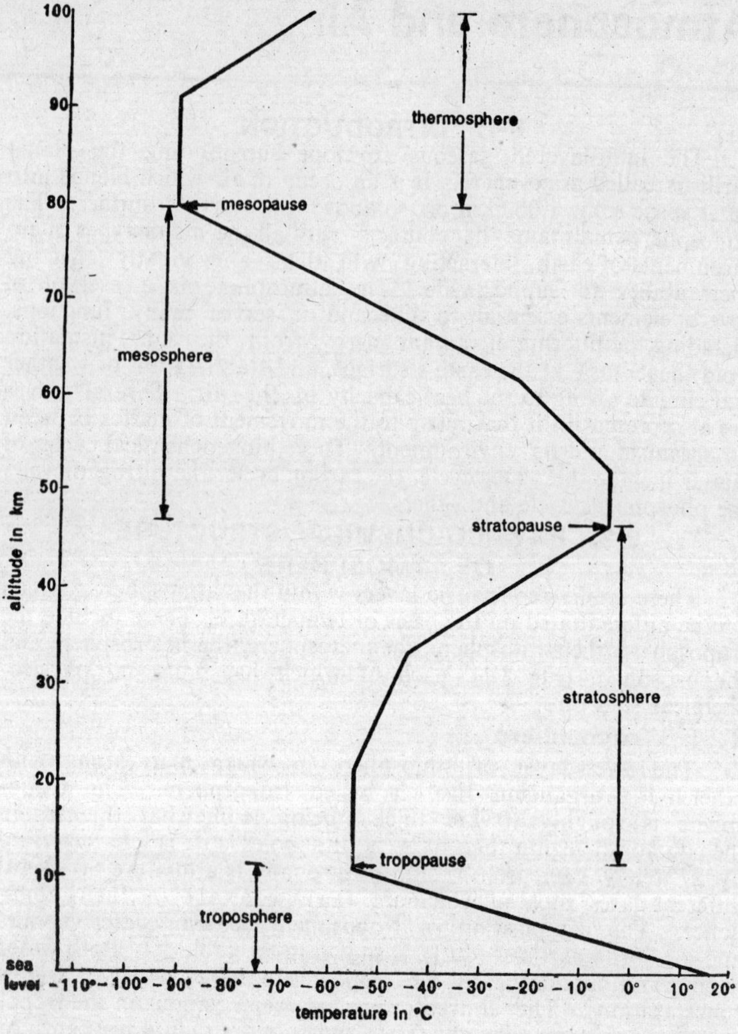

Fig. 8·1. Diagram showing the extent of different layers of atmosphere and temperature stratification of the atmosphere (after Clapham, Jr., 1973).

Troposphere is the layer of sulphates and is the region of strong air movements and cloud formation, *i.e.* it is the locus of most weather phenomena which affect different ecosystems of the biosphere.

Table 8·1. Composition of the dry atmosphere (air of troposphere) by volume (Clapham, Jr., 1973).

Constituent	Percentage
Nitrogen (N_2)	78·084
Oxygen (O_2)	20·9476
Argon (Ar)	0·934
Carbon dioxide (CO_2)	0·0314*
Neon (Ne)	0·001818
Helium (He)	0·000524
Methane (CH_4)	0·0002*
Krypton (Kr)	0 000114
Nitrous oxide (N_2O)	0·00005
Hydrogen (H_2)	0·00005
Xenon (Xe)	0·0000087
Ozone (O_3)	0·000001*

(*These components are highly variable.)

8-2·2. Stratosphere

The stratosphere can be defined as the air mass extending from the **tropopause**, or uppermost level of the troposphere, to the **stratopause** (*i.e.*, uppermost level of the stratosphere), about 50 km, above the surface of the earth. In stratosphere, the temperature rises from a minimum of about −55°C to a maximum of about 5°C.

The stratosphere exhibits several significant differences from the troposphere. Water vapour is virtually absent. The only clouds found in the stratosphere are very thin wispy clouds formed of tiny ice crystals. Ozone (O_3) is present there in significant quantities and there it forms a well marked ozone layer called **ozonosphere** within the stratosphere. Ozone is formed from oxygen by a photochemical reaction in which energy from the sun (here symbolized by hv) splits apart the oxygen molecule (O_2) to form atomic oxygen :

$$O_2 + hv \rightleftharpoons 2O$$

The atomic oxygen then combines with molecular oxygen to form ozone :

$$O_2 + O \rightleftharpoons O_3$$

Both of these reactions are reversible, but the ozone content of the stratosphere is sufficiently constant, *i.e.*, ozone is clearly in chemical equilibrium with the rest of the air, which means that ozone is being produced from oxygen as fast as it is broken down to molecular oxygen. This is important because ozone absorbs ultraviolet radiation from the sun. In fact, the reason that the stratosphere becomes warmer with increasing distance from the earth is that the ultraviolet energy absorbed by the ozone is transformed into heat (**Craig,** 1968). For the ecosystem, the ozone umbrella is of vital importance because its absorption of ultraviolet radiation prevents the latter from reaching the surface of the earth, where it would be lethal to most living organisms. Moreover, because of its extra heat ozonosphere also acts like a blanket that reduces the cooling rate of the earth and thus adds to the effect of water vapour.

8-2·3. Mesosphere

Above the stratosphere is the **mesosphere** which is characterized by cold temperatures and very low atmospheric pressure. In fact, the temperature reversal starts from the stratopause, *i.e.*, temperature begins to drop, reaching a minimum about −95°C at a level some 80−90 km above the earth's surface. This level is termed the **mesopause**.

8-2·4. Thermosphere

Above the mesosphere is the **thermosphere** which extends upto 500 km above the earth's surface and is characterized by steady temperature increase with height from mesopause. The thermosphere includes the region in which ultraviolet radiation and cosmic rays cause ionisation of molecules like oxygen and nitric oxide. This region is called the **ionosphere**. In the thermosphere molecules of air are so widely spaced that high frequency audible sounds are not carried by the atmosphere.

8-2·5. Exosphere

The region of atmosphere above the thermosphere is called **exosphere** or **outer space** which lacks atoms except that of hydrogen and helium and extends upto 32,190 km from the planet. Exosphere has very high temperature due to solar radiations. The earth's magnetic field becomes more important than gravity in the distribution of atomic particles in the exosphere.

All the layers of the atmosphere are of interest to the ecologists since together they form the total blanket of air in the biosphere. Air is an important ecological factor and also serves as a medium of living for some organisms.

Moreover, sometime the ionosphere is effected by magnetic storms on the sun, with an increase in the number of free electrons present during periods of intense activity. Population cycles have been attributed to these storms and subsequent electrical activity in the ionosphere. **A. N. Moen** (1973) has found that white-tailed deer seems to respond to unidentified factors, with an increase in nervousness, activity and heart rate.

8-3. AIR AS AN ECOLOGICAL FACTOR

The gaseous mixture of troposphere is called **air**. The air has a relative constant composition of gases (see Table 8·1) and is utilized by most organisms in respiration to liberate chemical energy from food during oxidation.

Importance of three constituent gases of air, namely oxygen, carbon dioxide and nitrogen is well known. These keep on cycling in nature between organisms and their atmosphere through the biogeochemical cycles. Oxygen is usually available in adequate amounts in land environments. It becomes scarce only at high altitudes (where the atmosphere is thin), deep in ground or in water-soaked soil.

Aquatic organisms make use of oxygen dissolved in the water. Some of this oxygen is absorbed from the air, some is released by plants growing in the water. The amount of oxygen dissolved in water varies greatly with the temperature and depth of the water. Warmer water has a lower capacity for dissolved oxygen than colder water. The amount of oxygen in the water limits the range of water organisms in many cases.

Plants obtain carbon from the gas carbon dioxide, which they absorb in the the process of photosynthesis. Atmospheric nitrogen is directly used by only some nitrogen-fixing bacteria of roots of some plants.

8-3·1. Effect of Air on the Plants

Plants differ in their requirements of aeration. Hydrophytes germinate and show best development under conditions that are almost unsuitable for mesophytes and xerophytes. Reduced aeration of soil bring about a number of morphological and physiological effects on plants. Excess of carbon dioxide in soil air may result in the production of such toxins as hydrogen sulphide, bicarbonates of iron and manganese, acetic, oxalic, and other organic acids. Even carbon dioxide sometimes has toxic and lethal effects on plants and some soil organisms.

In aquatic habitats, the medium is generally deficient in oxygen content, because much of the dissolved oxygen in water diffuses in the atmosphere. The solubility characteristics of oxygen and carbon dioxide in water are also different, for carbon dioxide is about 200 times more soluble than oxygen. Moreover, both temperature and salinity have pronounced effects on the solubility of these two gases ; with either increased temperature or increased salinity, there is a decrease in their solubility in water (Table 8·2). Thus, the oxygen content of lakes and deep ponds are closely related to their thermal stratifications and salinity.

Table 8·2. Coefficients of saturation of CO_2 in water of different temperatures and salinities (ml/litre in equilibrium with 760 mm Hg) (Vernberg and Vernberg, 1970).

Temperature, °C	Salinity, %		
	0	28·91	36·11
0	1715	1489	1438
12	1118	980	947
24	782	695	677

The plants growing in conditions of suboptimal aeration exhibit certain morphological and physiological adaptations. For example, they develop shallow root systems, special aerating tissues and special aerating organs. They have a low requirement of oxygen and have an ability to respire anaerobically.

8-3·2. Effect of Air on the Animals

The atmospheric composition is not always the most important factor to organism ; rather, it is the partial pressure of a gas, especially oxygen and CO_2, which influences the existence of life, as illustrated by altitudinal studies on organisms. The **partial pressure** of a gas is the product of the total barometric pressure times the concentration of gas in dry air. For every 18,000 ft., the barometric pressure is reduced by half, but the **composition of air** is essentially unchanged. To illustrate this effect on animals, the following critical altitudes have been identified for most humans ; above 10,000 ft., oxygen must be added to inspired air ; at 35,000 ft., 100 per cent oxygen must be inhaled ; and at 50,000 ft., little oxygen reaches the alveoli even if 100 per cent oxygen is breathed (see **Vernberg** and **Vernberg**, 1970).

If gases are to be functionally important to an organism, there must be a mechanism whereby gases can enter and leave the body. Apparently the passage of oxygen across a membrane is accomplished by **diffusion**. The rate of diffusion can be calculated by **Fick's law**, which is based on concentration coefficient, amount of surface thickness of the membrane, and time, and is expressed as the diffusion coefficient (**Davson**, 1959). The diffusion coefficient value is not the same for different animal tissues. For example, some representative values of diffusion coefficient of oxygen are : air, 11·0 ; water, 0·000034 ; muscle, 0·000014 ; chitin, 0·000013 ; connective tissue, 0·000011. Not only is the diffusion rate different in various tissues, but the behaviour of each gas also differs. For example, carbon dioxide diffuses through water and animal tissues 25 times faster than oxygen.

8-4. AIR AS A MEDIUM FOR LIVING ORGANISMS

Air is not an easy and suitable medium to support life (biota) and actually no organism ever originated in air, though, certain aquatic and terrestrial organisms have become secondarily adapted for aerial existence. Air has so much less **buoyancy** than water that organisms emerging from water or land, are immediately subjected to a stronger pull of **gravity** which holds them to the earth. They are also exposed to **evaporation** of water from their bodies, which threatens loss of water from the protoplasm and consequent death due to desiccation. Changes in temperature are much more drastic in air and there is a greater danger of chilling or overheating. Further, periods of light exposure are much longer and much more intense in air than in water or upland. Even more, the problem of **mineral supply** becomes acute in air.

Only a few micro-organisms, plants and animals have invaded the air. The essential function of dispersal of fruits and seeds of certain flowering plants is performed by the air currents. These fruits and seeds have various kinds of adaptations for such wind dispersal. Various animals of different groups exhibit adaptations for the aerial mode of life. The animals which can fly in air, are said to have **flight** or **volant adaptations**.

8-4·1. Volant Adaptations of Animals

Animals which use atmosphere as the medium of living, have twofold adaptations, one kind of adaptations for aerial mode of life (*i.e.*, volant adaptations) and others for trees, earth or aquatic environment, to which, they have to return for rest, feeding, reproduction and other essential activities of life, since, no animal can **permanently** dwell in air. Flight or volant adaptations fall into following two types:

(a) **Adaptations for gliding type flight**. Gliding is characterized by leaping from a high point and held up by certain sustaining organs, then to glide to lower level. Such gliding flight is performed by following animals: fishes like *Exocoetus pecilopterus*, *Dactylopterus*, *Gastropelecus* and *Pegasus volitans*, amphibians like *Rhacophorus* (flying frog), reptiles like *Rhamphorhynchus* (extinct flying dinosaur called pterodactyl of mesozoic era), flying lizards like *Draco dussumieri*, *Draco volans* and *Ptychozoon*; birds like *Archaeopteryx* (extinct bird), *Struthio* (ostrich), etc., and mammals like flying phalangers (*Petaurus*), rodents (*Anomalurus*, *Pteromys*, ***Sciuropterus*, and *Eupetaurus***), and lemurs.

Morphological adaptations for gliding include (i) Development of a fold of skin called **patagium** between fore-limbs and hind-limbs. This patagium provides sustaining surface for gliding. Patagia of both sides can be folded like a fan against the sides of the body when not in use. (ii) In flying fish, pectoral fins become enlarged in the form of parachutes and are highly inserted on the body. (iii) In flying frog (*Rhacophorus pardalis*) the feet are webbed which sustain it in the prolonged leaps. Its digits terminate in adhesive pads which help in adhesion to the trees.

(b) **Adaptations for true flight**. In true aerial flight, the power (*i.e.*, muscular power) is implied and the movement in air is sustained. Adaptations for true flight occur in various insects like; dragon flies, locusts, moths, butterflies, wasps, flies, bees, etc., most birds like vultures, kites, parrots, owls, crows, house-sparrows etc., and some mammals like insectivorous bats (*Eptesicus*, *Desmodus* and *Magaderma*), frugivorous bats or flying foxes (*Pteropus medius*, and *Cynopterus*).

Volant adaptations for true flight include following morphological and physiological adaptations: 1. Development of **feathers** in birds which provide many advantages during flight. 2. In birds, the fore-limbs are modified into flight structures, the **wings**. In bat wing, humerus bone is well developed, radius bone is long and curved ulna is vestigial. The thumb or pollex is free and clawed for crawling and climbing. In insects, wings are the lateral membranous folds of exoskeleton. They are attached with the thorax and are made up of thin, light but tough and elastic chitin.

3. The bones of birds are hollow, spongy and contain many air-cavities. Such pneumatic bones reduce the weight of body and also provide buoyancy during flight. Birds lack teeth and instead of them have a much lighter beak. 4. Birds also possess **air-sacs** which have air for respiration and are buoyant. 5. All flying animals have a compact and sleek body build-up and also have well developed flight muscles. 6. Nervous system and sense organs of flying animals are well developed. 7. For the reduction of body weight, female birds have only one ovary.

8-5. WIND AS AN ECOLOGICAL FACTOR

Wind or air current is an important ecological factor of the atmosphere, as it affects life of living organisms (plants and animals) mainly on flat plains, along sea coasts and at high altitudes in mountain. Meterologists describe wind flow in terms of an average velocity over a vertical profile. For estimating wind speeds and their effects on land surface, the **Beaufort scale** is used (Table 8·3).

Table 8·3. Beaufort scale for estimating wind speeds and their effects on land (Subramanian and Radhakrishnan, 1980).

S. No.	Beaufort description	Speed (m.p.h.)	Land effect
0	Calm	Under 1	Smoke rises straight up
1	Light air	1–3	Direction shown by smoke dirt
2	Light breeze	4–7	Wind felt on face leaves rustle
3	Gentle breeze	8–12	Leaves in constant motion, wind extends light flag
4	Moderate breeze	13–18	Raise dust and loose paper ; small branches move
5	Fresh breeze	19–24	Small trees in leaf begin to sway
6	Strong breeze	25–31	Large branches in motion ; whistling of telegraph wires
7	Moderate gale	32–38	Whole tree in motion ; difficulty in walking
8	Fresh gale	39–46	Breaks twigs of trees
9	Strong gale	47–54	Some damage to vegetation and structures
10	Whole gale	55–63	Trees uprooted, considerable damage
11	Storm	64–75	Widespread damage
12	Hurricane, cyclone and Typhoon	Over 75	Total damage

8-5·1. Causes of Air Movements

The air is always in a state of flux. It flows under the influence of forces generated by heat from the sun, and the centrifugal forces resulting from the earth's rotation. Thus, the **wind** may be defined as the motion of air caused by gravity, by deflective forces due to earth's rotation from west to east and by centrifugal forces due to the curvature of the wind path (see **Subramanian** and **Radhakrishnan**, 1980). These forces are opposed by other forces which arise from friction and viscosity.

The air never flows with a perfectly smooth and streamline motion, but always with horizontal and vertical fluctuations which when sudden and relatively brief, are called **gusts**.

The circulatory pattern of wind flow on the planet earth occurs by following ways: The earth spins on its axis from west to east. This spinning produces a **Coriolis force** that deflects the winds and prevents a simple direct return flow of air from the poles to the equator. In the Northern Hemisphere, the winds are deflected to the right ; in the Southern Hemisphere to the left. Friction produced by the irregular land masses and oceans on earth's surface modifies the simple flow pattern of air. The result is a series of belts of prevailing **east winds** in the polar regions, the **polar easterlies**, and near the equator, the **easterly trade winds**. In the middle latitudes is a region of **west winds** known as the **westerlies**. These belts break the simple flow of air towards the equator and the flow aloft towards the poles into a series of cells.

The flow of air is divided into three cells in each hemisphere. The air that flows up from the equator forms an equatorial zone of low pressure. The equatorial air cools, loses its moisture and descends. By the time the air has reached 30° latitude it has lost enough heat to sink, forming a cell of semi-permanent high pressure encircling the earth and a region of light winds known as the **horse latitudes**. The air, warmed again, picks up moisture and rises once more. Some of it flows toward the poles, some toward the equator. Meanwhile, at each pole another cell builds up and flows outward in a southerly and northerly direction to meet the rising warm at approximately 60° latitude, flowing towards the poles. This convergence produces a semi-permanent low pressure area at about 60° N and S latitude.

Further, land heats and cools more rapidly than the oceans. Because the masses of land and water on the surface are not uniform, the surface of the earth experiences uneven heating and cooling. Consequently, at any given time the temperature changes are much drastic over the continental areas than over oceans. Oceans act as heat reservoirs and that are continents which affect the circulation of winds. In winter the west coasts of the continents are warmer than the east coasts because the air reaching the west coasts has travelled over warmer ocean areas. The interaction of winds and heating produces more or less permanent high-pressure cells called **Subtropical Highs** in the Atlantic and Pacific oceans ; winds and cooling produce low-pressure cells such as the Aleutian and Icelandic Lows. The highs are more pronounced during the summer months, the lows in the winter months.

There are also produced the monsoon winds and dry winds. The dry winds blow from continental interiors to the oceans in summer and winds heavy with moisture (*i.e.*, monsoon winds) blow from the oceans to the interiors in winter. Lastly, there are moving air masses with their cyclonic and anticyclonic frontal systems. These major air circulations are responsible for the changing swirls or cloud patterns over the earth.

The circulatory patterns of air flow and resultant moisture regimes are responsible for the forests, grasslands and deserts of earth, which in turn reflect the climatic pattern of the planet (**Smith**, 1974).

8-5·2. Jet Stream, Tornado, Hurricanes and Eddies

At altitudinal heights between 8000 metres and 13000 metres a tubular current of air may form, with a very high wind speed, called a **jet stream**. Another high speed wind called **tornado** is characterized by a vertical spiral of very high speed wind with a relatively short life.

Hurricanes or **typhoons** are tropical depressions (Depression has a belt of rain or snow surrounding the warm air. In a depression, the cold air will force itself into a less dense warm air. Thus, the centre will be at low pressure and the outer winds will be at higher pressure), smaller in area but more severe than the depression of temperate zone. Hurricanes originate over warm oceans and derive energy from the latent heat of evaporation of water sucked into low pressure centre. Their paths are entirely random and are characterized by high tidal surges. Tornadoes and hurricanes are cyclonic storms but, most cylones are neither tornadoes nor hurricanes.

Eddies are created in air when they meet a sharp discontinuity on the earth's surface, such as building, tree or mountain. Eddies travel for long distances and when superimposed on the mean airflow, give rise to variable winds. When the wind reaches a peak, it is called a gust and at its minimum it is called a **lull**.

8-5·3. Ecological Effects of Wind on Organisms

Wind modifies the water relations, temperature and light conditions of a particular area. Birds confine themselves to shelter on days when the wind is strong. The transportation of microbes, plants and animals by hurricanes to **hitherto** unoccupied land may change the nature of the whole biotic community. Wind brings about following physical as well as anatomical and physiological effects on plants :

8-5·3A. Physical effects. (i) The wind of high velocity may cause the breaking of living branches of trees and sometimes even **causes** their complete uprooting. Usually such breakage occurs in soft woods of such plants as cotton woods and river maple. The uprooted single or groups of forest trees are called **wind throws** or **wind falls**. (ii) Strong winds from a constant direction often cause permanent alteration in the form and position of the plant shoots. Sach wind-caused deformations are common in trees growing on ridges and along coasts. (iii) Violent winds often cause the flattening or lodging of the herbaceous plants like wheat, maize, oat, sugarcane, etc., against the ground. (iv) Particles of soil or ice carried by strong winds may act as abrasive force, by which the buds, leaves and other delicate parts of plants may be eroded away. Crops grown on sandy soils are usually suffered by such damages. (v) Strong winds cause soil erosion and soil deposition and make a land unsuitable for plant growth. (vi) Along sea-coasts, the salts of the water are carried along by strong winds and such salt-sprays have injurious effects on some plants growing in the vicinity of the ocean.

8-5·3B. Anatomical and physiological effects. (i) The wind deformation often causes the development of a dense, reddish type of xylem, called **compression wood**, on the compressed side of the trees. The wind-deformed organs of the herbaceous plants contain more collenchyma than the normal plant organ. (ii) Strong winds cause an increase in the rates of evaporation and transpiration in plants, both of which disturb the normal water balance of the plants and thus, result the **desiccation** and death of plants. However, some plants with crowded branches escape the desiccation due to wind effect and they develop cushion-form root system and are called **cushion plants**. (iii) Plants growing on sea-coasts, arctic or alpine timberlines, are subjected to drying winds and consequently are suffered from dehydration and consequent loss of turgidity. Under such circumstances, their organs become dwarfed.

8-6. REVISION QUESTIONS

1. Define following terms : environment, atmosphere, ozonosphere, ionosphere, air, wind, volant adaptation.
2. Describe physico-chemical nature of the atmosphere.
3. Describe, how does air act as a ecological factor ?
4. How do living organisms use the air as the medium of living ? Describe.
5. Describe the different kinds of winds ? How are they caused ? What ecological changes are caused by wind in plants ? Describe.

9

Lithosphere and Soil

9-1. THE LITHOSPHERE

The earth is a cold, spherical, solid planet of the solar system, which spins on its axis and revolves around the sun at a certain constant distance. The solid component of earth is called **lithosphere.** The lithosphere is multilayered and includes following three main layers : **crust**, **mantle** and **outer** and **inner core** (Fig. 9·1). The **core** is the central fluid or vapourized sphere having diameter of about 2500 km from the centre and is possibly composed of Nickel-Iron. The **mantle** extends about 2900 km above the core. This is in a molten state. The **crust** is the the outermost solid zone of the earth and it is about 8 to 40 km **above the mantle. The crust** is very complex and its surface is covered with the **soil** supporting rich and varied biotic communities, for living organisms find in the soil an environment providing food, shelter, anchorage and concealment from the predators (**Wallwork**, 1970).

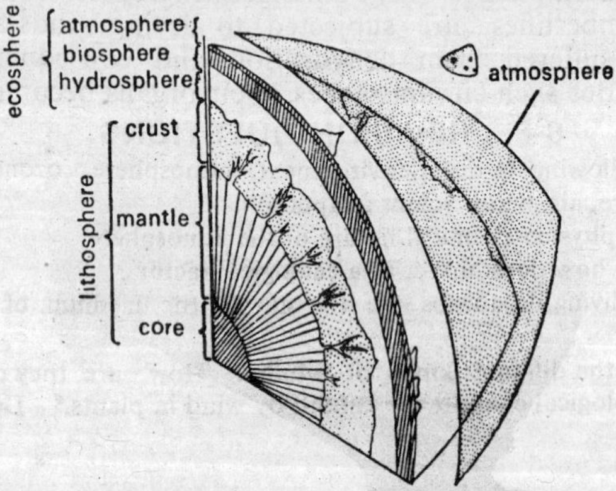

Fig. 9·1. Cross section of the planet earth showing its different layers (after Smith, 1974).

9-2. THE SOIL

The soil is one of the most important ecological factor (called **edaphic factor**) and also is the most characteristic feature of terrestrial environments. It is the loose, friable, unconsolidated top layer of earth's crust and is the mixture of weathered rock materials (*i.e.*, minerals) and organic detritus, both of which are formed through the physical, chemical and biological processes occurring slowly and slowly for a long period at earth's surface. Besides being the source for the entry of nutrients and water into plants, the soil is the medium for the detritus food chain : nutrients released in detritus are decomposed by various soil microbes like bacteria, algae, fungi, protozoa, etc., bound in or on soil particles, and taken back into plants through their roots. Soil (mud) is also the main source of nutrients for all aquatic plants (hydrophytes), rooted or submerged or free-floating. In addition, soil is the means of support for all terrestrial organisms : plants are anchored to the soil by their root systems ; animals walk upon it and are supported by it, as many animals like nematodes, polychaetes, arachnids, insects, rodents, etc., live under the soil.

The word **soil** is derived from a Latin word '*solum*' meaning earthy material in which plants grow. The science which deals with the study of soil is called **soil science**, **pedology** (*pedos*–earth), or **edaphology** (*edaphos*–soil). The study of soil is helpful in practices of agriculture, horticulture, forestry, etc., such as cultivation, irrigation, artificial drainage and use of fertilizers. Pedology is also useful in sciences like geology, petrology, minerology, palaeobotany and palaeozoology.

A fully developed or **mature soil** is that state of soil that has assumed the profile features (*i.e.*, succession of natural layers), characteristic of predominant soils on the smooth uplands within the general climatc and botanic regions in which it is found (**Marbut**, 1926). According to **Dokyachev** (1879), the first soil scientist, the soil is a result of the actions and reciprocal influences of parent rocks, climate, topography, plants, animals and age of the land.

9-3. ORIGIN, DEVELOPMENT OR FORMATION OF SOIL (Pedogenesis)

Soil is a stratified mixture of inorganic and organic materials, both of which are decomposition products. The mineral constituents of soil are derived from some parent material, the **soil forming rocks** by fragmentation or weathering, while, organic components of soil are formed either by decomposition (or transformation) of dead remains of plants or animals or through metabolic activities of living organisms present in the soil. Before discussing the soil-forming processes, a brief discussion about the nature of soil forming rocks will be highly desirable.

9-3·1. Soil Forming Rocks

Basically, there are three kinds of soil forming rocks—**igneous rocks**, **sedimentary rocks** and **metamorphic rocks**. The main characteristics of these rocks have been summarized in table 9·1.

Table 9·1. Classification, characteristics and mineral composition of the principal soil-forming rocks (Wallwork, 1970).

Class of rock	Mode of formation of rock	Type	Description of rock-types	Principal minerals of soil-forming rocks
A. Igneous rocks	Formed by cooling of molten megma (lava)	1. Granite	Usually light in colour, coarse to medium grain	Quarts, feldspar, mica, amphibole, iron oxides
		2. Diorite	Grey to dark in colour, coarse to medium grain	Feldspar, amphibole, iron oxide, biotite
		3. Basalt	Dark to black in colour, dense to fine grain	Feldspar, pyroxene, iron oxide, biotite
B. Sedimentary rocks	Formed by deposition of weathered minerals which derived from igneous rocks	1. Shales	Light to dark in colour, thinly laminated	Clay minerals, quartz
		2. Sandstone	Light to red in colour, granular and porous	Quartz, clay minerals, iron oxides, calcium carbonate
		3. Limestone	Light grey, red, brown, or black in colour, fine grained and compact	Calcite, dolomite, iron oxide, clay minerals
C. Metamorphic rocks	Formed by change of pre-existing rock (*i.e.*, igneous or sedimentary rock) through heat and pressure	1. Gneiss	With light and dark bands	Much as in granite from which it is formed
		2. Schist	Foliated structure	Much as in basalt or shale from which it is formed
		3. Slate	Grey to black in colour, compact and uniform texture	Similar composition to shale from which it is formed
		4. Quartzite	Light to brown in colour, compact and uniform texture	Similar composition to sandstone from which it is formed
		5. Marble	Light, red, green or black in colour, compact, fine to coarse texture	Calcite, dolomite formed from lime stone

9-3·2. Chemistry of Minerals of Soil-forming Rocks

Rocks are the chemical mixture of numerous kinds of minerals. The chemical nature of certain most common and abundant minerals of soil forming rocks has been listed in table 9·2.

Table 9·2. Chemical composition of some common soil minerals.

Minerals	Chemical Constituents
A. Sand and silt minerals	
1. Quartz or silica	SiO_2
2. Feldspars	
a. Orthoclase	$K_2Al_2Si_6O_{16}$
b. Plagioclase	$NaAlSi_3O_8$
c. Calcium feldespar	$CaAl_2Si_2O_8$
3. Micas	
a. Muscovite	$K(OH)_2Al_2(AlSi_3)O_{10}$
b. Biotite	K, Mg, Fe, Al silicate
4. Pyroxene	$(Mg, Fe) SiO_3$
5. Amphibole	$(Mg, Fe)_7 (Si_4O_{11})_2 (OH)_2$
6. Olivine and serpentine	$(Mg, Fe)_2 SiO_4$
7. Calcite ; magnesite ; and dolomite	$CaCO_3$, $MgCO_3$; and ($CaCO_3$, $MgCO_3$)
8. Iron oxides	
a. Haematite	Fe_2O_3
b. Magnetite	Fe_3O_4
c. Limonite	$FeO (OH), xH_2O$
B. Clay minerals	
1. Kaolin	$Al_2O_3, 2SiO_2, 2H_2O$
2. Montmorillonite	$(Ca, MgO)Al_2O_3, 5SiO_3, 5H_2O$

Besides these tabulated minerals following minerals occur in low percentage in soil-forming rocks—Toarmalive (Boron silicas or Aluminium with alkali metals and Iron and Magnesium), Rutile (Titanium oxides), Zircon (Zirconium oxide), and Gauconite (hydrated silicon of iron and potassium).

9-3·3. Process of Soil-formation

The processes which are involved in the formation of mature soil can be studied under following heads :

9-3·3A. Weathering of soil-forming rocks. Soil formation is started by **disintegration** or **weathering** of parent rocks by some physical, chemical and/or biological agents, because of which the soil-forming rocks are broken down in small particles called **regoliths**. Regoliths are the basic materials which under the influence of various other pedogenic processes finally develop into mature soil.

(a) **Physical weathering.** The physical weathering agents are primarily climatic in character, exerting a mechanical effect on the substratum with the result that fragments are comminuted into progressively decreasing particle sizes (*i.e.*, regoliths). Such climatic weathering of rocks does not cause any chemical transformation of rock-minerals and commonly occurs in deserts, in high altitudes, in high latitudes, and in localities with marked topographic relief and sparse vagetation cover. The agents which are involved in climatic weathering of rocks, are **temperature, water, ice, gravity** and **wind** (see Wallwork, 1970).

The temperature causes break down of those rocks which have heterogeneous structure, due to the fact of differential expansion and contraction coefficient of materials composing the rocks. Minerals composing the rocks have got different degrees of expansion. These minerals expand in the high temperature of day and contract when the temperature falls. The differential expansion and contraction of different minerals of rocks set up internal tensions and produce cracks in the rocks and consequently, the rocks weather into finer particles.

In its liquid state **water** causes mechanical weathering of rocks by following methods : (i) **Rain water**—Natural water falling either in the form of rain drops or hail storm on the surface of rock with beating effect bring about abrasion of massive rocks into smaller particles. (ii) **Torrent water**—Rapidly flowing water rolls the heavy rock masses such as rock boulders along the bottom of stream and grinds them into finer particles. (iii) **Wave action**—The wave actions are most effective in sea shores. The rapidly striking water waves dislocate solid particles of varying diameters from sea shore rocks and the debris is then settled at the sea bottom to form **marine soil**. Water also acts as a mechanical carrier. Nearly 5 billion tons of mineral matter is annually carried away in solution from land to sea (see Pearse, 1939).

In its freezing and ice-melting states water causes rock-weathering by **frost action** and **glacier formation**. Water in the form of frost or ice, is an extremely effective physical weathering agent of rocks. It steeps into rock crevices, freezes due to sudden fall of temperature of rocks, expands about nine per cent of its original volume, exerts a pressure (expansion force) of approximately 150 ton/ft^2 and eventually cracks the rock into smaller pieces. Likewise, in summer when ice at mountain tops starts melting and glaciers (huge sliding masses of ice) move downwardly on the slope, then, during the glacier movement, the rock over which they move is gradually worn down to produce fine particles which are deposited as **drift** or **till**, when glacier finally retreats.

Gravitation weathering action is most effectively demonstrated by land slides and rock slippages caused by earth quakes and faulting during which the rock is fragmented by abrasion and the forces of impact. Lastly, the rapid stormy **wind** carrying suspended sand particles causes the abrasion of exposed rock. It acts like a mechanical carrier in moving the particles over the surface of earth **as dunes** or **drifts** and in transporting large quantities of fine suspended particles long distances.

(b) **Chemical weathering**. The physical disintegration produces a greater surface area of rock exposed to the **chemical weathering**, which occurs simultaneously with physical weathering and continues much beyond that. During chemical weathering, a chemical transformation or decomposition of parental mineral materials into new mineral complexes or secondary products occurs. For example, primary minerals that contain aluminium and silicon, such as feld-spar, are converted to secondary mineral such as clay. Because chemical weathering requires the presence of moisture and air as essential factors, therefore, chemical weathering is not effective in deserts. It includes following reactions :

1. **Solution.** Water is the most potent chemical weathering agent in most kinds of rocks through its solvent action. Solution helps in the removal of water-soluble minerals of weathered rocks. For example, soluble rocks like gypsum, limestone and those with a calcareous content get weathered by the solvent action of water which increases in the presence of CO_2 and organic acids formed by decay of organic remains of plants and animals. The solution of these minerals may become absorbed on the surface of negatively charged colloidal particles or are removed by leaching.

2. **Hydrolysis.** Hydrolysis involves an exchange of constituent parts between water and minerals such as the chemical union of water with strong bases producing hydroxides of iron, magnesium, calcium, aluminium, etc. Hydrolysis mostly acts in combination with other reaction such as oxidation/reduction or carbonation as in the production of secondary kaolinitic clay minerals from primary orthoclase :

$$\underset{\text{Orthoclase}}{K_2Al_2Si_6O_{16}} + 2H_2O + CO_2 \rightarrow \underset{\text{Kaolin}}{Al_2O_3 . 2SiO_2 . 2H_2O} + 4SiO_2 + K_2CO_3$$

The release of calcium, magnesium, potassium, sodium, and silicates into the soil solution, as a result of hydrolysis, enhances the availability of these ions to plants.

3. **Oxidation**. Oxidation results when oxygen is added to minerals to produce oxides which when dissolve in water weaken the rock and bring about weathering. It occurs in well-aerated, well drained soil. Because iron is especially reactive with water and oxygen, iron-bearing minerals are prone to rapid decomposition. For example, due to oxidation iron is oxidised in the red ferric state.

$$\underset{\text{Ferrous oxide}}{4FeO} + O_2 \rightarrow \underset{\text{Ferric oxide}}{2Fe_2O_3}$$

Further, oxides and sulphides of Iron, Aluminium, Manganese, etc., are easily oxidised and cause into the weathering of rocks.

4. **Reduction**. Reduction occurs in wet, badly drained, poorly aerated sites such as deep zones of earth crust. It means removal of oxygen from minerals. For example, red ferric oxide may be reduced to the gray ferrous state as follows :

$$2Fe_2O_3 \rightarrow 4FeO + O_2$$

5. **Carbonation.** In carbonation water and carbon dioxide combine to form carbonic acid which reacts with hydroxides of calcium, magnesium and other minerals of rock to form carbonates and bicarbonates in following manner :

(i) $$CO_2 + H_2O \rightarrow \underset{\text{Carbonic acid}}{H_2CO_3}$$

(ii) $$\underset{\text{Calcium hydroxide}}{Ca(OH)_2} + CO_2 \rightarrow \underset{\text{Calcium carbonate}}{CaCO_3} + H_2O + CO_2 \rightarrow \underset{\text{Calcium bicarbonate}}{Ca(HCO_3)_2}$$

The slightly soluble carbonates of these minerals either accumulate deeper in the rock material or are carried away, depending upon the amount of water passing through. The solubility of these carbonates, leaves gaps in rocks and consequently makes them weak enough for weathering process.

6. **Hydration.** During hydration water molecules become chemically attached to particular rock mineral, due to which the volume of parental material increases and hydrated minerals become soft and more readily weatherable :

$$\underset{\text{Hematite}}{2Fe_2O_3 + 3H_2O} \rightarrow \underset{\text{Limonite}}{2Fe_2O_3.3H_2O}$$

(c) **Biological weathering.** Though the surface of bare rock is unsuitable for many forms of life, even then a number of microorganisms (bacteria, protozoans, fungi, nematodes, etc.), lichens and mosses can gain a foothold. These early colonizers transform the rock into a dynamic system, storing energy and synthesizing organic material (**Jacks**, 1965). Their activities alter the mineral composition as well as the physical structure of the rock. For example, lichens are present in the initial stages of biological succession and their growth may cause cracking or flaking, exposing greater area of rock to further weathering. The lichens and mosses extract mineral nutrients such as P, S, Ca, Mg, K, Na, Fe, Si, and Al, from the rock. These elements are combined with organic complexes, and eventually return to the developing soil when the vegetation decomposes.

Products of Weathering and Soil Types

The soil which is formed by weathering of soil-forming rocks is called **embryonic** or **primary soil**. It may mature into following types of soils :

(i) **Residual soil or sedimentary soil.** This is the mature soil lying immediately over the parent rocks.

(ii) **Skeletal soil or immature soil.** This is only partly weathered material in which maturation has not occurred.

(iii) **Secondary soil or transported soil.** The transported soils are those in which the weathered parent material has been shifted to different places by the agency of glacier (**moraine soils, glacial drift** and **till**), streams and rivers (**alluvial soil**), the gravitational forces as land slides (**colluvial soil**), wind (**aeolian soil**), sand storms (**sand dunes**), standing water and wave action (**lacustrine soil**) and the oceanic waves (**marine soil**).

On the basis of profile development, the soils are also classified into zonal, intrazonal and azonal soils. Some soils are called **zonal**, because climate is the foremost control over the process of soil formation and fully developed or mature soils of a given climatic zone tend to be very similar. **Intrazonal soils** result from local geological conditions that preclude the normal pattern of soil formation. One type of intrazonal soils are **hydromorphic soils**, which are formed in areas where the soil becomes waterlogged during formation. Waterlogged soils are not aerated, and the normal aerobic breakdown of detritus cannot take place ; thus hydromorphic soils are rich in organic materials. A **halomorphic** is another type of intrazonal soil and it is formed in arid and semiarid regions due to waterlogging. Its chief characteristic is the high salt contents. Thus, intrazonal soils show some inherited and some acquired characteristics. In contrast, the characteristics of **azonal soils** are entirely inherited. Alluvial soil is best example of azonal soil which is rich in colloids and because it is not leached, it may be rich in nutrient materials, but lacks humus (For soil classification see section 9-10).

9-3·3B. **Addition of organic matter, humification and mineralization.** Organic matter is continuously added to the embryonic soil during soil formation by various living organisms. Plants contribute their deciduous parts at regular intervals and when each plant dies it adds its whole body to the soil. Likewise, animal living in and on the soil add their excreta regularly and contribute their own bodies to the substratum when they die. All organic plant debris fallen recently to the ground is called **litter**. The litter is composed of dead leaves, twigs, woods, dead roots and various plant products. In thick forests, during each season, there is much accumulation of thousands of tons of such litter on the soil surface.

Just beneath the fresh litter, often occurs the material derived from the preceding season's litter in which decay or microbial decomposition has been in progress and this is called **duff**. The litter is decomposed by combined efforts of soil microbes like bacteria, actinomycetes and other fungi. The resultant products of decomposition include various kinds of inorganic and organic plant nutrients. They all are incorporated into the mineral particles of soil which now becomes dark in colour. The residual, finely divided, amorphous, incompletely decomposed black coloured organic matter which is added to the mineral matter of soil, is called **humus** and the process of its formation is called **humification**. In fact humus includes two kinds of organic matter : the partially decomposed organic matter that has been derived from litter and excreta of soil animals like centipedes, millipedes, earthworms, mites, springtails, grasshoppers, etc., which feed on litter or fresh plant material. In a gradual way, all humus is completely decomposed or reduced into simple compounds like carbon dioxide, water, minerals and salts by a process called **mineralization**.

The rate of humification and humus accumulation are chiefly determined by the organisms involved in the decaying process and the temperature, moisture, nitrogen supply, aeration, and pH of the soil. The decomposition period of litter in the tropical forests varies from 2·5 to 7·2 months (see **Vyas** and **Golley**, 1975). Normally, the nutrient contents of leaf litter closely approximates the annual release of nutrients through decomposition. However, during dry season, the floor of tropical deciduous forests turns bare, but still the forest soils store considerable amounts of essential nutrients; according to **Misra** (1972), about 13,301 kg/ha of calcium, 9378 kg/ha of nitrogen, 551 kg/ha of phosphorus are available for free growth within a half meter of soil depth.

Sources of litter. The main sources of litter are forests, grasslands and aquatic plants. In deciduous forests huge amounts of litter are produced annually as a crop, whereas in evergreen forests the leaf fall is gradual with simultaneous decomposition and the amount of litter is not so pronounced at a time. The amount and chemical nature of litter depend upon the type of vegetation of the forest. For example, the tree litter of forests chemically consists of mainly carbohydrates in the form of cellulose, hemicellulose, lignin, and some fats, resins, waxes and proteins. The inorganic constituents of litter are calcium, potassium, magnesium, iron, manganese, silicon, copper, aluminium, phosphorus, nitrogen, etc. Coniferous litter and most evergreen

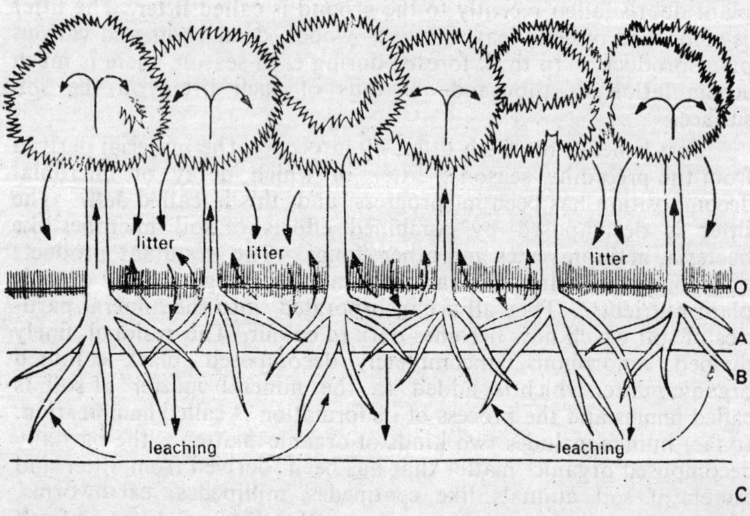

Fig. 9·2. Nutrient cycling within the soil (after Clapham Jr., 1973).

types of litters have comparatively lower amounts of calcium than the deciduous litter. Indian tropical forests of moderate density and growth produce approximately 5 tons of litter per hectare per year. In a deciduous Indian forest, main bulk of litter was found to be contributed by following forest trees : *Shorea robusta*, *Terminalia tomentosa*, *Diospyros melanoxylon*, *Tectona grandis*, and *Butea monosperma* (**Vyas** and **Golley**, 1975).

Thus, the mineral capital of the forest ecosystem operates in a dynamic state through a series of inputs and outputs for essential elements. Autotrophic forest organisms and the soil are subsystems and serve as storage compartments. Atmosphere is the open reservoir. Fluxes of nutrients keep on tree to soil pathway (*i.e.*, detritus food chain) via litter fall and soil to tree pathway through root activity; these pathways are supported by atmosphere-soil channel, composed of minor routes such as stem-flow, through fall and crown wash. The output of elements from the system materializes through soil drainage, harvest by man, fire and grazing by herbivores. The mobility of elements through plant-soil-atmosphere continuum constitutes a **nutrient cycle**.

Types of humus. Depending upon the organic matter content, soils are generally classified as **mineral soils**, rich in mineral particles ; **peat** and **muck**, rich in organic matter accumulation in wet areas, **mor**, low in basic minerals and **mull**, with higher base contents. Accordingly **Muller** (1879, 1884) has recognized two kinds of humus—**mor** and **mull**. **Mor humus is acidic and supports** an abundant fungal growth and low number of soil bacteria. Fungal mycelia may help to bind together particles of humus and decomposition litter into matted layers and in a well developed mor, three such layers can be distinguished, namely a surface litter or **L layer**, comprising undecomposed leaves and twigs, below which is the **fermentation** or **F layer**, in which decomposition has proceeded same way toward the development of humus, and beneath this is the **humus** or **H layer**, in which degraded humus fractions accumulate. Mor has low calcium **content** and developed on sandy soils under conifers. **Mull humus** is neutral or slightly alkaline and contain rich microflora of bacteria. It lacks distinct layering of the mor, largely because the presence of calcium compound favours the development of a rich earthworm fauna which promotes a greater mixing of organic and mineral materials. The mull humus develops in brown forest soils under tree species having a relatively high calcium content, such as alder, elm, bass wood, aspen, and few conifers (red cedar). Intermediate between the two types of humus is the **'moder'** which has a richer and varied fauna.

In order to emphasize the part played by living plants and animals in soil formation, **Taylor** (1930) proposed the following formula :

$$S = M(C + V + VA + A)\, t + D$$

where, A signifies animals, ; C=climate ; D=erosion or deposition ; M=parent material ; S=soil ; t=time and V=vegetation.

C. Formation of Organo-Mineral Complexes

During final stage of pedogenesis, colloidal particles which are formed due to weathering humification and mineralization accumulate and may aggregate into **crumbs** or **concretions.** Some colloidal humus particles may become associated with mineral particles to form organo-mineral complexes. According to **Wallwork** (1970) the crumbs or organo-mineral complexes are formed by following two mechanisms—**electro chemical bonding** and **cementing.** In electrochemical bonding method of crumb formation, aggregation of negatively charged colloidal clay and/or humus particles of water molecules and metallic ions, particularly calcium, takes place. The cementing mechanism of crumb formation involves the action of substances absorbed on the surface of soil particles which effectively glues them together.

The crumbs increase the total pore space in the soil, allowing good aeration and drainage. Eventually, a characteristic profile develops under the influence of climate, vegetation, parent material and the activities of the soil communities. The mature soil, thus, becomes a complex system of living and non-living materials not inert but active.

9-4. THE SOIL PROFILE

Soil profile is the term used for the vertical section of mature soil generally upto the depth of 2 meters or upto the parent material to show different layers or horizons of soil for the study of soil in its undisturbed state. It is made up of a succession of horizontal layers or **horizons,** each of which varies in thickness, colour, texture, structure, consistency, porosity, acidity and compositon.

In general, soils have following four horizons : an **organic** or **O-horizon** and three **mineral (A, B, C) horizons.** Some workers recognized a **D-horizon**, in which rocks are in active weathering state, in between C and R-horizons. **R-horizon** is the consolidated bed rock on which a soil profile rests. A and B-horizons form the **true soil** or **solum**. Each horizon of soil profile is further subdivided. Horizon subdivisions are indicated by a series of letters with arabic numbers as subscripts, *e.g.*, O_1, O_2, A_1, A_2, etc., (Fig. 9·3). Different layers of soil profile have following characteristics :

9-4·1. O Horizon

The uppermost horizon of soil profile is called **O horizon** or **litter zone**. It is present in soils of forests but absent in the soils of deserts, grasslands and cultivated fields. It includes following two sublayers :

9-4·1A. O_1 horizon (Aoo or L horizon). It is the top layer of soil consisting of freshly fallen litter (*i.e.*, dead leaves, twigs, bark, flowers, fruits and animal excreta and remains). In O_1 horizon, original form of plant and animal residues can be recognized with the naked eye and in it, the decomposition has not yet started. The A_1 level may be seasonal in nature : it is thickest in a deciduous forest immediately after leaf fall, when the forest floor is covered with fresh leaves and is virtually gone by the end of the following summer, when the leaves have largely decomposed.

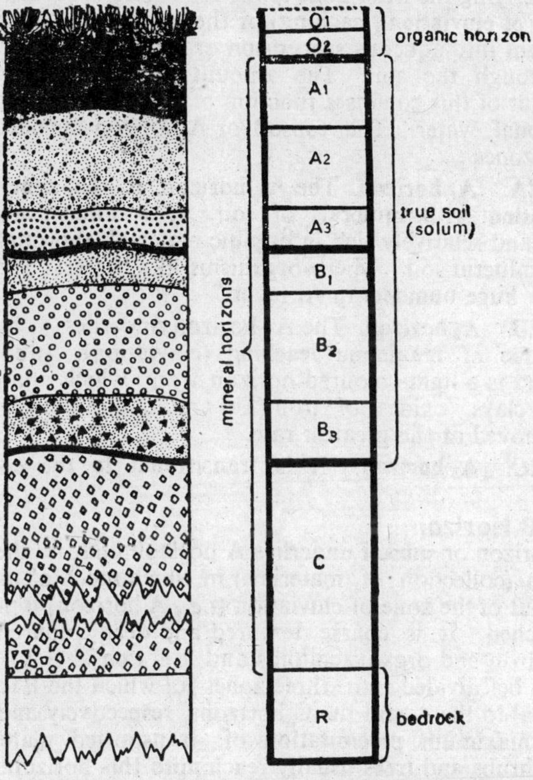

Fig. 9·3. A generalized profile of soil. O_1 : Loose leaves and organic debris; O_2 : organic debris partly decomposed or matted; A_1 : A dark coloured horizon with a high content of organic matter mixed with mineral matter; A_2 : A light coloured horizon of maximum leaching; A_3 : Transitional to B but more like A than B; B_1 : Transitional to B but more like A than B; B_2 : A deeper coloured horizon of maximum accumulation of clay minerals or of iron and organic matter; B_3 : Transitional to C; C : Weathered material (regolith); R : Consolidated bedrock (Smith, 1974).

9-4·1B. O_2 horizon (Ao or H horizon). O_2 horizon underlies the O_1 or litter horizon and contains blackened unrecognizable decomposed litter. The upper portion of O_2 horizon contains partially decomposed detritus, the **duff**, so is called **duff layer**. Its lower part contains completely decomposed, light and amorphous organic matter, the **humus** and is called **humus** or **H layer**. Insects and other small animals are abundant in this layer.

9-4·2. A Horizon

Underlying the litter zone is the A horizon or **topsoil**. It is the **zone of eluviation** (leaching) or the horizon in which materials are brought into aqueous suspension or solution and move downward through the soil. The amount of material that is actually leached out of this zone is a function of the amount of percolating gravitational water. The topsoil or A horizon includes following three subzones :

9-4·2A. A_1 horizon. The A_1 horizon is the **zone of humus incorporation** with minerals of soil. It is almost always dark coloured and relatively rich in organic materials thoroughly mixed with the mineral soil. Micro-organisms like bacteria and fungi are present in huge numbers in A_1 layer.

9-4·2B. A_2 horizon. The A_2 horizon underlies A_1 horizon and is the **zone of maximum leaching** (eluviation). It contains less humus and is a light-coloured horizon from which materials like silicates, clays, oxides of iron (Fe) and aluminium (Al), etc., are being removed at the greatest rate.

9-4·2C. A_3 horizon. It is transitional to the subjacent B horizon.

9-4·3. B Horizon

B horizon or **subsoil** underlies A horizon and is the **zone of illuviation** (collection of materials) in which much of the material leached out of the zone of eluviation (*i.e.*, A horizon) is precipitated and enriched. It is coarse textured and deep coloured with aluminium, iron and organic colloids and it is rich in clay. B horizon can also be divided into three zones of which the B_1 and B_3 are transitional to the A and the C horizons, respectively and B_2 is the zone of maximum precipitation of transported material. The roots of shrubs and trees usually reach upto this horizon.

9-4·4. C Horizon

Underlying the B horizon is the weathered rock or sediment that serves as the parent material for the mineral fraction of the soil. It is called **C horizon** or **regolith**. It is a light-coloured and is virtually lacking in organic materials.

9-4·5. R Horizon

C horizon is underlain by unweathered bedrock which is called **R horizon.**

The relative thickness and importance of the major horizons are highly variable. However, the concept of the soil profile is of great value because it provides a single genetic model by which all zonal soils can be compared.

9-5. SOIL TEXTURE

The soil texture refers to the proportion of mineral particles of various sizes by weight. Depending upon their size, the mineral particles may be of the following types :

Name of the mineral particles	Diameter (mm)
1. Clay	Less than 0·002
2. Silt	0·002—0·020
3. Fine sand	0·020—0·200
4. Coarse sand	0·200—2·000
5. Fine gravel	2·000—5·000
6. Coarse gravel	Above 5·000

Most soils are mixture of these mineral particles. Based on the proportions of the various particles contained in them, soils can be grouped into 5 textural groups : **(1) Coarse textured soils** are loose, consist mainly of sand and gravel. They retain very little moisture and supply some plant nutrients. **(2) Moderately coarse soils** include sandy loam to very fine sandy loam. **(3) Medium textured soils** are mixture of sand, silt and clay high enough to hold water and plant nutrients. **(4) Moderately fine textured soils** are high in clay. They are moderately sticky and plastic when wet, they may form a crust on the surface if organic matter is low. They have a high moisture-holding capacity. **(5) Fine textured soils** contain

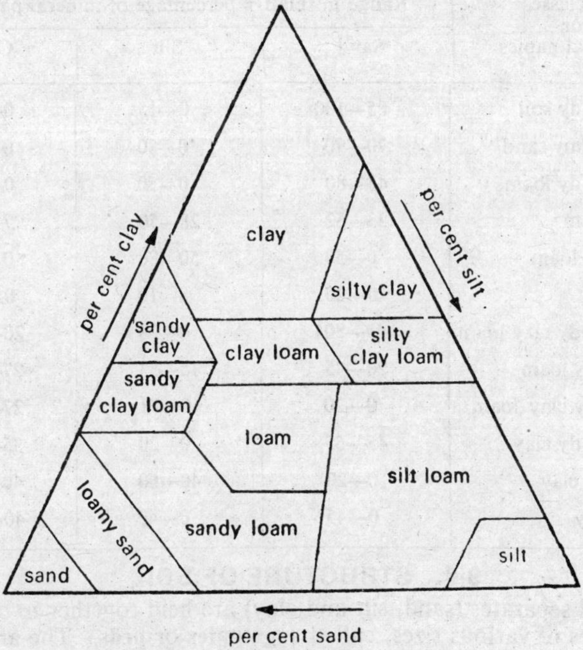

Fig. 9·4. Classification of soil texture.

more than 40 per cent clay, may be sticky and plastic when wet and hold considerable water and plant nutrients, but may have restricted internal drainage. All these 5 textural soil-groups include 12 textural classes (Table 9·3) which have been recognized on the basis of relative percentage of soil particles (*i.e.*, sand, slit and clay).

The larger soil particles are inert and the amount of water that can be held in the spaces between these particles is very little. The sandy soils cannot support a good vegetation though the roots grow easily and branch profusely. The clay particles are colloidal in nature and being small in size, leave very thin capillary spaces between the particles. The clayey soils have therefore a high water-holding capacity. The cations of minerals or nutrients get easily absorbed on the clay particles and become available to plants. Their only disadvantage is that when wet the clay particles imbibe water, swell and become sticky. On drying the particles remain tightly stuck together and large cracks develop in the soil, leading to injury to the plant roots. However, loamy soils being mixture of sand and clay combine the properties of the two particles and are better suited for plant growth.

Table 9·3. Textural classes of soils.

Soil classes or Textural names	Range in relative percentage of mineral particles		
	Sand	Silt	Clay
1. Sandy soil	85—100	0—15	0—10
2. Loamy sand	70—90	0—30	0—15
3. Sandy loam	43—80	0—50	0—20
4. Loam	23—52	28—50	7—27
5. Silt loam	0—50	50—88	0—27
6. Silt	0—20	8—10	0—12
7. Sandy clay loam	45—80	0—28	20—35
8. Clay loam	20—45	15—53	27—40
9. Silty clay loam	0—20	40—73	27—40
10. Sandy clay	45—65	0—20	35—45
11. Silt clay	0—20	40—60	40—60
12. Clay	0—45	0—40	40—100

9-6. STRUCTURE OF SOIL

Soil separates (sand, silt and clay) are held together in clusters or shapes of various sizes, called **aggregates** or **peds**. The arrangement of these peds in earth crust is called **soil structure**. Like texture, there are many types of soil structure. Soil peds may be classified as **granular**, **crumb-like**, **plate-like** or **platy**, **blocky prismatic** and **columnar**. The properties of different soil peds have been tabulated in table 9·4.

Table 9·4. Characteristics of peds.

Types of ped	Properties
1. Granular	Small, spherical and non-porous.
2. Crumb-like	Small, porous and spheroidal.
3. Agranular blocky	Block-like with shrap ends, one end may be pointed
4. Sub-granular blocky	Block-like but bounded by other aggregates.
5. Platy	Plate-like, sometimes plates are overlapped.
6. Prismatic or columnar	Prism-like but without rounded surface.

Structureless soil can be either single-grained or massive soil aggregates (peds) tend to become larger with increasing depth. Structure is influenced by texture, air, moisture, organic nutrients, micro-organisms, root-growth and soil chemical status.

9-7. SOIL COLOUR

Soils exhibit a variety of colour. Soil colour may be inherited from the parental material (*i.e.*, **lithochromic**) or sometimes it may be due to soil forming processes (**acquired** or **genetic colour**). **Though colour has little direct influence on the functions of a soil, but is important in the identification of soil type. In temperate regions dark-coloured soils generally are higher in organic matter** than light coloured ones. Well-drained soil may range anywhere from very pale brown to dark brown and black, depending upon the organic matter-content. Red and yellow soils are the result of iron oxides, the bright colours indicating good drainage and good aeration. Other red soils obtain their colour from parent material and not from soil-forming processes. Well-drained yellowish sands are white sands containing a small amount of organic matter and such colouring matter as iron oxide. Red and yellow colours increase from the cool regions to the equator. Quartz, kaolin, carbonates of lime and magnesium, gypsum and various compounds of ferrous iron give whitish and grayish colours to the soil. The grayest are permanently saturated soils in which the iron is in the ferrous form.

Soil colour and temperature. Soil colour influences the soil temperature. The dark coloured soils absorb heat more readily than light coloured soils. Ramdas *et al.*, (1936) showed that black cotton soils of Poona absorbed 86% of the total solar radiations falling on the soil surface as against 40% by the grey alluvial soil.

9-8. PHYSICAL PROPERTIES OF SOIL

Soil possesses many characteristic physical properties like density, porosity, permeability, temperature, water and atmosphere, each of which can be studied under following separate headings :

9-8·1. Soil Density

Average density of soil is 2·65 gms per ml. Density of soil varies greatly depending upon the degree of weathering.

9·8·2. Porosity

The spaces present between soil particles in a given volume of soil are called **pore spaces**. The percentage of soil volume occupied by pore space or by the interestitial space is called porosity of the soil. Porosity of soil depends upon the texture and structure compactness and organic contents of soil. Porosity of the soil increases with the increase in the percentage of organic matter in the soil.

The pore spaces are of two types—(1) **Micro-pore spaces** (capillary pore spaces) and (2) **Macro-pore spaces** (non-capillary pore spaces). Capillary pore spaces can hold more water and restrict the free movement of water and air to a considerable extent, whereas macro-pore spaces have little water holding capacity and allow free movement of moisture and air in the soil under normal conditions.

9·8·3. Permeability of Soil

The characteristic of soil that determines the movement of water through pore spaces is known as **soil permeability**. Soil permeability is directly dependent on the pore size, therefore, it is **higher for the loose soil with large number of macro-pore spaces than it is for compact soil with numerous micro-pore spaces.**

9·8·4. Soil Temperature

Soil gets heat energy from different sources such as solar radiation, decomposing organic matter, and heat formed in the interior of earth. The temperature of soil is affected by its colour, texture, water content, slope, altitude of the land and also by climate and vegetational cover of the soil. Evaporation of water from soil makes it cooler. Black soils absorb more heat than white soils. Sandy soils absorb more heat and radiate it out quickly at night than clay or loam soils.

The soil temperature greatly affects the physico-chemical and biological processes in the soil. For example, the germination of seeds, normal growth of roots and biological activity of soil-inhabiting micro- and macro-organisms require proper and specific temperature.

9·8·5. Soil Water

In soil, water is not only important as a solvent and transporting agent but in various ways it maintains soil texture, arrangement and compactness of soil particles and makes soil habitat livable for plants and animals. It comes in soil mainly through infilteration of precipitated water (dew, rain, sleet, snow, and hail) and irrigation. Water may be chemically **combined** or **uncombined**. The most important types of chemically combined water are the **water of crystallization** of mineral grains and **water of hydration** of clay mineral particles. Both kinds of these waters are not at all available to living organisms. **Uncombined water** is not chemically combined with soil particles and held in soil by **adhesion** (attraction of water molecules to solid particles) and **cohesion** (attraction of water molecules to each other). Following four kinds of uncombined soil water have been differentiated on the basis of their tendency to be retained in soils (**Bouyoucos** 1970 ; **Clapham, Jr.** 1973).

9-8·5A. Gravitational water. In a well-saturated soil, the accessory (extra) amount of water displaces air from the pore spaces between soil particles and percolates downwardly under gravitational influence and finally it is accumulated in the pore spaces. This accumulated excess water of large soil spaces is called **gravitational water**. When this gravitational water further percolates down and reaches to the level of parent rock, it is called **ground water**. Both kinds of these soil water are ecologically important in the leaching of nutrients.

9-8·5B. Capillary water. The water which is held by capillary forces (*i.e.*, surface tension and attraction forces of water molecules) in smaller soil channels, when the gravitational water and ground water have been drained, is called **capillary water**. Capillary water occurs as a thin-film around soil particles in the capillary spaces and represents the normal available water to the plants. It remains in soil for long periods and carries with it nutrients in solution. Humus has more capillary water than soil minerals.

9-8·5C. Hygroscopic water. Hygroscopic water is held very tightly by the small particles of the soil. Plants cannot absorb the hygroscopic water.

9-8·5D. Water vapour Some uncombined soil water occurs as moisture or water vapours in the soil atmosphere.

Further, the total amount of water present in the soil is called **holard**. The quantity of water that plant-roots can obsorb out of holard is called **chresard** and that amount of soil water which cannot be absorbed by plant-roots is called **echard**.

Retention of water in soil and its availability to plants. The retention of water in a soil depends on the nature of the soil and physical forms of soil water. For example, if a soil is absolutely saturated with water, so that water fills all the pores between soil particles and there is no air space, the soil is said to be at its maximum retentive capacity, or **saturation** (Fig. **9·5A**). If any more water is added, it runs off on the surface. When water is allowed to drain from a saturated soil under the influence of gravity, the gravitational water is lost, and the larger pores refill with air, leaving only the small pores with capillary water. The amount of water that can be retained in a soil by capillary attraction when it is free to percolate under the influence of gravity is termed the **field capacity** (Fig. 9·5B). If the soil is dried out so that all capillary water is lost from even the smallest pores, leaving only hygroscopic water, the amount of water remaining in the soil is called **hygroscopic coefficient** (Fig. 9·5D). If a soil is dried out to a point where the tension at which the remaining water is held precisely equal to the maximal ability of the plants to extract it, the amount of water in the soil is referred to as the **permanent wilting point** (Fig. 9·5C) This is always intermediate between the field capacity and the hygroscopic coeffieient. In general, the water which is most useful to plants is capillary water (**Clapham Jr.**, 1973).

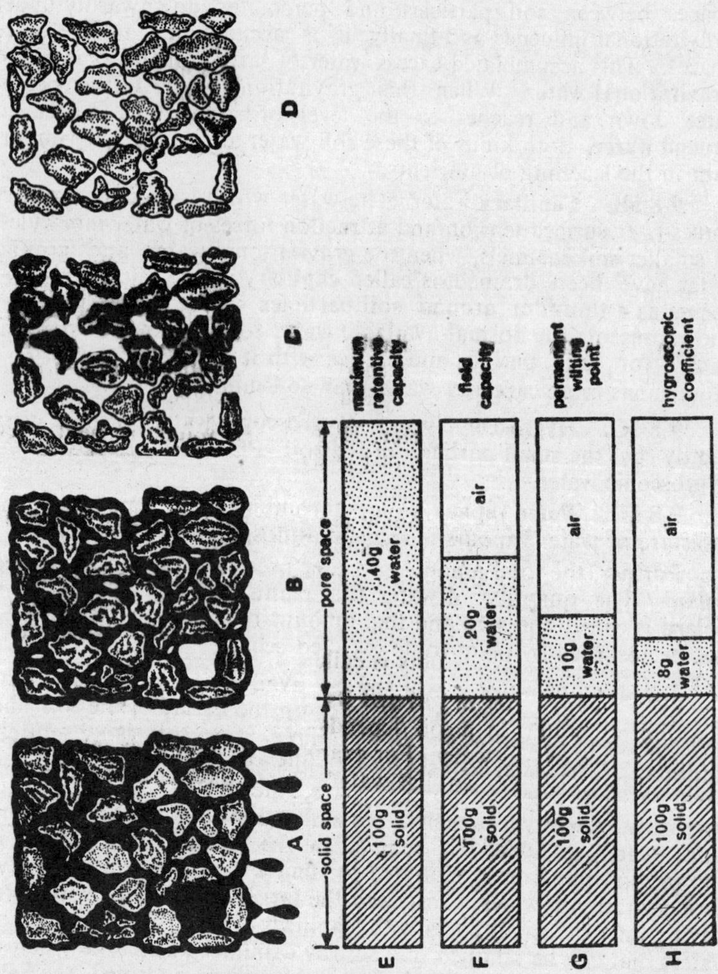

Fig. 9·5. Illustrations of the water contents of a soil show it at saturation (A, E), at field capacity (B, F), at its permanent wilting point (C, G), and in a state of extreme dessication, the hygroscopic coefficient (D, H) (After Clapham Jr., 1973).

9·8·6. Soil Atmosphere

Gases found in pore spaces of soil profiles form the **soil atmosphere**. The soil atmosphere contains three main gases namely O_2, CO_2, and N_2. Soil air differs from atmospheric air in having more of moisture and CO_2 and less of O_2. The soil atmosphere is affected by temperature, atmospheric pressure, wind, rain-fall, etc. Loam soils with humus contain a normal proportion of air and water (about 34% air and 66% water) and, therefore, are good for majority of crops.

9·9. CHEMICAL PROPERTIES OF SOIL

Soil is a mixture of various inorganic and organic chemical compounds and exhibits certain significant chemical properties, all of which can be discussed as follows :

9·9·1. Inorganic Elements and Compounds of Soil

The chief inorganic constituents of soil are the compounds of following elements—Al, Si, Ca, Mg, Fe, K and Na. Soil also contains smaller amounts of compounds of following inorganic elements—B, Mn, Cu, Zn, Mo, Co, I, F, etc. Most of these inorganic salts exists in soil in the form of weak solution. Soil solution may contain complex mixtures of minerals as carbonates, sulphates, nitrates, chlorides and also organic salts of calcium, magnesium, sodium, potassium, etc. The chemical nature of nutrient solution depends on the nature of the parent matter through which water has percolated and climatic conditions of the region and it affects the types of vegetation of that region. For example, temperate soils with high rainfall has H ions in abundance, due to which leaching of basic nutrient ions like Ca, Mg and K occurs and fertility of soil is greatly reduced. Soils with suboptimal concentration of nutrients are called **oligotrophic** and those with more or less optimal concentrations of nutrient solutes as **eutrophic**. The oligotrophic soils cause many physiological disorders in plants which have very poor root system and fruiting. To compensate the efficiency of nutrients, fertilizers containing N, P and K salts are added to such soils.

9·9·2. Organic Matter of Soil

The chief organic component of soil is **humus** which chemically contains—amino acids, proteins, purines, pyrimidines, aromatic compounds, hexose-sugars, sugar alcohols, methyl sugars, fats, oils, waxes, resins, tannins, lignin and some pigments. Further, humus is black coloured, odourless, homogeneous complex substance.

9·9·3. Colloidal Properties

As soil is composed of crystalloids and colloids, therefore, it exhibits all the physico-chemical properties which are related with these two soil particles. Colloids for example exhibit absorption, electrical properties, coagulation, Tyndal phenomenon, Brownian movement, dialysis, etc.

9·9·4. Soil pH

Many chemical properties of soils centre around soil reaction. As regards their nature, some soils are **neutral**, some are **acidic** and some **basic**. The acidity, alkalinity and neutrality of soils are described in terms of **hydrogen-ion concentrations** or **pH** values. A pH value of 7·0 indicates neutrality, a value above this figure (7·1-14·00) indicates alkaline condition and a value below (0—6·9) indicates acid conditions. Normally, pH value of soils lies between 2·2 and 9·6. Surface soils are more acidic. Warm, dry climate soils are strongly basic. In India, acidic soils (pH below 5·5 to 5·6) occur in the high rainfall areas of Western ghats, Kerala, Eastern Orissa, West Bengal, Tripura. Manipur and Assam. The saline, alkali or basic soils (called '*Usar*', contain pH upto 8·5) of India, occur in U. P., West Bengal, Punjab, Bihar, Orissa, Maharashtra, Madras, M. P., A. P., Gujrat, Delhi and Rajasthan.

pH value of soil controls the concentration and availability of other minerals also. For example, certain plants require considerable amounts of calcium (**calciphytes**) and thus grow on basic soils. Plants requiring low calcium amounts are called **oxylophytes**. At low pH, generally copper, iron, zinc, manganese and aluminium become toxic. Highly acidic and highly saline or alkaline soils are often injurious for plant growth and micro-organisms, etc. Neutral or slightly acidic soil, however, are best for the growth of majority of plants.

9·10. CLASSIFICATION OF SOILS

Soils are of various types as regards to their physical, chemical and biological properties. Soils are generally classified by following two methods ; the **zonal soil classification** (Table 9·5) and **comprehensive classification** (Table (9·6), developed by the U. S. Department of Agriculture (U.S.D.A., 1960). Both systems have many advantages and disadvantages. The comprehensive system stresses the properties of the soils as they are found in the field and allows a great precision in classifying them, however, it does not relate a soil to the ecosystem in which it is formed. The zonal soil classification on the other hand, relates soils to the characteristics of the ecosystems in which they are formed, but it is not quite as precise as a tool for describing and classifying soils.

9·10·1. Climate and Soil Types

The development and types of soils are affected by many factors, of which the parent rocks, vegetation and climate are most significant. The depth and chemistry of the soil is greatly determined by the nature of parent rocks. Calcium-rich rocks form soils which are rich in exchangeable bases. The siliceous rocks weather very slowly forming shallow soils. The climatic factors like temperature and precipitation influence the soil development (Fig. 9·6) both directly and indirectly as they affect physico-chemical **processes**, type of vegetation and rate of decomposition

Table 9·5. Zonal soil classification.

Soil order	Soil suborder	Great soil groups
A. Zonal soils	1. Humid region	(i) Podozolic yellow soil with black humic top soils.
		(ii) Red yellow podozolic soils.
		(iii) Non-laterised red soils.
		(iv) Laterised red soils.
		(v) Degraded grey and black clay soils.
		(vi) Alluvial soils.
	2. Arid region (Desert soils)	(i) Red desert soils
		(ii) Dark soils
B. Intrazonal soils	1. Calcimorphic (calcareous soils)	(i) Red soil (developed over limestone).
		(ii) Black soil (developed over limestone).
		(iii) Red soil (developed over marls).
		(iv) Black soil.
	2. Halomorphic	(i) Saline soil.
	3. Hydromorphic	(i) Ground water lateritic soil
		(ii) Planosols.
		(iii) Organic soils.
C. Azonal soils	1. Lethosols	
	2. Regosols	
	3. Alluvial soil	

Table 9·6. Compreshensive soil classification of U.S.D.A., 1960.

Soil order	Important characteristics	Old equivalent type
1. Entisols	No stratification	Azonal soils
2. Vertisols	Excess of clay	Grummusols
3. Inceptisols	Usually moist, incomplete weathering, no clay, Fe and Al oxide saturated layers absent	Humic clay soils, brown forest soils
4. Aridisols	Low organic matter, more than six months dry in a year	Desert soils, sierozems, solonchak
5. Mollisols	Almost black, organic upper horizon, high base content	Chestnut, chernozemes, rendzinas, prairie soils, brown forest soils, solonetz
6. Spodosols	Amorphous substances excessively present in middle horizons	Podosol, brown podosol, ground water podosol
7. Alfisols	Brown upper layer, high base content, excess clay in middle layers, usually moist	Brown soil, chernozem, half bog soil, planosols, grey-brown podosols, noncalcic brown soils
8. Ultisols	Normally moist, low base content, clay content high	Reddish yellow posol, reddish brown laterite
9. Oxisols	Fully weathered, excess of Fe and Al hydroxides, and clay in middle layers	Laterite, latosols
10. Histosols	Excess of organic matter	Bog soils

of the organic matter. Those soils that form in wet regions are extensively leached, while desert soils are not leached at all. The soils of cold regions are both thinner and less iron-rich than those of hot regions, Similarly, the type of vegetation also affects the soil type. The chemical composition of litter and its rate of de-

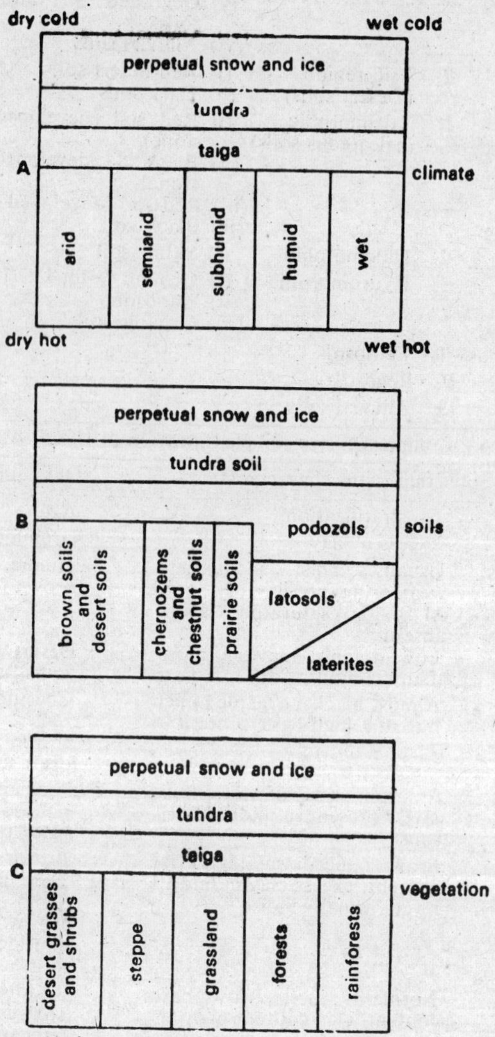

Fig. 9·6. Schematic representation of the relationship between the rainfall and temperature, and climate (A), the major zonal soils (B) and the major biome types (C), (after Clapham Jr., 1973).

composition determine the amount of organic matter, pH and total base exchange capacity of the soils.

There are following six major classes of soils, each of which has many subclasses and widespread geographical distribution.

1. Soils of cold zone : tundra soils.
2. Light-coloured soils of temperate forests : podzols.
3. Soils of the wet and tropical subtropical forests : latosols.
4. Soils transitional between wet and dry regions ; prairie soils.
5. Soils of semiarid forests and **grasslands** : chernozems and chestnut soils.
6. Light-coloured soils of arid regions : brown soils and desert soils.

9-10·1A. Tundra soils. Tundra soils (Fig. 9·7A) are very thin, irregular soils which tend to be very wet, because of lack of drainage through the permafrost. The subsoil is compact and the parent material is poorly weathered and frozen. The organic content of the topsoil is high.

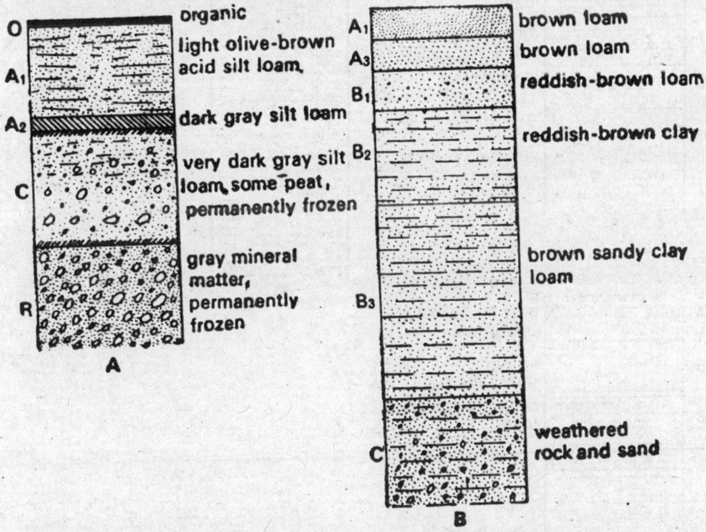

Fig. 9·7. Soil profile of tundra soil (A) and transition-zone soil (B).

9-10·1B. Podzolic soils. The podzols (Fig. 9·8A) are brown to gray in colour, with horizons distinctly separate and some very highly coloured. Podzols form in areas of temperate climate with considerable rain. The typical vegetation is a deciduous forest. These soils are highly leached and acid. Litter layer is thick and A horizon has pH of 4·0–4·5. Alumina and iron are leached from the A horizon, leaving it an ashy gray silicious layer. The B horizon is typically thick and well-developed.

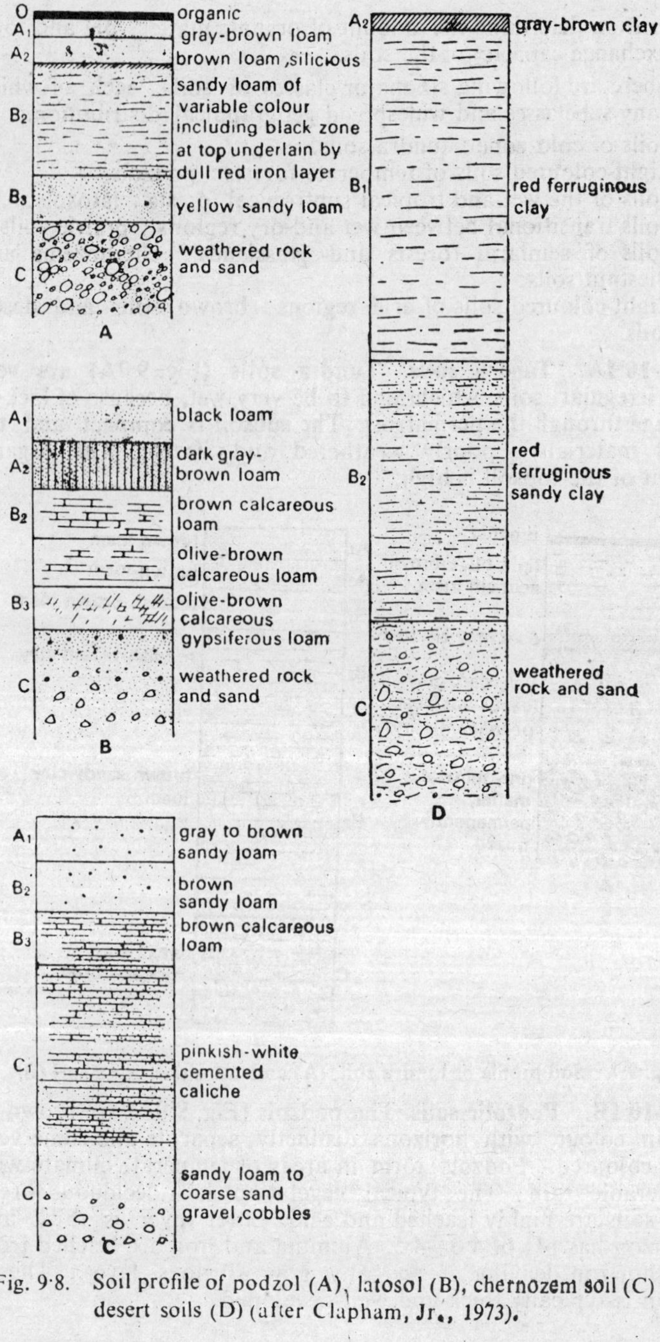

Fig. 9·8. Soil profile of podzol (A), latosol (B), chernozem soil (C) and desert soils (D) (after Clapham, Jr., 1973).

9-10·1C. Latosols and Lateritic soils. (Fig. 9·8 B). Latosols are also strongly leached like podzols. They occur in tropical and subtropical soils. In latosols, the A horizon is a very thin brown or gray layer which is easily lost to erosion. Oxides of iron and aluminium accumulate in a very thick B horizon, but silica is commonly removed from the soil profile in its entirety. Due to abundance of iron oxide clay content, the colour of latosols tends to be orange-red. Organic matter is very uncommon and detritus decomposition may be complete. Latosols are very thick (upto 20 m) and are commonly covered with lush vegetation such as rainforests or savannas.

Laterite is a kind of latosol which has the texture of soil when shaded by a forest so that it does not dry out during dry periods. But if the forest is cut and the soil dries, it hardens irreversibly into a cement-like substance that has been used to construct roads and buildings.

Latosols are nutrient-poor and only certain types of crops such as rubber, bananas and cassava can be raised successfully on them. One of the chief reasons for the intense poverty of tropical South America, the West Indies, Central Africa, India, South-east Asia and Oceania, is the fact that these areas occur on latosols and cannot sustain a sufficiently high-quality (*i.e.*, protein rich) agriculture to allow the areas to feed themselves and industrialize at the same time.

9-10·1D. Transitional and Prairie soils (Fig. 9·7B). The soils of the transition zone between wet and dry climates show leaching and their B horizons have soluble cations like sodium and potassium. These soils exhibit **calcification**. In a leached soil, calcium tends to be leached out of the soil altogether ; in calcified soils, calcium carbonate is deposited in the B horizon. Transitional soils are represented by prairie soils of United States and Canada.

9-10·1E. Chernozems and grassland soils (Fig. 9·8C). Grassland soils are black chernozem soils. Potential evapotranspiration and rainfall are fairly close, and the available water is sufficient to maintain a rich flora of mixed tall and short grasses. This results in sufficient productivity to produce an abundance of organic detritus combined with a minimum of leaching, so that the A horizon is unusually thick and rich in organic colloids. Because of the abundance of organic materials with their very high cation exchange capacity and a low level of leaching, the soils are rich in all mineral nutrients, as well as nitrogen and sulphur. Calcification is typical of the grassland soils and a layer of calcium carbonate generally occurs at the base of the B horizon. Sometimes, a layer of more soluble calcium sulphate (gypsum) is also present. These most fertile and highly productive lands occur in the Great Plains of the United States and Canada, the steppes of the Ukraine and Southern Asiatic USSR and part of South America.

9-10·1F. Desert and Brown soils (Fig. 9·8D). Desert and brown soils develop in areas where the potential evapotranspiration is much greater than the rainfall. They are thin, and a layer of cemented calcium carbonate often lies at the surface of the soil. In hot deserts, the soils may be red in colour. Although there is little leaching in desert and brown soils, the productivity of the areas in which they are developed is so low that detritus production and organic colloid content are also extremely low and such soils are not very useful for agriculture purposes without extensive irrigation.

9-11. MAJOR SOIL TYPES OF INDIA

Geographically soils of India are divided into the following three groups :

1. Mature soils of Peninsular India (include red soils, black soils, lateritic soils). 2. Alluvial soils of Indogangetic plain. 3. Scanty soil of Himalayas. Further, in India the alluvial soils occur in the belt of the submountian regions running along the base of the Himalayas (in Punjab, Uttar Pradesh, Bihar, West Bengal) and valleys of Narmada, Tapti (Madhya Pradesh), Godavari, Krishna and Cauvery. The black cotton soils occur in West Madhya Pradesh, Maharashtra, parts of Andhra, Bihar, West Bengal, Assam, Tamilnadu, Orissa, etc. Red soils occur in large areas of Madras, Mysore, North East Andhra and other regions. Laterites and laterite soils occur on the summit hills of Deccan, Madhya Pradesh, Eastern Ghats, parts of Orissa, Assam, Mysore and Kerala. Besides these there are mountain soils, desert soils peaty and other organic soils. Peaty soils are mostly found in Kerala. They are black and acidic (pH 3·9).

9-12. SOIL AS A HABITAT FOR ANIMALS

Soil is used by various animals as a refuge from enemies, to escape desiccation and extremes of temperature ; as material for abodes, and as a highway. Clay is cold, dense, poorly aerated and though it holds much water, does not give up water readily to plants and animals. Sand is dry, loose (cave in readily) and has a variable temperature. Rock heats and cools quickly, has little available water and practically no air, and is usually too resistant to allow penetration by burrowing animals. Alkaline soils are "**physiologically dry**" because the salts present hold the water so that it is not readily available to organisms. Humus has a great water capacity and gives up water readily, is usually well-aerated and furnishes nitrogen as well as mineral salts and water. It remains rich in biota. Peaty soils are raw humus and are uninhabitable for most animals because they are poorly aerated and have an acid reaction. In general, the character of the vegetation cover serves as a good index of the habitability of soils for animals ; conditions that make plants flourish are usually favourable for animals.

9-13. SOIL FAUNA AND SOIL FLORA

Soil supports a wide array of organisms of different body-sizes and taxonomic groups. Generally, soil organisms are classified into three major groups namely **micro-fauna and micro-flora, meso-fauna** or **meio-fauna,** and **macro-fauna**. Mesoflora and macroflora because occur above the surface of soil (land-surface), therefore are excluded from this discussion :

1. Microfauna. It includes animals with body size within the range of 20μ to 200μ. It includes all Protozoa and small-sized mites, nematodes, rotifers, tradigrades and copepode Crustacea. Soil inhabiting **protozoans** like amoeba, ciliates, zoomastigine flagellates occur near the surface soils, while the testate forms like *Thecamoeba*, *Euglypha* and *Difflugia*, have a wider vertical distribution. The common terrestrial **polyclad** is *Bipalium*. The **nematodes** such as *Rhabditis*, *Diplogaster*, *Tylenchus*, *Heterodera*, *Aphelenchoides*, *Mononchus*, *Pratylenchus*, *Xiphinema* and *Criconemoides*, abound by as much as 1–3 million in raw humus soils to 20 million /m_2 in grassland soils.

2. Microflora. The microflora of soil includes bacteria, soil fungi, soil actinomycetes, blue green algae and algae. In soil, microflora bacteria form about 90 per cent of the total population. Fungi and algae together represent one per cent and actinomycetes cover only 9 per cent.

Soil bacteria grow fairly well in the neutral soils richly supplied with organic nutrients. Soil inhabitant bacteria fall into two categories namely—**autotrophic bacteria** and **heterotrophic bacteria**. The autotrophic bacteria derive their energy from the oxidation of simple carbon compounds or from inorganic substances and their carbon from the atmospheric CO_2. The common autotrophic bacteria of soil are nitrifying bacteria, hydrogen bacteria, sulphur bacteria, iron bacteria, manganese bacteria, carbon monoxide bacteria and methane bacteria. Most of soil bacteria are heterotrophic bacteria depending upon the organic matter of soil for their energy source and are primarily concerned with the decomposition of cellulose, and other carbohydrates, proteins, fats and waxes. They bring about mineralization of organic matter of soil and release considerable amount of nitrogen, phosphorus and other nutrients for plants. The common nitrogen-fixing bacteria of soil are *Rhizobium* (occurs in root nodules of leguminous plants) ; *Azobacter* and *Clostridium pasteurianum* (the latter two are free occurring in soil).

Majority of **soil fungi** are found in acidic soils. They may be parasitic, saprophytic and symbiotic. **Parasitic fungi** of soil infect roots of plants and cause plant diseases such as cotton root rot and many kinds of wilts, rusts, blights and smuts. Certain wilt-forming fungi produce toxins which are harmful, for example, *Fusarium lini*, which causes wilt of flax (*Alsi*) and secrets HCN and *Fusarium udum*, a fungus causing wilt of pigeon pea (*Arhar*) secretes fusaric acid in the roots of host plants. However, certain parasitic fungi produce growth stimulating substances for host plant. *Fusarium* sp., for example, have been found to secrete Gibberellin and Gibberellic acid ($C_{19}H_{22}O_6$). **Symbiotic fungi** of soil live on the roots of certain plants and both fungus and plants are benefitted. **Saprophytic fungi** depend on dead organic matter of soil and derive the energy from decomposition of the latter. They break down cellulose, lignin, gum, sugars, starch, protein, etc., into simple gradients to be utilized by higher plants as nutrients.

Actinomycete fungi prefer saline soils and bring about the decomposition of organic matter such as cellulose. They produce a variety of antibiotics of great economic significance for man. The most important blue green algae of soil are those which fix nitrogen in soil. *Anabaena*, *Nostoc*, *Microcystis*, *Tolypothrix*, *Oscillatoria*, etc., are important nitrogen fixing blue green algae of soil. They also make soil aggregates because of having mucilage.

3. **Mesofauna.** Mesofauna include animals with body size within the range 200μ to 1 cm. The micro-arthropods Acarina (mites) and Collembola (spring tails) are important members of this group which also includes the larger nematodes, rotifers, and tardigrades, together with most of the isopods, Arachnida (spiders), Chelognathi (pseudoscorpions), Opiliones (harvestmen), Enchytraeidae (potworms) insect larvae and small millipedes (Diplopoda), isopods and molluscs.

Among **annelids** the microscopic enchytraeids are represented by *Enchytraeus fridericia* and *Achaeta lumbricellus*, which are more abundant in organic soils and forests than in grass land. *Oniscus*, *Porcellio*, and *Armadillidum* are the most common isopods (**crustacean**) of the tropics in the humid zone. Among the soil **arachnids**, mites are the most predominant. Mites flourish in moist organic soils and certain mites such as *Galumna*, *Cepheus*, *Hemorobates* occur in lichens and mosses. Certain mites, such as, *Schelorbates* and *Brachychthornus* live in humus. The mites are saprophagous, predatory and phytophagous and are involved in the process of organic decomposition and its resultant processes. Certain arachnids like scorpions, *Thelyphonus*, *Galeodes*, and some spiders are crepuscular, hiding under rocks or in crevices in soil and in loose litter, and has no ecological significance in decomposition like other arachnids. Many **opiliones** or harvestmen occur in forest litter, frequently preying upon soil organisms.

Besides mites, only the pseudoscorpions or **chelognathi** occur in surface soils and most decaying vegetations. Of the xerophil litter inhabitants are *Stenatemnus indicus*, *Dhanus indicus*, *Fealla indica* and the hygrophil inhabitants living in the litter and under stones are—*Comsaditha indica*, *Tyrannochthonius madrasensis*, *Tyrannochithonius chelatus* and *Hygrochelifer indicus* (Murthy, 1964), feeding on Collembola. enchytraeids, etc.

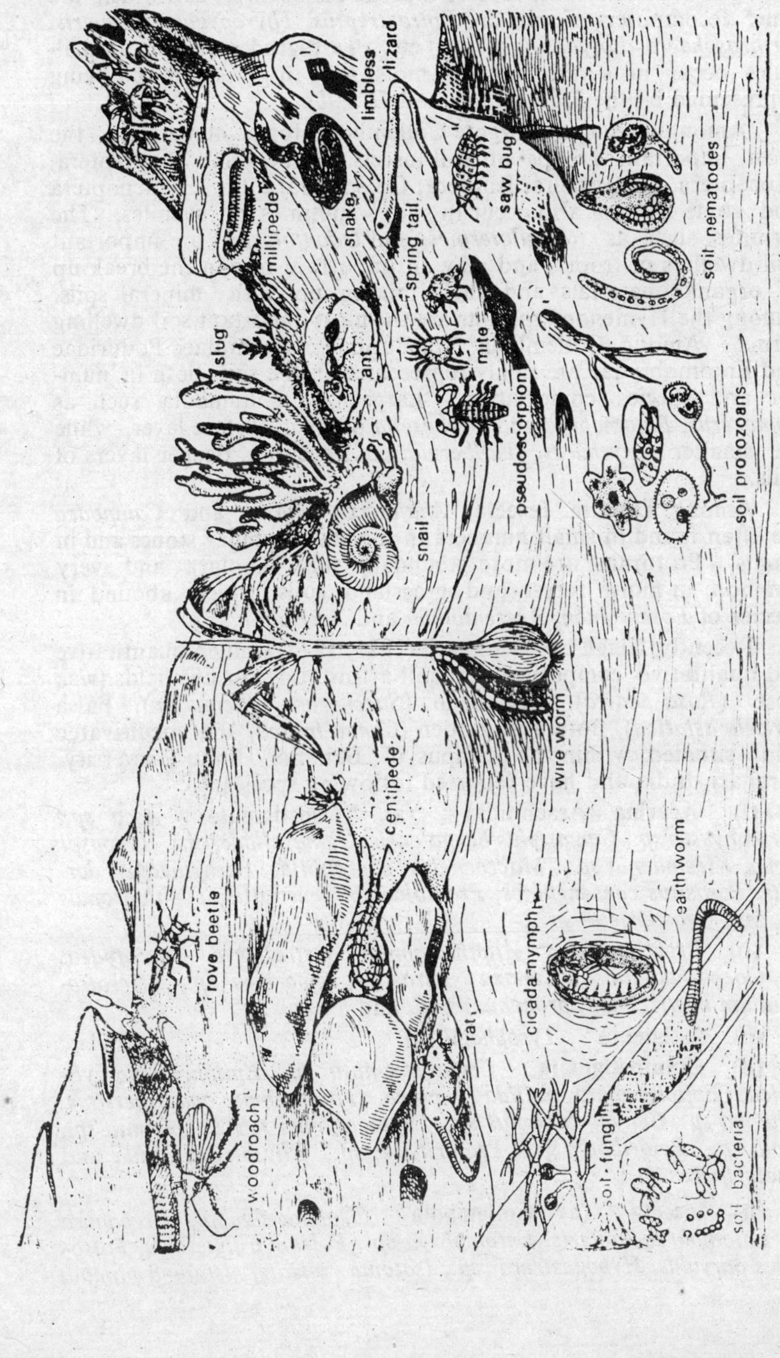

Fig.9·9. Certain most common living organisms (biota) inhabiting the soil (Smith, 1974).

The common millipedes or diplopodes of forest soil which are chief decomposers of soils are *Spirostreptus*, *Thyropygus*, *Glomeris*, *Arthrosphaera*, *Polydesmus*, *Iulis*, etc. **Tardigrades** or bear animalcules occur in surface layer of moist soils in grassland and being represented by the *Macrobiotus* and *Hypsibius*.

Among **insects**, apterygote Collembola form numerically the most important groups of soil insects. Others like Dermaptera, Psocoptera, Dictyoptera, Isoptera, Coleoptera, a few Hymenoptera and some Diptera also occur in soil, sometimes as juveniles. The termites such as *Reticulotermes* and *Odontotermes* are important soil-dwellers of tropics and play an important role in the break up of organic materials and their mixing up with mineral soils. Among the Hymenoptera, ants are the most important soil dwelling forms. Among Collembola, Onchuridae, Isotomidae, Poduridae and Entomobrydae are richly represented in the soil both in number and species composition. Large-sized Collumbola such as *Tomocerus*, *Entomobrya* and *Orchesella* occur in surface layer, while the smaller *Onychurus*, *Tullbergia*, etc., occur in deeper layers of soils.

Among Diplura **Insecta**, *Anajapyx*, *Japyx* and *Campodea* are often found in small numbers in moist soils under stones and in humus. Proturans are more abundant than Diplura and very common in moist forest and grassland soils which abound in species of *Eosentomon*, *Acerentomon* and *Acerentulus*.

Recently, **Singh** and **Mukharji** (1971, 1973) studied quantitative and qualitative composition of soil arthropods in five fields (*viz.*, Rose (*Rosa indica*), sugarcane (*Saccharum officinarum*), Falsa (*Grewia asiatica*), Botanical garden (*Canna indica*) and uncultivated field) situated within the campus of Banaras Hindu University, Varanasi, India and have reported following species :

(i) **Acarina** (Arachnida) : (a) **Mesostigmata** : *Asca sp.*, *Dermanyssus sp.*, *Gamasiphis bengalensis*, *Gamasellodes sp.*, *Hypoaspis vacua*, *Kleemannia sp.*, *Macrocheles coprophila*, *Pachylaelaps dorsalis*, *Parasitus consanguines*, *Proctolaelaps curtipilis*, *Typhlodromus sp.*, and *Urosternella sp.*

(b) **Prostigmata** : *Allothrombium australiense*, *Coccotydeus sp.*, *Coccorhagidia sp.*, *Cunaxa setirostris*, *Imparipes sp.*, *Microtrombidium hystricinum*, *pronematus sp.* and *Tydeus sp.*

(c) **Astigmata** : *Tyrophagus sp.*

(d) **Cryptostigmata** : *Cryptostigmata sp.*, *Epilohmannia cylindrica*, *Epilohmannia pallida pacifica*, *Fosseremus quadripertitus*, *Galumna sp.*, *Gehypochthonius sp.*, *Leptogalumna sp.*, *Lohmannia sp.*, *Oppia sp.*, *Peloribates sp.*, *Protoribates sp.*, *Rhysotritia ardua* and *Scheloripates sp.*

(ii) **Insecta** : (a) **Collembola** : *Cyphoderus sp.*, *Drepanura sp.*, *Entomobrya santeris*, *Entomobrya sp.*, *Folsomia fimetaria*, *Folsomides parvulus*, *Hypogastrura sp.*, *Isotoma viridis*, *Isotoma pinnata* *fasciata*, *Isotomurus*, *palustria*, *Isotomina thermophila*, *Isotomides sp.*, *Isotoma* (*Desoria*) *sp.*, *Isotomurus punctiferus*, *Isotomina pontica*, *Neanura muscorum*, *Onychiurus armatus*, *Proisotoma sp.*, *Pseudachorutes*, *Sminthurus virdis annulatus*, *Salina indica*, *Seira* ***biformis***, *Seira indica*, *Tullbergia sp.*, *Xenylla sp.*

(b) **Diplura** : *Anajapyx sp.*, *Campodea sp.*, *Heterojapyx sp.*, *Japyx sp.*, *Projapyx sp.*

(iii) **Arachnida** (a) **Pseudoscorpionida** : *Chelifer sp.*, *Neobisium muscorum.*

(b) **Palpigradi** : *Koenenia sp.*

(c) **Uropygi** : *Thelyphonus sp.*

(iv) **Symphyla** : *Scutigerella sp.*, *Scolopendrella sp.*

(v) **Pauropoda** . *Eurypauropus sp.*, *Decapauropus sp.*, *Pauropus sp.*

(vi) **Schizomida** : *Nyctalops sp.*, *Trithyreus sp.*

4. **Macroflora of soil. Warming** (1909) has recognised following five ecological classes of plants on the basis of soil characteristics on which they grow : 1. **Oxylophytes** : Plants usually found on acid soils, *e.g.*, *Rhododendron*, *Rumex sp.* 2. **Halophytes** : **Plants** usually found on saline soils, *e.g.*, *Salsola foetida*, *Salicornia* and *Suaeda fruticosa*. 3. **Psammophytes** : These plants grow on sand 4. **Lithophytes** : Plants which grow on rock surfaces. 5. **Chasmophytes** : Plants which grow in rock crevices.

Soil supports a variety of plants like herbs, shrubs, trees and thalloid plants, all of which have specific adaptations for their particular soil types. For instance, in areas of thick soil, roots may be thick and deep ; in areas of thin soil cover, roots may be broad and flat, in areas with water logged swampy soils, roots may be swollen buttresses (**Clapham Jr.**, 1973).

5. **Macrofauna of soil**. Macrofauna of soil includes those animals whose body size is greater than 1 cm. Here belong the majority of Lumbricidae, the Mollusca, the large-sized chilopods, arachnids and insects and the soil-dwelling or fossorial vertebrates.

Earthworms usually occur in abundance in alkaline and moist soils and sparse in acid soils. They have been proverbial for their influence on the process of breaking up litter fragments, decomposition of organic materials and mixing them thoroughly with mineral soils resulting in the formation of organic soils. Some of the common Indian annelidan species of soil are *Megascolex*, *Pheretima*, *Octochaetus*, *Drawida* and *Moniligaster*. Among chilopods the carnivorous *Scolopendra* and *Lithobius* are common in moist soils feeding on leaf litter inhabitants. Land snails are abundant in soils rich in lime. Among insects, *Gryllotalpa* and *Forficula* are adapted for fossorial life in soil.

Among soil vertebrates following animals are well adapted for fossorial or burrowing life in soils : *Ichthyophis*, *Cacopus systema*, *Breviceps* (Amphibia) ; *Sphenodon*, *Uromastix*, limbless lizards, snakes (Reptilia) ; burrowing owl (Aves) and monotremes, kangaroo-rats, moles (*Talpa*), prairie dogs, hedgehog, hares, ground squirrels, otter, badgers and rodents (Mammalia).

Adaptations of soil animals. Animals which are adapted for digging the burrows and for subterranean mode of life are called **fossorial animals**. These animals may dig either for their food or simply for retreat. Zoologically they are primitive, defenseless and unambitious animals. They have following adaptations : 1. The body contour is either cylindrical (*e.g.*, *Ichthyophis*, limbless lizards, snakes, earthworms, *Scolopendra*, etc.), or spindle-shaped or fusiform (*e.g.*, *Talpa*, *Echidna*, etc.) as to offer least resistance to subterranean passage. 2. The head tapers anteriorly to form a sort of snout for burrowing. 3. The tail is short or vestigial. 4. The eyes tend to become vestigial as they are of no use in dark habitat. 5. The external ears also tend to disappear since they would be obstructing in burrowing. 6. For digging, many structures may be found in different fossorial animals. For example, hands are well adapted for digging. In the insect *Gryllotalpa*, the fore-leg's are modified for digging purpose.

The ground-dwelling animals which may be **cursorial** (running), like ostrich, rhea, ungulates, wolves, cats, bears, hyaenas, etc ; **saltatory** (jumping) like rodents, rabbits, wallabies, kangaroos, etc., or **graviportal** (heavy) like turtles, armadillos, elephants, hippopotami, etc., **exhibit** different kinds of adaptations for different kinds of soils. For example, if the soil is firm and hard, the large animals inhabiting the ecosystem tend to have small hooves or paws ; if the soil is wet and spongy, they tend to have broad hooves or paws.

9-14. REVISION QUESTIONS

1. What is soil ? Describe the formation of soil.
2. What is soil profile ? Describe the soil profile of a typical soil.
3. Describe the 'Nutrient cycle' in a deciduous Indian forest.
4. Discuss soil as an ecological factor.
5. Explain two basic methods of soil classification and also describe different types of soils.
6. Describe the influence of climate, vegetation and parent rock on soil development.
7. Give an account an 'soil organisms in Indian ecosystems'.
8. Explain the following : Humus, soil water, field capacity, lithosphere.

10
Hydrosphere And Water

10·1. INTRODUCTION

Liquid water covers about three quarters of the earth's surface, either as oceans (salt water) or as fresh-water and forms the **hydrosphere**. Most of the earth's surface water (99·9 per cent) is in oceans, which contain about thirty-five parts per thousand (‰) dissolved salts. Of the remainder, most is freshwater with a salt content under 1/5‰ found in either lakes and ponds (still water) or in rivers and streams (running water). Fresh-water also becomes available to earth in the form of rain fall, snow, dew, etc.

Water is essential to all life. First life was originated in water. Means of obtaining and conserving it have shaped the nature of terrestrial life. Means of living within it have been the overwhelming influence on aquatic life. Water is one of the main agents in pedogenesis and is also the medium for several different ecosystems. Its significance as a potent ecological factor can be appreciated only by considering its structure and physical and chemical properties which are as follows :

10·2. STRUCTURE OF WATER

Water is a simple inorganic compound composing of hydrogen (H) and oxygen (O) elements. Because of the physical arrangement of its hydrogen atom and hydrogen bonds, liquid water consists of branching chains of oxygen tetrahedra. The physical state of water, whether **liquid**, **gas** or **solid**, is determined by the speed at which hydrogen bonds are being formed and broken. Heat increases that speed ; hence weak hydrogen bonds cannot hold molecules together as they move faster. The thermal status of water in a liquid state is such that hydrogen bonds are being broken as fast as they form. At low temperatures, the tetrahedral arrangement is almost perfect; when water freezes, the tetrahedral arrangement becomes an almost perfect lattice with considerable open space **between** ice crystals and thus, a decrease in density. This is the reason why ice floats on water body.

As the temperature of frozen water is increased, this molecular arrangement becomes looser and more diffuse, resulting in random packing due to continuous breaking and reforming of hydrogen bonds and contraction of molecules. The higher the temperature rises, the more diffuse the pattern becomes, until the hydrogen bonds and whole structure break down and as a result, water melts. Upon melting, water contracts and its density increases up to a temperature of 3·98°C. Beyond this point the loose arrangement of the molecules means a reduction in density **again**, The existence of this point of maximum density at 3·98°C or 4°C is of fundamental significance to aquatic organisms (**Smith**, 1977).

However, the behaviour of seawater is somewhat different. The density (or mass per unit, volume) of sea water (salinity of 24·7‰, and higher) or, rather, its specific gravity relative to that of an equal volume of pure water (*sp gr* = 1) at atmospheric pressure, is correlated with salinity. At 0°C, the density of sea water with a salinity of 35‰, is 1·028. The lower its temperature, the greater the density of sea water, the higher the temperature, the lower the density. No definite freezing point exists for sea water. Ice crystals begin to form at a point on the temperature scale that varies with salinity. As pure water freezes out, the remaining unfrozen water becomes higher in salinity and lower in its freezing point until finally a solid block of ice crystals and salt is formed.

10-3. PHYSICAL PROPERTIES OF WATER

Water is an extremely inert body in relation to most other **chemical substances. It contains following unique physical properties which have ecological relevance :**

10-3·1 Specific Heat

Water is capable of storing tremandous quantities of heat energy with a relatively small rise in temperature. In this property, it is exceeded only by ammonia, liquid hydrogen and lithium. Thus water is said to have a high **specific heat** which can be defined as *the number of calories necessary to raise one gram of water one degree centigrade.* The specific heat of water is given the value of 1.

Because such great quantities of heat are absorbed before the temperature of natural waters such as ponds, lakes and seas can be raised one degree centigrade, they warm up slowly in the spring and cool off just as slowly in the fall. This tends to maintain a relatively constant temperature of aquatic habitats and prevents wide seasonal fluctuations in the temperature of ponds, lakes and seas and also moderates the temperatures of local or worldwide environments. The property of high heat capacity of water is functionally important to aquatic organisms.

10-3·2 Latent Heat

Water possesses the highest heat of fusion and heat of evaporation, collectively called **latent heat**, of all known substances that are liquid at ordinary temperatures. Thus, large quantities of heat energy must be removed before water can change from a liquid to a solid state, and conversely, it must absorb considerable heat before ice can be converted to the liquid state. It takes approximately 80 calories of heat to convert one gram of ice to a liquid state when both are at 0°C. This is equivalent to the amount of heat needed to raise the same quantity of water from 0°C that 80°C. The value of latent heat of melting of water is thus 80.

Evaporation occurs at interface between air and water at all ranges of temperature. It also requires considerable amounts of heat, as 536 calories of heat are needed to overcome the attraction between molecules and convert one gram of water at 100°C into vapour. This is as much heat as is needed to raise 536 grams of water 1°C. Thus, the value of latent heat of water is 536.

The properties of latent heat of water are important not only because they moderate the temperature of the biosphere, but also because they play a basic role in the evaporation of water and its precipitation (condensation) as rain and as dew in the **hydrological (water) cycle.**

10·3·3. Thermal Conductivity

Although water is a poor thermal conductor compared to metals, among the common liquids it is excellent : for example, the conductivity value for silver is 1·10 ; for water, 0·0125 ; for alcohol, 0·00048 and for benzene, 0·00033.

10·3·4. Expansion Before Freezing

The unusual relationship between temperature and density of water (section 10·2) has a great ecological significance. At the freezing point (0°C) water expands markedly and ice always floats on the top of a lake or stream, and it is very unusual for an aquatic system ever to freeze solid, unless it is very small. This phenomenon protects the aquatic biota from sudden freezing and consequent death due to it.

10·3·5 Viscosity

The viscosity of water is also high because of the energy contained in the hydrogen bonds. Viscosity of water can be visualized best if one observes it or any other liquid flowing through a glass tube or a clear plastic hose. The liquid moving through the tube behaves as if it consisted of a series of parallel concentric layers flowing one over another. The rate of flow is greater at the centre, but because of the amount of internal friction between layers, the flow decreases toward the sides of the tube. This phenomenon can be observed along the side of any stream or river with uniform banks. The water along the banks is nearly still, while the current in the centre may be swift. This resistance between the layers is called **lateral** or **laminar viscosity.** In another kind of viscosity, called **eddy viscosity,** water masses pass from one layer to another and creates turbulence both horizontally and vertically.

Viscosity is the source of frictional resistance to objects including organisms moving through the water. Since this resistance is 100 times that of air, animals must expend considerable muscular energy to move through the water. A mucous coating on fish, frogs and certain other aquatic organisms reduces surface tension. Streamlining of body does likewise. In fact the body forms of some aquatic organisms have evolved under the stresses of viscosity. The faster an aquatic organism moves through the water, the greater the stress placed on the surface and greater the volume of water that must be replaced in a given time. Replacement of water in the space left behind by the moving animal adds additional drag on the body. An animal streamlined in reversed direction, with a short, rounded front and a rapidly tapering body and meets least resistance in the water. The acme of such streamlining is the sperm whale.

The viscosity of water allows organisms to swim using relatively simple movements. It also protects the aquatic animals and plants from the mechanical disturbances.

10-3·6 Surface Tension

Within all substances including water particles of the same matter are attracted to one another. Molecules of water below the surface are symmetrically surrounded by other molecules. The forces of attraction are the same on one side of the molecule as on the other. But at the water's surface the molecules exist under a different set of conditions. Below is a hemisphere of strongly attractive similar water molecules ; above is the much smaller attractive force of the air. Since the molecules on the surface are drawn into the liquid, the liquid surface tends to be as small as possible, taut-like the rubber of an inflated balloon. This is **surface tension**, which is greatest among all common liquids, except mercury. The surface tension of water is important in the lives of aquatic organisms.

For example, the water surface is able to support small objects and animals, such as water striders and water spiders that run across the pond surface. To other organisms surface tension is a barrier, whether they wish to penetrate into the water below or escape into the air above. For some, the surface tension is too great to break ; for others it is a trap to avoid while skimming the surface to feed or to lay eggs. If caught in the surface tension, the insect may flounder on the surface. The imagoes of mayflies and caddisflies find surface tension a handicap in their efforts to emerge from the water. Slowed down at the surface, these insects become easy prey for trout.

Likewise, surface tension is the force that draws liquids through the pores of the soil and the conducting network of plants. Aquatic insects and plants have evolved structural adaptations that prevent the penetration of water into the tracheal systems of the former and the stomata and internal air spaces of the latter.

10-3·7 Solvency

No other compound can be compared to water as a **solvent**. So many different substances can be dissolved in it that it is known as the **universal solvent**. More things, in fact, can be dissolved in water than in any other liquid. This is especially true for inorganic chemicals which split, or dissociate to form electrically charged entities termed **ions**. Ionization influences most electrical phenomena and many chemical phenomena of solutions. It is probable that all natural elements are soluble in water, at least in trace amounts, and that they are all found in natural water at some place or other on the earth's surface. In addition, many organic chemicals are water-soluble. Thus, water is the main medium by which chemical constituents are transported from one part of an ecosystem to the other. It is the only medium by which these constituents can pass from the abiotic portion of the ecosystem into the living portion. Even in the driest of terrestrial environments, nutrient materials pass into the roots of plants in aqueous solution; when air is breathed by animals, oxygen is dissolved in water at the surface of the lung before it can cross the mucous membrane and be absorbed by the blood.

10-3·8. Buoyancy

Water is a buoyant medium. Organisms can exist in it without specialized supportive structures such as are needed by organisms that inhabit terrestrial environments.

10-3·9. Light Penetration of Water

Water is a transparent medium. Its transparency enables the penetration of light to the depths where it is ultimately absorbed. Water absorbs light, transforming radiant energy into heat. But it does not absorb all wave lengths equally (Fig. 10·1). Both ends of the visible light spectrum, especially the red end, are selectively absorbed, so that as one descends into a body of water

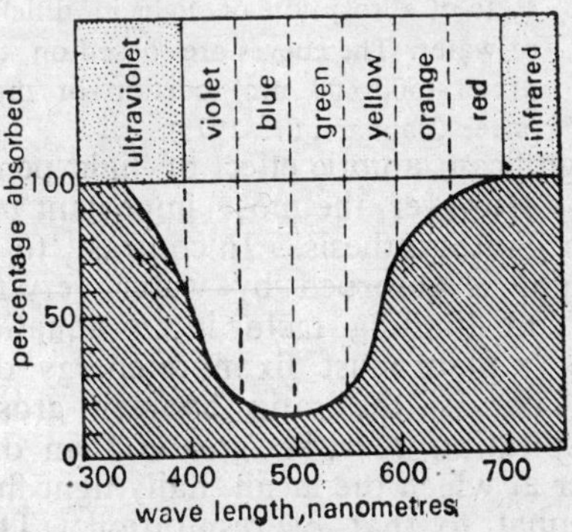

Fig. 10·1. Absorption of different wave lengths of light by 10 metres of pure water (after Clapham, Jr., 1973).

the colour changes from white to bluish to a dull blue-green. Light may be diffused or absorbed by particles in the water such as sediments, detritus, animals and plants. The more particles in the water, the greater the degree of diffusion and absorption. Figure 10·2 indicates the amount of light transmission in different bodies of water under various conditions of turbidity. The zone upto which light rays penetrate is called **photic zone** and below this zone there is complete darkness.

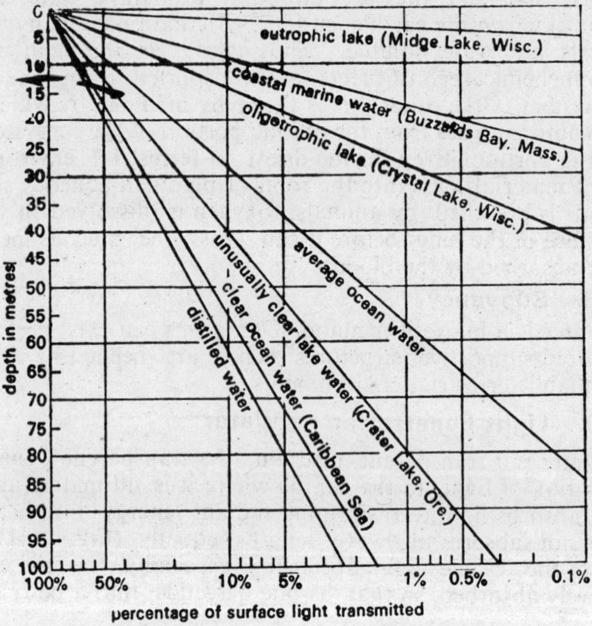

Fig. 10·2. Rate of absorption of light in different bodies of water. The curves are based on the yellow-green (500–600 nm) portion of the spectrum (after Clapham, Jr., 1973).

The most significant abiotic effect of light penetration is heating of the water. However, the most important biotic function of light is its role in photosynthesis. In contrast to the air of terrestrial systems, light is absorbed by water very much faster, so light is an important limiting factor in an aquatic ecosystem. In order to survive, a plant must fix more energy than it utilizes in its respiration. That is, the ratio between gross primary production and respiration must be greater than one. There is a level in the water at which the mean daily light flux allows a rate of production equal to that of respiration. This is termed the **light compensation level.** Above the compensation level, autotrophs can produce sufficient food for themselves and the animals associated them. Below the compensation level, all energy must be derived from organisms living and feeding above the compensation level and sinking below it, either as part of their normal behaviour or after death. The compensation level fluctuates under various circumstances. It tends to be higher in winter than in summer because the total radiation flux is lowest in the winter. In highly turbid areas, due to light-scattering and great diffusion of light, the compensation level raises upto the surface of water.

10-3·10. Pressure

Organisms living at sea level experience a pressure of about 15 psi, which is defined as 1 atm (760 mm of Hg). Pressures increase with increased depth of water at the rate of one atmosphere (1 atm) for every 10 meters of descent. Organisms inhabiting the floor of deep-sea areas at depth of 10,500 meters are exposed to hydrostatic pressure of about one ton per square centimeter. Pressure influences solubility, ionic dissociation, and surface tension, and water is slightly compressible with increased pressures.

10-3·11. Salinity

Salinity has been defined as "the total amount of solid material in grams contained in one kilogram of the water, when all the carbonate has been converted into oxide, bromine and iodine replaced by chlorine and all organic matter completely oxidised." All types of natural waters contain various amounts of different salts (ions) such as Na, K, Mg, Cl, SO_4, PO_4, CO_3, HCO_3, NO_3, etc., and all these salts are responsible for the **saltiness, salinity** or **salt content** of water. The salinity of marine water is rather constant being about 3·5%. The salinity of fresh water varies greatly. Some salt lakes may have a salinity of 25% to 30% which greatly restricts life in them.

Table 10-1. Comparison of some of the principal ions **found** in different kinds of water.

Water	Na	K	Ca	Mg	Cl	SO_4	CO_3	Total per litre
1. Soft water	0·016	—	0·01	0·0005	0·019	0·007	0·012	0·065
2. Hard fresh	0 021	0·016	0·62	0·014	0·041	0·025	0·116	0·301
3. Sea	10 5	0·30	0·40	1·27	18·98	2·65	0 71	34·85
4 Great salt lake	65·54	3·76	0·065	4·47	110·08	13·04	—	197·51

Salinity of water acts as an important limiting factor for the distribution of a number of species of plants and animals. Certain animals such as, spider crab, *Maia*, etc., can tolerate only narrow fluctuations in salinity of water and are known as **stenohaline** animals. While some animals such as *Mytilus*, *Aplysia*, etc., can withstand wider ranges of salinity and are called **euryhaline** animals. However, there are certain animals, such as, *Anguilla*, *Salmon*, etc., **which are both stenohaline and euryhaline.**

10-4. CHEMICAL PROPERTIES OF WATER

Water consists primarily of a single compound, H_2O. It is a universal solvent and most chemical compounds ionize readily in water and provide many radicals and considerable versatility in the rearrangement of chemical substances. It has following chemical properties.

10-4·1. Solubility of Gases in Water

Most gases dissolve readily in water, most notably those that are essential for life. The concentration of any gas in water generally varies between zero and a theoretical maximum or **saturation**. The latter is the amount of gas that can be dissolved in water when the atmosphere and the water are in equilibrium with one another. Except for water falls and very turbulent streams, the water in natural ecosystem is seldom in equilibrium with the atmosphere. A gas may show a **deficit**, if it is being utilized in the ecosystem faster than it is going into solution across the air water interface, or it may be **supersaturated**, if it is being produced in the ecosystem faster than it is being released from solution across the interface. The concentration of important gases may vary widely in any ecosystem, both horizontally and vertically.

The saturation level of any gas in water depends on several variables, most notably temperature, salinity, the concentration of the gas in the atmosphere, and its relative solubitiiy in water. The greater the concentration of a gas in the atmosphere, the greater its concentration in the water will tend to be, depending on its relative solubility in water.

10-4·1A. Oxygen. One of the most critical factor in an aquatic environment is the amount of oxygen in the water because most living organisms (excepting anaerobic forms) require this gas for respiration. In contrast to atmosphere, the oxygen becomes limiting factor for aquatic animals as the saturation concentration of oxygen in water is governed by temperature and salinity. As is evident in table 10·3, the lower the temperature, the greater the oxygen retaining capacity of water, whether it is fresh-water or sea water (Fig. 10·3).

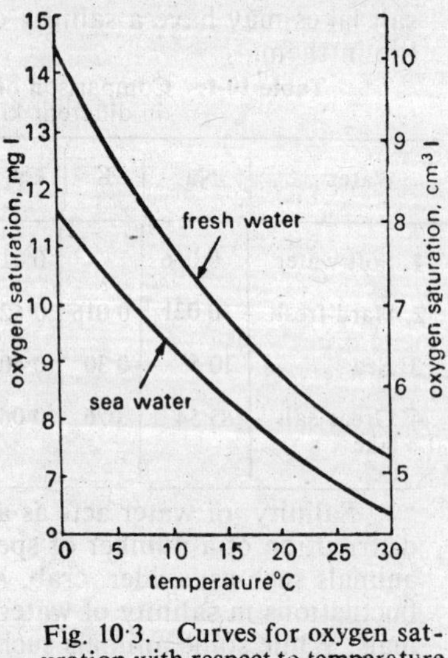

Fig. 10·3. Curves for oxygen saturation with respect to temperature in fresh and sea water.

In fresh-water areas of any depth and in the salt water habitat, there are three recognizable zones with regard to oxygen concentration. A surface stratum, where the oxygen tends to be in equilibrium with the atmosphere above. Below this surface stratum of variable depth is an intermediate stratum where oxygen values fluctuate in accordance with existing factors. Respiration, decomposition of organic materials (stagnant ponds) and stream pollution all tend to reduce the amount of available oxygen, while photosynthetic activity will often balance or more than balance oxygen loss. The deepest layers of water will usually have a very low oxygen concentration in the deeper lakes and oceanic areas because the continual decomposition of organic debris, the respiration of organisms inhabiting these deeper water, and the complete absence of photosynthetic activity in these lower strata will tend to deplete the oxygen concentration. The deep stratum is entirely dependent upon the slow transport of oxygen from the overlying intermediate layer.

Table '0·2. Comparison between equilibrium concentrations of important gases in the atmosphere and in water (Clapham, Jr., 1973).

Gas	Atmospher c concentration	Saturation in water
Oxygen	210 cc./1·(21%)	7 cc./1·(32·9%)
Nitrogen	780 cc./(1·(78%)	14 cc./1·(65·7%)
Carbon dioxide	0·3 cc./1·(0·03%)	0 3 cc./1·(1·4%)

Because oxygen is needed for the respiration of all heterotrophs, it tends to be taken out of solution continuously by the organisms. The only means by which it can enter the water are via solution at the air-water interface or through photosynthesis by aquatic plants. Oxygen is transferred to deeper water by diffusion or through circulation of the water. The rate of oxygen solution at the surface is highly variable, depending mainly on the turbulence of the water and the presence of waves spray and foam.

Table 10·3. Comparison of the saturation concentration of oxygen in fresh-water and salt water environment with varying temperatures.

Water	Temperature. °C	Saturation point in millilitre per litre
Fresh water	0	10·27
Salt water	0	8 08
Fresh water	30	5·57
Salt water	30	4·52

Some organisms (*e.g.*, aquatic beetles, true bugs, fresh water snails, larvae of mosquitoes, whales, seals, porpoises, alligator, crocodiles, etc.) are independent of the oxygen concentration in water because they move to the surface and obtain atmospheric oxygen. Ingenious devices such as air tubes or the ability to trap air bubbles in surface hairs or under wings cover are utilized by some of these forms to insure submersion for an interval of time. Some bacteria and animals (*e.g.*, *Teredo*, *Mytilus edulis*, etc.) can live anaerobically by breaking down glucose, glycogen and other carbohydrates and obtain oxygen from this source.

10·4·1 B. Nitrogen. Nitrogen is significantly less soluble in water than oxygen. But because it constitutes 78 per cent of the atmosphere, it still accounts for about 65 per cent of the dissolved gases at equilibrium. It is fairly inert chemically and does not react with water, although some bacteria, fungi, blue-green algae, and so on, can use it to satisfy their nitrogen requirements, and other bacteria can produce it through reduction of nitrate under conditions of very low oxygen concentration.

10-4·1C. Carbon dioxide. The decomposition of organic matter and the respiratory activity of aquatic plants and animals produce carbon dioxide. This gas is one of the essential raw materials necessary for photosynthetic activity by green plants. Carbon dioxide combines chemically with water to produce carbonic acid (H_2CO_3) which influences the hydrogen ion concentration (pH) of water. Carbonic acid dissociates to produce hydrogen (H^+) and bicarbonate (HCO_3^-) ions. The bicarbonate radical may undergo further dissociation forming more hydrogen (H^+) and carbonate (CO_3^-), as represented in following reactions :

$$\underset{\text{atmospheric}}{CO_2+H_2O} \underset{\text{air-water interface}}{\rightleftarrows} \underset{\text{dissolved}}{CO_2+H_2O} \rightleftharpoons \underset{\text{carbonic acid}}{H_2CO_3} \rightleftharpoons \underset{\text{bicarbonate}}{H^+ + HCO_3^-}$$

$$\downarrow\uparrow$$

$$\underset{\text{carbonates}}{2H^+ + CO_3^-}$$

The amount of free or uncombined carbon dioxide in water is of ecological importance : it governs the precipitation of calcium in the form of calcium carbonate ($CaCO_3$). Calcium precipitates when temperatures and salinity are high and the amount of uncombined carbon dioxide is low. This means more carbonate (CO_3^-) is present to combine with the calcium cation (Ca^{++}). These conditions exist in shallow tropical waters, where evaporation is high. This raises the salinity and photosynthetic activity of plants and reduces the quantity of free carbon dioxide in water. The precipitation of calcium carbonate in tropical areas as the Bahamas explains the preponderance of thick calcareous shells of shallow water

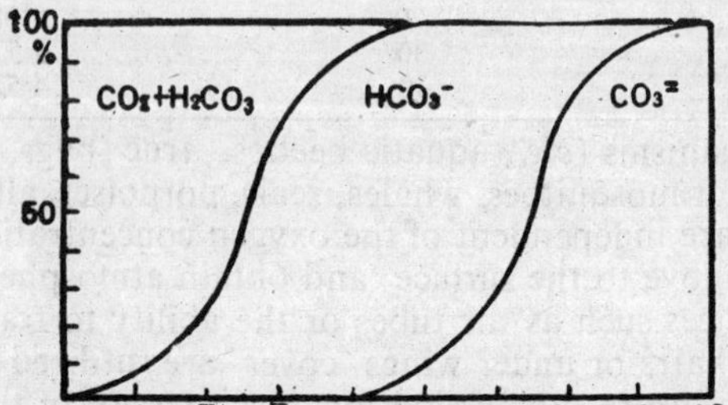

Fig. 10·4. Proportions of various forms of carbon dioxide over the range of pH commonly found in ecosystems.

tropical molluscs, plankton and algae. In deep oceanic water, temperatures are low and there are no photosynthetic plants, consequently, the carbon dioxide content of the water is high. Deep water fauna (molluscs, crustaceans) possess very fragile skeleton because the precipitation of calcium carbonate is minimum.

10-4·1D. Hydrogen sulphide. The deeper layers of many bodies of water, including ponds, lakes, and some estuaries, may contain significant amounts of the toxic gas, hydrogen sulphide, which is released by decaying organic matter. If concentration of the gas builds up, all life but anaerobic bacteria excluded from the area (*e.g.*, deeper strata of Black sea).

10-4·2. Hydrogen-ion Concentration (pH)

One property of natural water is their acidity or alkalinity. About one water molecule in every 10,000,000 splits into two, producing a hydrogen ion (H^+) and a hydroxyl ion (OH^-) :

$$H_2O \rightleftharpoons H^+ + OH^-$$

In pure water, there are equal numbers of H^+ and OH^- ions, it therefore, has a **neutral reaction**. Some natural waters, however, acquire an excess of H^+ and are **acidic**, while others, with an excess of OH^- are **alkaline**. For example, in fresh-water habitats such as acid bogs, swamps and drainage streams carrying water from these areas are acidic, contain pH value as low as 1·4. Despite the high acidity, rich populations of acidophilic flora and fauna thrive under such conditions. On the other hand, certain lake waters may be quite alkaline, particulraly in lime stone areas, where the pH may range from 10 to 12. Basophilic plants and animals are found in these areas. In a vast majority of fresh-water localities pH value ranges between 5·5 and 8·5. Oceanic areas exhibit little change in pH values over vast areas because of the effcetive buffering action of anions such as HCO_3^- and $CO_3^{=}$. Surface waters of open ocean have pH values of 8·0 to 8·4, but deeper water is close to neutrality (pH 7·4 to 7·9). There is often a wider range of pH values in shallow marine water, estuaries and tide pools.

10-5. HARD AND FRESH-WATERS

Water containing soluble salts of calcium and magnesium such as chlorides, sulphates and bicarbonates, is called **hard water**. The hardness of water may be of two types—**temporary hardness** and **permanent hardness**. Temporary hardness is caused by dissolved calcium bicarbonate or magnesium bicarbonate. Temporary hard water can be converted into soft water by boiling, as at higher temperature soluble bicarbonates of calcium and magnesium are converted into insoluble carbonates of calcium and magnesium. Permanent hardness of water is caused by dissolved chlorides or sulphates of calcium and magnesium. This hardness cannot be removed by boiling and requires some specific chemical treatments.

10-6. HYDROLOGIC CYCLE

Water continuously circulates between the atmosphere and the earth's surface (lithosphere) in what is known as the **hydrologic**

cycle. The energy for driving the cycle and thus ensuring a constant supply of fresh-water on land comes from the sun. Solar heat evaporates water from the ocean which is the great reservoir of water. A lesser amount of water is also evaporated from the surface of the land and from plants, a process known as **evapotranspiration**. All these vaporized water forms clouds which moved by winds, may pass over land where they are cooled enough to precipitate the water as rain or snow. Some of the precipitated water soaks into ground, some runs off the surface into streams and goes directly back to the sea. The ground water is returned to the surface by springs, by pumps and by transpiration—the movement of water in plants from roots to leaves. Water inevitably ends up back in the sea, but it may become incorporated into the bodies of several different organisms, one after another, en route.

10-7. EFFECT OF FACTOR OF AQUATIC ENVIRONMENT ON ORGANISMS

The aquatic environment is subject to water movements ranging from small vertical circulations to strong currents. The streams have a unidirectional movement and in seas the movement is reversible. Many aquatic animals have accordingly taken to sedentary or sessile lives depending on water movements. Radial symmetry is a characteristic of such animals. Transformation from a sessile to a locomotive existence favours a bilateral symmetry. The currents of water often abrade the inhabiting flora and fauna and varied modifications are encountered to withstand this abrasive action. Thick scales, strong shells and many attachment devices like the holdfasts and suckers all are the results of this environmental stress. The ability to breathe air dissolved in water, at times even resorting to anaerobic existence, the modification of various senses to respond to stimuli characteristic of aquatic environments, the phenomenon of osmoregulation, and above all the phenomenon of external fertilization are other spectacular physiological adaptations to live in an exclusively aquatic medium.

10-7·1. Atmospheric Humidity and its Ecological Effects

The amount of water vapours or moisture in the atmosphere is referred to as **absolute humidity**. However, of great significance is the **relative humidity** which is the ratio of the actual amount of water vapours in the atmosphere to the amount that can be held in the air at a particular temperature and pressure. As the air warms up, the relative humidity drops (as long as the moisture content of the air is constant) because warm air can hold more moisture than cool air. As the relative humidity drops, the water pressure deficit—the difference between the partial pressure of water at saturation and the prevailing vapour pressure of the air —increases, and increased evaporation takes place.

The relative humidity varies during a twenty-four hour period and from place to place, depending on the topography. Generally relative humidity is lower by day and higher by night. The daily range of humidity is greater in the valley and less at high elevations.

Both excessive and deficient moisture can be detrimental to organisms. Usually moisture becomes a problem only in two environments, the marine and saltwater environment and the desert. The former are **physiological deserts**, in which the concentration of salts outside the organism can osmotically dehydrate it. In the latter, an absolute lack of moisture exists. Most organisms are adapted to moisture extremes in one way or another.

Most of the plants cannot make use of atmospheric humidity, however, several lichens, mosses, filmy ferns, and epiphytic orchids can absorb moisture directly from the air. Some beetles can also absorb moisture from the air through their body surface. Other plants, except the submerged aquatic plants, obtain their water supply by absorption through roots. Animals obtain water and solutes by anyone of following methods : (i) by absorbing it through their skin from contact with some damp ground, (*e,g.*, by osmosis in amphibians) ; (ii) by drinking ; (iii) directly from their food ; or (iv) from water produced by metabolism. They can lose water and solutes through urine, faeces, evaporation over the skin and respiration. If the animal is not to dehydrate, the input of water must equal to the losses.

Since water makes up a large proportion of the bodies of plants and animals (*e.g.*, cytoplasm holds 70–90 per cent of water), it affects all life processes directly. In plants, the rate and magnitude of photosynthesis, respiration, absorption of nutrients, growth and other metabolic processes are influenced by the amount of water available. Low relative humidity increases water loss through transpiration and affects plant growth. Conversly, plants in regions with high moisture reduce transpiration. They survive the physiologically dry periods of winter by shedding leaves. The germination of seeds and establishment of seedlings are directly affected by water. In lower plants, water is essential for fertilization and among higher plants, pollination and seed dispersal are effected through the agency of water in many cases. The moisture and temperature, acting together, determine in large measure the climate of a region and the distribution of plant and animal life (see **Smith**, 1977).

The growth and reproduction of animals are also effected by water factor. Many lower animals show preferences for certain moist levels. The distribution of several species of spiders in the forest was found to be directly related to humidity levels (**Todd**, 1949). A number of nematodes, annelids, isopods, chilopods, diplopods, and insects do not survive at humidities below 100 per cent. In many insects and mites the reproduction, fecundity and speed of development is directly governed by soil moisture. If the air is too dry, the eggs of some locusts and other insects may become quiescent. There is an optimum humidity at which nymphs develop fastest. Some insects lay more eggs at certain relative humidities than above or below that point. The mosquitoes cannot withstand high temperatures under low humidity levels, and the malarial parasite cannot be transmitted at a temperature below 15·5°C and mean relative humidity under 65 per cent (see **Gopal** and **Bhardwaj,** 1979). Many insects and nematodes can survive extreme desiccation. **Ramamurty** (1966) demonstrated that the colouration of animals (*e.g., Labidhura ripania*, Insecta) is influenced by humidity.

Too much water may be as determinative as too little. High water tables result in shallow-rooted trees that are easily toppled by the wind and are sensitive to draught and frost. Terrestrial plants subject to prolonged flooding, particularly during the growing season, will die from lack of oxygen about the roots. Similarly, long periods of dry weather that result in soil drought reduce plant growth, causing dieback of plants or outright death. Drought-injured plants are susceptible to outbreak of insects and highly susceptible to fire.

Heavy rains and prolonged wet spells cause widespread death among mammals and birds from drowning, exposure and chilling. Excessive moisture and cloudy weather kill insect nymphs, inhibit insect pollination, and spread parasitic fungi, bacteria and viruses among both plants and animals.

10-7·2. Precipitation

The moisture falling on an area in liquid, vapours or frozen form is designated as **precipitation**. Precipitation depends upon season, wind, air, pressure, and temperature. **Rainfall** is the most common form of precipitation which has ecological effects on organisms. **Light drizzle** is of the little importance as very little moisture penetrates the soil because much of it evaporates rapidly. Gentle steady rain is most effective because much of it penetrates the soil. However, torrential rains on land are most disastrous because they leads to flooding, soil erosion, destruction of vegetation and of animals.

Hail, sleet and **snow** are forms of frozen rain. Hail is in the form of balls of frozen moisture. These may be of variable size and weight and may cause considerable damage to agriculture. Sleet and snow form destructive glaze on trees, shrubs and grass. The weight of snow causes destruction and damage to plants.

10-8. ECOLOGICAL GROUPS OF ORGANISMS AND ECOLOGICAL ADAPTATIONS

Plants and animals exhibit considerable variation in their requirements of water. **Warming** (1909) classified plant communities on the basis of nature of soil and their water requirements as follows :

1. Hydrophilous plants. Hydrophytes (in water) ; **Helophytes** (in marsh).

2. **Xerophilous plants. Oxylophytes** (on acid soil); **Psychrophytes** (on cold soils); **Halophytes** (on saline soils); **Lithophytes** (on rocks); **Psammophytes** (on sand and gravel); **Chersophytes** (on waste lands); **Eremophytes** (deserts and steppe); **Psilophytes** (on savannah); **Sclerophyllous formations** (bush and forests) and **Coniferous formations** (forests).

3. **Mesophilous plants. Mesophytes**. Thus, on the basis of water requirements following three broad ecological groups of plants can be recognised: (a) **Hydrophytes** (plants living in water and require large quantities of water); (b) **Xerophytes** (terrestrial plants which can tolerate extremely dry conditions and pass long periods without water) and (c) **Mesophytes** (terrestrial plants requiring moderate amounts of water).

Likewise, there exists a broad classification of animals into following three main ecological groups, depending on the requirement of water:

1. **Hydrocoles.** Aquatic animals which live in and require large quantities of water.
2. **Xerocoles.** Terrestrial animals which can tolerate extremely dry conditions and pass long periods without water.
3. **Mesocoles.** Terrestrial animals requiring moderate amounts of water.

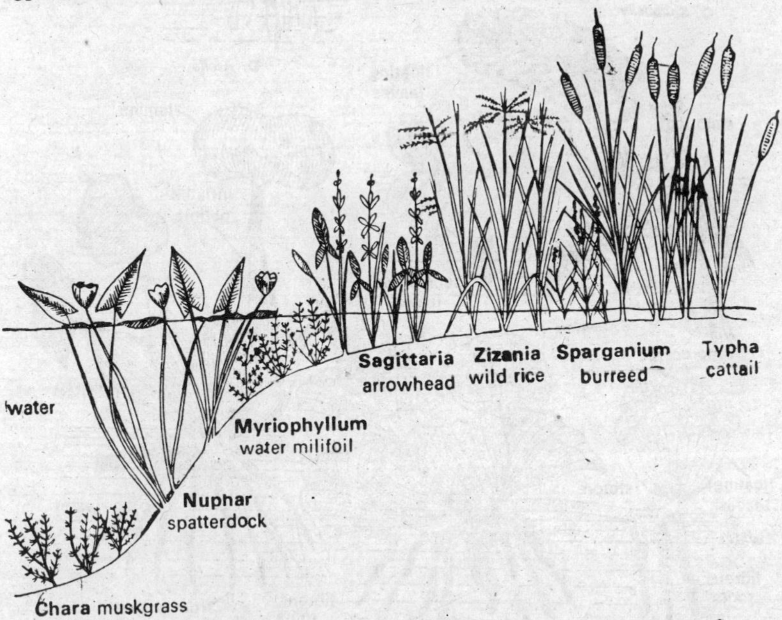

Fig. 10·5. Zonation of aquatic vegetation (hydrophytes) along a pond and along river banks. Note the changes in vegetation with water depth (after Smith, 1974).

Both the plants and animals living in water or dry habitats exhibit a number of special features or adaptations in their morphology, anatomy and physiology which enable them to live under

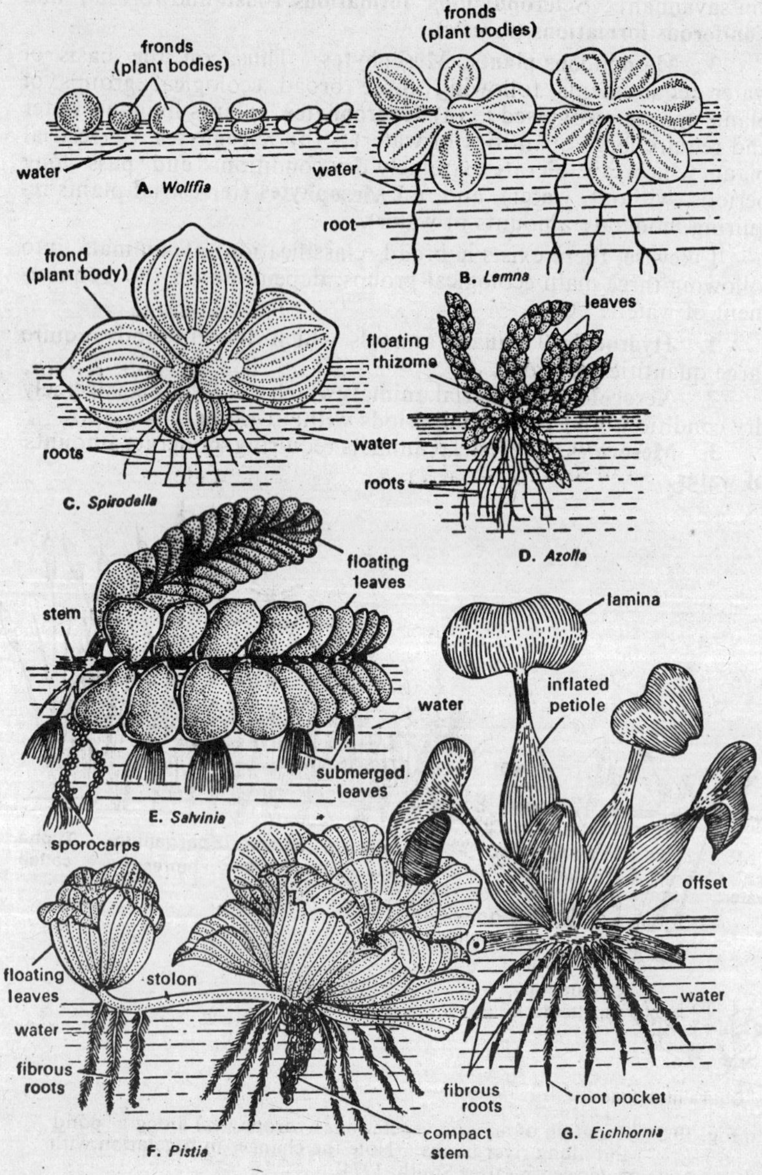

Fig. 10·6. Free-floating hydrophytes.

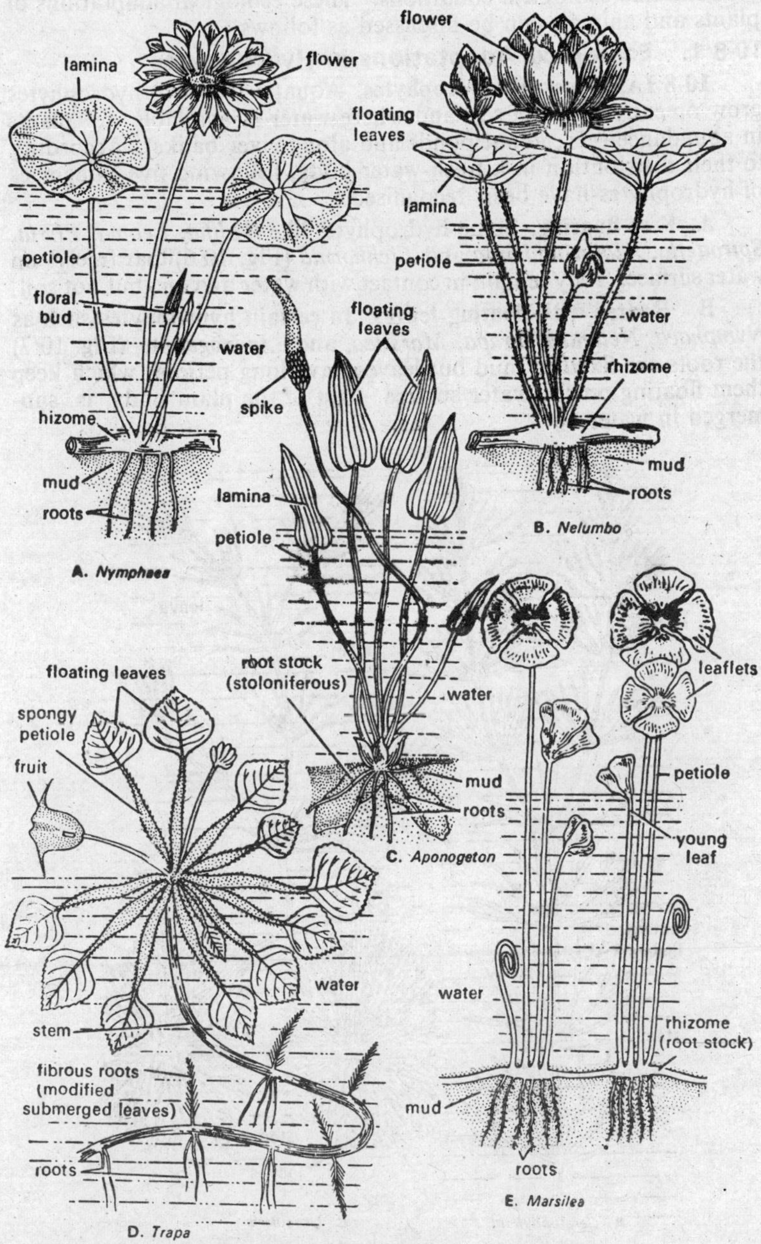

Fig. 10·7. Rooted hydrophytes with floating leaves

such extreme ecological conditions. These ecological adaptations of plants and animals can be discussed as follows :

10-8·1. Ecological Adaptations in Hydrophytes

10-8·1A. Types of Hydrophytes. Aquatic plants or hydrophytes grow on extremely wet soil and where water is available to plants in abundance (*viz.*, about ponds and along river banks). According to their distribution in a fresh-water body, following five categories of hydrophytes have been recognised :

A. **Free-floating.** Some hydrophytes like *Wolffia, Lemna, Pistia, Spirodella, Azolla, Salvinia* and *Eichhornia* (Fig. 10·6)float freely on water surface. They remain in contact with water and air, but not soil.

B. **Rooted with floating leaves.** In certain hydrophytes such as *Nymphaea, Nelumbo, Trapa, Marsilea,* and *Aponogeton,* (Fig. 10·7) the roots are fixed in mud but leaves have long petioles which keep them floating on the water surface. Rest of the plant body is submerged in water.

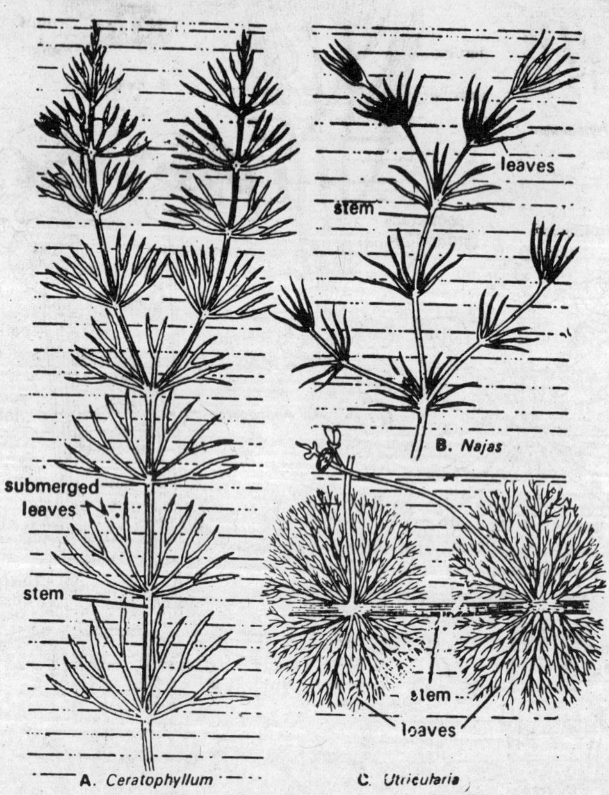

Fig. 10·8. Submerged floating hydrophytes.

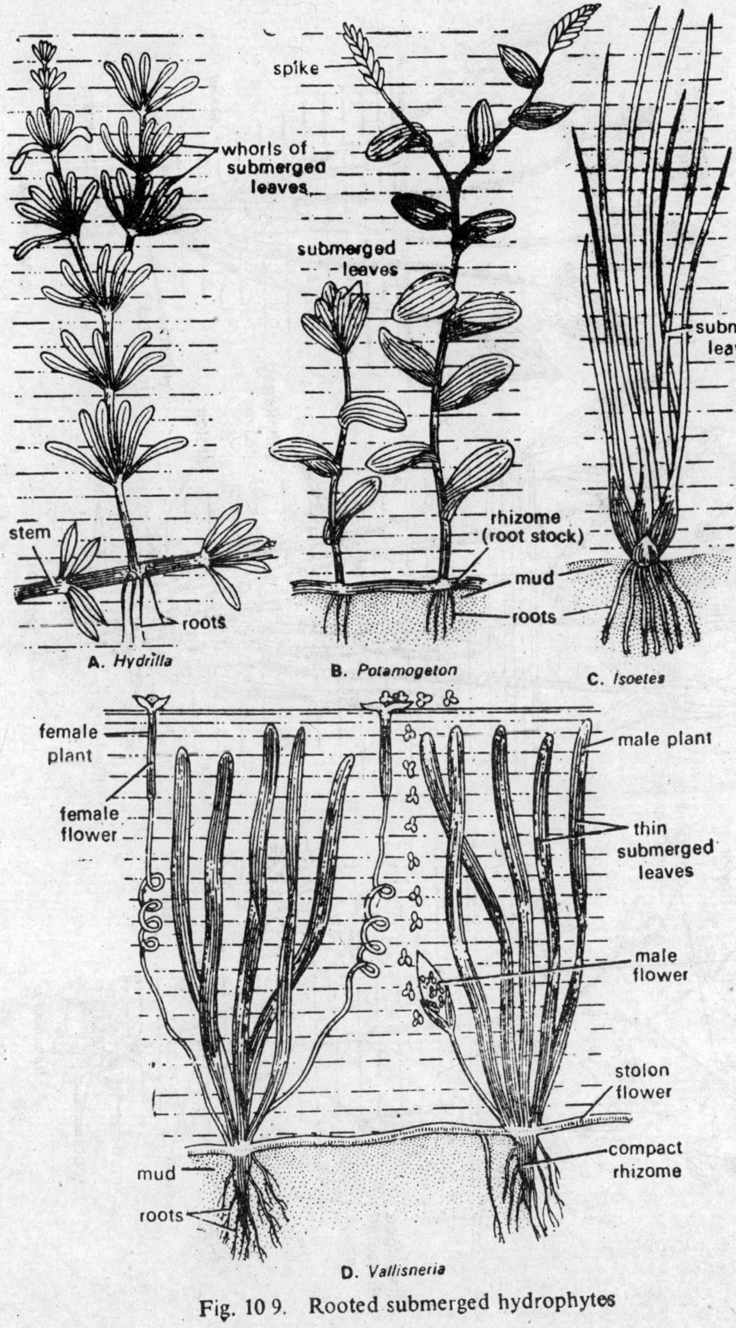

Fig. 10 9. Rooted submerged hydrophytes

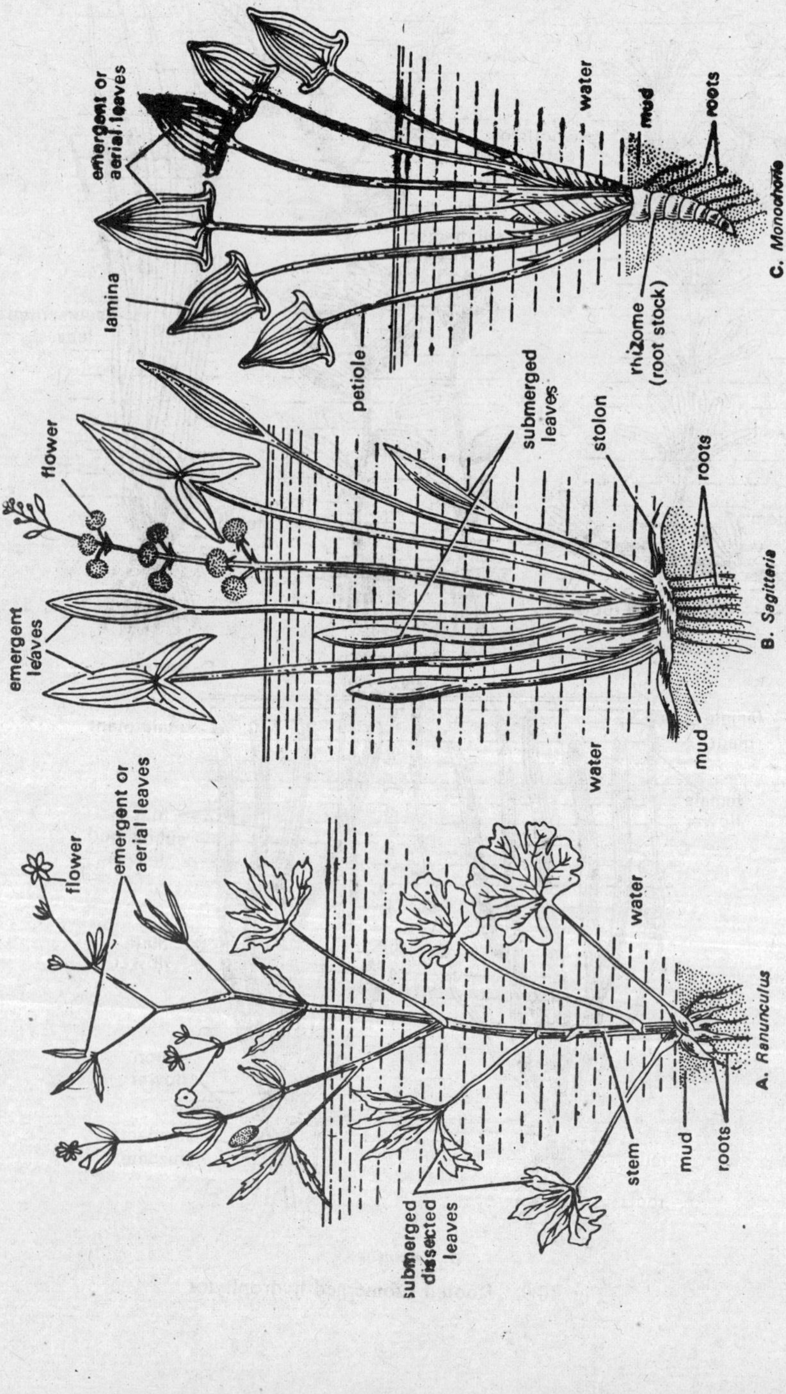

Fig. 10·10. Rooted em rgent hydrophytes.

C. Submerged floating. Hydrophytes like *Ceratophyllum*, *Utricularia*, *Najas*, etc., (Fig. 10·8) remain in contact only with water. They are not rooted in the mud and being completely submerged in water.

D. Rooted submerged. Some hydrophytes such as *Hydrilla*, *Potamogeton*, *Vallisneria*, *Chara* and *Isoetes* (Fig. 10·9) remain completely submerged in water, but rooted in the soil.

E. Rooted emergent. These hydrophytes grow in shallow waters and require excess of water but their shoots are partly or completely exposed to air. The root system is completely under water, fixed in soil. In some of these hydrophytes such as *Ranunculus*, *Sagittaria*, *Limnophila*, and *Monochoria* (Fig 10·10), shoots are partly in water and partly exposed out of water in air. Some other hydrophytes of this category such as *Cyperus*, *Rumex*, *Scirpus*, *Typha*, etc., (Fig. 10·11) are called **marsh plants** and their shoots are completely exposed to air, like the land plants.

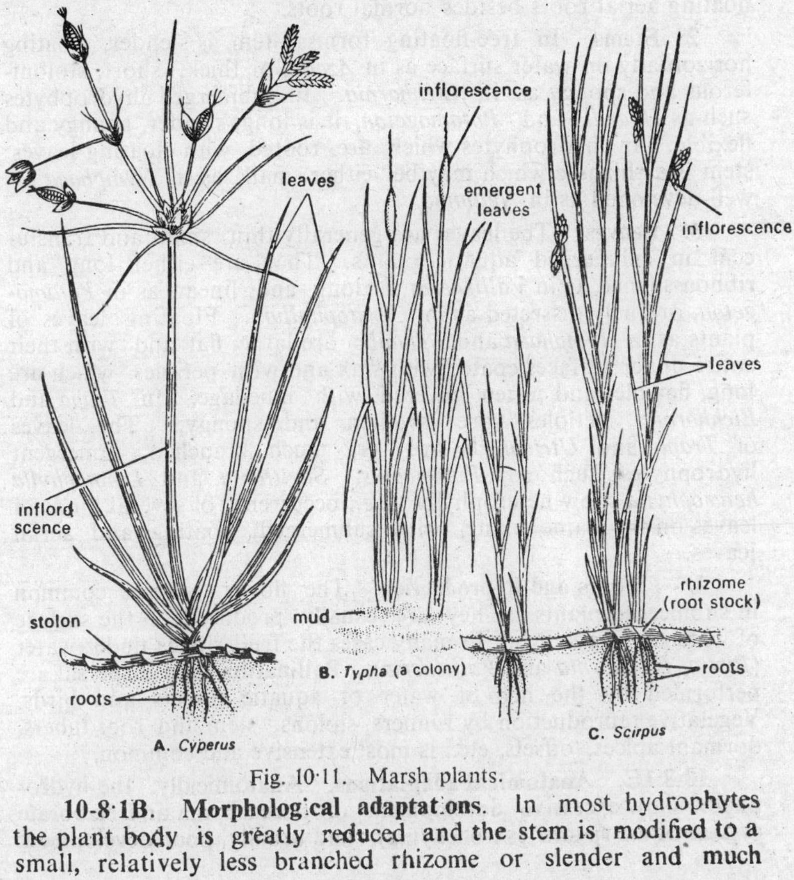

Fig. 10·11. Marsh plants.

10·8·1B. Morphological adaptations. In most hydrophytes the plant body is greatly reduced and the stem is modified to a small, relatively less branched rhizome or slender and much branched system which does not offer resistance to the water currents. Plants like *Lemna*, *Spirodela* and *Wolffia* show extreme reduction in the size of plant body. Different parts of various hydrophytes exhibit following specialized features :

1. **Roots. Due to availability of water in plenty, roots, the principal organs of water absorption, in such plants become of** less significance. Roots may be either entirely absent as in *Wolffia*, *Salvinia* and *Ceratophyllum* or are poorly developed as in *Hydrilla*. However, the emergent forms like *Ranunculus* and *Typha* contain well-developed and branched roots with distinct root caps. Root hairs and root caps are usually absent. However, in free-floating plants the root-hairs and root-caps develop prominently. In *Eichhornia*, root caps are replaced by "**root pockets**".

Roots of hydrophytes are generally fibrous, adventitious, short, unbranched or poorly branched. In *Lemna* roots act simply as balancing and anchoring organs. *Jussiaea repens* possesses floating aerial roots besides normal roots.

2 **Stems.** In free-floating forms, stem is slender, floating horizontally on water surface as in *Azolla*, or thick, short, stoloniferous and spongy as in *Eichhornia*. In submerged hydrophytes such as *Hydrilla* and *Potamogeton*, it is long, slender, spongy and flexible. In hydrophytes which are rooted with floating leaves, stem is a rhizome which may be either small as in *Nymphaea* or well-developed as in *Nelumbo*.

3. **Leaves.** The leaves are generally thin, small and translucent in submerged aquatic plants. They are either long and ribbon-shaped as in *Vallisneria* or long and linear as in *Potamogeton*, or finely dissected as in *Ceratophyllum*. Floating leaves of plants as in *Nymphaea* and *Nelumbo* are large, flat and with their entire upper surfaces coated with wax and with petioles which are long, flexible and often covered with mucilage. In *Trapa* and *Eichhornia* petioles are swollen and spongy. The leaves of *Trapa* and *Utricularia* are very much branched. Emergent hydrophytes such as *Ranunculus*, *Sagittaria* and *Limnophylla heterophylla* show heterophylly (*i.e.*, occurrence of several kinds of leaves on the same plant) with submerged, floating and aerial leaves.

4. **Flowers and Reproduction.** The flowers are not common in submerged plants. They are usually produced on the surface of water or above it but in many cases the fruit ripens under water (*Pistia*, *Eichhornia* and *Vallisneria*). Pollination and dispersal are performed by the help of water or aquatic insects and birds. Vegetative reproduction by runners, stolons, stem and root tubers, dormant apices, offsets, etc., is most extensive and common.

10-8·1C. Anatomical adaptations. Anatomically, the hydrophytes **show excessive development of parenchyma and elaborate system of air spaces (aerenchyma), and usually poor development**

of vascular and mechanical tissues. The cuticle is absent or poorly developed. Stomata are completely absent in submerged leaves (*e.g.*, *Anacharis* and *Potamogeton*) ; in floating forms as *Nymphaea*, they are confined only to the upper surface of leaf. The chlorophyll is found in all the tissues throughout the plant body. Even the roots of some plants as *Azolla* contain chlorophyll. Mucilage cells and mucilage canals are present which secrete mucilage to protect the plant body from decay under water. The reserve food is in the form of starch grains which occur abundantly in pith and cortex. Cystoliths (sclereids) of various shapes are also common in leaves and other tissues. These provide some mechanical support to the tissues.

10·8·1D. Physiological adaptations. Physiologically, the aquatic plants exhibit a low compensation point and low osmotic concentration of cell sap. In submerged plants, the nutrients are absorbed through the plant surface. The gases are also exchanged from the water through epidermis. The gases produced during photosynthesis and respiration are partly retained in the air chambers of aerenchyma to be utilized when required. There is no transpiration from the submerged hydrophytes, however, emergent plants and free-floating hydrophytes have excessive rate of transpiration.

10·8·2. Ecological Adaptations in Hydrocoles

Aquatic animals usually have streamlined and flexible bodies which are provided with structures like flagella, antennae or fins which help in movement. They show following two kinds of adaptations for aquatic environment :

10·8·2A. Primary aquatic adaptations of hydrocoles. Such adaptations occur only in those animals such as fish which had never lived on land and their ancestors were also aquatic. They live permanently in water and have evolved from more primitive aquatic forms. They exhibit following adaptations :

1. The compression of head, body and tail into an elongated stream-lined form. There is no protuberance over the spindle-shaped body which would retard the swift passage of the animal through water. 2. Development of paired and unpaired fins for swimming occurs. A fish bears two pairs of paired fins (**pectoral fins** and **pelvic fins**) and the unpaired median fins (one or two **dorsal fins**, an **anal fin** and a **caudal fin**). The pectoral fins and dorsal fin act as stabilizers or balancers and the caudal fin provides a forward push to the body. 3. Body muscles of fish are arranged in the form of bundles separated by myocommata. Their contraction produces lateral undulations of the flexible body and helps in locomotion. 4. They respire or breath by gills which are specialized breathing organs to use gases dissolved in water. 5. Many teleost fishes have air-bladder or swim-bladder which is filled with gas or air and acts like an accessory respiratory as well as hydrostatic organ. By controlling the gas contents in the swim-bladder, fishes are able to maintain

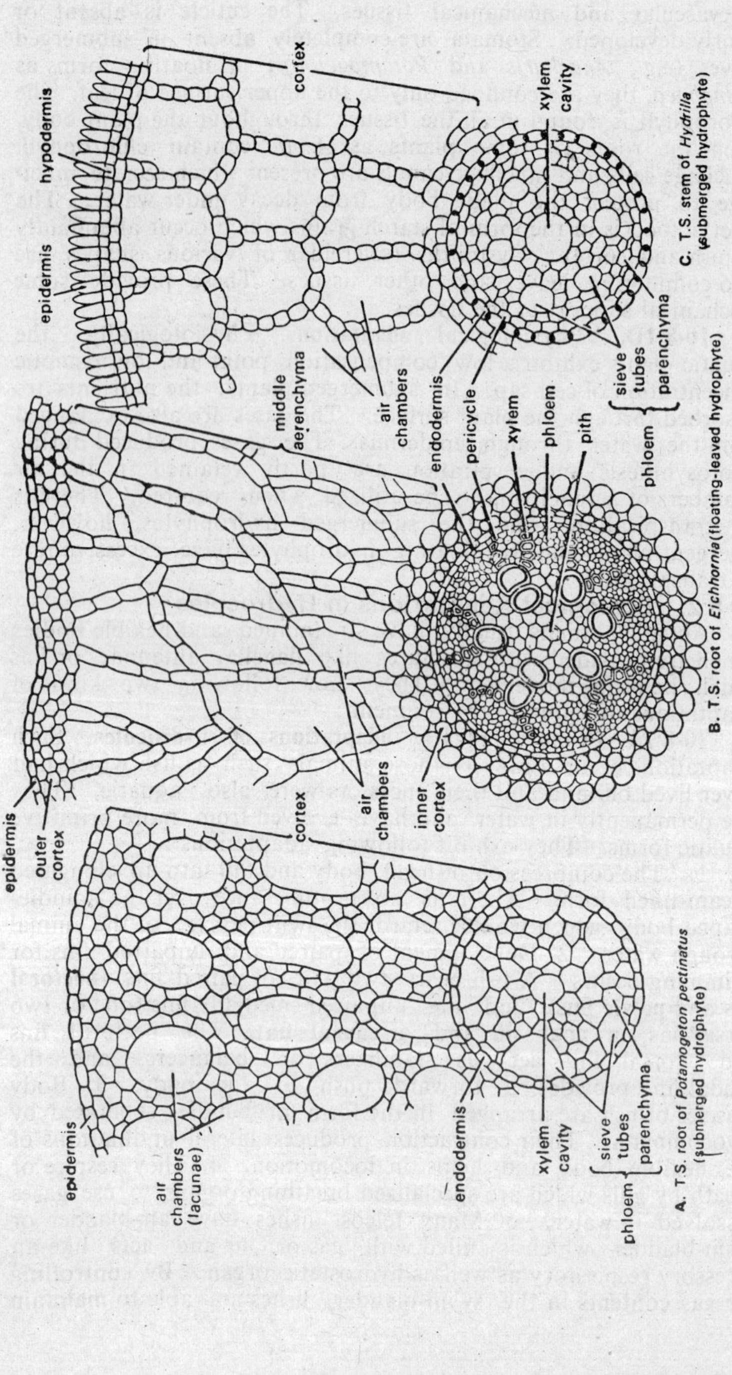

A. T.S. root of *Potamogeton pectinatus* (submerged hydrophyte)

B. T.S. root of *Eichhornia* (floating-leaved hydrophyte)

C. T.S. stem of *Hydrilla* (submerged hydrophyte)

Fig. 10·12. Anatomical features of some hydrophytes.

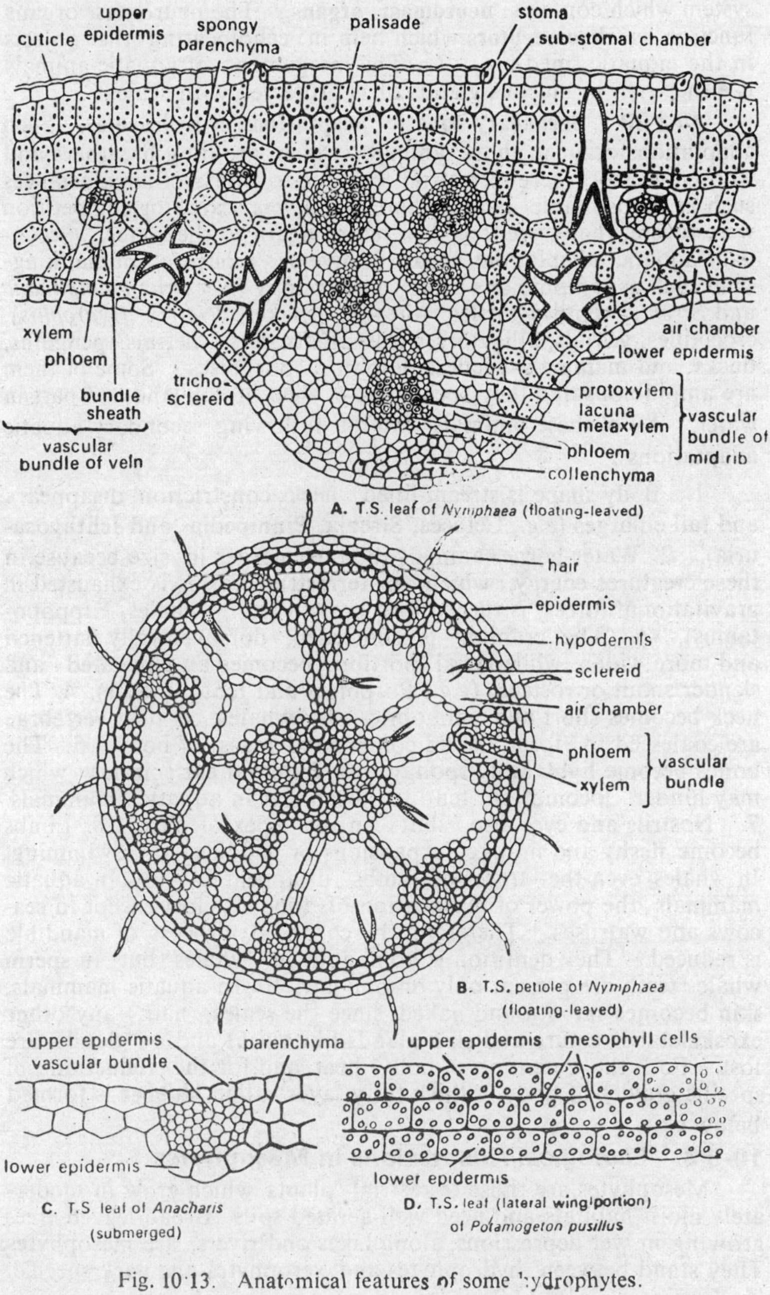

A. T.S. leaf of *Nymphaea* (floating-leaved)

B. T.S. petiole of *Nymphaea* (floating-leaved)

C. T.S. leaf of *Anacharis* (submerged)

D. T.S. leaf (lateral wing portion) of *Potamogeton pusillus*

Fig. 10·13. Anatomical features of some hydrophytes.

themselves at a particular depth. 6. Fishes possess lateral-line system which contains **neuromast organs**. The neuromast organs function as **rheoreceptors** which help in echolocating the objects in the aquatic medium. 7. The integument of aquatic animals is either rich in mucous glands or is protected with scales.

10-8·2B. Secondary aquatic adaptations of hydrocoles. Such adaptations occur in those animals which are lung breathers and whose ancestors were terrestrial but due to stress of circumstances such as inhospitable lands, where food was scarce or competition were severe, they were forced to return again to the water. Secondary aquatic adaptations occur in diverse groups of animal kingdom such as amphibians–frog, mud puppy (*Necturus*), *Amphiuma* and *Siren* ; reptiles–turtles, tortoise, snakes (*Natrix*, *Hydrophis*), crocodiles, gavials, alligators; birds–albatrosses, petrals, penguins, ducks; and mammals–whales, dolphins, porpoises. Some of them are amphibious and they spend part of their life on land and part in water. All of these animals exhibit following secondary aquatic adaptations :

1. Body shape is stream-lined, neck constriction disappears and tail enlarges (*e.g.*, Cetacea, Sirenia, Pinnipedia and Ichthyosauria). 2. Water-borne animals become larger in size because in these creatures energy, which in terrestrial forms is exhausted in gravitational forces, is turned into growth (*e.g.*, whales, Hippopotamus). 3. The cranium is shortened, dorsoventrally flattened and more wider, while facial portion becomes an elongated and slender snout or rostrum (*e.g.*, Porpoises and Ichthyosaurs). 4. The neck becomes short and immobile. In whales cervical vertebrae are coalesced to form a solid compressed mass of bone. 5. The bones become light and spongy. 6. External ear (pinna) which may hinder locomotion tend to disappear in aquatic mammals. 7. Nostrils and eyes also shift on the apex of face. 8. Limbs become fleshy and fin-like expansions or paddles for swimming. In whales even the traces of limbs disappeared. 9. In aquatic mammals, the power of mastication of food is lost except in sea-cows and walruses. Therefore, the coronoid process of mandible is reduced. The dentition is homodont in whales but in sperm whales teeth are present only on one jaw. 10. In aquatic mammals, skin becomes smooth and naked, since the scales, hair, any other exoskeletal structure and skin glands like sweat and oil glands are lost. For the retention of body heat and for the reduction of specific gravity of body, a thick fatty layer called **blubber** is formed below skin.

10-8·3. Ecological Adaptations in Mesophytes

Mesophytes are those terrestrial plants which grow in moderately moist habitats and need well-aerated soils. Broad-leaved trees growing in wet depressions, along lakes and rivers, are mesophytes. They stand between hydrophytes and xerophytes and lack specific.

adaptations of them. Some of the significant morpho-anatomical features of mesophytes are as follows : 1. Root system is well developed. Roots are fairly branched and contain root caps and root-hairs. 2. Stems are gererally aerial, solid and fairly branched. 3. Leaves are generally large, broad, thin and varied in shapes. They are green and lack hairy or waxy coatings. 4. In all aerial parts, cuticle is moderately developed. 5. Epidermis is well developed and has no chloroplasts. 6. Stomata generally present on both the surfaces of leaves. 7. Mesophyll in leaves is differentiated into palisade and spongy parenchyma, with many intercellular spaces. 8. Well developed vascular and mechanical tissues are well differentiated. 9. Mesophytes may exhibit temporary wilting during noon hours.

10-8·4. Ecological Adaptations in Mesocoles

Water economy is one of the major problems confronting terrestrial and semiterrestrial (amphibious) animals. The amount of water necessary to maintain life varies from species to species. It also varies with the stage in the life cycle of a single species. Many mechanisms have evolved which enable animals to maintain the critical level of water balance, thereby permitting some species to live in deserts, some in water and others in habitat intermediate in moisture content.

Adaptations of animals to meet the problem of dryness or desiccation are behavioural, anatomical and physiological. Two of the main effects of desiccation are of loss of body water and the drying out of the respiratory epithelium. The body wall of some animals, such as land planarians, earthworms and amphibians, is an especially sensitive epithelium, which also functions as a respiratory membrane ; if this membrane is subjected to drying, the cells are permanently injured. Therefore, to survive, these animals must always remain in moist, humid conditions. To maintain a moist situation, many animals secrete a mucus coating and in unusually dry situations they may resort to behavioural adaptations such as burrowing, coiling or forming aggregates.

Arthropods have an external covering over the respiratory surfaces, and some of the predominantly terrestrial forms of marine Crustacea have an actual reduction in the respiratory surfaces (Gray, 1957). Insects and myriapods have a particularly well-developed covering over the respiratory surface as well as a tracheal system. Vertebrates have stratified skin with several cellular layers in addition to a well-protected respiratory surface. Many of the structural modifications that prevent water loss are devices for effectively closing off the animals from the external environment. This is well illustrated particularly by the operculum of gastropod molluscs. The serpulid polychaetes have one or more of the gill filaments enlarged at one end, which are utilized to close the mouth of the tube when the animal withdraws into it during periods of stress. In insects a wax layer (in epicuticle) makes the integument relatively impermeable. The integument of terrestrial Crustacea is less permeable to water than that of predominantly littoral Crustacea.

Some animals have the ability to regulate the rate of loss of body water, an example is the amphibians, that are able to slow the rate of water loss when they are first exposed to dry air (**Thorson**, 1955, 1956). In the crab *Carcinus maenas*, the rate of water loss through transpiration is reduced by increased calcification of the exoskeleton. In intertidal-zone animals, the loss of water is accompanied by osmotic adjustments and regulation.

Some animals obtain needed water from food as partial replacement of water lost through excretion. The excretory process has also evolved differently for the terrestrial animals. Aquatic species tend to excrete more non-protein nitrogen than do terrestrial species. With the approach to terrestrialism, there also appears to be a general suppression of nitrogen metabolism rather than conservation of water, and the species shifts from ammonotelic to uricoletic excretion so that little or no water is excreted with uric acid. In the more terrestrial species the faecal pellets tend to be drier (see **Vernberg** and **Vernberg**, 1970).

10-8·5. Ecological Adaptations in Xerophytes

Xerophytes are the plants which grow in dry habitats. **Daubenmire** (1950) has defined xerophytes as "plants which grow on substrata that usually become depleted of growth water to a depth of at least 2 decimeters during a normal season". Thus, all plants of arid zone which are not confined to the margins of streams or lakes have been considered as xerophytes. Likewise, in regions of heavy rain fall, xerophytes are represented by shallow-rooted plants of sandy soils, by plants of dry ridgetops, and by algae, mosses and lichens which grow on tree barks or rock surfaces, etc.

10-8·5 A. Types and adaptations of xerophytes. Xerophytes are classified into following three categories :

1. Ephemerals. They are short-lived annual plants of arid zones which complete their life-cycle within a very small period when sufficient moisture is available. They are also called '**drought evaders**" or "**drought escapers**". Since these plants avoid dry seasons and thus escape dryness in external and internal environments. In this case, the seed germination is soon followed by flowering without undergoing a phase of vegetative growth. The dry phase of the year is passed in the form of seed, *e.g.*, *Cassia tora*, *Argemone mexicana*, *Solanum xanthocarpum*, *Tephrosia purpurea*, etc.

2. Succulents. These xerophytes suffer dryness in external environment only. Their succulent, fleshy organs (stems, leaves, roots) serve as water-storage organs which accumulate large amounts of water during brief rainy seasons. In these xerophytes, the succulence, *i.e.*, development of water-storing parenchymatous tissues may occur in roots (*Ceiba parvifolia*, *Asparagus*), or leaves

(*Aloe*, *Begonia*, *Peperomia*, *Salsola*, *Agave*, *Bryophyllum*, *Yucca*, *Tredescantia*, etc.,) or stems (*Opuntia* and other cacti, *Euphorbia splendens*). The rest of the plant body other than the succulent organs is generally very much reduced. The roots system is shallow and roots are fibrous and help in the absorption of traces of moisture from the soil surface (as dew and small amounts of rain). Among cacti and Euphorbias, the leaves are absent or modified to spines. In *Opuntia*, stem becomes fleshy, green, leaf-like (**phylloclade**) covered with spines, The cuticle is thick and vascular system is usually ill-developed.

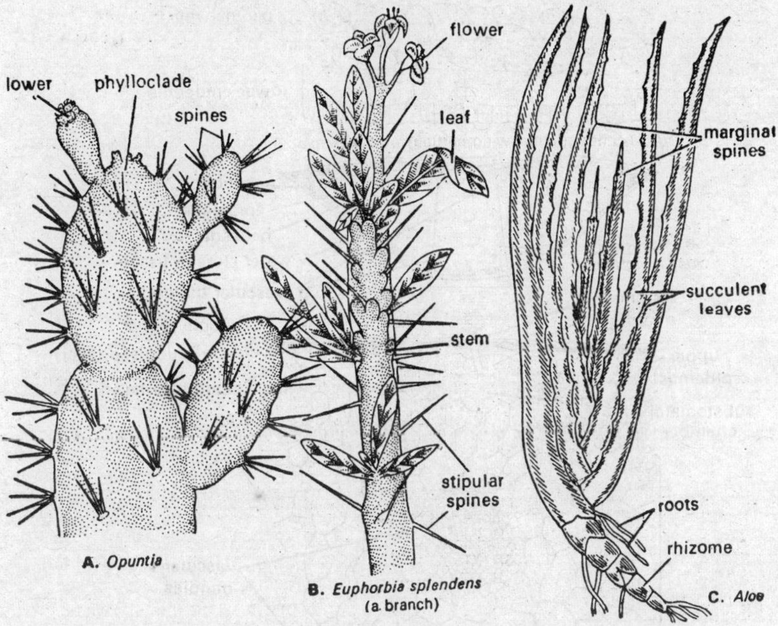

Fig. 10·14. Some succulent xerophytes

These succulents have an advantageous physiological adaptation : they can utilize carbon dioxide during night through the CAM pathway (crassulacean acid metabolism). This adaptation is in response to the fact that the stomata remain closed during the day to avoid water loss and gaseous exchange can occur only during night when stomata are open.

3. True xerophytes or drought-resistants. The plants which have the ability to maintain growth under critical dry conditions and high temperature are true xerophytes. Thus, these plants suffer from dryness both in their internal as well as external environments. The plants are woody trees, shrubs or herbs. The common examples are *Acacia nelotica*, *Calotropis*

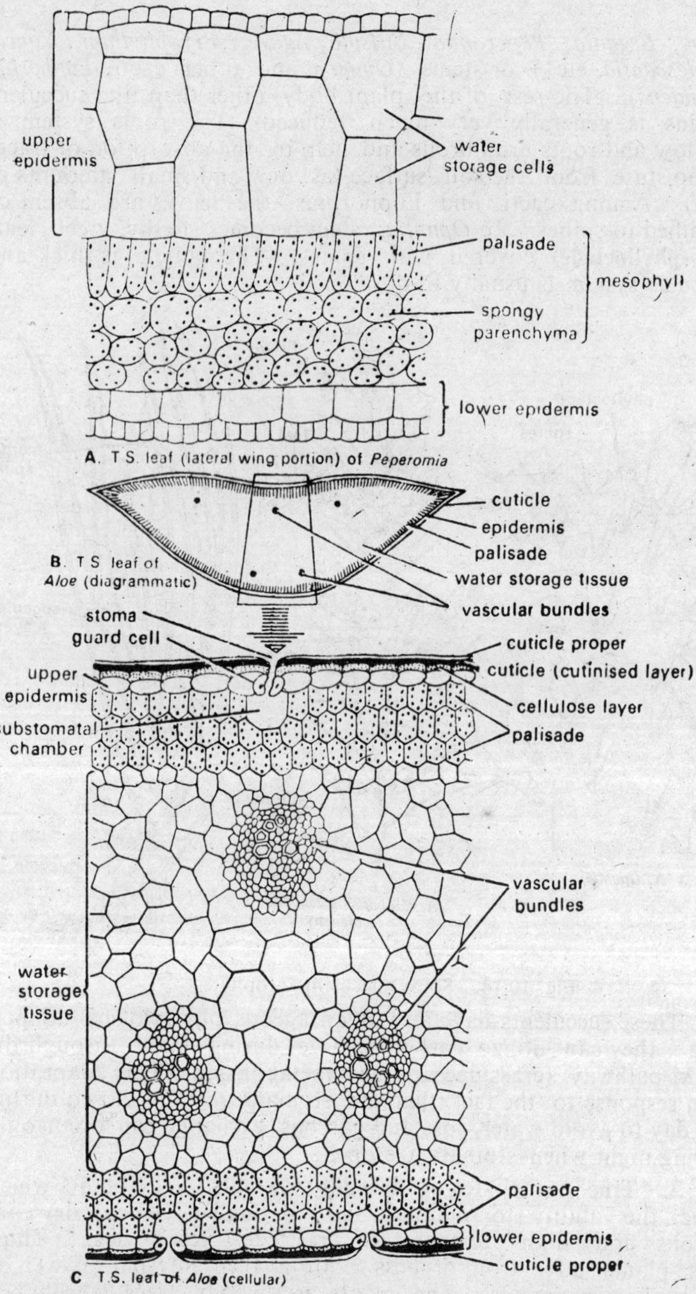

Fig. 10·15. Anatomical features of certain succulents. A—T.S. leaf of *P·peromia* ; B. C—T. S. leaf of Aloe.

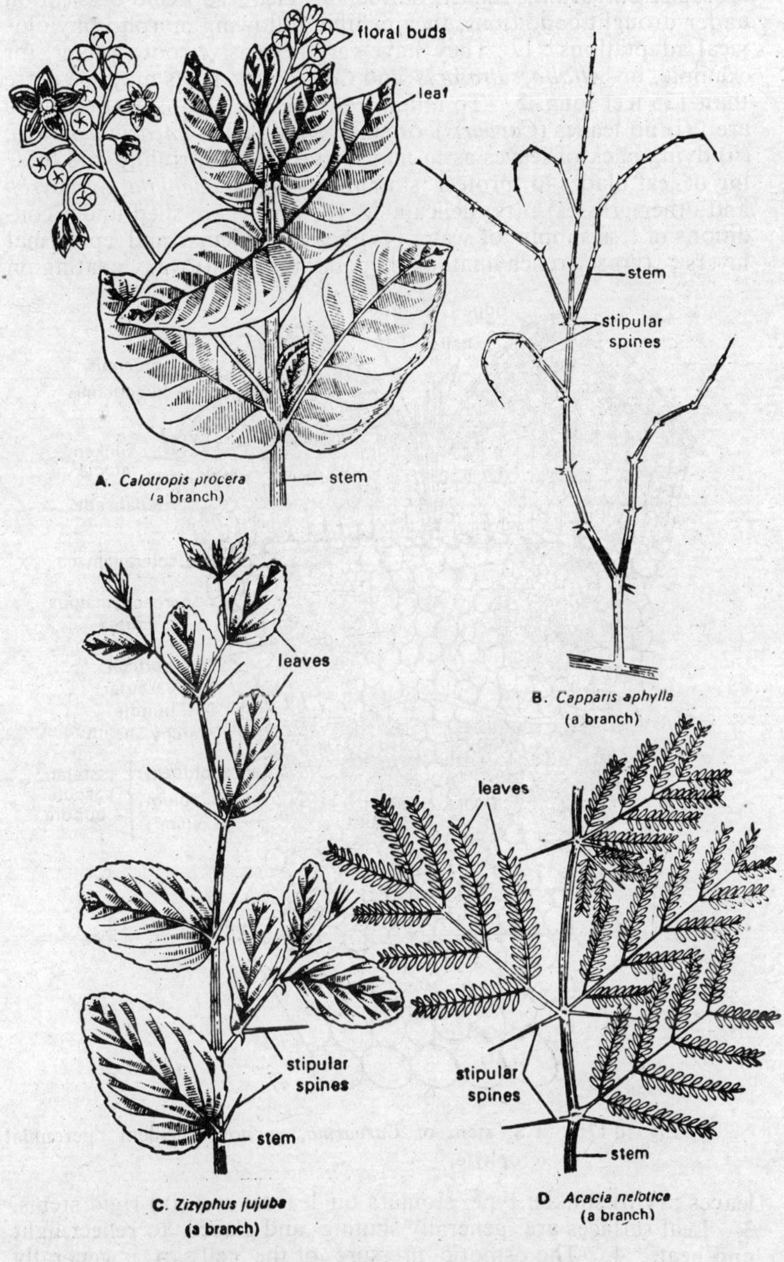

Fig. 10·16. Some common true **xerophytes.**

procera, *Zizyphus jujuba*, *Capparis aphylla*, *Saccharum*, *Nerium*, *Prosopis*, *Casuarina*, *Alhagi*, *Salvadora*, etc. To avoid dessication under drought conditions, they exhibit following morpho-phyiological adaptations : 1. They have very extensive root system ; for example, in *Alfalfa*, *Prosopis* and *Calotropis*, roots may be more than 125 feet long. 2. To minimize the rate of transpiration there are : (i) no leaves (*Capparis*) or small-sized leaves (*Acacia nelotica*); (ii) dying back of leaves as in many grasses ; (iii) rolling and folding of leaf blades to protect stomata (*Poa*, *Ammophila*, *Agropyron* and other grasses) ; (iv) delicate leaves which are shed under conditions of less supply of water ; (v) heavy cuticular and epidermal layers ; (vi) sclerenchymatous hypodermis ; (vii) waxy coating on

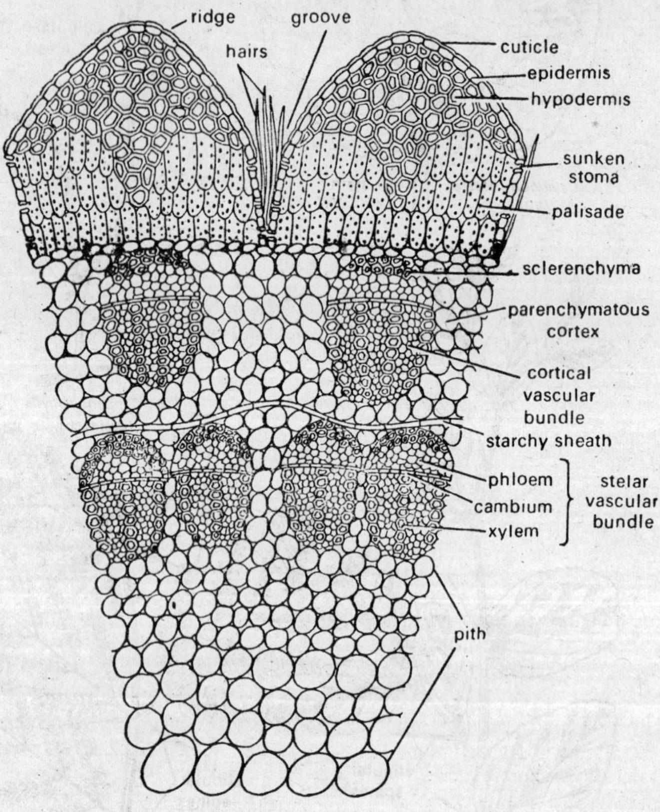

Fig. 10 17. T.S. stem of *Casuarina*, a non–succulent perennial xerophyte.

leaves ; (viii) sunken type stomata on leaves; and (ix) rigid stems. 3. Leaf surfaces are generally shining and glazed to reflect light and heat. 4. The osmotic pressure of the cell sap is generally

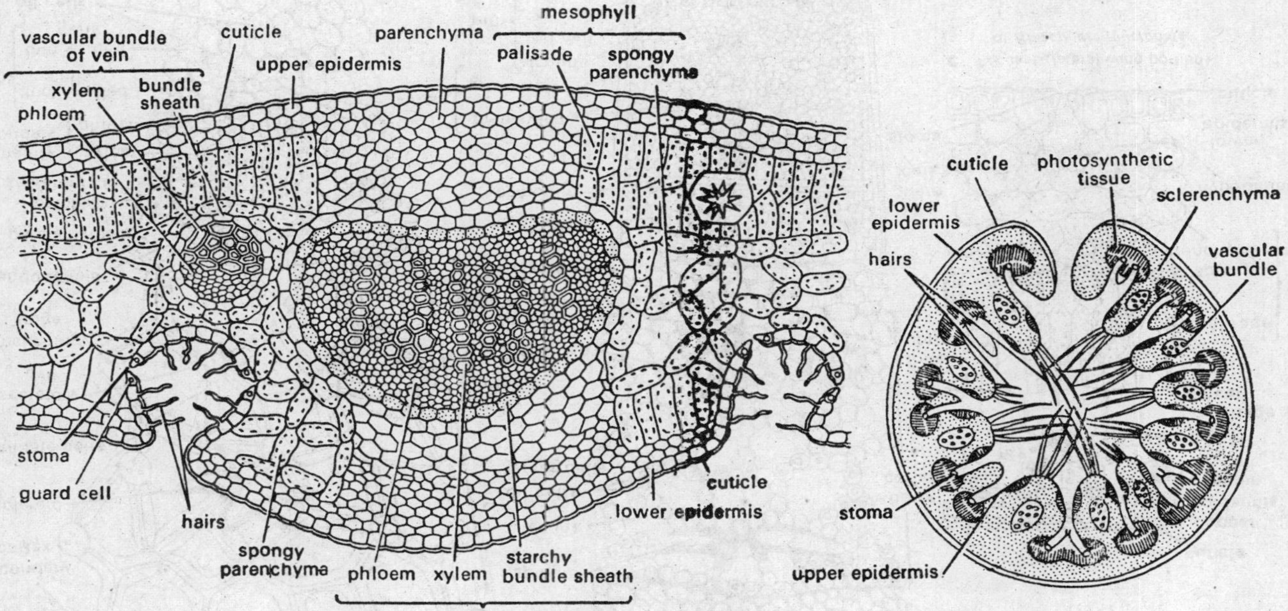

A. T.S. leaf of *Nerium* (non-succulent perennial)

B. T.S. rolled leaf of *Agropyron sp.*

Fig. 10·18. Anatomical features of some true xerophytes.

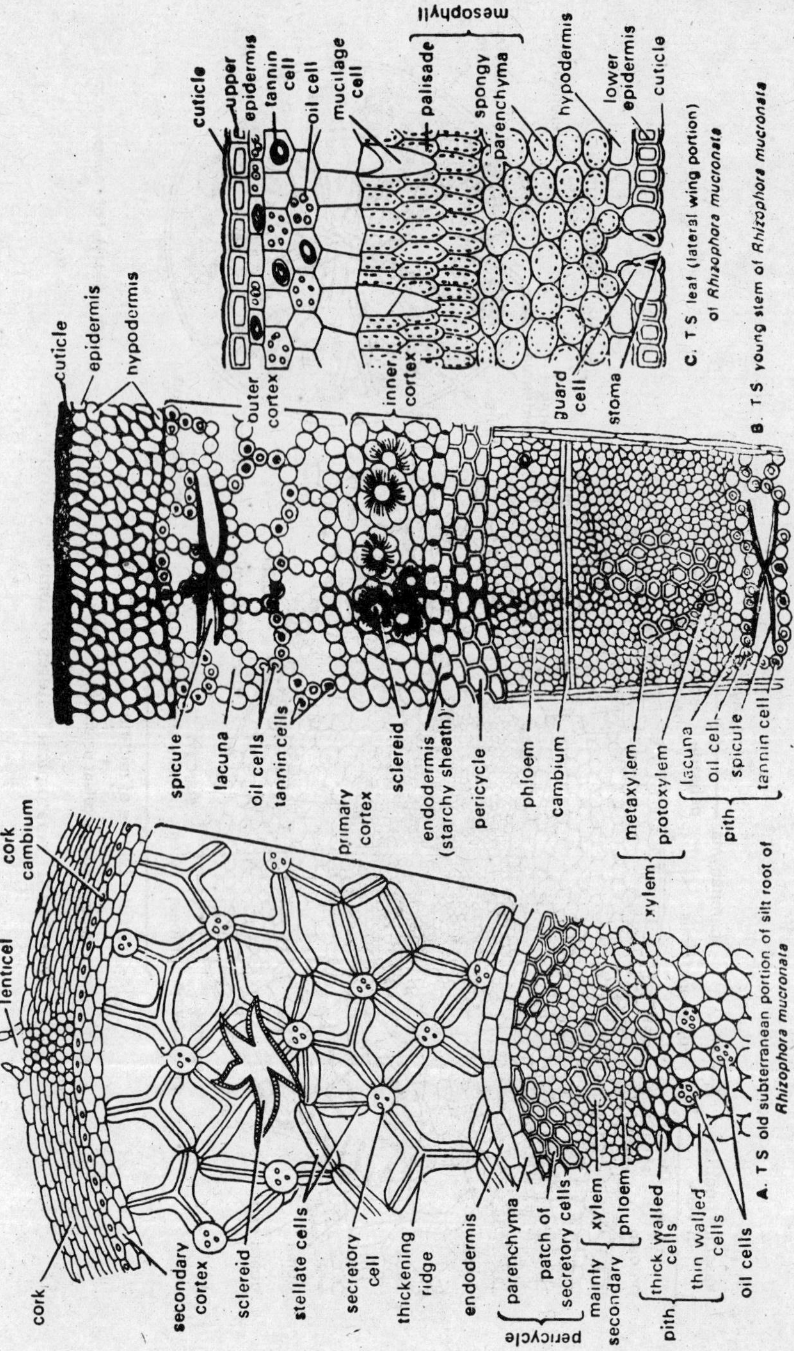

A. T S old subterranean portion of silt root of *Rhizophora mucronata*

B. T S young stem of *Rhizophora mucronata*

C. T S leaf (lateral wing portion) of *Rhizophora mucronata*

Fig. 10·19. Anatomical features of a halophyte of mangrove vegetation : *Rhizophora mucronata*.

high. 5. The plasmodesmata are absent. 6. Plants contain excess of sulphahydryl molecules and a variety of sclereids in leaf tissues.

10-8·5 B Halophytes. Halophytes are special type of xerophilous plants, which grow on physiologically dry soils such as saline soils with high concentration of such salts as sodium chloride, magnesium chloride and magnesium sulphate. Warming (1909) have classified halophytes into following four categories : 1. **Lithophilous**, 2. **Psammophilous**, 3. **Pelophilous**, and 4. **Helophilous** types, according to the substratum of rock and stones, of sand, of mud and of swamp, respectively. Helophilous halophytes include two types of vegetation : (i) **salt-swamp** and **salt desert** and (ii) **littoral swamp-forest (mangroves)**. In India, mangroves are common on seashores of Bombay and Kerala and in Andaman and Nicobar islands. They grow on water-logged soil, where habitat is characterized by sandy, swampy and saline soils, high precipitation, high atmospheric humidity and almost no fluctuation in temperature throughout the year. Common examples of mangrove plants are *Sonneratia* and *Rhizophora mucronata*.

Xeromorphic characters of halophytes are following : 1. Most of them are succulents, their leaves are evergreen, thin, small and leathery with water storage tissues, thick cuticle and well-developed palisade tissues. 2. Most halophytes produce special type of branched, negatively geotropic roots that come out of the mud surface to encourage the entry of oxygen gas. These roots are called **pneumatophores** and they possess breathing pores for gaseous exchange. 3. Some halophytes like *Rhizophora* produce stilt roots which have well-developed cork, lacunar primary cortex with sclereids and secretory cells and well developed mechanical and vascular tissues. 4. Their osmotic pressure values are very high (150 atms. in *Salicornia*). 5. Some halophytes as *Rhizophora* show **vivipary**, *i.e.*, germination of seeds before they are shed from parent plant. 6. Psammophilous halophytes **are** *Suaeda fruticosa* and *Salsola foetida*.

10-8·6. Ecological Adaptations in Xerocoles (Desert Animals)

Like plants, desert animals are also adapted to conserve moisture and escape heat of shimmering sun. They fall into following two categories :

10-8·6 A. Drought evaders. Drought evader xerocoles, like ephemeral plants, adopt an annual life style or go into aestivation or some other stage of dormancy. For eight or nine months, perhaps even several years, the eggs of insects and other invertebrates and insect pupae lie dormant. When the rains arrive and plants flourish, the deserts swarm with insects—crickets, grasshoppers, ants, bees, wasps, butterflies, moths, beetles, etc. Young bees emerge from underground cells at the very time when the particular flowers on which they feed are in flower. Some amphibians like spade-foot toads (*Scaphiopus*) a estivate for eight or nine months in an underground cell lined with a gelatinous substance that reduces evaporative losses through the skin. They appear on ground when the rain fall saturates the earth, move to the nearest puddle, mate and lay eggs. Young tadpoles hatch in 1½ to 2 days, rapidly mature (15 to 45 days) and metamorphose into functioning adults capable of diging their own retreats in which to aestivate until the next rainy season.

Likewise, birds nest during the rainy season, when food is most abundant for the young. If extreme drought develops during the breeding season, some birds do not reproduce. **Keast** (1959) has reported that among some desert birds the endocrine control of reproduction depends on rainfall rather than day length. A few birds such as swifts, poorwills and humming birds become torpid when food is scarce. Small rodents such as kangaroo or pocket mice, aestivate during periods of most severe drought.

10-8·6 B. Drought resisters. Drought resister xerocoles are active all the year round and have evolved ways of circumventing aridity and heat through morpho-physiological adaptations or by modifying their feeding and activity patterns (*i.e.*, behavioural adaptations). Some animals simply avoid the heat by adopting nocturnal habits and remaining underground or in the shade during the day. Some desert rodents that are active by day periodically seek burrows and passively lose heat through conduction by pressing their bodies against the burrow walls. Some birds such as poorwill (*Phalaenoptilus nuttallii*) and humming birds and bats go into a daily torpor.

The large ears of some desert animals such as jackrabbits (*Lepus*) and kit fox (*Vulpes velox*) may serves to reduce the need of water evaporation to regulate body heat. The ears may function as efficient radiators to the cooler desert sky, which on clear days may have a radiation temperatture of 25°C below that of the animal. By seeking shade, where the ground temperatures are relatively low and solar radiation is screened out, and sitting in depressions, where radiation from the hot ground surface is obstructed, the *Lepus* could radiate 5kcal a day through its two large ears (400 cm²). Such a radiation heat loss minimizes the heat loss through water evaporation (sweat glands) (**Schmidt-Nielsen**, 1964).

The kangaroo rat, which seals its burrow by day and thus keeps its chamber moist, can live throughout year without drinking water. Similar adaptation occurs in the jerboas and gerbils of Africa and the Middle East and the marsupial kangaroo mice and pitchi-pitchi of Australia. These animals feed on dry seeds and dry plant material even when succulent green plants are available. The kangaroo rat obtains water from its own metabolic processes and from hygroscopic water in its food. To conserve water the animal remains in its burrow by day, it possesses no sweat glands, its urine is highly concentrated and its faeces is also dry.

Large desert animals such as the camel can use water effectively for evaporative cooling through the skin and respiratory system because their low surface-area-to-body-size ratio and lower internal heat production result in slower accumulation of heat. The camel not only excretes highly concentrated urine **but can with stand dehydration upto 25 per cent of body weight, and it loses** water from body tissues rather than from blood (**Schmidt-Nielsen,** 1959). Body temperature of camel is labile, dropping to 33·8°C over night and rising to 40·6°C by day, at which point the animal begins to sweat. The camel accumulates its fat in the hump rather than over the body. This speeds heat flow away from the body and its thick coat prevents the flow of heat inward toward the body.

More extreme adaptation to aridity occurs in some African antelope such as *Oryx*. Many African ungulates migrate to escape the heat and dryness, but the *Oryx* does not migrate. Its water requirements are low because it stores heat in its body during the day. This causes a substantial rise in body temperature (**hyperthermia**) (**Taylor,** 1969). The *Oryx* further reduces daytime evaporative loss by suppressing heating. It pants only at very high temperatures. It reduces its metabolic rate and conserves water by decreasing the rate of internal production of calories, thus reducing the amount of evaporative cooling. These are daytime adaptations.

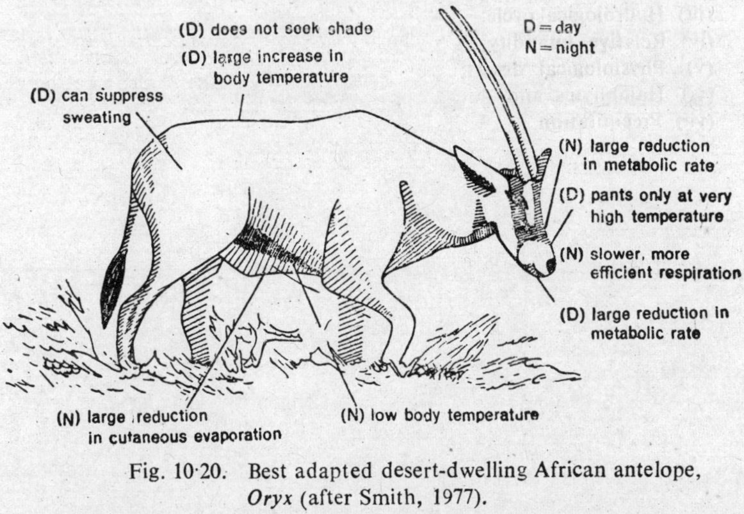

Fig. 10·20. Best adapted desert-dwelling African antelope, *Oryx* (after Smith, 1977).

By night the *Oryx* decreases its non-sweating cutaneous evaporation by 60 per cent and its metabolic rate by 60 per cent. Its respiratory rates are proportional to its respiratory efficiency and inversely proportional to body temperature. A cool animal breathes more slowly than a warm one, thus using a greater portion of inspired oxygen. With a lowered night-time body temperature, the saturation level for water vapour in the exhaled air is lower. The animal normally does not drink water ; it exists on metabolic water and by feeding on grasses and shrubs, many of them are succulent.

Some desert birds, like marine birds, utilize a **salt gland** to help in the maintenance of water balance. Some desert birds like black-throated sparrow (*Amphispiza bilineata*) of North America, budgerigar and zebra finch of Australia, and grey-backed finch lark of Africa feed on dry seeds and when have some access to green vegetation, they do not drink water. These birds can survive indefinitely without drinking by reducing the water content of its excreta from about 81 per cent to 57 per cent (**Smyth** and **Bartholomew**, 1966 ; **Dawson** and **Bartholomew**, 1968 ; and **Serventy**, 1971).

10-9 REVISION QUESTIONS

1. Describe physical and chemical properties of water and show, how they affect the organisms ?
2. Discuss water as an ecological factor.
3. Describe different ecological adaptations of plants and animals in relation to water.
4. Write short notes on the **following** :
   (i) Latent heat;
   (ii) Viscosity;
   (iii) Hydrological cycle;
   (iv) Relative humidity;
   (v) Physiological desert;
   (vi) Halophytes, and
   (vii) Precipitation.

11
Abiotic Environmental Factors

(Limiting Factors, Temperature, Light and other Factors)

11-1. INTRODUCTION

Every species has its own peculiar and perhaps unique set of environmental requirements for survival. Unless these needs are met, extinction is certain. However, these requirements are not totally inflexible, for they may vary throughout the lifetime of the individual and the history of species. The environmental factors which influence the growth, distribution, abundance, behaviour and ultimate survival of organisms are of two basic types: **non-living** or **abiotic environmental factors** which determine the interactions between the population and the abiotic environment, and **living** or **biotic environmental factors** which include interactions between different populations and instinctive control mechanisms that are internal to the population itself (**Clapham, Jr.**, 1973).

The abiotic environmental factors are of two basic kinds : 1. **Physical abiotic factors** such as temperature, light, pressure, geomagnetism, gravity, wind, precipitation, topography, habitat, etc. 2. **Chemical abiotic factors** such as pH, salinity (salt content), oxygen availability, nutrient availability, etc. In chapters 8, 9 and 10, some of these abiotic factors like wind, precipitation, pH, salinity, availability of oxygen and nutrients, habitat, etc., have already been discussed. In these three chapters, also have been discussed three fundamental media namely air, soil (land) and water which provide necessary conditions for sustaining life, serve as **macrohabitats** for different organisms and also act as abiotic environmental factors. In the present chapter will be discussed law of limiting factors and only some important abiotic factors such as temperature, light, etc.

11-2. NATURE OF RESPONSE OF ORGANISMS TO ABIOTIC ENVIRONMENTAL FACTORS

The environmental factors are not constant. They fluctuate temporally and geographically, and the stresses on animals resulting from these environmental alterations affect the complex biotic structure of our biosphere. Probably the first response of any organism to a change in the environment is **physiological** (**Kendeigh**, 1974). Different abiotic environmental factors influence organisms physiologically in various ways (**Fry**, 1947). These effects may be classified as follows :

1. **Lethal.** Causing death ; for instance, extreme heat or cold, lack of moisture, and so forth.

2. **Masking.** Modifying the effect of some other factor. Low relative humidity increases the rate of evaporation of moisture from body surfaces so that warm-blooded animals are able to survive at otherwise intolerably high air temperatures.

3. **Directive.** Producing an orienting response in relation to some environmental response so that the organism gets itself into favourable conditions.

4. **Controlling.** Influencing the rate at which some processes function, but not entering the reaction. Temperature, pressure, and viscosity, for instance, affect metabolism, secretion, and locomotion.

5. **Deficient.** Some abiotic environmental factors curtail an activity of an organism because some essential ingredient, such as salt, oxygen, or the like, is absent or at unfavourably low concentration.

Further, the same abiotic environmental factor may produce different effects at different times and under different conditions. Temperature may be lethal, if extreme ; masking, as when cold reduces the demand of cold-blooded organisms for food ; directive, by inducing a search for more favourable locations; or controlling, as a modifier of the rate of metabolism.

11-3. ESSENTIAL ELEMENTS AND LIMITING FACTORS

The individual organisms of species population, in order to grow and multiply, must be supplied with certain essential materials. Of the hundred four (*i.e.*, 104) naturally occurring chemical elements on the earth, all living organisms are believed to utilize only 16 different chemical elements for their survival. These are called **essential elements.** Several other elements are needed in small quantities by some species. They are summarized in table 11·1.

Table 11·1. Essential elements (Clapham, Jr., 1973).

Macronutrients : Elements used in relatively large quantities.	Micronutrients : Elements generally needed in relatively small quantities.	Micronutrients : Elements needed by certain species in relatively small quantities
Carbon	Iron	Sodium
Hydrogen	Manganese	Vanadium
Oxygen	Boron	Cobalt
Nitrogen	Molybdenum	Iodine
Phosphorus	Copper	Selenium
Calcium	Zinc	Silicon
Magnesium	Chlorine	Fluorine
Sulphur		Barium

Some of these tabulated elements are used in relatively large quantities as the fundamental building blocks of organic tissues and are called **macronutrients**. Others,the **micronutrients**, are used in much smaller quantities. The micronutrients are also known as **trace elements**. For instance, a crop such as corn will remove over 45 kg. of nitrogen per acre from the soil but only 10–15 g. of boron. In addition to the nutrients required by all species, some organisms have special nutrient requirements.

11-3·1. Liebig-Blackman Law of Limiting Factors

An organism is seldom, if ever, exposed solely to the effect of a single factor in its environment. On the contrary, an organism is subjected to the simultaneous action of all factors in its immediate surroundings. However, some factors exert more influence than do others, and the attempt to evaluate their relative roles had led to the development of the law of the minimum by the German biochemist **Justus Liebig** in 1840.

Liebig while investigating the relationship between the available amounts of essential elements and plant growth, discovered that crop yield was frequently limited by elements other than those utilized in the largest quantity. Freely translated, a part of his statement on his experimental results is that "*growth is dependent on the amount of food stuff that is present in minimum quantity*". This statement has come to be known as **Liebig's law of minimum.** It is now usually **incorporated** with a **law of limiting factors** developed by a British physiologist **F. F. Blackman** (1905), who at the beginning of this century investigated the factors affecting the rate of photosynthesis. He listed five factors involved in controlling the rate of photosynthesis : amount of CO_2 available, amount of H_2O available, intensity of solar radiation, amount of chlorophyll present, and temperature of the chloroplast. **Blackman** discovered that the rate of photosynthesis is governed by the level of the factor that is operating at a limiting intensity. Further work on limiting factors and substances has shown that a high level of one factor will modify the limiting effect of a second, a process described as **factor interaction**. The same principle of limiting factors applies well to animal functions.

Since the rate of a process may be controlled by too great an amount of a substance ; such as heat, as well as by too small an amount, and since the presence or abundance of an organism may be limited by a variety of environmental factors, biotic as well as abiotic, and since the limiting effect may be due to two or more interacting factors rather than a single isolated one (**Shelford**, 1932) the law of minimum has been restated by **Taylor** (1934) in broad ecological terms as follows : "*The functioning of an organism is controlled or limited by that essential environmental factor or combination of factors present in the least favourable amount. The factors may not be continuously effective but only at some critical-*

period during the year or perhaps only during some critical year in a climatic cycle.

11-3·2. Threshold and Rate

Every environmental factor varies through a wider range of intensity than any single organism can tolerate. Characteristically, there is for each individual organism a lower and an upper limit in the range of an environmental factor between which it functions efficiently. For any one factor, different organisms find optimal conditions for existence at different points along the range, hence their segregation into different habitats.

The **threshold** is the minimum quantity of any factor that produces a precipitable effect on the organism. It may be the lowest temperature at which an animal remains active, the least amount of moisture in the soil that permits growth of a plant, the minimum intensity of light at which photoreceptor is stimulated, and so forth. Above the threshold, the rate of a function increases more or less rapidly as the quantity of heat, moisture, light or other environmental factor is increased, until a maximum rate is attained. Above the maximum, there is usually a decline in the rate of a process either because of deleterious effect produced, the interference of some other factor or exhaustion.

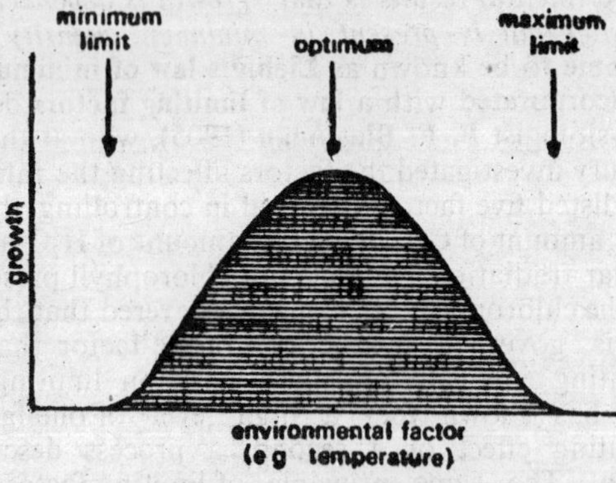

Fig. 11·1. An idealized tolerance curve showing that growth (or survival) of an organism can be limited by both too much or too little of something in the environment.

11-3·3. Shelford's Law of Tolerance

Organisms may be limited in their growth and their occurrence not only by too little of an element or too low an intensity of a factor but also by too much of the element or too high intensity of the factor. For example, carbon dioxide is necessary for the growth of all green plants, small increase in concentration of carbon dioxide in the atmosphere will, under certain circumstances, increase the rate of plant growth, but very considerable increases become toxic. Likewise, small additions of arsenic to the human diet actually have a tonic effect, further increase in the dosage, however, soon proves fatal.

The idea that factors could be limiting at their maximum as well as minimum quantities was incorporated in **law of tolerance** formulated by **V. E. Shelford** in 1913. This law postulates that each ecological factor to which an organism responds has maximum and minimum limiting effects between which lies a range or gradient that is now known as the **limits of tolerance.**

Between the **lower** and **upper** limits of tolerance lies a broad middle sector of a gradient which is called the **zone of compatibility,** the **zone of tolerance,** the **biokinetic zone** or the **zone of capacity adaptation.** The region at either end of the zone of compatibility is called the **lethal zone** or the **zone of resistance** or **zone of intolerance.** The zone of compatibility too includes a broad **range of optimum** and narrow **zones of physiological stresses** in between the range of optimum and lethal zones.

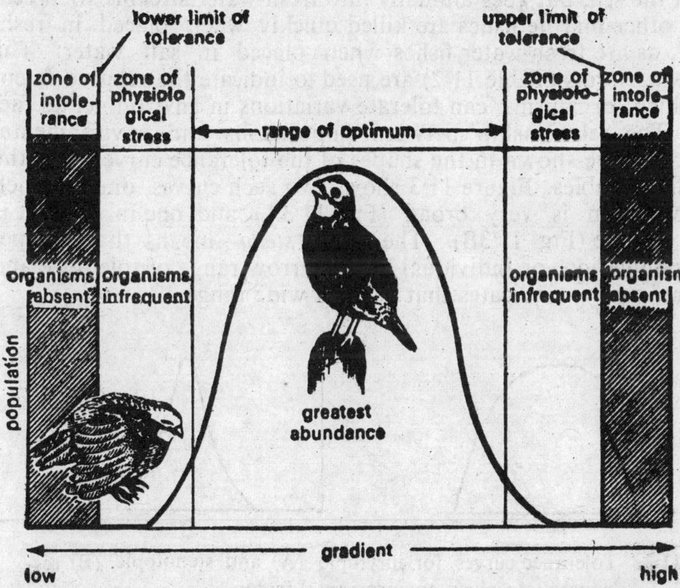

Fig. 11 2. Graphic illustration of the law of tolerance (after Smith, 1977).

Upper and **lower limits of tolerance** are intensity levels of a factor at which only half of the organisms can survive. These limits are sometimes difficult to determine, as for example with low temperature, organisms may pass into an inactive, dormant, or hibernating state from which they may again become functional when the temperature rises above a threshold At high temperatures, there may be similar inactivation or aestivation before the lethal level is attained. Even without dormancy occurring, there are normally **zones of physiological stresses** before the limits of tolerance are reached.

The species as a whole is limited in its inactivities more by conditions that produce physiological discomforts or stresses than it is by the limit of tolerance themselves. Death verges on the limits of tolerance, and the existence of the species would be seriously jeopardized if it was frequently exposed to these extreme conditions. Therefore, in retreat before conditions of physiological stress there is a margin of safety, and the species adjusts its activities so that limits of tolerance are avoided. There is a variation in hardiness of individuals within a species, so that some hardy individuals find existence possible under conditions that disrupt other individuals. The population level of a species becomes reduced before the limits of its range are actually reached.

Further, species vary in their limits of tolerance to the same factor. For example, the Atlantic salmon spends most of its adult life in the sea, but goes annually into fresh-water streams to breed. Most other marine fishes are killed quickly when placed in freshwater, as are fresh-water fishes when placed in salt water. The following terms (Table 11·2) are used to indicate the relative extent to which organisms can tolerate variations in environmental factors. The relationship between populations and environmental factors can be shown in the shapes of the tolerance curves for the specific variables. Figure 11·3 shows two such curves, one in which the optimum is very broad (Fig. 11·3A) and one in which it is quite definite (Fig. 11·3B). The prefix *steno*—means that the species, population, or individual has a narrow range of tolerance and the prefix *eury*—indicates that it has a wide range.

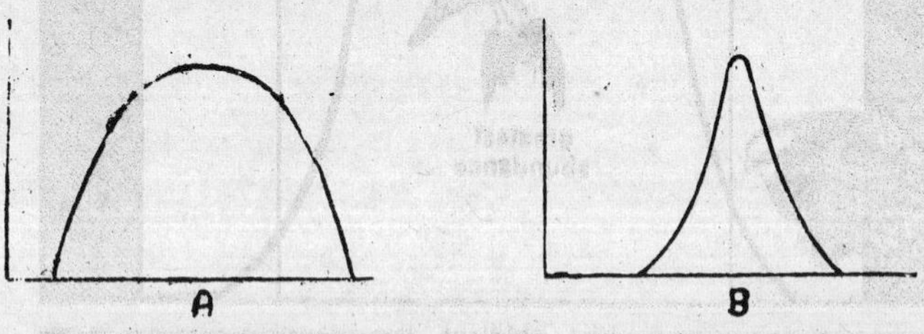

Fig. 11·3. Tolerance curves for eurytopic (A) and stenotopic (B) responses to a given enviornmental factor.

Table 11·2. Terminology for ranges of certain abiotic environmental factors.

Stenothermal—Eurythermal	Pertaining to temperature
Stenohaline—Euryhaline	Pertaining to salinity
Stenohydric—Euryhydric	Pertaining to water
Stenophagic—Euryphagic	Pertaining to food
Stenobathic—Eurybathic	Pertaining to depth
Stenoecious—Euryecious	Pertaining to niche or habitat selection

Moreover, a plant or animal may have a wide range of tolerance for one factor in the environment, but a relatively narrow range of tolerance for another condition. Thus, we find that some species of fresh-water fishes are eurythermal but they are stenohaline.

11-4. TEMPERATURE

Temperature is one of the essential and obvious changeable environmental factor. It varies not only between climatic regions, but also in the temporal changes of all habitats. It is one of the most extensively studied environmental factor. It penetrates into every region of the biosphere and profoundly influences all forms of life by exerting its action through increasing and decreasing some of the vital activities of organism such as metabolism, behaviour, reproduction, embryonic development and death. Temperature has an universal influence and acts as a limiting factor for the growth and distribution of plants and animals. The interaction of temperature with certain other abiotic environmental factors such as humidity, air, etc., cause into many other climatic changes which influence the living organisms in one way or another.

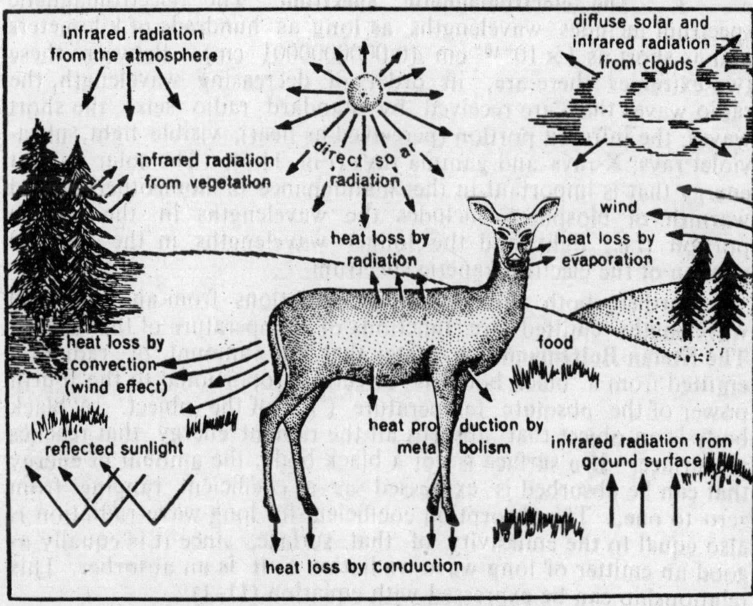

Fig. 11·4. Thermal energy exchange between an animal and its environment includes radiation, conduction, convection and evaporation as the four basic modes of heat transfer (after Moen, 1973).

11-4·1. Nature of Temperature

Temperature is a measure of the intensity of heat in terms of a standarized unit, and is commonly expressed as degrees either in the **Fahrenheit** or **Celsius scale (centigrade).** Heat is a form of energy and is called **thermal energy.** The same influx of energy on which photosynthesis depends is also the source of thermal energy that characterizes the physical environment of life.

Thermal energy is exchanged between animal and environment by radiation, conduction, convection and evoporation (Fig. 11·4). These four basic modes of heat transfer occur within the organism, in the interface between the organism and its environment (a thermal boundary layer), and between the environment and the thermal surface of an organism.

11-4·1A. Radiant energy exchange. Radiant energy exchange or radiation occurs between two surfaces as each surface emits energy at wavelengths that are dependent on the temperature of the emitting surface. Radiant energy travels through those media that are transparent, neither reflecting nor absorbing the radiation. A complete vacuum presents no obstructions to radiant energy exchange. The radiant heat exchange can be understood well by discussing its following basic characteristics :

1. The electromagnetic spectrum. The electromagnetic spectrum includes wavelengths as long as hundreds of kilometers and as short as 1×10^{-10} cm (0·0000000001 cm). Between these two extremes there are, in order of decreasing wavelength, the radio waves that are received by standard radio sets, the short waves, the infrared portion (perceived as heat), visible light, ultra-violet rays, X-rays and gamma rays (Fig. 11·5). The solar radiant energy that is important in the maintenance of homeothermy and warmth of biosphere includes the wavelengths in the visible portion (*i.e.*, light) and the longer wavelengths in the infrared portion of the electromagnetic spectrum.

Further, both the amount of radiations from an object and wavelengths emitted are functions of temperature of the object. The **Stefan-Boltzmann law** states that the amount of radiation emitted from a black body is directly proportional to the fourth power of the absolute temperature (°K) of the object. A black body is an object that absorbs all the radiant energy that reaches its surface. If a surface is not a black body, the amount of energy that can be absorbed is expressed as a coefficient ranging from zero to one. The absorption coefficient for long-wave radiation is also equal to the emissivity of that surface, since it is equally as good an emitter of long wave radiation as it is an absorber. This relationship can be expressed with equation (11–1).

$$Q_r = \epsilon\sigma T_8^4 \qquad \ldots(11\cdot1)$$

where :

Q_r = radiant energy emitted in kcal m^{-2} hr^{-1}

ϵ = emissivity (range from zero to one)

σ = Stefan-Boltzmann constant = $4\cdot93 \times 10^{-8}$ kcal m^{-2} hr^{-1}

T_8 = surface temperature of the object in °K

The **Wien displacement law** states that the wavelength (λ) of maximum intensity that is emitted from the surface of a black body is inversely proportional to the absolute temperature of the body (equation 11-2).

$$\lambda_{max}\ (\mu) = 2897\ T^{-1} \qquad ...(11\cdot2)$$

Thus, a very hot surface emits shorter wavelengths, while a cooler surface emits longer wavelengths.

2. Solar radiation. The wavelength of maximum emission from the sun is 0·5μ, which is in the visible portion of the spectrum (**Sellers** 1965). **Sellers** points out that 99% of the sun's radiation is the wavelength range of 0·15μ to 4·0μ, including 9% in the ultraviolet (<0·4μ), 45% in the visible (0·4μ to 0·74μ) and 46% in the infrared (>0·74μ).

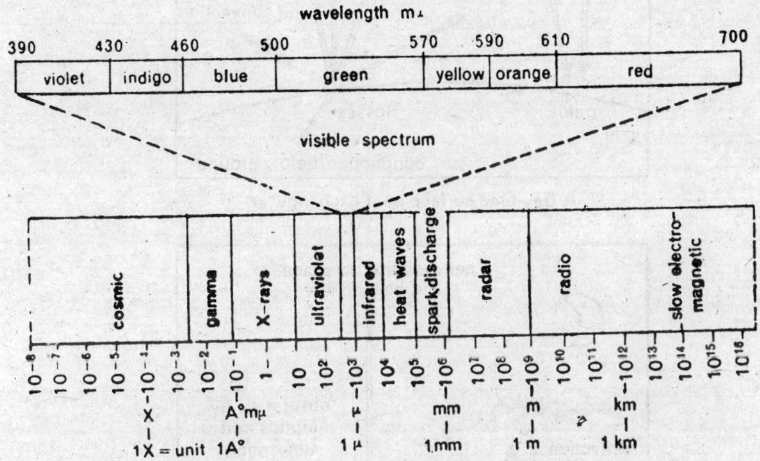

Fig. 11·5. The electromagnetic spectrum of radiant energy (after Vernberg and Vernberg, 1970).

The amount of solar energy reaching a surface perpendicular to the rays of sun at the outer limits of the earth's atmosphere is called the **solar constant**. The value of solar constant is 2·00 calories per square centimeter per **minute**. Of this quantity about 1·00 $cal/cm^{-2}/min^{1}$ reaches the earth's surface.

The amount of solar radiation that reaches a place depends on the moisture, concentration of ozone and dust particles in the atmosphere and altitude and location on the planet. On a clear day, a high percentage of the solar radiation is transmitted through the atmosphere with a completely overcast sky, no direct solar radiation penetrates the cloud cover. Likewise, as the altitude above sea level increases, the amounts of water vapour, dust, air molecules and carbon dioxide in the path of the sunlight diminish. Thus at higher altitudes solar radiation is more intense. The

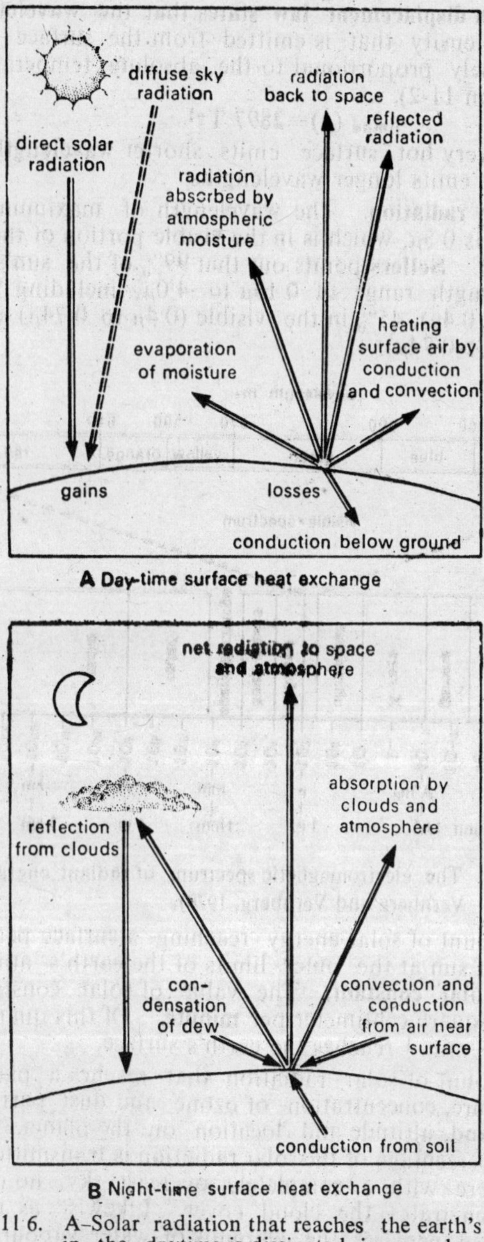

Fig. 11 6. A–Solar radiation that reaches the earth's surface in the daytime is dissipated in several ways, but heat gains exceed heat losses. B–At night there is a net cooling of the earth's surface, although some heat is returned by various processes (after Smith, 1974).

solar radiation absorbed as short wave energy by the earth and reradiated as longwave energy is responsible for heating the earth's atmosphere. Longwave radiation, prevented from escaping into space by the ozone layer of the stratosphere, warms the upper atmosphere. Thus the atmosphere receives most of its heat directly from the earth and indirectly from the sun.

Solar radiation also affects climate of earth. Sunlight does not strike the earth uniformly. Because of the earth's shape and tilt on its axis , the sun's rays strike it more directly on the equatorial than on the polar regions. The lower latitudes are heated far more than the polar ones. Air heated at the equatorial regions rise until it eventually reaches the stratosphere, where the temperature no longer decreases with altitude. There the air, whose temperature is the same or lower than that of stratosphere, is blocked from any further upward movement. With more air rising, the air mass is forced to spread out north and south toward the poles. As the air masses approach the poles they cool, become heavier and sink over the arctic regions. This heavier cold air then flows towards the equator, displacing the warm air rising over the tropics.

3. Solar radiation and organisms. There are three possible pathways for radiant energy to take once it reaches a plant or animal (**Moen** 1973). It may be reflected from the surface, it may be absorbed by the surface, or it may be transmitted through the material (Fig. 11·7). Energy that is reflected from the surface is of no thermal benefit to an animal or plant. Reflected energy in the visible portion of the spectrum is detected as shades of gray or ash colour, depending on the receptors of the organism detecting the light energy. Transmitted solar energy is of no value to an animal, and all animals except the smallest protozoans are essentially opaque. The leaves of plants, however, transmit some solar energy. Absorbed energy becomes a part of the thermal and physiological regime of an organism, and the quantity and distribution of absorbed radiant energy is of interest to the physiologist and the ecologist.

For an instance the amount of energy absorbed by the hair surface of a mammal depends on the spectral characteristics of the hair and the angle at which the solar energy strikes the surface. **Riemerschmid** and **Elder** (1945) measured absorption coefficients for cattle and found that white coats absorb less and reflect more solar energy ; black coats absorb the most solar energy. Further, the greatest amount of energy is absorbed when the solar radiation strikes perpendicular to the surface and no absorption takes place when the rays are parallel to the surface. Inclination of hair, the smoothness or curliness of the coat, and seasonal changes in the characteristics of the coat have little effect on the absorptivity. The distribution of solar radiation on the surface of an animal is called **the solar radiation profile (Moen**, 1973).

Moreover, the effect of direct solar radiation on the radiant temperature over the entire surface of a plant or animal is not uniform because both the colour of this surface and the angle of the rays striking the surface are important in determining just how much solar energy is absorbed. For example, a deer bedded in the sun (Fig. 11·8) might have 40 per cent of its surface exposed to direct solar radiation, 80 per cent exposed to indirect solar radiation, and 80 per cent exposed to infra-red radiation. These differences result in different radiant temperatures, but they act in combination with the distribution of tissue metabolism beneath the hair surface as well as with the distribution of blood. The radiant temperature distribution of an animal exposed to only infra-red radiation at night is much simpler since the hair surface is almost a black body and infra-red energy is much more uniformly distributed in the environment than is solar energy.

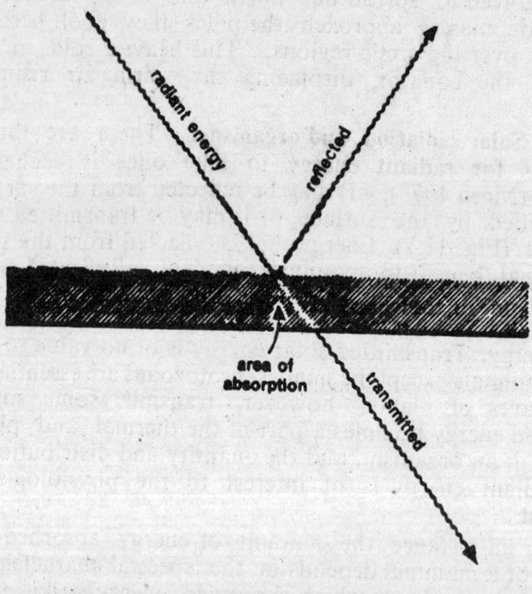

Fig. 11·7. The three pathways for radiant energy reaching a plant or animal (after Moen, 1973).

In fact, radiant temperature and air temperature are much more similar to plants than to animals because only a small amount of metabolic heat is released inside the leaf. In homeothermic animals, however, there is heat flow from the many exothermic reactions inside the animal to the outside through a layer of insulating hair or feathers. Thus the surface temperature of an animal is dependent on blood flow beneath the skin, local tissue metabolism and on the quality of the insulating pelage. In a plant, the surface temperature is dependent primarily on the interactions between thermal parameters alone, with virtually no input from metabolic reactions within the leaf.

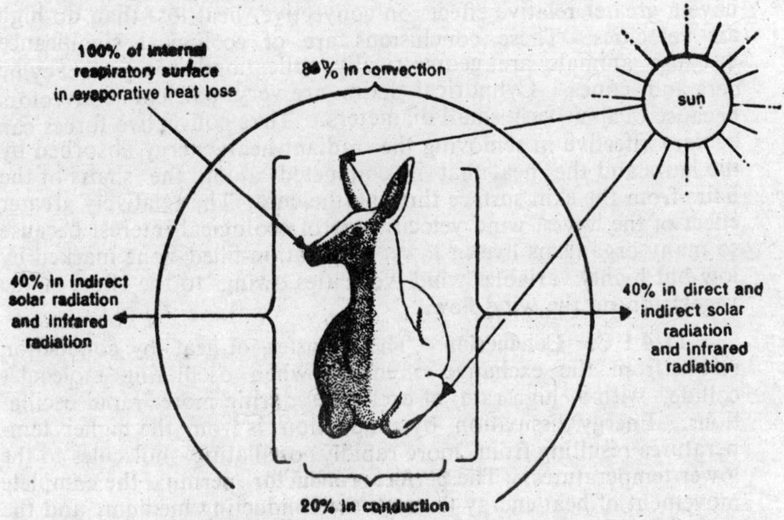

Fig. 11·8. **The relative proportions of the surface area of a deer bedded in the sun light (After Moen, 1973).**

11-4·1 B. Convection. Thermal energy may be removed from the surface of an object by fluid or air flowing over the surface. This process is called **convection.** Two types of convection occur : **free convection** or **natural convection** and **forced convection.** Free convection occurs when temperature differences in the boundary layer of air surrounding an object cause a movement of the air in response to changes in air density. Forced convection occurs when external pressure differences cause wind to blow past the object.

Convective heat loss. The amount of heat that is removed **from** an object by convection is a function of the factors expressed by the following equation (11-3) :

$$Q_c = h_c\, At\, (T_s - T_a) \qquad \ldots(11{\cdot}3)$$

where

Q_c=calories of heat transferred by convection

h_c=convection coefficient

A=area

t=time

T_s=temperature of the surface of the convector

T_a=temperature of the air (fluid).

Following two kinds of conclusions have been drawn from experiments which were performed to determine convective heat loss from cylinders of different diameters : (i) small cylinders are more efficient convectors than large ones, and (ii) low air velocities have a greater relative effect on convective heat loss than do high air velocities. These conclusions are of ecological significance because animals are, geometrically, collections of imperfect cylinders and cones. Cylindrical hairs are very efficient convectors because of their very small diameters. Thus convective forces can be very effective in removing the radiant heat energy absorbed by the hairs and the heat that is conducted along the shafts of the hairs from the skin surface through the coat. The relatively greater effect of the lower) wind velocities is of ecological interest because so many organisms live in lower vegetation-filled zone marked by low but highly variable wind velocities owing to the effect of the vegetation on the wind flow.

11-4·1 C. Conduction. The transfer of heat by conduction results from the exchange of energy when oscillating molecules collide, with a high rate of exchange during more rapid oscillations. Energy dissipation by conduction is from the higher temperatures resulting from more rapidly oscillating molecules to the lower temperatures. The **perfect conductor** permits the complete movement of heat energy through the conducting medium, and the **perfect insulator** pervents all movement of heat energy through the medium. The thermal conductivity (k) of a material is an expression of the rate of heat flow by conduction through the material under a specified set of conditions.

The amount of heat flow by conduction is expressed by following equation (11·4).

$$Q_k = \frac{kA\,t\,(T_1 - T_2)}{d} \qquad \ldots(11\cdot4)$$

where

Q_k=calories of heat transferred by conduction
k=thermal conduction coefficient
A=area
t=time
T_1=temperature of first surface
T_2=temperture of second surface
d=distance between the surfaces.

Conduction through the hair layer includes heat flow through the hair shafts themselves and through the air trapped between the hairs. Since air is a better insulator than hair, the important function of hair is the stabilization of the trapped air. This trapped air is a more important thermal barrier than the hair shafts themselves. The same relationship is true for feather surfaces

11-4·1 D. Heat loss by evaporation. Thermal energy or heat is lost from the surface of a plant or an animal by evaporation because energy is absorbed as liquid water is changed to a gaseous state. At 100°C the heat of vaporization is 540 kcal per gram, and at 0°C it is 595 kcal per gram. There is a linear relationship between air temperature and the heat of vaporization of water. Thus, if one knows the amount of water evaporated from an organism, the amount of energy removed by vaporization can be determined:

Evaporative heat loss comes from two sources : evaporation from the skin surface in the form of perspiration from the animal and evaporation from the lungs and lining of the nasal passages.

11-4·1 E. Heat budget. The temperature at the earth's surface is governed by the brightness of the sun ; the constancy of brightness of sun has remained virtually unchanged for about 3 billion years. In fact, average terrestrial temperatures now probably do not differ rapidly from those at the earth's beginning (**Schwarzs Hild,** 1967). The total amount of heat entering the biosphere from the sun must balance the amount lost per unit time if temperatures are to remain unchanged, since the flow of geothermal heat from the interior of the earth is small by comparison and probably has been negligible for at least the past 500 million years. The estimate of this energy flow is referred to as the **heat budget** (**Vernberg and Vernberg,** 1970). Heat budget has been projected for the total surface of earth as well as for special environments.

11·4·2. Temperature Fluctuations in Different Environments

Environmental temperatures fluctuate both daily and seasonally. The temperature of any one area will vary from sunlight to shade and from daylight to dark. The surface temperature of soil may be 30°C higher in the sunlight than in the shade and upto 17°C higher during the day than during the night. On the desert this spread may be as high as **40°C.** The Thar Desert of South Rajasthan (India) shows a diurnal change of 20—**30°C.** for all seasons. Differences in latitudes often causes variation in the annual temperature cycles. With the increased altitude for every 150 meters, the decrease in temperature of 1°C takes place. Further, different environments such as fresh-water, marine and terrestrial environments, are subjected to varied, reponses to fluctuating temperature. Temperature fluctuations are comparatively less in the aquatic environment than in the terrestrial environment. The increase in depth of aquatic medium often increases the temperature fluctuations. There exists a distinct difference in the response of living organisms of the fresh-water and sea to temperature fluctuations because of the presence of dissolved salts in sea water. The minimum temperature in the sea is −**3°C,** while in fresh water pond it never goes below 0°C. The maximum temperature of ocean generally goes upto 36°C, but in shallow pools of fresh-water and tide of pool's littoral zone, temperature may go higher. In deeper bodies of water, heating and cooling are restricted to the surface strata. But the deeper layers also get a lot of heat as a result of what is usually termed **vertical circulation,** where in, due to circulatory movement of water, surface waters are brought to the deeper regions, and vice versa. Studies on the vertical changes of temperature have led to the hypothetical classification of the freshwater media into three strata. The superficial layer of freshwater is constantly stirred by wind and is called **epilimnion** (Fig. 11·11). It is

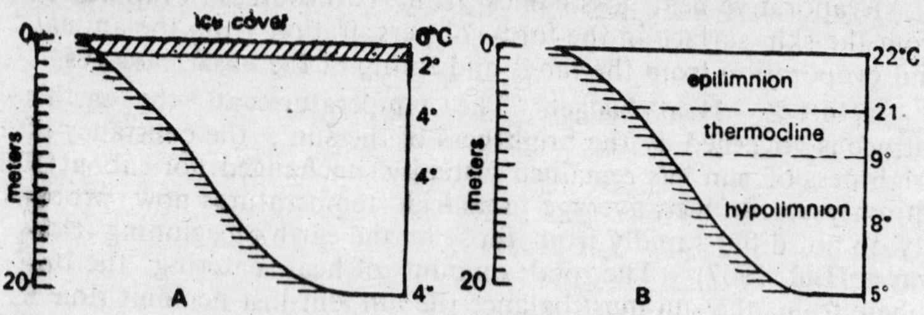

Fig. 11·9. Thermal stratification of a deep lake of the temperate region. A—During winter. B—During summer.

the layer of warmer water and its temperature may rise upto 27°C during summers. The stagnant water of the bottom constitutes the **hypolimnion.** The hypolimnion has temperature of 5°C only. In between epilimnion and hypolimnion occurs an intermediate zone called **thermocline** or **metalimnion**, which has rapid vertical temperature changes. During winter, the temperature of epilimnion of a fresh-water lake becomes 0°C and the lake becomes ice covered. The process of differentiation of fresh-water habitat into these three strata is called **thermal stratification.**

In the terrestrial environment fluctuations in temperature are varied and marked. The lowest temperature recorded for any land mass is −70°C (Siberia in 1947). Higher temperatures may likewise go often 85°C as in certain deserts at noon. However, the water in hot springs and geysers may approach 100°C, and even higher temperatures occur sporadically in the very special situation presented by volcanic areas. In Rajasthan the highest temperature exceeds **50°C.**

The differential responses of land and sea to temperature, bestow upon the terrestrial and aquatic organisms certain special adaptations to withstand the rapid changes of temperature. Because of, narrow range of temperature fluctuations in the aquatic media, the relative stability of the aquatic media has led to the adaptation of aquatic animals to a more or less constant temperature. But this constancy in temperature of aquatic media has its certain own drawbacks as a slight alteration in temperature often brings about mass killings. For example, torrential insects habitually inhabiting waters with temperature range of 2—4°C at altitudes of 1000—1400 meters are adversely affected by even a slight increase in the temperature of water (Mani, 1968 a).

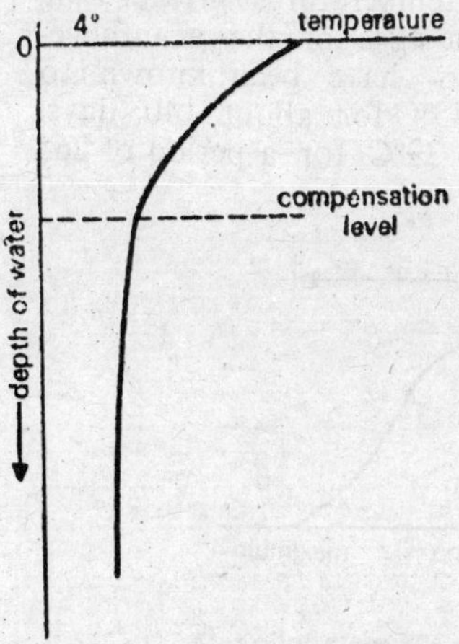

Fig. 11·10. Theoretical distribution of temperature in a lake during the summer if there were no mixing of water by the wind (after Clapham, Jr., 1973).

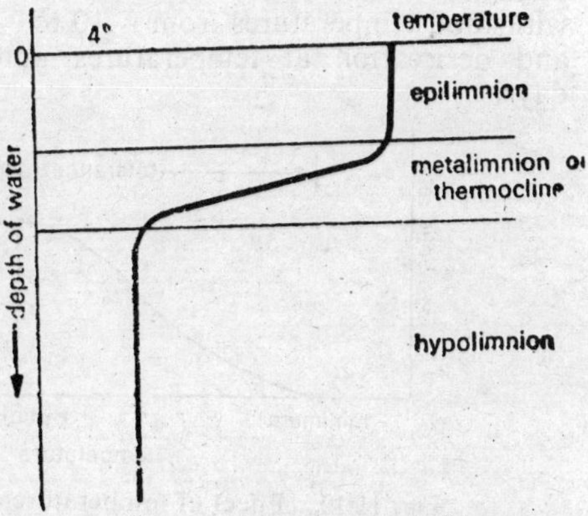

Fig. 11·11. Typical pattern of temperature distribution in a deep lake in the temperate zone during the summer (after Clapham, Jr., 1973).

11·4·3. Range of Temperature Tolerance

In organisms all metabolic processes necessary for life start at a certain **minimum temperature** and increase with rise in temperature until they reach the maximum level at a temperature called **optimum temperature**. Further rise in temperature beyond optimum brings about decrease in metabolic rate, until it ceases at a temperature called **maximum temperature** (Fig. 11·12). Thus, the favourable temperature range for any particular species is determined by the prevailing temperature at which normal physiological activities of the animals take place. For example, rotifer *Keratella procura* is known to appear in the ponds of **Pilani,** Rajasthan (India), when the temperature is below 24°C and to disappear when temperature rises above 24°C. Its frequency of distribution reaches the peak during the months of October to March (**Nayar,** 1970).

The ability to withstand extremes in temperature varies widely among plants and animals, but there are temperatures above and below which no life can exist. A temperature of 52°C is about as high a temperature as any animal can withstand and still grow and multiply. However, larvae of chironomids and certain other Diptera are found to thrive at temperature near 55°C. Praying mantis is found to live on bare ground at a temperature of **62°C** in deserts. Likewise, among plants, some hot-spring algae can live in water as warm as 73°C under favourable conditions and some arctic algae can complete their life cycles in places where the temperaturec barely rise above 0°C. Non-pathogenic bacteria inhabiting hot springs can actively grow at temperatures greater than **90°C Bott** and **Brock**, 1969). Further, the eggs of the acanthocephalan *Macracanthorhynchus hirudinaceus* have been known to withstand temperatures from −10 to −45°C for about 140 days, and desiccation at temperatures upto 39°C for a period of 265 days.

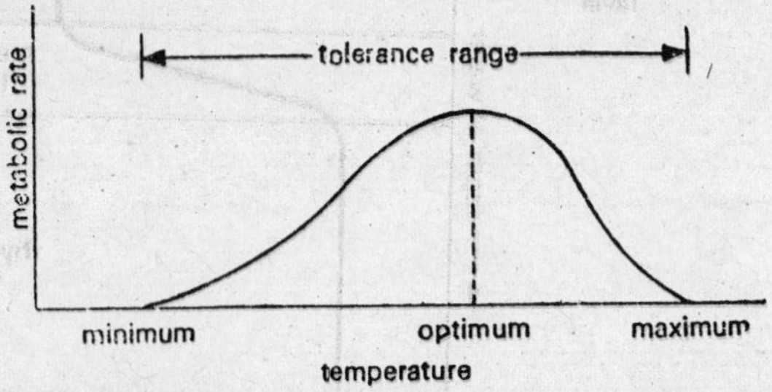

Fig. 11·12. Effect of temperature on the metabolic rate.

The organisms which can tolerate very large fluctuations in temperature are called **eurythermal organisms**. For example, cyclops, toad, wall-lizard, grass-snake, man, etc., are the eurythermal animals. The organisms which can tolerate only a small variation in temperature are termed **stenothermal organisms**. The common stenothermal organisms are fishes, snails, coral reefs, etc.

In fact, most animals can tolerate extreme temperatures only to a limited extent. Even with a very slow increase in temperature, there reaches an **absolute upper lethal limit** for the particular animal beyond which it cannot adjust to a further increase. Similarly, the **absolute lower lethal limit** is that point beyond which no further decrease in temperature is tolerated. For example, when all organisms including insects are kept at unfavourable low temperature for a long time, they are killed. The death at the sub-zero temperature is caused by the freezing of the body tissues. The freezing point of most insects of cold regions is between −10°C and −2°C when a **cold-hardy** insect is chilled below 0°C, its body do not freeze immediately but are supercooled beyond the freezing point. Thus, cold-hardiness is the quality of an insect of a cold region to resist freezing at sub-zero temperatures at which tropical and subtropical insects would readily freeze and die. The tropical grain pests lack this capacity (**undercooling**, the capacity of cooling beyond the freezing temperature) and if subjected to low temperatures beyond 0°C, die without freezing.

The best example of cold-hardiness is furnished by mosquito larvae, which in Alaska normally pass the winter frozen in the ice of water pools and are known to tolerate repeated freezings and thawings. In these larvae about 90 per cent water freezes to ice at **−15°C**. Many nematodes, rotifers, and tardigrades withstand cooling to −272°C without ill effects. Many lower organisms have been cooled in liquid O_2 to a temperature of $-183°C$ and in liquid helium to $-269°C$ and have survived. In cold-hardy insects supercooling has some advantages. For example, *Cephus* survives supercooling to $-23°C$ to -30 C as compared with only -5 to $-8°C$ by its less cold-hardy species. During the supercooling of *Bracon*, an insectan parasite on sawflies to $-47{\cdot}2°C$, an increase in proportion of "bound" to "free" water was observed. According to **Salt** (1950) glycerol found normally in the haemolymph of insects like *Bracon* and *Loxostege* plays an important role in increasing the supercooling. Glycerol lowers the freezing point of these cold-hardy insects and thus protect their tissues.

Vertebrates are limited in their tolerance of freezing. For example, arctic fishes do not survive total freezing but may withstand freezing of the body surfaces. However, deep sea fishes live in a permanent state of supercooling.

Upper limit of the range of lethal high temperatures varies from species to species and also within a species from season to season. The winter population would die at a much lower lethal higher temperature than the summer population. An insect with a cosmopolitan distribution will be more resistant to heat in those areas where the average temperatures remain higher. A high body temperature is an adaptation to a warm climate. Many tropical mammals are genetically adapted to heat. So, **heat-hardiness** is the quality of an animal of the hot region to resist heating beyond its normal temperature without serious effects like heat shock **(hyperpyrexia)** or heat death. Desert insects and mammals (*e.g.*, camel) are best examples of heat-hardy animals.

11·4·4. Classification of Organisms According to Temperature Tolerance and Regulation

Although both plants and animals experience and exist within the same external energy environment, there exist fundamental differences. Plants cannot move to escape adverse effects of heat or cold, and their metabolic heat is derived from absorbed solar rays. Plants has been divided into the following four categories on the basis of their heat tolerating capacity : (i) **Megatherms** : plants growing in regions where high temperatures prevail throughout the year, *e.g.*, desert vegetation and tropical rain forests. (ii) **Mesotherms** : plants of the regions where high temperature alternating with low temperature, *e.g.*, tropical deciduous forests and aquatic plants. (iii) **Microtherms** : plants of the regions where low temperature prevail throughout the year, *e.g.*, mixed coniferous forests. (iv) **Hekistotherms** : plants growing in regions with very low temperature, *e.g.*, alpine vegetation.

Except for the time they sprout from seeds or vegetative cuttings. plants receive all of their energy from radiation in the environment and from convection. To balance the input, plants lose heat by radiation, **convection and evapotranspiration.** They have some control over the latter by opening and closing the stomata and by changing the shape and position of the leaf.

Further, the ability to endure low temperature extremes varies among plants. They are not equally resistant at all stages of the life cycle. Flowers are more sensitive to low temperatures than are fruits and leaves, and young leaves are more resistant than old ones. Trees may be more severly injured than herbaceous plants. In general, adaptations among plants to endure low temperatures are primarily protoplasmic (**Smith**, 1977)

On the basis of temperature tolerance, fungi have also been classified into the following three kinds : **thermotolerant, thermophilic** and **mesophilic fungi** (**R. Emerson,** 1968). Thermophilic fungi require optimum temperature 45°C for growth and they occur in man-made self-heated habitats and also in natural habitats where geothermal heat, body heat of warm blooded animals, microbial metabolism and perhaps sun heating provide elevated temperature required for their growth. Thermophilic fungi has been reported in nesting material of American alligator (**Tansey**, 1973) and of birds. **T. Satyanarayana, B. N. Johri** and **S. B. Saksena** (1977) have reported the occurrence and seasonal variation of mycoflora of nesting material of the following Indian birds : crow (*Corvus splendens*), house sparrow (*Passer domesticus*), pipit (*Anthus rufulus*), bee-eater (*Merops superviliesus*) and crow-pheasant (*Centropus sinensis*). In the nest of these birds, they reported few thermotolerant fungi (that have optimum growth temperature 45°C and are constantly found throughout the year), *e.g., Aspergillus fumigatus*, and *A. flavus* ; many thermophilic fungi such as *Chaetomium thermophile*, *Humicola grisea*, *Mucor pusillus*, *Rhizopus microsporus*, *R. rhizopodiformis*, *Sporotrichum thermophile*, *Thermoascus aurantiacus*, *Thermomyces lanuginosus* and *Thielavia minor* ; and many mesophilic fungi (having optimum growth temperature 28°C) such as *Penicillium citrinum*, *Trichoderma viride*, four species of *Aspergillus* and one species each for *Mucor* and *Rhizopus*.

In contrast to plants and fungi, animal's response to energy environment is more complex. Not only do animals produce a considerable amount of heat by their own metabolism, but they are also able to move about to seek a favourable temperature regime and in other ways, both behaviourally and physiologically, to maintain some control over their body temperature.

As far as temperature control is concerned, animals can be divided into the following three groups :

11-4·4A. Poikilothermic or Ectothermic animals. Poikilothermic means 'having a variable temperature'. Poikilothermic animals are described as "**cold-blooded**", their body temperature changing with fluctuations in the environmental temperature. If the environment is cold, so is their blood.

Poikilothermic animals such as most invertebrates and most chordates excluding birds and mammals exhibit high rates of thermal conductance and low rates of heat production (usually less than 2°C). Consequently these animals exploit sources of heat energy other than metabolism (solar radiation, reradiation, etc.). Thus, poikilothermic animals are **ectothermic** which means that they cannot regulate their body temperature by physiological means but have gain heat from the environment, *i.e.*, from outside the body.

Within the range of temperatures that poikilotherms can tolerate, the rate of metabolism and thus, oxygen consumption rises with temperature. Lacking any homeostatic devices, poikilothermic animals of terrestrial environments in particular must depend on some behavioural control over body temperatures. Most aquatic poikilotherms generally encounter temperature fluctuations of lesser magnitude and usually have more poorly developed behavioural and physiological capabilites than do terrestrial forms.

11-4·4B. Homeothermic or Endothermic Animals. Homeothermic is a Greek word meaning having the same temperature. Homeothermic animals are popularly described as **'warm blooded'** ; their body temperature is independent of environmental temperature so that in cold conditions their blood is at a temperature higher than that of their surrounding.

The homeothermic animals are birds and mammals which possess a sufficiently high rate of oxidative metabolism and a sufficiently low rate of thermal conductance, that is, the body temperature is the product of the animal's own oxidative metabolism (thermogenesis). Due to this fact, homeotherms are also called **endothermic animals**, since they generate heat from within the body, and keep it there.

The oxygen consumption of homeotherms decreases linearly with increasing temperature to some critical point where it is independent of environmental temperature. In other words, the lower the temperature, the more metabolic work the animal does to maintain a uniform temperature. But within a certain range of temperatures, the amount of energy needed to maintain body temperatures is minimal. At all ambient temperatures, homeotherms maintain their metabolic rates and body temperatures by means of a closely regulated feedback system mediated by the hypothalmus. The metabolic costs of such a system, however, are high. Homeotherms use some 80 to 90 per cent of their oxidative energy to maintain thermal homeostasis.

Heterotherms. Some animals such as monotremes and **some** marsupials have a limited power of temperature regulation. They respond to temperature extremes by aestivating and hibernating.

Both poikilotherms and homeotherms have following methods of thermo-regulation :

11-4·4C. Thermo-regulation in poikilotherms. Certain poikilothermic animals avoid both heat and cold by undergoing into a state of dormancy during a period of environmental stress. Many species of insect as well as certain crustaceans, mites, and snails, enter **diapause**, a state of dormancy and arrested growth. Eggs, embryonic larvae, or pupal stages may be involved. In addition to

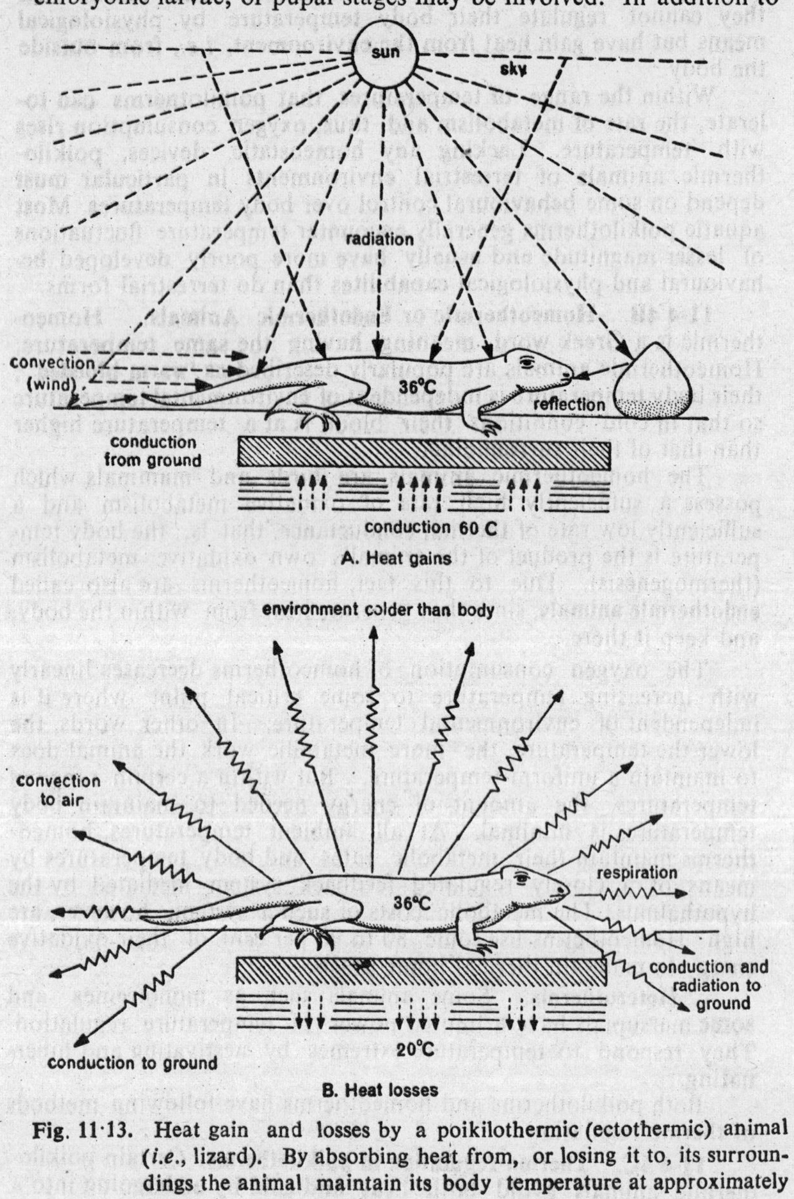

Fig. 11·13. Heat gain and losses by a poikilothermic (ectothermic) animal (*i.e.*, lizard). By absorbing heat from, or losing it to, its surroundings the animal maintain its body temperature at approximately 36°C (after Roberts, 1971).

tiding the animal over unfavourable periods, diapause also synchronizes the life cycle of a species with the weather and ensures that active stages will coincide with the climatic conditions and food supply that favour rapid development and high survival.

Some poikilothermal animals become dormant during periods of temperature extremes. Amphibians and turtles bury themselves in the mud of pond bottoms ; snakes and lizards seek burrows and dens in rocky hillsides. There they remain in a suspended animation until the temperature warms or cools again. Winter dormancy is called **hibernation** ; summer dormancy is **aestivation**. Most poikilotherms become inactive when the temperature of their surroundings goes below 8°C or rises to 42°C.

Like the poikilothermic animals, certain homeothermic animals, such as bats, ground squirrels, woodchucks and jumping mice also undergo hibernation and aestivation. During hibernation, they too have a state of cold torpor (*i.e.*, slow metabolic rate, low oxygen consumption, slow body movement and letharginess), like the poikilotherms, but there is a difference. When reptiles and amphibians are exposed to cold, the animal cools because it has no way to stay warm. As a poikilotherm's system becomes cooler, the heart rate and metabolic rate decline and animal become torpid. But when a mammal enters hibernation, first the heart rate and metabolic rate decline and then the body temperature drops (**Smith**, 1977).

A few exceptional poikilothrems, especially insects, certain amphibians and reptiles, exercise a degree of thermo-regulation by either physiological or behavioural mechanisms. For example, Hawk-moths can rise the temperature of their flight muscle to 32°—36°C by vibrating the wings before take-off and gregarious butterfly larvae may raise their temperature 1½—2°C when clustered together. Locust, and grasshoppers may increase their temperature 10°C by basking sideways in the sun. Ants move their larvae to warm or cool places within the nest and bees maintain temperatures within their hives between 13° and 25°C by fanning with their wings to evaporate water droplets when it is too hot, or releasing body heat through increased metabolic activity, when too cold.

When temperature drops, lizards bask in the sun to achieve the desired body temperature (Fig. 11·13) ; once this thermal level is attained, they will divide their time between sun and shade to maintain it. Desert lizards tend to be most active in the early morning and evening when it is neither too cold nor too hot. These are the times when one sees them scurrying about for food. In the heat of day they retire to a shady place ; and at night, when it can be extremely cold, they may burrow or seekout a crevice in which to build up a warm atmosphere from the heat generated by their metabolism.

Some poikilotherms such as frogs and reptiles lower their body

temperature slightly by evaporative cooling through the skin or via the respiratory tract by panting.

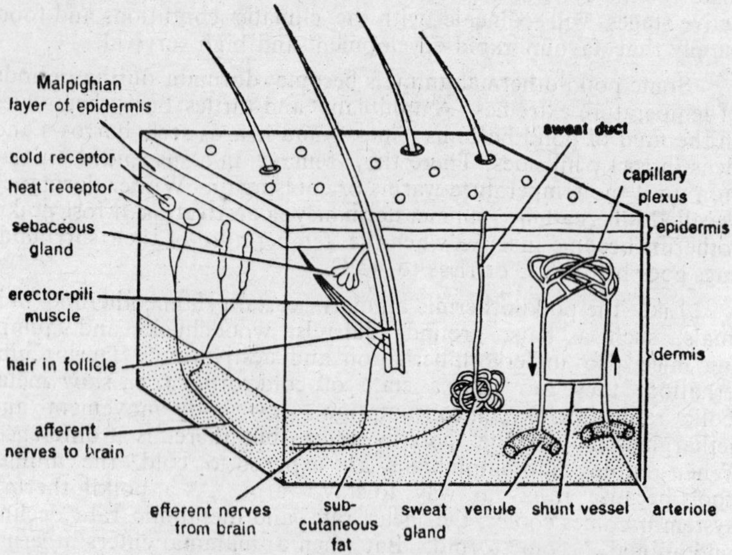

Fig. 11·14. V. S. of mammalian skin stereogram showing structures involved in temperature regulation (after Roberts, 1971).

11-4·4D. Thermo-regulation in homeotherms. Although many different systems co-operate in the process, the key structure involved in the temperature regulation of any endotherm is the **skin**. Literally all the structures included in the skin (Fig. 11·14) play some part in temperature **regulation**. A homeothermic or endothermic animal will have different responses to cold or heat as follows :

A. Response to cold by an homeothermic animal. When a mammal, a typical homeothermic animal is subjected to severe cold, following physiological processes and thermal adaptations may save it from excessive cooling and ultimate death :

1. **The subcutaneous fat** in the dermis of the skin serves as an **insulator** and reduces heat loss from the body. In fact, animals living in very cold habitats, the polar bear and seal for example, have a particularly thick layer of sub-cutaneous fats.

2. **The hair is raised** and brought up into a more-or-less vertical position by contraction of the **erector-pili muscles.** The advantage of this is that air gets trapped in the spaces between the hairs. This air is warmed by the body and being a poor conductor of heat it serves as an insulatory layer around the animal. In birds the same function is performed by the feathers. Mallard and black ducks (*Anas platyrhynchos* and *Anas rubripes*) with oil on their feathers show higher metabolic rates as the oil

concentrations are increased (**Hartung**, 1967). The oil, thus increases the insulating capability of the feathers.

3. **The superficial blood vessels in the skin constrict** so that blood is diverted from the surface to the deeper layers. This reduces loss of heat from the blood to the surrounding atmosphere. In exposed structures such as the ears there are specia **shunt vessels** interconnecting the arterioles and venules that take blood to and from the superficial capillaries of the skin. In cold conditions these shunt vessels dilate so that blood by-passes the surface of skin. This is aided by a general reduction in the total volume of circulating blood, achieved by some of the blood being taken up into reservoirs such as spleen. In prolonged conditions of extreme cold the blood may be diverted from the surface to such an extent and for so long that the cells die, resulting in frost-bite.

4. **Extra heat is produced** by an increase in the **metabolic rate**, particularly of the liver and muscles. This starts as a general increase in muscular tone but may be followed by rhythmical involuntary contractions of the skeletal muscles (**shivering**) and contraction of the smooth muscles in the skin (**goose flesh**). Heat production increases when an animal is active. For example, the heat produced while standing compared with lying to be 15 per cent greater for sheep and 17 per cent greater for cattle (**Ritzman** and **Benedict**, 1931).

Homeotherms also have the physiological capability of increasing their heat production without overt activity. **Summit metabolism** is one example of this ; newborn lambs can increase their heat production up to five times the basal rate without muscular activity. Postnatal tissue (*i.e.* **brown fat**) which is typically located between the shoulder blades and contains numerous mitochondria, produces large quantities of heat. This initial source of thermogenesis is followed by the lamb's increasing dependence on milk and forage for energy (**Moen**, 1973).

Out of these four techniques of endothermic animals to cope with environmental cold fall, first three are **physical mechanisms** designed to conserve heat by increasing the body's powers of insulation ; the fourth is a **chemical mechanism** in which body actively generates heat.

B. **Response to heat by an endothermic animal.** The response to excessive environmental heat involves the reverse of the above processes. Heat production is cut down and heat loss encouraged by following methods :

1. **Animals living in hot climates have comparatively little subcutaneous fat.** The fat deposits tend to be localized so as not to impede the loss of heat. Thus the camel's fat is stored in hump, and in the buffalo and bison it is located on top of the neck.

2. **The hair is lowered** by relaxation of the erector-pili muscles so that it lies flat against the surface of the body. With no

space between the hairs no air can be trapped against the skin. Insulation is reduced and heat can be more readily lost by radiation and convection.

However, when the environmental temperature becomes **higher than the body temperature, heat cannot be lost in this way. Under such circumstances the hair becomes important in insulating the body** against excessive heat uptake. **This is important in** large animals like the camel which finds it difficult to **escape the** heat of the sun.

3. The superficial blood vessels are dilated so that blood is brought up near the surface from which it can lose heat to the surrounding atmosphere. The shunt vessels are constricted and the total blood volume is raised, thereby further increasing the flow of blood to the surface.

4. Sweating or panting occur. Sweating by **sweat glands** of mammalian skin and evaporation of the sweat from the surface of the body cools the skin and the blood flowing through it. The evaporating power of the atmosphere is greatly enhanced by air movements.

In the dog and cat families there are no sweat glands (except in the pads of the paws) and heat is lost by **panting**. This greatly speeds up evaporation from the lungs with concomitant cooling of the blood. It also facilitates loss of heat from the blood as it flows through the pulmonary capillaries.

5. The metabolic rate falls in hot conditions so that less heat is generated by the body. That is why animals are generally less active in hot weather than in colder conditions.

Thus, when the temperature of the environment changes, the body takes the necessary steps to maintain its own temperature at a constant level. Figure **11·15** depicts the summary of the structures involved in the reflex control of body temperature in a mammal.

Thus in mammals the hypothalmic centre function as a **thermostat**. It is sensitive to temperature changes of the blood flowing through it and responds by sending nerve impulses through efferent nerves to the appropriate effectors. If the temperature of the blood is slightly higher than it should be, the thermo-regulatory centre of hypothalamus of brain detects this and sets into motion process that collectivelly encourage heat loss (Fig. 11·16). On the other hand, if the temperature falls below the optimum level, the centre initiates processes that produce and conserve heat (Fig. 11·16). The skin receptors detect temperature at the surface and enable the animal to feel whether the external environment is hot or cold. This information, acting via the thermo-regulatory centre, initiates voluntary activities such as taking muscular exercise in severe cold or moving into the shade if it is very hot.

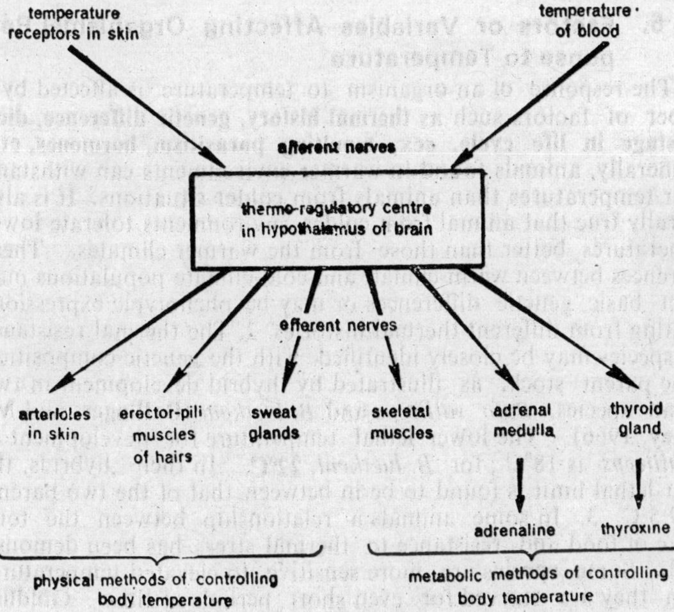

Fig. 11·15. **Summary** of structures and processes involved in the reflex control of body temperature in a mammal. Changes in the general **metabolic** rate are brought about by the hormones adrenaline and thyroxine which are secreted in extra large amounts in cold conditions (after Roberts, 1971).

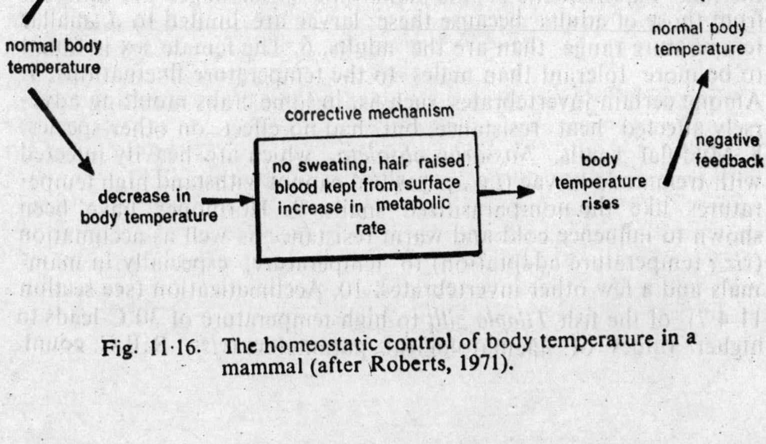

Fig. 11·16. The homeostatic control of body temperature in a mammal (after Roberts, 1971).

11-4·5. Factors or Variables Affecting Organismal Response to Temperature

The response of an organism to temperature is affected by a number of factors such as **thermal history, genetic difference, diet, size, stage in life cycle, sex, moulting, parasitism, hormones,** etc. 1. Generally, animals found in warmer environments can withstand higher temperatures than animals from colder situations. It is also generally true that animal from colder environments tolerate lower temperatures better than those from the warmer climates. These differences between warm-climate and cold-climate populations may reflect basic genetic differences or may be phenotypic expressions resulting from different thermal histories. 2. The thermal resistance of a species may be closely identified with the genetic composition of the parent stock, as illustrated by hybrid development in two anuran species, *Bufo valliceps* and *B. luetkeni* (**Ballinger** and **Mc Kinney**, 1966). The lower lethal temperature for development of *B. valliceps* is 18°C ; for *B. luetkeni*, **22°C**. In their hybrids, the lower lethal limit is found to be in between that of the two parents at 19·5°C. 3. In some animals a relationship between the total intake of food and resistance to thermal stress has been demonstrated. Some species are more sensitive to elevated temperatures when they are starved for even short periods of time. Goldfish showed increased resistance to high temperature when placed on a high fat diet (**Hoar** and **Cottle**, 1952). 4. The size of animal body is found to have some correlation with thermal lethal limits of animals. In some species, the smaller animals are more resistant to higher temperatures than the larger ones ; in some species the reverse is noted, and in other species size is not an apparent variable. Further, some smaller animals die faster at low temperatures than do larger animals of the same species, but at higher temperatures body size is not a factor. 5. Many species of animals have exceedingly complex life cycles during which the larval stages are not only morphologically dissimilar from the adult but also occupy different ecological niches. In the wharf crab, *Sesarma cinereum*, thermal requirements of the planktonic zoeal stages are different from those of adults because these larvae are limited to a smallar temperature range than are the adults. 6. The female sex is found to be more tolerant than males to the temperature fluctuations. 7. Among certain invertebrates, such as, in some crabs moulting adversely affected heat resistance, but had no effect on other species. 8. Mud-flat snails, *Nassarius obsoleta*, which are heavily infected with trematode larvae (*i.e.*, parasites) cannot withstand high temperatures like the non-parasitized snails. 9. Hormones have been shown to influence cold and warm resistance as well as acclimation (*viz.*, temperature adaptation) to temperature, especially in mammals and a few other invertebrates. 10. Acclimatization (see section 11-4·7) of the fish *Tilapia zilli* to high temperature of 30°C leads to higher values of haematological parameters (*viz.*, R.B.C. count,

W.B.C. count, Hb concentration and Haematocrit, etc.), while low temperature causes a depression of these values (**Farghaly** *et al.*, 1973). Similar changes in blood parameters due to temperature have been observed in many other fishes such as *Heteropneustes fossilis* (**Pandey**, 1977), *Salmo gairdneri* (**De wilde** and **Houston,** 1967), *Abramis brama* and *Lucioperca lucioperca* (**Molnar** *et al.*, 1960).

11-4·6. Effect of Temperature on Plants and Animals

Temperature has been found to affect the living organisms in various ways, for example it has significant role on the cells, morphology, physiology, behaviour, growth, ontogenetic development and distribution of plants and animals. Some of well-studied effects of temperature on living organisms are following :

1. **Temperature and cell.** The minimum and maximum temperatures have lethal effects on the cells and their components. If too cold, cell proteins may be destroyed as ice forms, or as water is lost and electrolytes become concentrated in the cells ; heat coagulates proteins (**Lewis** and **Taylor**, 1967).

2. **Temperature and metabolism**. Most of metabolic activities of microbes, plants and animals are regulated by varied kinds of enzymes and enzymes in turn are influenced by temperature, consequently increase in temperature, upto a certain limit, brings about increased enzymatic activity, resulting in an increased rate of metabolism. For instance, the activity of liver arginase enzyme upon arginine amino acid, is found to increase gradually and gradually, with the simultaneous increase in the temperature from 17°C to 48°C. But an increase in temperature beyond 48°C is found to have an adverse effect on the metabolic rate of this enzymatic activity which retards rapidly.

In plants, the absorption rate is retarded at low temperature. Photosynthesis operates over a wide range of temperature. Most algae require lower temperature range for photosynthesis than the higher plants. The rate of respiration in plants, however, increases with the rise of temperature, but beyond the optimum limit high temperature decreases the respiration rate. The rate of respiration become doubled (like in animals) at the increase of 10°C above the optimum temperature, provided other factors are favourable (**Vant Hoff's law**). However, optimum temperature for photosynthesis is lower than that for respiration. When temperature drops below the minimum for growth, a plant becomes dormant even though respiration and photosynthesis may continue slowly. Low temperatures further affect the plant by precipitating the protein in leaves and tender twigs and by dehydrating the tissues.

3. **Temperature and reproduction**. The maturation of gonads, gametogenesis and liberation of gametes takes place at a specific temperature which varies from species to species. For example, some species breed uniformly throughout the year, some only in

summer or in winter, while some species have two breeding periods, one in spring and other in fall. Thus, temperature determines the **breeding** seasons of most organisms.

Temperature also affects fecundity of animals. **Fecundity** of an animal is defined as its reproductive capacity, *i.e.*, the total number of young ones given birth during the life time of the animal. For example, females of the insect, acridid *Chrotogonus trachypterus* became sexually mature at 30°C and 35°C than at 25°C, and the highest number of eggs per female was laid at temperatures of 30°C. The number of eggs decreased from 243 to 190 when the temperature was raised to 30—35°C (**Grewal** and **Atwal**, 1968). Likewise, in grasshopper species—*Melanoplus sanguinipes* and *Camnula pellucida* when reared at 32°C produce 20—30 times as many eggs than those reared at 22°C (see **Ananthakrishan** and **Viswanathan**, 1976). On the other hand, the fecundity of certain inseets such as cotton stem weevil (*Pempherulus affinis*) was found to decline with an increase in temperature beyond **32·8°C** (**Ayyar** and **Margabandhu**, 1941).

4. Temperature and sex ratio. In certain animals the envirommental temperature determines the sex ratio of the species. For example, the sex ratio of the copepod *Macrocyclops albidu* is found to be temperature dependent. As the temperature rises there is a significant increase in number of males. Similarly in plague flea, *Xenopsylla cheopis*, males outnumbered females on rats, on days when the mean temperature remains in between 21—25°C. But the position becomes reverse on more cooler days.

5. Temperature and ontogenetic development. Temperature influences the speed and success of development of poikilothermic animals. In general complete development of eggs and larvae is more rapid in warm temperatures. Trout eggs, for example, develop four times faster at 15°C than at 5°C. The insect, chironomid fly *Metriocnemus hirticollis*, requires 26 days at 20°C for the development of a full generation, 94 days at 10°C, 153 days at 6·5°C, and 243 days at 20°C, (**Andrewartha** and **Birch**, 1954). However, the seeds of many plants will not germinate and the eggs and pupae of some insects will not hatch or develop normally until chilled. Brook trout grows best at 13°C to 16°C, but the eggs develop best at 8°C. In the common forest ground beetle *Pterostichus oblongopunctatus* development from egg to mature beetle takes 82 days at 15°C, whereas at 25°C it takes only 46 days. In pine lappet, *Dendrolimus pini* rate of development and mortality of various developmental stages are effected by temperature.

6. Temperature and growth. The growth rates of different animals and plants is also influenced by temperature. For example, the adult trouts do not feed much and do not grow until the water is warmer than 10°C . Likewise, in the oyster *Ostraea virginica*, the length of the body increase from 1·4 mm to 10·3 mm when temperature is increased from 10°C to 20°C. The gastropod *Urosa-*

lpinx cinerea and sea urchin *Echinus esculentus* show maximum size in warmer waters. Corals flourish well in those waters which contain water below 21°C.

7. **Temperature and colouration**. The size and colouration of animals are subject to influençe by temperature. In warm humid climates many animals like insects, birds and mammals bear darker pigmentation than the races of some species found in cool and dry climates. This phenomenon is known as **Gloger rule.** In the frog *Hyla* and the horned toad *Phrynosoma*, low temperatures have been known to induce darkening. Some prawn (crustacean invertebrates) turn light coloured with increasing temperature. The walking stick *Carausius* has been known to became black at 15°C and brown at 25°C.

8. **Temperature and morphology**. Temperature also affects the absolute size of an animal and the relative properties of various body parts (**Bergman's rule**). Birds and mammals, for example attain a greater body size when they are in cold regions than in warm regions, and colder regions harbour larger species. But poikilotherms tend to be smaller in colder regions. Body size has played a significant role in adaptation to low temperature because it has influenced the rate of heat loss. According to **Brown** and **Lee (1969)**, larger wood rats have a selective advantage in cold climates, apparently because their surface to air ratio and greater insulation permit them to conserve metabalic heat. For opposite reasons small-sized animals are favoured in deserts.

The extremities of mammals like tail, snout, ears, andlegs are relatively shorter in colder parts than in the warmer parts (**Allen's rule**). For example, there occurs difference in the size of ears of arctic fox (*Alopex lagopus*), red fox (*Vulpes vulpes*) and the desert fox (*Megalotis zerda*) (**Fig.** 11·17). Since heat is lost through the surface, the small ears of the arctic fox help to conserve the heat ; while, large ears of the desert fox help in heat loss and evaporation. Similarly, *Gazella picticanda* of Himalayas has shorter legs, ears and tail than *Gazella benetti* found in the plains of Himalayas, though both of them are of the same body size. Likewise, Eskimos have shorter arms and legs in proportion to their trunk size, which is comparalively larger than in any other contemporary group. The mice reared at 31°C to 33·5°C have longer tails than those of the same strain reared at 15·5°C to 20°C. All these **examples** of Allen's rule are clearly showing adaptive significance of short extremities in reducing heat loss from body in cold climate.

The races of birds with relatively narrow and more acuminate wings tend to occur in colder regions, while those in warmer climates tend to be broader (**Rensch's rule**). Temperature also influences the morphology of certain fishes and is **found** to have some relation with the number of vertebrae (**Jordon's rule**). Cod which hatches off New Foundland at a temperature between 4° and

8°C has 58 vertebrae, while that hatches East of Nantucket at a temperature between 10° and 11°C has 54 vertebrae.

Fig. 11·17. Heads of arctic fox (*Alopex lagopus*), red fox (*Vulpes vulpes*) and desert fox (*Megalotis zerda*) showing gradation in size of ears and illustrating Allen's rule (after Clark, 1954).

9. **Temperature and cyclomorphosis.** The relation between seasonal changes of temperature and body form is manifested in a remarkable phenomenon **termed cyclomorphosis** exhibited by certain cladocerans like *Daphnia* during the warm months of summer (Fig. 11·18). These crustaceans show a striking variation in the size **of their helmet or head projection between the winter and summer** month **(Coker, 1931). The helmet develops on the *Daphnia* head in spring, it attains its maximum size in summer and disappears al**together in winter to provide usual round shape to the head. Such a kind of cyclomorphosis in the terms of size of the helmet is clearly showing a correlation to the degree of warmth of different seasons. These prolongations of the helmet have been interpreted as an adaptation aiding flotation since the buoyancy of water becomes reduced as the temperature increases (the **buoyancy hypothesis**). According to other interpretation (*viz.*, **stability hypothesis**), the helmet acts like the rudder and gives greater stability to the animal. Besides temperature, such structural polymorphism can be caused by other environmental factors including the food.

10. **Temperature and animal behaviour**. Temperature generally influences the behavioural pattern of animals. In temperate waters the influence of temperature on the behaviour of wood borers is profound. For example, in the winter months in general, both *Martesia* and *Teredo* occur in smaller numbers in comparison with *Bankia* ***campanulata*** whose intensity of attack is maximum during the winter months. Further, the advantage gained by certain cold blooded animals through thermotaxis or orientation towards a source of heat are quite interesting. Ticks locate their warm blood hosts by a turning reaction to the heat of their bodies. Certain snakes such as rattle snake, copper heads, pit vipers are able to detect mammals and birds by their body heat which remains slightly warmer than the surroundings. Even in the dark these snakes strike on their prey with an unnerving accuracy, due to heat radiation

coming from the prey. The arrival of cold weather in temperate zones causes the snakes to coil up and huddle together.

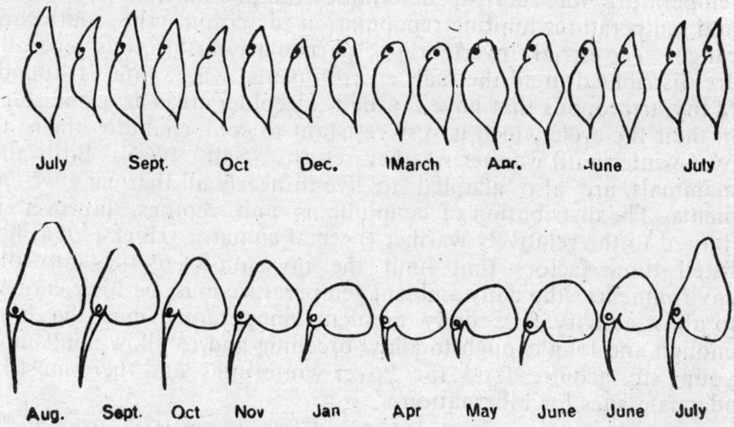

Fig. 11·18. Cyclomorphosis in *Daphnia cucullata* due to seasonal change in temperature (after Clarke, 1954).

11. Temperature and animal distribution. Because the optimum temperature for the completion of the several stages of the life cycle of many organisms varies, temperature imposes a restriction on the distribution of species. Generally the range of many species is limited by the lowest critical temperature in the most vulnerable stage of its life cycle, usually, the reproductive stage. Although the Atlantic lobster will live in water with a temperature range of 0° to 17°C, it will breed only in water warmer than 11°C. The lobster may live and grow in colder water but a breeding population never becomes established there. Not only temperature affects on breeding in the geographical distribution but also temperature affects on survivality (*i.e.*, lethal effect of temperature), feeding, and other biological activities are responsible in geographic distribution of animals. As noted earlier in this chapter, the animals from colder geographic regions are generally less heat tolerant and more cold tolerant than those animals from warmer regions ; for example, member of *Aurelia*, a jelly fish from Nova Scotia dies at a water temperature of 29–30°C, while *Aurelia* from Florida can tolerate temperatures upto 38·5°C. Thus, lethal limit of temperature may regulate the range of distribution of *Aurelia*.

Generally, the distribution of shallow-water marine species can be assigned to four types of zonation. In the first type, northward distribution is dependent on thermal lethal limits during the winter months, and southern distribution is dependent on summer temperature limits. In a second type, the thermal limits required for population determine the north to south distribution. In the third type of zonation, the thermal requirements for repopulation determine the poleward habitat in summer, and the maximum temperature determines the equatorward survival area. Finally, the minimum temperature for survival determines the poleward limit in winter and temperatures limiting repopulation determine the southward range. Terrestrial invertebrates, particularly arthropods generally are distributed in all thermal environments where life is found. Many arthropods that have invaded the colder areas have one stage in their life cycle which is very resistant to cold, enabling them to over-winter until warmer weather returns (**Salt**, 1964). Birds and mammals are also adapted to live in nearly all thermal environments. The distribution of amphibians and reptiles, however is limited to the relatively warmer thermal climates. **Hock** (1964) has listed three factors that limit the invasion of reptiles into cold environments : the daily ambient temperature must be high enough to allow activity, the daily ambient temperature must be high enough and long enough to allow breeding and to allow adults and young to acquire food for "over-wintering" and there must be adequate sites for hibernation.

12. Temperature and moisture. The differential heating of the atmosphere resulting from temperature variation over the earth's surface produces a number of ecological effects, including local and trade winds and hurricanes and other storms, but more importantly it determines the distribution of precipitation.

11-4·7. Thermal Adaptations of Plants and Animals

Most animals and plants of different ecological habitats have developed various sorts of thermal adaptations during the course of evolution to overcome the harmful effects of extremes of temperature. Some of the significant thermal adaptations of plants and animals are following :

1. Formation of heat resistant spores, cysts, seeds, etc. Some of the animals and plants produce heat resistant cysts, eggs, pupae, spores and seeds which can tolerate extremes of temperatures. *Amoeba* in encysted conditions, can tolerate temperature below 0°C. Similarly, *rye* seeds remain active even at 0°C and can germinate at that temperature. As an adaptation against frost the starch of plants changes to fats or oils in the autum. The fatty oils depress the freezing points and thus increase the power of resistance in plants against frost. Many leaves, that grow in the coldest lands, store fats. Pentosans mucilage and pectic substances which have high moisture retaining power are abundant in many plants. They decrease the danger of plants from desiccation during extremes of heat and save them from death.

2. Removal of water from tissue. Dried seeds, spores and cysts avoid freezing because there remains no liquid in them that can freeze. Due to removal of water from seeds, the cold resistance of seeds of certain plants increases upto the extent that their exposer for 3 weeks to 190°C, does not diminish their germinating capacity

3. Dormancy. Dormancy includes two already discussed phenomena namely **hibernation** and **aestivation**. During both kinds of dormancies metabolic rate becomes reduced, body temperature becomes low and heart beat rate is also reduced.

4. Thermal migrations. Thermal migrations occur **only in** animals. The journeys taken by animals that enable them to escape from extremely hot or cold situations are referred to as **thermal migrations**. For example, desert animals move to shaded places to avoid burning heat of noon and some animals like desert reptiles and snakes become nocturnal to avoid heat of the day. The frogs, toads, other amphibians, turtles, etc., make short trips into or out of water (or moist places) and this provides desired cooling and warming to the animal.

5. Acclimatization. Organisms can modify their response to an environmental stress even in their life time. Obviously such a change is not genetic, and it is called **acclimatization**. Nude **men,** exposed to moderately cold temperatures shiver violently. However, after several weeks of exposure, this response decreases and they become better able to withstand the cold. Within a few days after a move from low to high altitudes, people experience an increase in the number of red blood cells in their blood. Red cells carry oxygen, so these people can then tolerate the lower oxygen level in the air at high altitudes without signs of stress. Acclimatization is an important weapon in the arsenal of the organism trying to maintain homeostasis.

11-5. LIGHT

Light is one of the most vital and essential abiotic ecological factor without which no life can exist. On light depends the photosynthesis or synthesis of food by green plants which in turn support all animals and most microbial life on the earth. Light also influences the daily and seasonal activity patterns of plants and animals. It is a non-lethal limiting factor, both at the maximum and minimum levels.

The chief natural sources of light are **sunlight, moonlight,** starlight and the light from light-producing or luminescent organisms. However, only sunlight has greatest ecological significance.

Light can be defined as the visible part of the spectrum of solar radiant energy. In subsection 11-4·1A (1), it has already been discussed that visible light comprises radiations of wave-lengths ranging from 390 mμ to 700 mμ and constitutes about 48 per cent of the total energy received on the earth's surface. Visible light is made of a series of colours such as violet, indigo, blue, green, yellow, orange and red, constituting the **visible spectrum** (Fig. 11·5). Each colour of visible spectrum has different wave-length.

Further, light is highly directional. It differs in this respect from the temperature factor since heat often **reaches the living organism** from many different directions at the same time. Light is also extremely variable. Ecologically variations in the **quality** of light (wavelength or colour), the **intensity** of light (actual energy measured in gram-calories or foot candles) and the **duration** of light (length of day) are known to influence different ecosystems.

11·5·1. Light Variations in Different Environments

Light energy varies with different media. The transparency of air and water is important in **regulating** the amount and quantity of light that may be available in particular habitats. For example, the intensity of light reaching the earth's surface varies with the angle of incidence, degree of latitude and altitude, season, time of day, amount absorbed and dispersed by atmosphere and a number of climatic and topographical factors such as fog, clouds, suspended water drops, dust particles, etc., when the angle of incidence is smaller, light rays have to travel by a longer distance through the atmosphere, which resulting into relative reduction in intensity. Likewise, sun's altitude changes due to differences in latitude, changes in the season and in the time of day. When sun remains overhead, the intensity of sunlight over the earth's surface will be greatest. At higher latitudes the intensity of light becomes correspondingly reduced. The illumination or intensity of daylight is greatly diminished by moisture, clouds, and dust in the atmospere and also by forest vegetation. The direction and slope also affect light intensity. There will be no light on the one side of the slope.

The layers of vegetation bring about variation in light intensities reaching at various heights of a mountain. For example, in forests, major proportion of light intensity is absorbed by tree vegetation and the light reaching to the lower part of the ground vegetation is considerably reduced by 90—98 per cent of that in the exposed areas. The amount of light reaching the forest floor depends upon the height of canopy, crown development of trees, age of trees and phenological characteristics of the constituent species. Thus in a forest, the mature tallest trees receive full insolation, undershrubs receive subdued illumination, and herbs and especially epigenous cryptogams grow in lighter light conditions. When leaves of a tree are fully expanded, its canopy may reduce light to less than 1% of full solar radiation. However, in nature sun-flecks and movement of leaf canopy of trees play an important role in compensating for the reduced light and in making it available to the ground flora.

The light which enters in the aquatic media, comes from sun by passing through the atmosphere existing above the water surface and hence, that is subjected to all kinds of atmospheric factors like that of terrestrial media. About 10 per cent of the sunlight which falls over the water surface, is reflected back and rest 90 per cent of that pass downward in the water and is modified in respect to intensity, spectral composition, angular distribution (refraction) and time distribution. The phytoplankton, zooplankton, suspended organic and inorganic particles either reflect or absorb the light rays. Further, in water there is a selective absorption of light at various depths. The longer light rays are absorbed near the surface and in general the shortest light rays penetrate deepest. Thus, long heat and infra-red rays are absorbed in the upper layers of water (about 4 metres) ; red and orange rays are completely absorbed upto the depth of 20 metres ; yellow rays penetrate upto 50 metres and green and blue rays penetrate up to 80 to 100 metres deep. Violet and ultraviolet rays penetrate beyond 100 metres and no light ray penerates beyond 200 metres depth. Depending upon the **penetration** of light, oceans are divided into **euphotic zone** (upto 50 metres depth), **disphotic zone** (upto 80 to 200 metres depth) and **aphotic zone** (below 200 metres of depth). In the ocean, algae are distributed according to length of light rays that their colours are best suited to absorb and to utilize : green algae live in the intertidal zone, brown algae decend somewhat deeper and red algae are characteristic of deep oceanic water.

In general, light intensities are weaker at dawn, sunset and in water. At equator daylight prevails for about 12 hours out of every 24 hours, in both summer and winter. Towards the poles, in summer day length becomes longer than 12 hours.

11·5·2. Effects of Light on Plants

As an ecological factor light affects almost all aspects of plant life directly or indirectly. Thus, it controls plant's structure, form, shape, tissue differentiation, chlorophyll production, number and position of chloroplasts, leaf structure, stomata movement, transpiration, growth and development of flowers, seeds and fruits, locomotion, local distribution, etc. Some of well known effects of light on the plants are the following :

1. Light is an essential factor in the formation of **chlorophyll** pigment in chlorophyllous plants. It also has a very strong influence on the number and position of **chloroplasts**. The upper part of the leaf which receives full sunlight has larger number of chloroplasts that are arranged in line with the direction of light. In leaves of plants which grow under shade, chloroplasts are very few in number and are arranged at right angle to the light rays, thus increasing the surface of light absorption.

2. The effect of light intensity and quality on **photosynthesis** is well known. The photosynthetic rate of a plant leaf increases linearly with increasing light intensity up to the point of light saturation, usually one-tenth to two-tenth of full sunlight ; after this point the rate remains the same. The efficiency of the photosynthetic process, however, declines steadily with increasing light intensity. Thus a leaf exposed to full sunlight is not very efficient at utilizing light energy ; at best it utilizes about 5 per cent. At low light intensity the photosynthetic rate is lower, but efficiency

increases and may approach 20 per cent. However, the plant yield increases under high light intensity, since more light reaches the lower leaves and even the lower layers of chloroplast within the leaf (see **Smith**, 1977).

The light intensity at which plants can no longer carry on sufficient photosynthesis to maintain themselves is called the **compensation intensity**. This is the point at which photosynthesis balances respiration, where plants are just able to replace material lost in respiration night and day. Few green plants can live where the light intensity is less than 1 per cent full sunlight or about 100 foot-candles.

Further, the red light is absorbed by chlorophyll and only in certain marine algae other regions of light spectrum are utilized. For example, red algae possess supplementary pigments for the utilization of the energy of green light in photosynthesis and they occupy the zone of sea which contains green light of the spectrum. Certain plants can also utilize blue light in photosynthesis (see **Smith**, 1977). The photoperiod and periodic fluctuations in light intensities also affect the rate of photosynthesis.

The plants have been grouped according to their tolerance to light intensities into **shade loving**, **photophobic plants** or **sciophytes** and **sun plants**, **photophilous plants** or **heliophytes**. The shade plants maintain a high rate of photosynthesis in low light intensities, while the heliophytes are adversely affected by shade. There are some heliophytes, which grow best in sun, but can grow fairly well under shade, such plants are called **facultative sciophytes.** Similarly, there are **facultative heliophytes** which, although grow best at lower light intensities, but can grow well in full sunlight. Sciophytes and heliophytes exhibit the following morphological and physiological differences (Table 11·3).

3. At high intensity of light photo-oxidation of chlorophylls and other enzymes takes place, which consequently reduces the rate of synthesis of carbohydrates and proteins. However, high intensity of light results in the formation of **anthocyanin** pigments. **It** is for this reason alpine plants have beautiful coloured flowers.

4. Light affects the movement in plants. The effect of sunlight on the plant movement is called **heliotropism** or **phototropism.** The stems elongate towards light (**positive phototropism**) and roots are negatively phototropic. The leaves grow transversely to light.

The growth of shoot towards light has been found to be affected by blue light. Carotenoids and flavin are held responsible for mediating this light response. The blue light has also been found to affect the respiration, protein synthesis, cytoplasmic movement and opening of stomata (see **Gopal** and **Bhardwaj**, 1979). Red light is found to favourably affect several morphogenetic processes such as seed germination, seedling growth, leaf expansion and apical growth. As light inhibits the production of **auxins** or

growth hormones, therefore, it influences the shape and size of plants.

Table 11·3. Morphological and physiological characteristics of heliophytes (sun plants) and sciophytes (shade plants).

Characteristic	Heliophytes	Sciophytes
1. Stem	Thick, short internodes, profusely branched; well developed conducting tissues and mechanical tissues.	Thin, long internodes, sparsely branched ; poorly developed conducting and mechanical tissues.
2. Leaves	Small, thicker, leaf blades not flat, **less** compound, oriented obliquely to path of incident radiation; more hairy; thicker cuticle and cell walls, small-sized cells, few small sized chloroplasts well developed pallisade, poorly developed spongy parenchyma, less chorolophyll content.	Large, thinner, oriented at nearly right angles to incident radiation ; less hairy or glabrous; thin cuticle and cell walls, less developed pallisade, chloroplasts more and larger, well developed spongy parencyma, high chlorophyll content.
3. Roots	Longer, numerous, more branched, high root/shoot ratio. More nodules in legumes.	Small, few, less branched, low root/shoot ratio Less nodules.
4. Physiology	Low photosynthetic rate per unit surface, high respiration rate, low water content, higher concentration of salts and sugar and high osmotic pressure, decrease in acidity in cell sap, low potassium content, early appearance of flowers, high resistance to temperature injury, drought and parasites.	

5. Light affects opening and closing of stomata, influences the permeability of plasma membrane and has heating effect. All these in turn affect **transpiration** which in turn affects **absorption** of water. In many plants the respiratory rate increases with the increase in the light intensity (*e.g.*, *Canna*, *Nerium*, *Bougainvillea*). However, in certain plants respiration rate is decreased slightly in intense light.

6. The seeds of many plants are light sensitive, *i.e.*, either their germination requires light or is inhibited by light (Table 11·4). In most cases, the red light promotes germination and far-red light inhibits germination. Investigations regarding the understanding of the mechanism of seed germination in light have found the involvement of a pigment **phytochrome**. The pigment occurs in two reversible forms (*i.e.*, P_r and P_{fr}) developed under red and far-red light and the germination depends upon the balance between two forms in favour of the P_{fr} which develops under red light :

$$\underset{\text{(inactive)}}{P_r} \underset{730\text{ nm}}{\overset{660\text{ nm}}{\rightleftharpoons}} \underset{\text{(active)}}{P_{fr}}$$

The seeds of certain plants are found to **germinate** in blue light. However, in *Typha* species yellow light has been found to promote germination of seeds and also counters the inhibitory effect of blue light (**Gopal** and **Sharma**, 1978).

Table 11·4. Some light sensitive seeds.

Light necessary for germination	Light inhibitory to germination	No effect of light
Nicotiana tabacum	*Ailanthus glandulosus*	*Anemone*
Alisma plantago	*Aloe variegata*	*mamarosa*
Capparis spinosa	*Ephedra helvetica*	*Cystisus nigricans*
Colchium autumnale	*Lycopersicum*	*Datura*
Lepidium virginicum	*esculentum*	*stramonium*
Daucus carota	*Primula spectabilis*	*Hyacinthus*
Ocimum americanum	*Tulipa gessenreiana*	*candidans*
Anagallis arvensis	*Nigella damacaena*	*Linaria cymbalaria*
Taraxacum officinale	*Cistus radiatus*	*Pelargonium*
Sueda maritima	*Hedera helix*	*zonale*
Nasturtium officinale	*Mirabilis jalapa*	*Sorghum halapense*
Lactuca sativa	*Ranunculus crenatus*	*Tragopogon*
Salvia pratense		*paratensis*
Phacelia sp.		*Vasiccaria viscosa*

7. The development of **flowers, fruits** and seeds is greatly affected by light intensity. Diffused light or reduced light promotes the development of vegetative structures and causes delicacy. For example, vegetative crops like turnips, carrots, potatoes and beets give highest yield in regions with high percentage of cloudy days. Intense light favours the development of flowers, fruits and seeds.

8. The light not only affects the autotrophic plants but also other plants. In many bacteria and fungi such as *Psaliota campestris* and *Pilobolus microsporus* the spore development and germination are regulated by light factor.

9. **Photoperiodism in plants.** Actual duration or length of the day (**photoperiod**) is a significant factor in the growth and flowering of a wide variety of plants. The controlling effect of photo-

period on organisms is called **photoperiodicity**. Based on photoperiodic responses, plants can be classified **short-day plants, long-day plants** and **day-neutral plants** (Table 11·15).

Table 11·5. Classification of plants according to their photoperiodic requirements.

1. **Day-neutral plants.**
   Cucumis sativus, Gossypium hirsutum, Impatiens balsamina, Nicotiana tabacum. Phaseolus lunatus, Poa annua, Solanum tuberosum.
2. **Quantitative short-day plants.**
   Salvia splendens, Datura stramonium, Cannabis sativa, Andropogon virginicus, Cosmos bipinnatus.
3. **Quantitative long-day plants.**
   Brassica rapa, Nigella avroensis, Secale cereale, Sonchus oleraceum, Sorghum valgare.
4. **Qualitative short-day plants.**
   Bryophyllum pinnatus, Chenopodium rubrum, Ipomoea batatas, Lemna purpusilia, Coffea arabica, Calanchoe blosfeldiana.
5. **Qualitative long-day plants.**
   Agropyron smithii, Anagallis arvensis, Lolium temulentum, Avena sativa, Agrostis palustris.
6. **Intermediates.**
   Chenopodium album, Tephrosia candida, Saccharum officinarum.

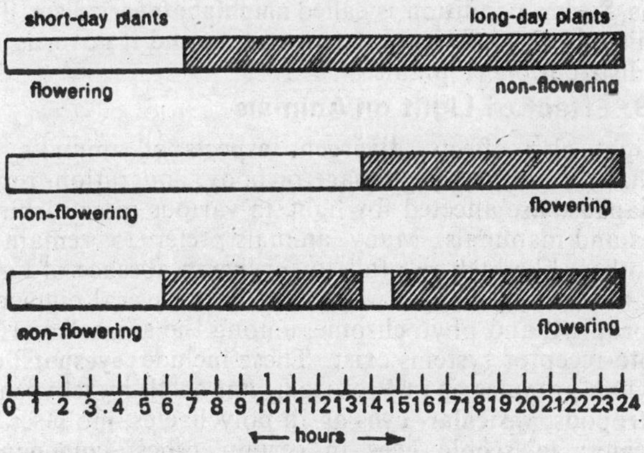

Fig. 11·19. The time of flowering in long-day and short-day plants is influenced by photoperiod. When plants are held under short-day and long-night conditions, the short-day plants are stimulated to flower and the long-day plants are inhibited. When day length is increased, flowering is inhibited in the short-day plants and stimulated in the long-day plants. If the dark period of the short-day and the long-day plants is interrupted, each reacts as if it had been exposed to a long day : the long-day plant flowers; the short-day plant does not (after Smith, 1977).

Day-neutral plants are those whose flowering is not affected by day length but, rather, is controlled by age, number of nodes, previous cold treatment and the like. Short-day and long-day plants are influenced by day length (Fig. 11·19). When the period of light reaches a certain portion of the 24-hour day, it inhibits or promotes a photoperiodic response. The length of this period, so decisive to the response, is called the **critical day length**. It varies among organisms but usually falls somewhere between 10 and 14 hours. Throughout the years plants "compare" this time scale with the actual length of day or night. As soon as the actual length of day or night is greater or smaller than the critical day length, the plant may flower or cease to flower, expand its leaves, or lengthen its shoots. **Short-day plants** are those whose flowering is stimulated by day lengths shorter than the critical day length. **Long-day plants** are those whose flowering is stimulated by day lengths longer than a particular value. The latter usually bloom in late spring and summer. Long-day plants and short-day plants are further classified into subgroups on the basis of quantity and quality of light (Table 11·5).

Further, **intermediate plants** are those which flower at neither too short nor too long photoperiods. However, in a plant, *Media elegans*, flowering occurs only at too small or too large photoperiods and the plant remains vegetative at intermediate day-lengths. Such a condition is called **amphiphotoperiodism.** The photoperiodism is also effected by temperature and it governs the latitudinal distribution of plants.

11·5·3. Effect of Light on Animals

Light also affects divergent aspects of animal's life. The growth, colouration of plumage or body, migration, reproduction and diapause are affected by light in various insects, birds, fishes, reptiles and mammals. Many animals prefer to remain in dark, while others like hydroids fail to survive in absence of light. While the plants respond to light with the help of several pigment systems as chlorophyll and phytochrome, among the animals various kinds of photo-receptor systems exist. These include '**eyespots**' consisting of amylum granules as in Protozoa; **flat ocelli** in jellyfish; **pit eyes** in gastropods; **vesicular eyes** as in polychaetes, molluscs and some vertebrates; **telescopic eyes** in certain fishes; **compound eyes** in Crustacea and insects; **simple eyes** or ocelli in other arthropods and dermal light receptors in other animals. Light has also been found to influence the development of these visual organs (**Tobias** 1976). For example, many inhabitant animals of caves or deep sea generally have either vestigial eyes or no eyes due to the absence of light in these environments. *Bathymicrops regis*, the deep sea fish (5000 metre sea depth) have no eyes. Some of the other important effects of light on animals are following :

1. **Effect of light on protoplasm**. Though the bodies of most

animals remain **protected** by some sort of body covering which save animal tissues from the lethal effects of solar radiations. But, sometimes sun rays penetrate such covers and cause excitation, activation, ionization and heating of protoplasm of different body cells. Ultraviolet rays are known to cause mutational changes in the DNA of various organisms.

2. **Effect of light on metabolism**. The metabolic rate of different animals is greatly influenced by light. The increased intensity of light results in an increase in enzyme activity, general metabolic rate and solubility of salts and minerals in the protoplasm. Solubility of gases however, decreases at high light intensity. Cave dwelling animals are found to be sluggish in their habits and to contain slow rate of metabolism.

3. **Effect of light on pigmentation.** Light influences pigmentation in animals. Cave animals lack skin pigments. If they are kept out of darkness for a long time, they regain skin pigmentation. The darkly pigmented skin of human inhabitants of the tropics also indicate the effect of sunlight on skin pigmentation. The skin pigment's synthesis is dependent on the sunlight. Light also determines the characteristic patterns of pigments of different animals which serve the animals in sexual dimorphism and protective colouration. Animals that dwell in the depths of the ocean where the environment is monotone, though pigmented do not show patterns in their colouration.

4. **Effect of light on animal movements.** The influence of light on the movement of animals is evident in lower animals. Oriented locomotory movements towards and away from a source of light is called **phototaxis**. Positively phototactic animals such as *Euglena*, *Ranatra*, etc., move towards the source of light, while negatively phototactic animals such as planarians, earthworms, slugs, copepodes, siphonophores, etc., move away from the source of light.

The light directed growth mechanisms are called **phototropisms** which occur in sessile animals. Phototropisms also include responsive movement of some body part of some active animal to the light stimulus, such as the movement of flagellum of *Euglena* towards light and movements of polyps of many coelenterates.

The velocity or speed of the movement of certain animals is also regulated by light. It has been observed that animals when responding to light reduce their velocity of movement and these movements which are non-directional are ealled **photokinesis**. Photokinesis may be a change in linear velocity (**rheokinesis**) or in the direction of turning (**klinokinesis**). During photokinesis when only a part of the body of an animal deviates always from the source of light, the reaction is termed **photoklinokinesis**. Larvae of *Musca domestica* show such movements. When animals are confronted with two lights of equal brightness they move towards or away to a position that is distance between the two lights. This is termed **phototropotaxis**. Attraction of males towards the flesh of the female is called **telotaxis**. Movement of animals at a constant angle towards the source of light is called **light compass reaction** or **celestial orientation**.

Celestial orientation. Some organisms, particularly arthropods, birds and fish, utilize their time sense as an aid to find their way from one area to another. To orient themselves, the animals use the sun, moon, or stars as a compass. To do this, they utilize both their biological clock and observations on the azimuthal position of the sun in relation to an established direction. The **azimuth** is the angle between a fixed line on the earth's surface and a projection of the sun's direction on the surface. Using the sun as a reference point involves some problems for animals because the sun moves. The target angle changes throughout the day. But animals which use the sun as a reference, correct their orientation somehow. Such celestial orientation has been observed in fishes, turtles, lizards, most birds, and such invertebrates as ants, bees, wolf spiders and sand hoppers.

5. **Photoperiodism and biological clocks.** Regularly occurring daily cycles of light (day) and darkness (night) have been known to exert a profound influence on the behaviour and metabolism of many organisms. Underlying such environmental rhythms of light and darkness are the movements of the earth relative to the sun and the moon. The earth's rotation on its axis results in alternation of night and day. The tilt of the earth's axis, along with the annual revolution around the sun produces the seasons. The response of different organisms to environmental rhythms of light and darkness is termed **photoperiodism**. Each daily cycle inclusive of a **period** of illumination followed by a period of darkness is called the **photoperiod**. The term **photophase** and **scatophase** are sometimes used to denote the period of light and the period of darkness respectively. Different animals have evolved different morphological, physiological, behavioural and ecological adaptations during the course of their evolution to varying photoperiods, which provide them with environmental information regarding the intensities of natural light.

(a) Daily responses : Circadian rhythms. Life evolved under the influence of daily and seasonal environmental changes, so it is natural that plants and animals would have some rhythms or pattern to their lives that would synchronize them with fluctuations in the environment. For years biologists have been intrigued over the means by which organisms kept their activities in rhythm with the 24 hours day, including such phenomena as the daily pattern of leaf and petal movement in plants, the sleep and wakefulness of animals and the emergence of insects from pupal cases (Fig. 11·20). At one time biologists thought that these rhythmicities were entirely **exogenous**, that is, the organisms responded only to external stimuli such as light intensity, humidity, temperature and tides. But now it is well investigated that most animals possess **internal**

or **endogenous** rhythms in synchrony with the external or exogenous rhythms of the environment, due to which they remain able to measure the length of day. The internal or endogenous rhythms are approximately of 24 hours duration, while the exogenous or environmental rhythms are exactly of 24 hours duration. The term **circadian** (from the Latin *circa*, about, and *dies*, daily) has been used to denote these daily rhythms. The period of circadian rhythm, the number of hours from the beginning of activity one day to the beginning of activity on the next, is called **free running.** Photoperiod plays a role in providing time signals, for adjustments of the animals concerned to these daily rhythms. Circadian rhythms apparently are internally driven or endogenous, are affected little by temperature changes, are insensitive to a great variety of chemical inhibitors, and are innate, not learned from or imprinted upon the organisms by the environment.

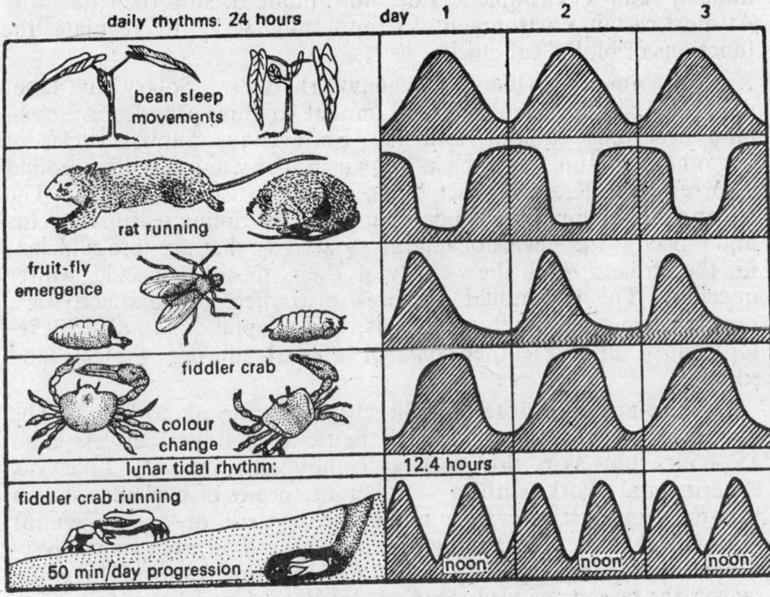

Fig. 11·20. Certain well-known examples of circadian rhythms (after Smith, 1977).

The innate character of circadian rhythm is demonstrated by several animals. When *Drosophila* are kept under constant conditions from the larval stage on, they will still emerge from pupae with a regular circadian rhythm. Eggs of chicken and lizards kept under constant conditions produce animals that later show regular circadian cycles. The circadian rhythms have been observed in zooplanktons, polychaete annelids, many insects (Lepidoptera, Diptera, Hymenoptera, Neuroptera, Coleoptera, Orthoptera, Odonata, etc.), most birds, and certain mammals.

Planktons of sea and lakes provides very interesting instance of circadian rhythms by showing diurnal changes in their vertical distribution. For example, numerous copepods and zooplanktons tend to swim toward the surface at night and to move downward to deeper layers during the day (see **Clarke**, 1954). Reverse is true with the phytoplanktons. Phytoplanktons of Dal Lake, Shrinagar exhibit diurnal movement in reverse order : they are abundant in the surface layer during day time and at the depth of 2·5 metre at mid-night (**Kant** and **Kachroo** 1975).

The possession of a circadian rhythm that can be entertained to environmental rhythms provides plants and animals with a **biological clock**, which is an integral part of cellular structure and is a chemosensory system very receptive to the environmental stimuli. The biological clocks of different animals run or oscillate continuously and environment does not initiate or stop their function. At most certain environmental stimuli may serve to regulate the functions of biological clocks.

(b) Annual rhythms : Circannual rhythms. Solary day, lunar day, tidal rhythms, monthly and annual rhythms are also of common occurrence among animals. Endogenous annuals cycles or circannual rhythms have been known in many animals like **ground** squirrels, warblers and other birds, some crayfishes and slugs. The circannual rhythms are of adaptive value for timing seasonal events and specify the levels of migratory activity that are just sufficient for the birds to reach the vicinity of their species—specific winter quarters. The circannual rhythms also affect gonadial activities, reproductive cycles, metamorphosis, and adaptations to cold (development of fur and feather coats of animals during winter), and so on.

The **diapause** in insects is directly related to photoperiod. The pupae of *Apatele rumicis* enter diapause at photoperiods less than 15 hours but skip this pause at 16 hour photoperiod. Likewise, experimental work with a number of species of birds has shown that the reproductive cycle is under the control of an exogenous seasonal rhythm of changing day lengths and an endogenous physiological response timed by a circadian rhythm. After the breeding season the gonads of birds studied to date have been found to regress spontaneously. This is the **refractory period**, a time when **light** cannot induce gonadal activity, the duration of which is regulated by day length. Short days hasten the termination of the refractory period ; long days prolong it. After the refractory period is completed, the **progressive phase** begins in late fall and winter. During this period the birds fatten, they migrate, and their reproductive organs increase in size. This process can be speeded up by exposing the bird to a long day photoperiod. Completion of the progressive period brings the birds into the **reproductive stage**. A similar photoperiodic response exists in the cyprinid fish, the minnows (see **Smith**, 1977).

Seasonal cycles of photoperiodism influence the breeding cycles of many mammals such as white-tailed deer (Fig. 11·21) and flying squirrel. For example, flying-squirrel has two peaks of litter production, the first in early spring, usually April, in the Northeastern United States, and the second in late summer, usually August.

6. **Effect of light on reproduction.** In many animals (*e.g.*, birds) light is necessary for the activation of gonads and in initiating annual breeding activities. The gonads of birds are found to become active with increased illumination during summer and to regress during shorter periods of illumination in winter.

7. **Effect of light on development**. Light in some cases (*e.g.*, *Salmon* larvae) accelerates development, whereas, in other (*e.g.*, *Mytilus* larvae) it retards it.

Further, occasionally the output of sunlight is increased by the development of **sunspots**. As a result of this excess energy is radiated to space and this naturally increases the output of **solar** energy near the earth. A direct consequence of this is the greater evaporation of water which results in cloud formation to prevent more exposure to sunshine and thus to equalize temperature and modifying climate.

Lunar periodicity. It can be defined as a biological rhythm in which the maxima and minima appear once or twice in every lunar month at the same time ; if the rhythm occurs once in 15 days (14·77 days) it is called as **semilunar** ; if it occurs once in 30 days, it is called **lunar**. The lunar cycle or periodicity controls many living activities. For example, marine algae, *Dictyota*, produces its gamets at the time of full-moon spring tide. The spawning of fish, *Leuresthes tenuis* follows a semilunal cycle. Certain polychaete worms also exhibit lunar periodicity.

11-6. PRESSURE

Barometric pressure have little effect on organanisms. An amoeba in motion in spite of stopping under slight pressure is insensitive to increased pressure on the whole body. But in an ocean, pressure is an unavoidable ecological factor. In general, an increase of hydrostatic pressure is inversely proportional to the pace of life. While as altitude increases atmospheric pressure decreases. Pressure influences solubility, ionic dissociation and surface tension and water is slightly compressible with increased pressures. Since incompressible bodies are influenced very little by increased pressure, they will sink in the sea, and apparently this factor is important in food settling to the deep-water benthic fauna.

11-7. GEOMAGNETISM AND GRAVITY

The earth may be described as a magnetized sphere in which magnetic axis is inclined about 11 degrses relative to the geographical axis. Geomagnetism and gravity are involved in the adaptation of an organism to its environment. Like light, temperature and pressure they are omnipresent ecological factors with which life steadily interacts. The living systems of planarians and snails have been found to be extraordinarily sensitive to magnetic fields of earth.

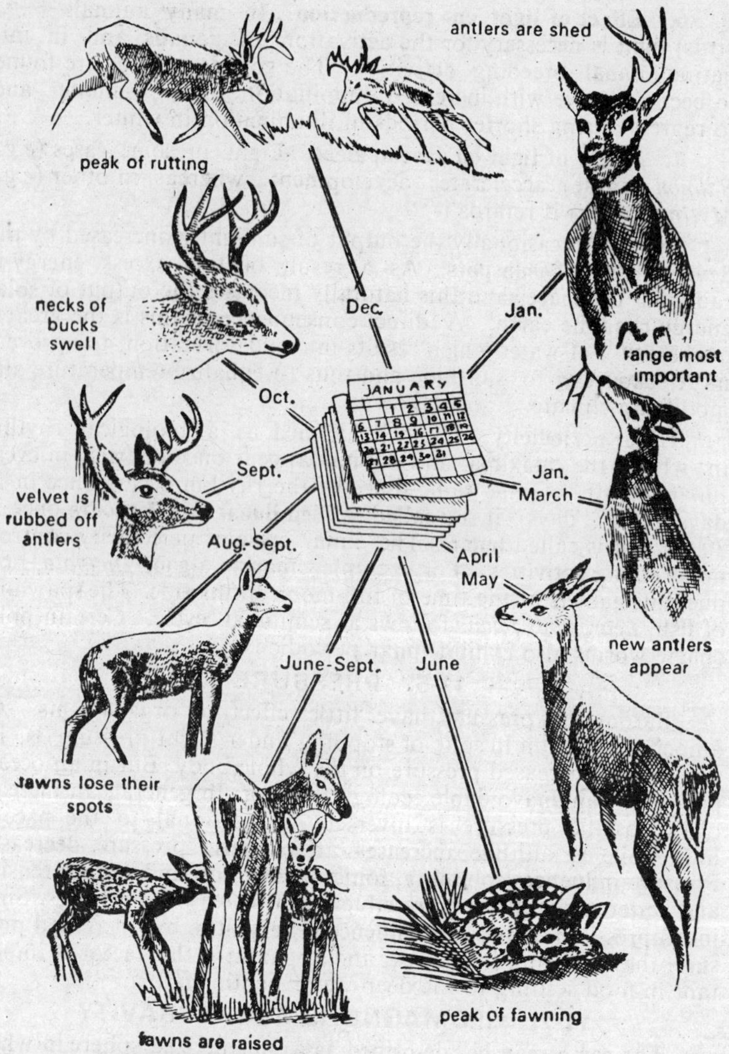

Fig. 11·21. The seasonal cycle of the white-tailed deer which begins with the breeding season in the fall. The annual breeding cycle of the white-tailed deer is attuned to decreasing day-length (after Smith, 1977).

11-8. TIDE

Tides affecting biota comprise a regular rhytmic rise and fall of oceanic water twice a day, as a result of which changes occur in the shore line. Tides are regulated by moon. At new moon and full moon the attraction of the sun and moon on tides reinforce each other and cause tides to rise higher and fall down without changing the mean sea level. This is called **spring tide**. During the first and last quarters of the moon the forces oppose each other and the daily fluctuations are much smaller due to neutral effects of the forces and this is the **neap tide**. Covering and uncovering of the tidal zone directly affect the biota by covering them with water and exposing them to air and sunlight, respectively.

11-9. FIRE

Fire is an interesting ecological force. It may be caused naturally by lightning or carelessness use of it by man. The forest fire is naturally controlled either by precipitation which often follows lightning or due to unavailability of accumulated litter. Fire causes immediate thermal effects because of high temperatures. Primarily it destroys an ecosystem, by destroying primary producers and by driving out consumers from the ecosystem. It also adversly affect the fertility of the soil by destroying organic binding agents in the soil, which results in erosin and leaching of soil. However, it releases nutrients as the organic material is reduced to ash (**Moen,** 1973).

11-10. CHEMICAL ABIOTIC FACTORS

Oxygen is primary source of life. It becomes a limiting factor in its complete absence, but, due to its abundance it very rarely becomes a limiting factor. High CO_2 concentrations become limiting to animals. pH value often becomes a limiting factor in different habitats. Non-availability of certain micronutrients such as calcium carbonate becomes a limiting factor for land snails (**calciphilic species**). *Sellana radiata*, the common limpet of Madras has been known to react to the change of sea water concentrations of various ions. For example, it responds to very small dilutions of the sodium chloride of 0·0001 per cent.

11-11. MICROCLIMATE

Untill now, we have discussed the influence of different ecological media and of different abiotic environmental factors on the living organisms. All these factors form the **macroclimate** which is measured by the meterologists. But the climate in which animals actually live is very much different from the gross environmental conditions. This immediate environment of animal is called **effective climate** or **microclimate** and relates to such of the habitats of animals as those living in the soil within crevices of barks of trees, inside sheaths of grasses, within burrows, in ant nests, beneath logs of wood or rock, within caves, on vegetation in sand dunes, within plant galls and a variety of habitats which are occupied by endoparasites. Woodlice, centipedes and millipedes, for example, remain restricted to dark microhabitats during the daytime to prevent rapid water loss or transpiration. When temperature falls and relative humidity increases at night, these animals emerge of the daytime microhabitats.

11·11·1 Causes of Fluctuations of Environmental Factors in Microclimates

Thus heat, moisture, air movement and light, all vary in microclimates of a given area. Such microclimatic variations result from differences in slop, soil and vegetation. On a summer afternoon the temperature under calm and clear sky may be 82°F at six feet, the standard level of temperature recording. But on or near the ground—at the 2-inch level—the temperature may be 5° lower than the standard level. The main reason for the great differences between the ground and six-foot level is solar radiation. During the day the soil, the active surface, absorbs solar radiation, which comes in short waves as light and radiates back as long waves to heat a thin layer above. Since air flow at ground level is almost non-existent, the heat radiated from the surface remains close to the ground. Temperature decreases sharply in the air above this layer and in the soil below. Thus, on a sunny but chilly spring or late winter day, one can walk in muddy ground while the air about him is cold.

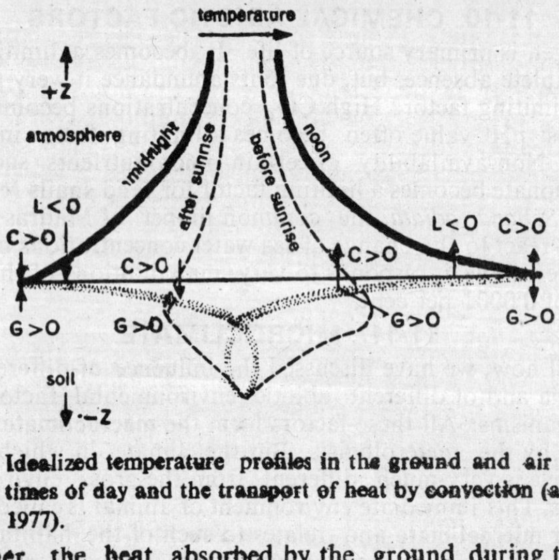

Fig. 11·22. Idealized temperature profiles in the ground and air for various times of day and the transport of heat by convection (after Smith, 1977).

Further, the heat absorbed by the ground during the day is reradiated by the ground at night. This heat is partly absorbed by the water vapour in the air above. The drier the air, the greater the outgoing heat and stronger the cooling of the surface of the ground and the vegetation. Consequently the ground and the vegetation are cooled to the dew point, and water vapour in the air may condense as **dew**. After a heavy dew a thin layer of chilled air lies over the surface, the result of rapid absorption of heat in the evaporation of dew.

Likewise, vegetation influences the microclimate of an area, especially near the ground by altering wind movement, evaporation, moisture and soil temperatures. Temperatures at the ground level under the shade are lower than in places exposed to sun and wind. On fair summer days a dense forest cover can reduce the daily range of temperatures at one inch by 20°F to **30°F**, compared with the temperatures in the soils of bare field. Vegetation also reduces the steepness of the temperature gradient and influences the height of the active surface, the area that intercepts the maximum quantity of solar insolation. In the absence of or in the presence of very thin vegetation, temperature increases sharply near the soil; but as the plant cover increases in height and density, the leaves of the plants intercept more solar radiation (Fig. 11·23). The plant crown then become the active surface and raise it above the ground. Consequently daytime temperatures are higher just above the dense crown surface and lowest at the surface of the ground.

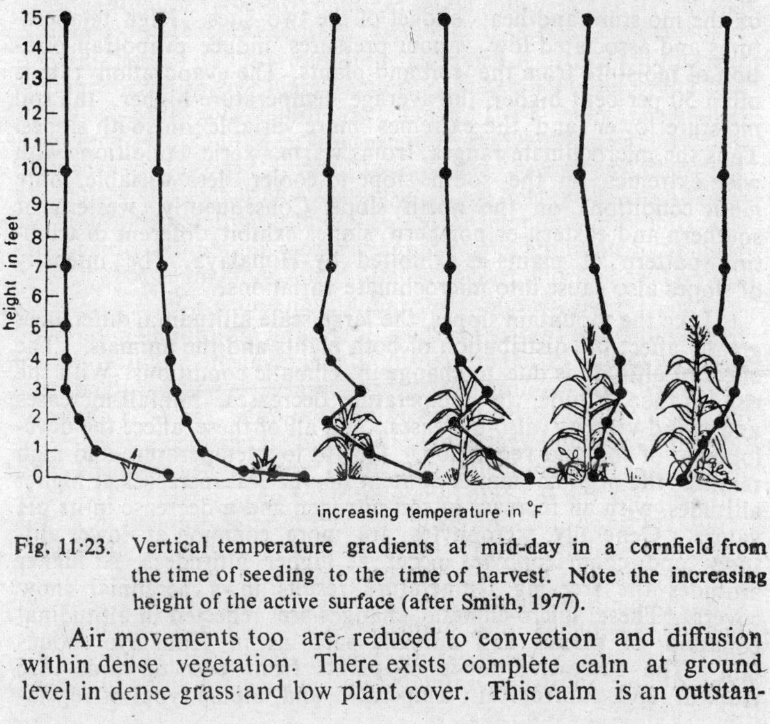

Fig. 11·23. Vertical temperature gradients at mid-day in a cornfield from the time of seeding to the time of harvest. Note the increasing height of the active surface (after Smith, 1977).

Air movements too are reduced to convection and diffusion within dense vegetation. There exists complete calm at ground level in dense grass and low plant cover. This calm is an outstanding feature of the microclimate near the ground, since it influences both temperature and humidity and creates a favourable environment for insects and other animals.

Even humidity differs greatly from the ground top. Since evaporation takes place at the surface of the soil or at the active surface of plant cover, the vapour content (absolute humidity) decreases rapidly from a maximum at the bottom to atmospheric equilibrium above. Relative humidity increases above the surface, since actual vapour content increases only slowly during the day, while the capacity of the heated air over the surface to hold moisture increases rather rapidly.

11-11·2. Topography and Microclimate

Topography and microclimate bear intimate inter-relationship. There exist great microclimatic differences between eastern and western or northern and southern slopes of the mountains or hills. For example, south-facing slopes receive the most solar energy, which is maximal when the slope grade equals the sun's angle from the zenith point. North-facing slopes receive the least energy, especially when the slope grade equals or exceeds the angle of sun ray inflection. At latitude North 41° mid-day insolation on a 20° slope is, on the average, 40 per cent greater on south slopes than on north slopes during all seasons. This has a marked effect on the moisture and heat budget of the two sites. High temperatures and associated low vapour pressures induce evapotranspiration of moisture from the soil and plants. The evaporation rate is often 50 per cent higher, the average temperature higher, the soil moisture lower and the extremes more variable on south slopes. Thus the microclimate ranges from warm, xeric conditions with wide extremes on the south slope to cooler, less variable, more mesic conditions on the north slope. Consequently, western or southern and eastern or northern slopes exhibit different distribution pattern of plants as exhibited by Himalaya. The intensity of slopes also cause into microclimate variations.

Like the mountain slopes, the large scale altitudinal differences greatly affect the distribution of both plants and the animals. The effect of altitude is due to change in climatic conditions. With the rise in the altitude, the temperature decreases, rainfall increases and wind velocity also increases, and all of these affect the development of soil and vegetation. Due to low temperature and high rainfall the organic matter content of the soil increases at higher altitudes, with an increase in soil nitrogen and a decrease in its pH values. Generally, xerophytes are more common at lower altitudes and chamaeophytes occur at higher altitudes. At higher altitudes the freezing temperature results in a perennial snow covers. These micro-climatic changes are reflected in altitudinal zonation of plants and animals such as in Himalaya mountain. The successive zones of plants from base upwards are tropical and subtropical, temperate, and alpine types of vegetation. The altitude at which the tree growth passes into shrub phase is called **timber line** or **tree line**. The tree line zone is characterized by following aspects which affect vegetation : lack of soil, desiccation of leaf in cold weather, short growing season, winter drying, excessive snow, high wind velocity, rapid heat loss at night, drought and high soil temperatures in the day.

Similar to vegetational zonation, zonation of animals is also observed with the altitude. Yaks and hares live at altitudes upto 5500 metres and insect dominates still higher altitude. Other animals of high altitude include the pelobatid frog, snow partridge, snow leopard, vole, Tibetan sheep, ibex, wildgoat and large variety of insects mostly coleopterans.

The environment of high altitude is also peculiar in having low air density, low oxygen content, high ozone content, greater insolation and high intensity of ultra-violet radiations. These environmental aspects make the life difficult. The ultra-violet solar radiations are largely responsible for the dwarfing of shrubs. The high humidity, lower temperature and the ultra-violet radiations result in dark pigmentation of high altitude insects (Mani, 1968). The high wind velocity at high altitudes is responsible for the dwarfing of trees and sometimes the formation of flag trees. As the winds blow in one direction, the growth of branches of trees on the side facing the wind is stopped and the development of branches only on one side gives the flag appearance of the trees.

11-11·3. Microclimate of Valleys

The widest climatic extremes occur in valleys and pockets, areas of convex slopes and low concave surfaces. All these places have much lower temperatures at night, especially in winter, and much higher temperatures during the day, especially in summer and a higher relative humidity. Protected from the circulating influences of the wind, the air of these places becomes stagnant. It is heated by insolation and cooled by terrestrial radiation, in sharp contrast to the wind-exposed, well-mixed air layers of the upper slopes. In the evening cool air from the uplands flows down the slope into the pockets and valleys to form a lake of cool air. Often when the warm air in the valley comes in contact with the inflowing cold air, the moisture in the warm air may condense as valley fog (Fig. 11·24). Such a distinct microclimate of valleys influence their plants and animals.

11-11·4. Microclimate of the City

The climate of rural areas markedly differs with the climate of the urban area or city (Table 11·6). The urban microclimate is a product of the morphology of the city and the density and activity of its occupants. In the urban complex stone, asphalt and concrete pavements and buildings with their high capacity for absorbing and reradiating heat replace natural vegetation with its low heat conductivity. Rainfall on impervious surfaces of city is drained away as fast as possible, reducing evaporation. Metabolic heat from masses of people and their domesticated animals and waste heat from buildings, industrial combustion and vehicles raise the temperature of the surrounding air. Industrial activities, building construction, power production and vehicles pour water vapour, gases and particulate matter into the atmosphere of the city in great quantities. The effect of this storage and reradiation of heat is the formation of a **heat island** above cities, large and small in which the temperature may be 6° to 8°C higher than the surrounding countryside (**Landsberg**, 1970).

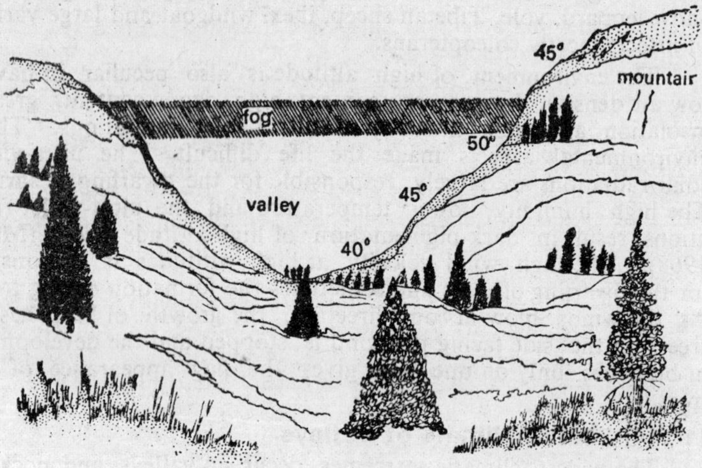

Fig. 11·24. Topography plays an important role in the formation and intensity of night-time inversions At night air cools next to the ground, forming a weak surface inversion in which temperature increases with height. As the cooling continues during the night, the layer of cool air gradually deepens. Simultaneously cool air descends down-slope. Both of these cause the inversion to become deeper and stronger. In mountain areas the top of the night inversion is usually below the main ridge. If air is cool and moist, fog may form in the valley. Smoke released in such inversions will rise only until its temperature equals that of the surrounding air. Then smoke flattens out and spreads horizontally just below the thermal belt (after Smith, 1974).

The heat islands are characterized by high temperature gradients about the city. The highest temperatures are associated with areas of highest density and activity, while temperatures decline markedly toward the periphery of the city. Such heat islands are common during the summer and early winter and are more noticeable at night than during the day, when heat stored by pavements and buildings is reradiated to air. The magnitude of the heat island is influenced strongly by local climatic conditions such as wind and cloud cover.

Table 11·6. Comparision of climates of urban and rural areas (Smith, 1977).

	Elements	Comparision with rural environment
1.	Condensation nuclei and particles	10 times more
2.	Gaseous admixtures	5–25 times more
3.	Cloud cover	5–10 per cent more
4.	Winter fog	100 per cent more
5.	Summer fog	30 per cent more
6,	Total precipitation	5–10 per cent more
7.	Relative humidity, winter	2 per cent less
8.	Relative humidity, summer	8 per cent less
9.	Radiation, global	15–20 per cent less
10.	Duration of sunshine	5–15 per cent less
11.	Annual mean temperature	0·5–1·0°C more
12.	Annual mean wind speed	20–30 per cent less
13.	Calms	5–20 per cent more

11·12. REVISION QUESTIONS

1. Describe the law of tolerance and law of minimum.
2. Discuss the role of temperature as an ecological factor.
3. Describe the methods of heat regulation in different organisms.
4. Discuss the role of light as an ecological factor.
5. What is microclimate ? Describe ecological effects of microclimates of mountain slopes, high altitude, valleys and city.
6. What is periodicity ? Describe different kinds of biological rhythms of plants and animals.
7. Write short notes on the following : threshold; heat budget; solar constant; thermal stratification; ectothermic and endothermic animals; biological clock; diapause; cyclomorphism.

12
Autecology of Species

12-1. COMPARISON OF AUTECOLOGY AND SYNECOLOGY

As discussed earlier in the introductory chapter, both plant and animal ecology are artificially divided into two parts, **synecology** and **autecology**. A brief comparison of these two ecological subdivisions will explain the basic significance of each of them. **Autecology** is concerned with the study of the interrelations of individual organisms with the environment. **Synecology** is concerned with the study of groups of organisms—the community, hence, it is also called **community ecology**. Both have developed independently, although the knowledge of both of them is necessary for the understanding of the individual, population, or ecosystem. Autecology is experimental and inductive. Synecology is philosophical and deductive. Autecology, because it is concerned with the relationship of an organism to one or more variables such as light, temperature, moisture, or salinity, is easily quantified and subject to experimental design both in the laboratory and in the field. Synecology is largely descriptive and not easily subject to experimental design. Autecology has borrowed techniques from physics and chemistry. Synecology, on the other hand, has entered a strong experimental phase in recent years, with the development of such tools as computers and radioactive tracers (**R. L. Smith**, 1977).

Autecology has great significance for the following economic biological disciplines : economic botany which includes agriculture, horticulture, silviculture and forestry ; and economic zoology that includes pisciculture, prawn fishery, pearl culture, animal husbandary, apiculture, sericulture, etc. It is also important for soil conservation and wild life **conservation**. Before one proceeds to cultivate any organism of economic or any other significance, he must be well aware to all the requirements of that species for its best possible growth in the environment. Further, autecology gives an idea of distribution, adaptation, speciation, etc., of a particular species. For instance, one finds in nature so many species living in various habitats. Some species are confined to limited areas due to limited characteristics of adaptation, whereas others may grow in different habitat conditions due to having a wide range of ecological adaptations. To find out the answers of some such questions, one has to study autecology of the species.

There exists an intimate interrelationship between the individual organism and its surrounding environment. For example, each stage in the life cycle of a plant is greatly influenced by a number of environmental factors and conversely, species does modify continuously the environment. In fact, the requirements of germination, growth, flowering, fruiting, leaf fall, etc., are met with at the same place, but at different times of the year. There **occurs** so much synchronisation in the phenological behaviour of the species and the different factors of the environment that the plants are often termed as **biological clocks** or **ecological clocks** (see chapter 11). The biological clocks are said to be regulated by external signals from the environment (external timing hypothesis), and such clocks are ecologically much advantageous since these couple environmental and physiological rhythms and enable the organism to anticipate daily, seasonal and other periodicities in light, temperature and, tides, etc.

In fact, autecology is based on the study of the **ecological life history** which is concerned with the activities of a species throughout its life cycle, and in relation to its adjustments to natural conditions (S.C. **Kendeigh**, 1974). The ecological life history of a given plant or animal species is studied by the following two slightly different methods :

12·2. OUTLINE FOR STUDY OF ECOLOGICAL LIFE HISTORY OF THE PLANT

According to **Stevens** and **Rock** (1952), the life history of a plant species should be studied under the following steps :

12·2·1. Introductory Conformation

(i) **Taxonomy**. Botanical and local names of the species ; chromosome number ; geographical distribution and history ; morphological variations, if any; **fossil evidences**, centre of origin and migration route.

(ii) **Field observations**. Location and general description of areas where plants grow under natural conditions, (*i.e.*, habitat). The climatic and general conditions of the habitats in which plants grow.

12·2·2. Ecological Relations

(i) **Natural distribution**. General distribution, altitude limits, effect of slope, lakes, low lying areas, etc.

(ii) **Soil relations**. Type of soil, humus contents, water holding capacity, wilting coefficient, pH range and other edaphic factors.

(iii) **Climatic relations**. Light (intensity, duration and quality) and temperature, wind and soil water, etc., which influence the vegetative growth of plant.

(iv) Plant associations. Inter-as well as intraspecific competitions at different growth stages.

(v) Modification of the species. Correlation between plant variation and changing environmental conditions, development of ecotype, biotypes, etc.

(vi) Phenology. Time of seedling appearance, time and rate of vegetative growth, time of flowering, fruiting, seed maturity, and fruit dispersal, etc.

12-2·3. Regeneration or Developmental History

This depends mainly upon the average seed output, viability of seeds, seed dormancy, reproductive capacity, seed dispersal seedling growth, vegetative propagation, vegetative growth and reproductive growth.

(i) Seed output. Collection of seeds, date, habit and weather of seed collection, weight of seed ano average production condition of seeds; percentage of seed production and seed germination. Average seed output of a species is calculated as follows :

$$\text{Average seed output} = \frac{\text{Total number of seeds}}{\text{Number of plants from which seeds are collected}}$$

(ii) Seed dispersal. Fruits, bulbs, bulbils, spores, shoot apices, and seeds are generally carried away from parent plants by such natural agencies as animals, wind and water. Thus availability of these dispersal agents at a proper period of life cycle is a very significant factor for successful dispersal of seeds.

(iii) Viability of seeds. The seeds usually have a longer period in their life before they lose their capacity to germinate. This period is called the **viability period**. Seeds are stored for long time in soil, water or mud to withstand adverse environmental conditions. The viability of seeds lying in the soil is generally influenced by depth, water content, temperature and microbial population of soil.

(iv) Seed dormancy Methods of break the seed dormancy

(v) Seeds germination and reproductive capacity. Normally all the seeds produced by a plant do not germinate due to various reasons. Reproductive capacity of any species indicates its pressure upon the environment. Species with high reproductive capacity is presumed to have better chances of its survival and dispersal. Reproductive capacity of a species is calculated as follows:

$$\text{Reproductive capacity} = \frac{\text{Average seed output} \times \text{percentage germination}}{100}$$

Light, temperature, water, and levels of oxygen and carbon dioxide concentrations are the chief environmental factors affecting the seed germination. From his extensive survey of the reproductive capacity of flowering plants, **Salisbury** (1946) concluded that seed size is determined by the length of time for which the seeding needs to be supported by nutrient reserves in the seed before it becomes photosynthetically self-supporting. **Garrett** (1973) extended Salisbury's conclusions to fungi with reference to spores of some fungi causing leaf spots, macroconidia and chlamydospores of root-infecting fungi (*Fusarium* Sp.), mycelial strands and rhizomorphs of fungi infecting tree roots and sclerotia of pathogenic root-infecting fungi.

(vi) **Seedling growth**. Seedling represents the juvenile stage of plants. Seedlings of trees in forests, annuals, shrubs, climbers, etc., differ in their requirements of seedling establishment, especially in light conditions, water relations, soil characteristics and other environmental parameters. Extremes of environmental factors such as light, temperature, humidity, pathogens and birds and grazing animals adversely affect the establishment of seedlings.

(v) **Vegetative growth**. The vegetative growth is influenced by various environmental, mostly edaphic and aerial factors, such as the intensity, duration and quality of temperature, light, water, pH, etc. In grasses and some weeds, the vegetative growth, as the length of shoot, root depth, number of nodes, internode length, number and size of leaves, stomatal frequency, thickness of cuticle on leaf, etc., are affected by the environmental conditions. In other plants, vegetative growth include study of root system, root-shoot ratio, at different growth stages and arrangement, type, shape, variation, surface leaf area, chlorophyll, etc., in relation to environment in various growth stages.

(vi) **Reproductive growth**. It includes the flowering, pollination and fruiting of a species. Most of the terrestrial plants, for their successful growth, reproduce sexually, *i.e.*, flower and fruit. Various environmental factors influence the flowering, pollination and fruiting of a plant species. Different species differ in their time of flowering and their light and temperature requirements for flowering. Various characteristics of flowers affect pollination and agencies involved in the process. Plant species also differ in structure, and number of fruits, time of their formation and agents which damage their fruits. However, aquatic plants generally reproduce by vegetative means.

12·2·4. Growth and Dry Matter Accumulation

Measurement of Net Assimilation Rate (NAR), Relative Growth Rate (RGR), Leaf Area Index (LAI), Net primary production, biomass, energy accumulation pattern, phytochemical composition and accumulation pattern with reference to nitrogen, phosphorus, and other nutrients.

12·2·5. Economic Importance of the Plant Species

(For further details of autecology see **R. F. Daubenmire**'s *Plants and Environment : A Text Book of Plant Autecology* (1959); and **R. Misra**'s *Ecology workbook* (1968).

12·3. OUTLINE FOR STUDY OF ECOLOGICAL LIFE HISTORY OF THE ANIMAL

In case of animals, the ecological life histories usually proceed with, first, analysis of the behaviour adjustments needed for the survival of the mature animal; then of its reproductive behaviour and lastly, of the development of behaviour and physiological adjustments of the young animal. A complete ecological life history of an animal that is needed for the autecological study includes the following items :

1. Phylogenetic and geological history.
2. Geographic and habitat distribution with an analysis of adjustments to the physical environment and of biotic interrelations within the community.
3. Variation in population, through time and in space.
4. Changes in seasonal activities and physiological states: breeding, migration, hibernation.
5. Food and enemies.
6. Parasites and diseases.
7. Reproductive potential, mortality, rate of population turnover.
8. Requirements for reproduction : home range, territory, nest-site, nesting materials, etc.
9. Breeding behaviour : mating, nesting, etc.
10. Development of offspring : rate, stages, generations per year, etc.

Thus, to study the ecological life history of a plant or animal species each stage of the life cycle is studied in detail both in field and cultures grown under controlled or known environmental conditions. Then, observations and data are put in proper sequence and the whole picture of autecology is drawn.

12-4. REVISION QUESTIONS

1. Define autecology and show, why it is necessary to study ?
2. Compare autecology and synecology.
3. How will you study the autecology of a plant species ?
4. Write in brief outline about the method of study of the life history of an animal species.
5. Write short notes on the following : Autecology; Synecology; Life history; and Economic significance of autecology.

Part 3

BIOTIC COMPONENTS OF ECOSYSTEM

(Synecology or Population Ecology and Community Ecology)

All the living entities of an ecosystem form a single biotic component, the **community**. All the organisms of a community live together, share same habitat and influence each other's life directly or indirectly. Various vital life processes such as growth, nutrition, reproduction and distribution depend very much upon the interactions between the individuals of same species (**intraspecific interactions**), or between those of different species (**interspecific interactions**). Pollination, seed and fruit dispersal, symbiosis, grazing, predation, parasitism, competition are some of such interactions.

Community ecology deals with the ecology of the groups of different kinds of populations in the area. While, **population ecology** is the study of individuals of the same species where processes as aggregation, inter-dependencies between individuals and various factors governing such processes are emphasised. The population ecology has an applied significance for various human population oriented socio-economic problems at national and international levels.

13

Biotic Community and Ecological Niche

13-1. DEFINITION AND TYPES

The most visible component of the ecosystem is the biotic, the vegetation and animal life. The assemblage of different plants, animals and micro-organisms in any given ecosystem or in any given physical environment is a **community**. Thus, the forest with various trees, shrubs, herbs, animals, birds and insects and fungi and other micro-organisms is recognised as a community. Even within the forest which is a major habitat, any microhabitat such as a piece of decaying log, leaf or fruit or a tiny pool of water held in the hollow of a tree may support a community of heterotrophs such as fungi, bacteria, small insects, etc. In a community the organisms are interdependent—some organisms, called autotrophs (*e.g.*, plants) are synthesising the food by fixing solar energy in the photosynthetic activity and all others, called heterotrophs (*e.g.*, animals and micro-organisms), are utilizing the energy stored in the plant material or in the autotrophs either directly (herbivores) or indirectly (carnivores and detrivores). The community is thus essentially a **biotic community**. Since the plants, animals and micro-organisms all interact with each other and cannot be separated. At this stage, the biotic community can be defined as *a naturally occurring assemblage of plants, animals and micro-orgnisms living in the same environment, mutually sustaining and interdependent, constantly fixing utilizing and dissipating energy* (**Smith**, 1977). **However, McNaughton** and **Wolf** (1973) have considered groups of populations with similar ecological functions coexisting in space and time as distinct communities. The forest thus consists of several communities of plants, animals and micro-organisms, all of which together constitute the biotic component of an ecosystem.

The biotic communities represent a higher order of biological organisation than populations, yet, since communities refer only to living organisms, they are not as inclusive as ecosystems. In other words, the study of communities involves the entire biology of an area, but strictly speaking, it does not involve the specific study of the interactions of plants, animals and micro-organisms with the non-living components of the environment the way ecosystem analysis does (**Southwick**, 1976).

Some authors such as **S. C. Kendeigh** (1974) and **R. L. Smith** (1977) have divided the biotic communities into the following two main types : **major** or **autotrophic communities** which together with their habitats, form more or less complete and self-sustaining units except for the indispensable input of solar energy ; and **minor** or **heterotrophic communities** which are secondary aggregations within the major communities and are not, therefore, completely independent units as far as circulation of energy is concerned, *i.e.*, they depend on major communities for their energy source. For example, any forest forms the major community, while the community existing on any microhabitat which is occurring within the forest is the minor community.

A biotic community has a series of characteristics such as structure, dominance, niche, species diversity, stability, development and a metabolic rate which are not present in its individual species component, but have meaning only with reference to community level of organisation. Some of the significant attributes of biotic community can be studied as follows :

13-2. SIZE AND STRUCTURE OF BIOTIC COMMUNITY

The biotic communities may be large or small. Large-sized communities such as forests, swamps, deserts, grasslands, large lakes and many others have dimensions in hundreds of kilometers ; whereas others, such as ponds, tide flats, rocky plateaus, rivers, meadows, etc., occupy only a limited area.

A biotic community has certain structural characteristics which are often used in distinguishing the diverse communities. The structural attributes of plant and animal communities have been investigated in considerable details but in different directions. The plant ecologists have devoted greater attention to the structural characteristics, classification and nomenclature of plant communities, while the animal ecologists have emphasised the functional inter-relations of organisms within a biotic community (*i.e.*, energy flow, food webs, etc.). In fact, the structure of an animal community is dependent upon the plant community and its structure (see **Gopal** and **Bhardwaj**, 1979).

The structural characteristics of a plant community can be divided into two categories : analytical and synthetic. The **analytical characters** are those which are directly observed in the field and they may be either **qualitative** or **quantitative** depending on whether they can be expressed and measured in quantitative terms. The **synthetic characters** are based upon the data obtained for the analytical characters. The study of plant communities is generally known as **phytosociology**.

13-2·1. Qualitative Characters of Community

The major qualitative characters of the plant community (and also of the animal community) are as follows :

13-2·1A. Physiognomy. Physiognomy is the external appearance of the plant community which may be described on the basis of dominant plants, density, height, colour, etc., of the plants. It does not emphasise any particular species or individuals. In fact, **Cain** and **Castro** (1959) have defined physiognomy as the form and structure of the vegetation, the appearance that results from the life-form of the predominant plants. (**Life-form** is a general term referring to the shape or appearance of an organism irrespective of how formed, **Daubenmire** 1947). The terms like forests, grasslands, savannahs, scrub, bog, etc., describe the physiognomy of various communities.

13-2·1B. Growth-forms and life-forms. Many plant ecologists have attempted to classify plants according to their form (*i.e.*, growth form and life form), habitat, or some other characteristics. Thus, on the basis of growth forms, **Pound** and **Clements (1900)** have classified plants into shrubs, trees and herbs and further subdivided these categories into needle-leaf evergreens, broad-leaf evergreens, evergreen sclerophylls (small, tough, evergreen leaves, as in chamise), broadleaf deciduous, thorn trees and shrubs, dwarf shrubs, ferns, grasses, forbs, mosses, liverworts, lichens, fungi, algae and so on. Of the other various classifications, the one proposed by the Danish botanist **Christen Raunkiaer** (1903, 1934) has gained wide recognition and has been extensively followed. Instead of considering the plant's growth form, **Raunkiaer** classified plant life by the relation of the **embryonic** or meristemic tissues that remain inactive over winter or a dry period (**perennating tissue**) to their height above ground. Such perennating tissue includes buds, bulbs, tubers, roots, and seeds. **Raunkiaer** recognized five principal life forms which are summarized in Table 13·1. All the species in a community can be grouped into the five classes and the ratio between them expressed as a percentage, providing a **life-form spectrum** of the area that reflects the plant's adaptations to the environment, particularly climates. For example, community with a high percentage of perennating tissue well above ground (phanerophytes) would be characteristic of warm climates. A community with most of its plants chamaephytes and hemicryptophytes, and a community dominated by therophytes would be characteristic of deserts.

Life-forms of animals. There have been several attempts to classify the life-forms of animals, but no definite system has resulted (**Remane,** 1952 ; **Krivolutskii,** 1972). The major life-forms more often agree with their taxonomy than do plants, but some **life-forms include representatives from several different taxonomic groups.** These can be recognized **encrusting forms,** such as the fresh-water bryozoan *Plumatella* and some sponges ; **coral forms,** including grass, leaf, or shrub forms ; **radiate forms,** such as coelenterates and echinoderms ; **bivalve forms** ; **snail forms** ; **slug forms** ; **worm forms** ; **crustaecean forms** ; **insect forms** ; **fish, snake, bird** and **four-footed forms.** Each of these major types may be divided into narrower strutuctral or behavioural types, for example, **Osburn.** *et al.*, (1903) subdivided four-footed mammalian forms into the following types :

1. **Aquatic (swimming)** : Seal, whale, walrus.
2. **Fossorial (burrowing)** : Mole, shrew, pocket gopher.
3. **Cursorial (running)** : Deer, antelope, zebra.
4. **Saltatorial (leaping)** : Rabbit, kangaroo, jumping mouse.
5. **Scansorial (climbing)** : Squirrel, opossum, monkey.
6. **Aerial (flying)** : Bat.

Table 13·1. Raunkaier's life forms (from R. L. Smith, 1977).

1. **Phanerophytes** (Gr., *phaeros*, visible). Perennial buds carried well up in the air and exposed to varying climatic conditions. Trees and shrubs over 25 cm ; typical of moist, warm environments.

   The woody lianas, epiphytes and stem succulents (*e g.*, cacti and euphorbias) are recognised as sub-types of phanerophytes.

2. **Chamaephytes** (Gr., *chamai*, on the ground). Perennial shoots or buds on the surface of the ground to about 25 cm above the surface. Buds receive protection from fallen leaves and snow cover. Plants typical of cool, dry climate, *e.g.*, *Thymus*, *Silene*, *Trifolium*, *Rhamnus*, *Draba*, *Grewia*, etc.

3. **Hemicryptophytes** (Gr., *krypos*, hidden). Perennial buds at the surface of the ground where they are protected by soil and leaves. Many plants are characterized by rosette leaves and are characteristic of cold, moist climate, *e.g.*, grasses and herbs like *Thalictrum*, *Fragaria*, *Primula*, etc.

4. **Cryptophytes**. Perennial buds buried in the ground on a bulb or rhizome where they are protected from freezing and drying. This type includes the hydrophytes (buds remaining under water), helophytes (marsh plants with rhizomes under the soil) and geophytes (terrestrial plants with underground rhizomes, tubers), etc.

5. **Therophytes** (Gr , *theros*, summer). Annuals with complete life cycle from seed to seed in one season. Plants (*e.g.*, annual herbs) survive unfavourable periods as seeds and are typical of deserts and grasslands.

13-2·1C. Phenology, periodicity or aspection. The form of the plant changes with the age, so the same species in different stages of their life cycle provide a different structure to the community. As already discussed in section 11-5·2, **periodicity** refers to the regular seasonal occurrence of various processes and their manifestations (as formation of leaves, flowering, seed shedding). **Aspection** refers to the appearance of the community as a whole at different seasons as its appearance in rainy season, summer spring or winter seasons. **Phenology** has been defined by **Lieth** (1970, 1974) as the art of observing life-cycle phases or activities of plants and animals in their temporal occurrence throughout the year. Each phase in the life-cycle is called **phenophase**. The phenophases have been described by various methods using diagrams and symbols.

13-2·1D. Stratification. Organisms are distributed unevenly throughout the biotic community. Spatial distribution may be considered in both vertical and horizontal planes. Some ecologists also add another kind of stratification to these, the temporal stratification.

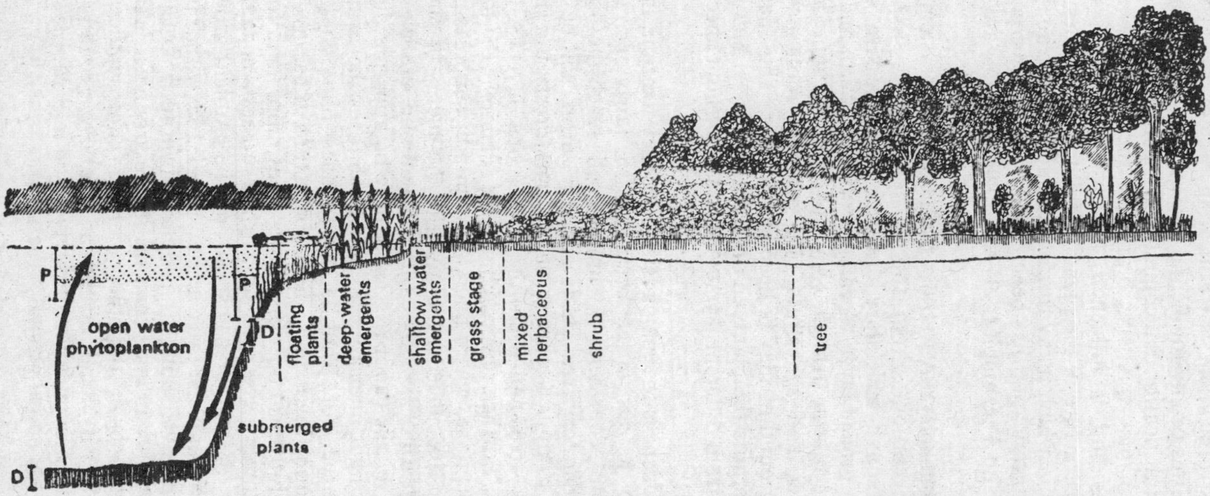

Fig. 13·1. A vertical sectional view of biotic communities from aquatic to terrestrial illustrates the general features of each. Both are structurally similar in that the zone of decomposition (D) and regeneration is in the bottom stratum and zone of energy fixation or production (P) is in the upper stratum. In succession from aquatic through the terrestrial stages, stratification and the complexity of the community become greater. In the aquatic community there is little storage of materials in biomass. However, in the terrestrial communities biomass storage increases as ecosystems becomes more mature (after Smith, 1974).

1. Vertical stratification. Stratification refers to vertical layering of organisms or environmental conditions within a biotic community. The stratification of a community is determined largely by the life-form of plants—their size, branching and leaves—which, in turn influences and is influenced by the vertical gradient of light. The vertical structure of the plant community provides the physical structure in which many forms of animal life are adapted to live. Vertical stratification has been observed in many communities and some distinct examples of this phenomenon are the following.

(i) Forest community. A well-developed forest ecosystem has several layers of vegetation. From top to bottom, they are **overstory stratum** or **canopy** ; **understory stratum** ; **transgressive stratum** or the **shrub** ; **ground layer, seedling stratum** or **herb**, and the **forest floor** or **subterranean stratum.** And one can continue down into the root layer and soil strata. For example, roots may be spread nearer the ground or may be deep penetrating. This spacing among the roots permits the plants to draw their water and nutrient requirements from different layers of the soil without affecting each other.

The canopy, which is the major site of energy fixation, has a major influence on the rest of the forest. If it is fairly open (*i.e.*, forest of dry climate), considerable sunlight will reach the lower layers and the understory tree strata and the shrub will be well developed. If the canopy is closed (*i.e.*, in rain-forest), the shrub and the understory trees and even the herbaceous layers will be poorly developed. Further, by reaching the canopy, the forest trees may gain advantage, where abundant sunlight supports photosynthesis, but the tree must spend much of the energy of photosynthesis in the growth of woody tissue of stem and branches to support the foliage in the canopy. There may be apparent disadvantages in the low light intensities in which the forest herbs must live, but the herbs need not spend its more modest photosynthetic profit on woody supporting tissue. Forest structure, thus, involves a gradient of growth-forms-upper and lower trees, upper and lower shrubs, upper and lower herbs, and soil-surface mosses—in adaptation to the gradient of light intensity. Along the gradient growth-form designs change from one extreme (the upper tree with foliage in full sunlight, massive supporting stem and branch structure, and a root system smaller in mass than the above ground structure) to herbs with adaptations at the other extreme (photosynthesis at low levels of light intensity, small investment in above ground supporting structure, and accumulation of reserve food in a root system more massive than the above ground structure).

The understory stratum consists of tall shrubs, understory trees and younger trees ; some are the same as those the crown, others are of different species. Species that are unable to tolerate shade and competition will die; others eventually reach the canopy after some of the older trees die or are harvested.

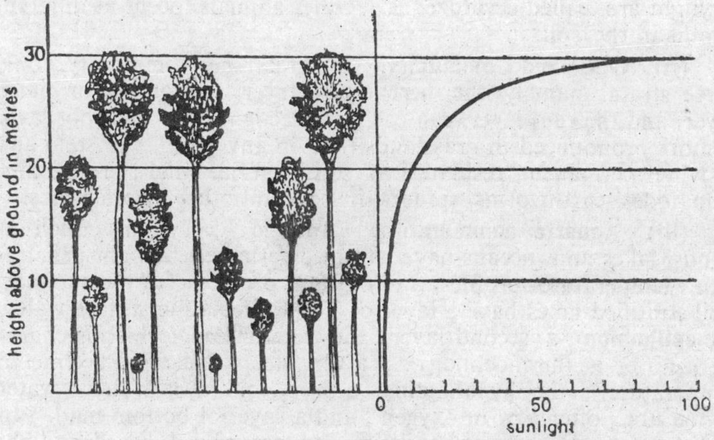

Fig. 13·2. Vertical stratification of a forest and light levels (after Herreid, II, 1977).

The shrub layer varies from forest to forest. The nature of the herb layer depends on the soil moisture conditions, slope position, the density of overstory and slope attributes, all of which vary from place to place through the forest. Beneath the herbs, small prostate plants, liverworts and mosses on the ground may form still another vegetation layer. The lichens and epiphytes grow at different strata on the tree trunk and branches. The final layer, the forest floor, is the site where the important process of decomposition of the forest litter takes place and where nutrients are released to the nutrient cycle.

Moreover, the variety of life in the forest is directly related to the number and development of the layers of the forest. For example, in the rain forests with greater species diversity, the stratification is more. While, in an agricultural field where weeds have been removed all the plants belong to the same strata. However, if certain layers are absent, then the animals they normally shelter and support are also missing. Thus a well-developed forest supports a rich diversity of life.

Vertical stratification in animals. As different plant species are adapted to different positions in this vertical gradient, so different animal species also occupy different levels in the forest. Different groups of bird species, for example, may be found feeding and nesting near the ground, in the shrub and small tree foliage beneath the canopy and in the canopy itself. Different arthropod species occur at different levels from the canopy downward to the herb stratum and to and below the ground surface. A group of animals—mites and spring tails, millepedes and centipedes, ground beetles, and so on—occur primarily in the leaf litter on the soil surface ; these animals, which are seldom seen on the surface by daylight are called **cryptozoans.** Other animals occur at different depths in the soil.

(ii) **Grassland Community.** In grassland community, only three strata, namely the **herbaceous layer,** the **ground** or **mulch layer,** and the **root layer,** can easily be recognised. The root layer is more pronounced in grasslands than in any other ecosystem and provides permanent residance to soil bacteria, fungi ; protozoans, nematodes, earthworms, spiders, insects and other invertebrates.

(iii) **Aquatic communities.** Aquatic ecosystems such as ponds, lakes and oceans have strata determined by light penetration, temperature profiles and oxygen profiles. In the summer, well-stratified lakes have a layer of freely circulating surface water, the **epilimnion** ; a second layer, the **metalimnion,** which is characterized by a thermocline (*i.e.*, a very steep and rapid decline in temperature) ; the **hypolimnion,** a deep, cold layer of dense water about 40°C, often low in oxygen ; and a layer of bottom mud. In addition two other structural layers are recognised, based on light penetration : an upper zone roughly corresponding to the epilimnion, which is dominated by phytoplankton and is the site of photosynthesis and a lower layer, in which decomposition is most active. The lower layer roughly corresponds to the hypolimnion and the bottom mud.

All communities, whether terrestrial or aquatic have similar **biological** or **trophic structures** (Fig. 13·1). They possess an **autotrophic layer,** concentrated where light is mostly available, which fixes the energy of the sun and manufactures food from inorganic substances. In forests this layer is concentrated in the canopy ; in grasslands, in the herbaceous layer ; and in lakes and seas, in the upper layer of water (*i.e.*, in phytoplanktons). The biotic communities also possess a **heterotrophic layer** that utilises food stored by autotrophs, transfers energy, and circulates matter by means of herbivory, predation in the broadest sense, and decomposition.

2. **Horizontal stratification.** Horizontal stratification or **dispersion** relates to the distribution of organisms, principally plants, in the horizontal space (*i.e.*, on the ground) or across the canopy. Like vertical stratification, it can influence the presence and absence of certain forms of animal life. The following three distribution patterns are observed for the organisms : (i) **Random distribution,** when each organism appears to be placed without any regard to where another organism occurs ; there is no negative or positive interaction between individuals. (ii) **Clumped** or **contagious distribution,** when several individuals are clustered together in various spots throughout the area. Ants are grouped into colonies, fish into schools and humans into cities rather than a

helter-skelter arrangement. (iii) **Regular** or **uniform distribution** which is produced due to negative effect of one organism on another ; each organism is spaced at certain intervals from others in the same species. Such distribution pattern is seen in certain desert plants, such as the creosote bush (*Larrea*), which are rather uniformly spaced. It is also seen in certain breeding birds that establish and defend territories from other individuals, thus parceling out the space in a regular way.

The plant distribution is mostly clumped to varying degrees. The horizontal distribution of plants and also of some animals is influenced by the following factors :

(i) **Seed distribution and vegetative reproduction.** Plants with airborne seeds may be distributed widely, while plants with heavy seeds or with pronounced vegetative reproduction by runners or rhizomes will be clumped near the parent plant. A similar dispersal process also explains clustering of certain sessile animals which release larvae that settle near the reproducing adult.

(ii) **Differences in environment.** Herbaceous plants of the forest may be clumped where pools of light reach the forest floor through the canopy. Further, there may be other differences, in the environment. Since the area is not homogeneous, some areas are more suitable for growth, reproduction and survival than others. The mosaic pattern of the environment leads to a patchy distribution of individuals within it. Clusters of individuals may occur not because of any attraction between them but because they are independently influenced by the same environment.

A pronounced form of horizontal stratification is **zonation,** which is caused primarily by differences in climatic or edaphic conditions that retard or inhibit rooted vegetation. This type of stratification is most conspicuous around ponds and bogs.

(iii) **Species interrelations.** There may be negative or positive interactions of species. Positive interactions lead to clustering of individuals and negative interactions lead to an even spacing between them.

3. **Temporal stratification.** Besides the differentiation of communities in space, there occur a differentiation of communities in time also. Environments of biotic communities are rhythmic : in most environments light, temperature and other environmental factors go through daily and yearly cycles. In some communities, especially those of ocean shores, there are also complex rhythms set by the rise and fall of tides. In general, there occur rhythms of function in biotic communities in adaptation to rhythms of environment.

Plankton population fluctuates rapidly in many water bodies, with species replacing one another in periods of days and weeks. A number of major species may occur as plankton dominants in the course of the year. Thus in freshwater lake species of diatoms and other yellow-green algae may predominate in the winter plankton

As water temperatures warm in later spring and early summer, these are replaced by desmids and other green algae. At peak summer temperatures blue-green algae may predominate or share dominance with the green algae ; as temperatures cool dominance shifts back toward the green and yellow-green algae. Each species has an inactive stage in which it survives the season unfavourable for its activity. Each species has its own place in the annual pattern, determined by its own response to the fluctuations of temperature and other environmental factors. The plankton community shows differentiation in time : different groups of species occur at different times during the seasonal cycle. The year-round total number of plankton species is much larger than the number present at a given time.

Seasonal and daily differentiation (stratification) occurs also in forests. One group of insects is active in the day time another group at night, and a third group may be active in the twilight transitions of morning and evening. In terrestrial communities flycatchers, warblers, and other insectivorous birds are active by day, bat at night, and nighthawks in the dusk. Progress of the seasons is marked by the appearance of different groups of plants in flower and different groups of insects visiting these flowers. In deciduous forests spring beauties, dogtooth violets, and other herbs develop their foliage and flower early, before the trees are in leaf.

13-2·1 E. Species abundance and species diversity. Among the array of species that make up the community, relatively few are abundant, and most of them are rare. **Abundance** or **species richness** of a species in the community represents its relative distribution in it. It is related to density but is a qualitative estimate. Large number of individuals occurring at one place in the community will not be referred to as abundant while the same number of individuals spread throughout the community may appear to be abundant. The species richness or abundance is expressed in five degrees on an arbitrary scale such as follows : (i) rare (*r*) or very sparse ; (ii) occasional (*o*) or sparse ; (iii) frequent (*f*) or not numerous ; (iv) abundant (*a*) or numerous ; and (v) very abundant (*va*) or very numerous.

Two parameters of species distribution within the community, species richness and species evenness, are useful in measuring an attribute of the community **species diversity**. A community that contains a few individuals of many species will have a higher diversity than will a community containing the same number of individuals but with most of the individuals confined to a few species. In order to quantify species diversity for the purpose of comparison, the **Shannonwiener index** is used. This index measures diversity by the following formula :

$$H = -\sum_{i=1}^{s} (P_i)(\log_2 P_i)$$

where, H = the diversity of species

S=the number of species
P_i=proportion of individuals of the total sample belonging to the ith species.

This index takes into consideration the number as well as the relative abundance of species.

Besides being a measure of comparison of similar communities within a given region, species diversity is also useful in examining global ecosystems. If one were to leave the tropics and travel north through the Temperate zone to the Arctic zone, one would find the numbers of plants and animals decreasing on a latitudinal gradient. For example, **Fisher** (1960) found that species of nesting birds approach 1395 in Colombia, drop to 1100 in Panama, to 143 in Florida, to 118 in Newfoundland, and to 56 in Greenland. One can follow the same pattern in mammals (**Simpson**, 1964), lizards (**Pinaka**, 1967), trees (**Monk**, 1967), and fish (**Lowe-McConnell**, 1969). Species diversity increases from cold to warm climates.

However, in the oceans species diversity increases from the continental shelf, where food is abundant but the environment is more changeable, to the deep, cold water where food is less abundant but the environment is stable. Likewise mountain areas in general support more species than do flatlands, and peninsulas have fewer species than do adjoining continental areas. Small or remote islands have fewer species than do large islands and those nearer continents (**Smith**, 1977).

The following hypotheses have been proposed to explain why the tropics should hold a greater abundance of species than the temperate region or why one island should hold more than another:

1. Time theory. Fischer (1960) and **Simpson** (1964) have related species diversity to evolutionary time. It is found that older communities hold a greater diversity than young communities. Communities in the tropics evolve and diversify faster than temperate or arctic communities, in part because the environment is more constant and climatic catastrophes less likely. Some palaentological evidences support this theory.

2. Spatial heterogenity theory. Simpson (1964) also believed that **more complex and heterogeneous the physical** environment, the more complex and diverse would be its flora and fauna. The greater variation in topographic relief, the more complex the vertical structure of the vegetation, and the more types of microhabitats the community contains, the more kinds of species it will hold. For example, the more complex the vertical stratification of a community, the more species of birds it holds (**MacArthur**, 1972).

3. Climatic stability theory. Fischer (1960) proposed that the more stable the environment, the more species would be present. Through evolutionary time, the tropics, of all regions of the earth, has probably remained the most constant and has been relatively free from severe environmental conditions that could have catastrophic effects on a population. Under tropical conditions, natural selection is strongly influenced by the competition of individuals against members of other species.

4. Productivity theory. Connell and **Orias** (1964) held that the level of diversity of a community is determined by the amount of energy flowing through the food-web. The rate of energy flow is influenced by the limitation of the ecosystem and by the degree of stability of the environment.

5. **Competition theory. Dobzhansky** (1951) and **Williams** (1964) proposed that in mild climatic regions like tropics biological competition becomes more important in the evolution of species and in the specialization of niches than in environments of high physical stress, such as Arctic.

6. Predation theory. Paine (1966) proposed that the more diverse communities such as tropics support more predators and these predators regulate the abundance of prey species, so that competition in prey species is greatly reduced.

7. **Stability-time hypothesis. Sanders** (1968) have suggested that two contrasting types of communities exist, the physically controlled and the biologically controlled. The species numbers diminish continuously along a stress gradient. The greatest diversity occurs among the predominantly biologically accomodated communities.

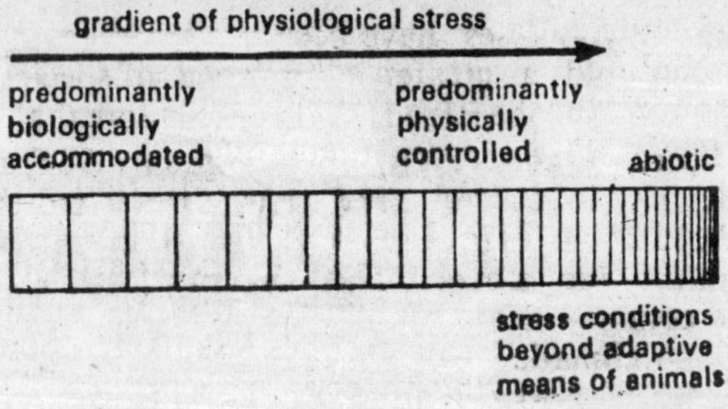

Fig. 13·3. The bar-graph representation of the stability time hyposthesis (after Smith, 1974).

13-2·1F. Sociability. The sociability expresses the relation of individuals to each other and indicates the closeness between individuals, within the community. **Braun Blanquet** (1932) have recognised the following five arbitrary types of sociability : S_1, plants growing singly; S_2, plants growing in small groups; S_3, plants occurring in small patches; S_4, plants forming large patches; and S_5, plants occurring in essentially continous populations.

Whitford (1949) expressed sociability of plants in quantitative terms of the following equation :

$$\text{Sociability} = \text{Density} \times 100/\text{Frequency}$$

13-2·1G. Vitality. Vitality refers to normal growth and reproductive ability of the plant species which help it in maintaining its position in the community. Plant ecologists have often recognised the following five categories of vitality : V_1, plants which germinate but die soon without reproducing; V_2, plants which linger after germination but cannot reproduce; V_3, plants reproducing but only vegetatively; V_4, plants reproducing sexually but rather feebly; and V_5, plants reproducing very well sexually.

13-2·1H. Disseminule type. In recent years the type of the dispersal organ or **disseminule** produced by a plant species has also been used as a community characteristic. **Dansereau** and **Lems** (1957) have recognised the following disseminule types of plants (Table 13·2) :

Table 13·2. Disseminule types of plants.

Disseminule type according to agent	Disseminule type according to morphology	Examples
1. Autochore	Expulsive	*Vinca, Ruellia*
(Self dispersed)	Stoloniferous	*Fragaria*
2. Anemochore	Pulmose	*Tridax, Taraxacum*
(wind dispersed)	Minute	Orchids, ferns
	Winged	*Tilia, Pinus*
	Tumbleweed	*Physalis*
	Catapult	*Papaver*
3. Zoochore	Fleshy	*Rubus*
(animal dispersed)	Nutlike	*Ficus, Psidium*
	Adhesive	*Xanthium, Cenchrus*
4. Hydrochore	Buoyant	*Eichhornia, Hydrilla*
(water dispersed)	Splash cup	*Marchantia*

13-2·2 Quantitative Characters of Community

13-2·2A. Density. Density represents the numerical strength of the species in the community. Plants grow at varying distance in respect with each other. The number of plants in an unit area gives an idea of its density. The density of various species in a community varies in time and space and affects the community structure. Density gives an idea of competition. If density is more, it means there is more degree of competition between the individuals of the species. Density is expressed as number of individuals per unit area and is calculated as follows :

$$\text{Density} = \frac{\text{Total number of individuals of the species in all the sampling units}}{\text{Total number of sampling units studied}}$$

13-2·2B. Frequency. Frequency is expressed as the percentage occurrence of individuals of a species in a number of observations. It indicates the homogeneity of dispersion of the individuals of a species in the community and is determined as follows :

$$\%\ \text{frequency} = \frac{\text{Number of sampling units in which the species occurred}}{\text{Total number of sampling units studied}} \times 100$$

Thus, the estimate of frequency depends upon the size of the sampling unit. A very small sampling unit underestimates the frequency of widely spaced individuals, while a large sampling unit overestimates the frequency.

After determining the percentage frequency of the various species recorded, these are distributed among **Raunkiaer's** (1934) five frequency classes—A, B, C, D and E, depending upon their percentage frequency as follows :

A=0 –20 per cent;
B=21–40 per cent;
C=41–60 per cent;
D=61–80 per cent;
E=81–100 per cent.

Raunkiaer proposed a "**law of frequency**" which states that the percentage of species falling under five classes of frequency will be in order A > B > C ⩾ D < E. He also proposed a normal frequency diagram in which per cent values of different frequency classes were : A=53, B=14, C=9, D=8 and E=16. Frequency distribution also varies greatly with the species diversity.

13-2·2C. Species dominance. The nature of biotic commnities is controlled either by physical or abiotic conditions such as substrate, the lack of moisture, and wave action, or by some biological mechanism. Biologically regulated communities are often influenced by a single species or by a group of species that modify the environment. These organism are called **dominants**, *e.g.*, in forest community large-sized trees are dominants, while shrubs and herbs are mostly subdominants.

It is not easy to describe a dominant or to determine the dominant species. The dominants in a community may be the most numerous, possess the highest biomass, preempt the most space, make the largest contribution to energy flow or nutrient cycling, or by some other means control or influence the rest of the community. Thus, in a forest community, trees are dominant. They decrease light intensity, increase the relative humidity, interecept precipitation, monopolize most of the moisture and nutrients in the soil, decrease wind velocity and furnish shelter and food for animals. Grasses play a similar, though less conspicuous, role in prairie communities; sedges, rushes and cattails in marsh communities; sagebrush in the arid habitat of the Great Basin ; mussels and barnacles on a rocky seashore ; and so forth.

Though some ecologists have given the role of the dominant organisms to those that are numerically superior, but numerical abundance alone is not sufficient. For example, a species of plant can be widely distributed over the area and yet exert little influence on the oommunity as a whole. In the forest the small or understory trees can be numerically superior, yet the nature of the community is controlled by a few large trees that overshadow the smaller ones. In such a situation the dominant organisms are those with the greatest biomass or **that preempt** most of the canopy space and thus, regulate the distributlon of light. Ecologists measure such dominants by **biomass or basal area (*i.e.*, cover).**

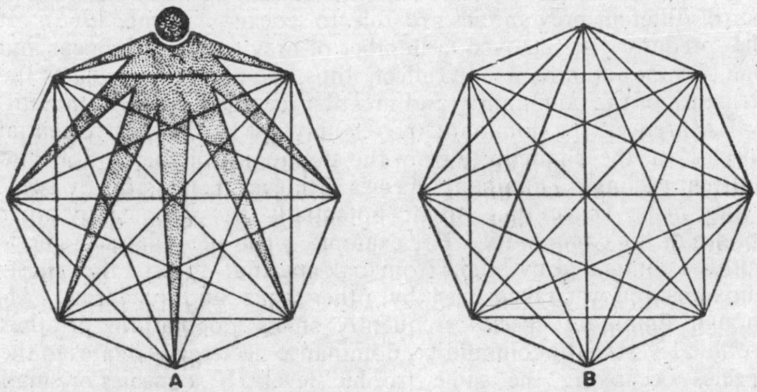

Fig. 13·4. Diagrammatic representation of the effects of dominance on a relatively simple community : (A) the dominance relationships are clear; the dominant species exerts a disproportionate influence on the community as a whole ; (B) no clear dominance relationship exists ; species interactions are more uniform (after Clapham, Jr., 1973).

(i) **Community mass.** The community mass is biomass which refers to the weight of organic material, living or dead. In terrestrial communities the biomass is correlated with the weight and complexity of strata of the community. Maximum biomass is observed in dense tropical forests growing on river flood plains. These forests may weigh upto 1000 metric tons dry weight per hectare (**Vyas** and **Golley**, 1975). **R. Mistra** (1972) has estimated the biomass of an Indian 60-year old forest near Varanasi as 239 metric tons per hectare. Grassland at Varanasi had a biomass as high as 33 metric tons per hectare. Biomass declines from the optimum for the community with decreased moisture and/or temperature and with increased biotic disturbance. In terrestrial communities the animal biomass is usually less than 10 per cent of the plant biomass. However, in aquatic communities the biomass of the plants may be much smaller than that of the animals, where the production is in the form of algae or where the animals feed on imported organic matter.

(ii) **Basal cover or canopy cover.** Cover is the area occupied by a plant and can be expressed as basal cover or canopy cover. The **basal cover** is the land area occupied by the cross section of the stem. The **canopy cover** is the total land area under the canopy of a plant. In fact, the basal area can only be a small fraction of the total land area in a community but the canopy cover of a single species, the dominant species, may be several times the total land area because of overlapping canopies.

Further, the dominant organism may be relatively scarce yet by its activity control the nature of the community. For example, the predatory starfish *Piaster* preys on a number of associated species and reduce competitive interaction among them, so a number of different prey species are able to coexist (**Payne**, 1966). If the predator is removed, a number of prey species disappear and one becomes dominant In effect, thus, the predator regulates the structure of the community and must be regarded as the dominant.

Moreover, the dominant species may not be the most essential species in the community from the standpoint of energy flow and nutrient cycling. Dominant species achieve their status by occupying niche space that might potentially be occupied by other species of the community. For example, when the American chestnut was eliminated by blight from oak-chestnut forest, the chestnut's position was taken over by other oaks and hickories. Although dominant species frequently shape populations of other trophic levels in the community, dominance necessarily relates to the species occupying the same trophic level. If a species or small group of species is to achieve dominance, it must relate to a total population of species, all of which have similar ecological requirements. One or several become dominants because they are able to exploit the range of environmental requirements more efficiently than other species in the same trophic level. The subdominant species exist because they are able to occupy a niche or portions of it that the dominants cannot effectively occupy. Dominant organisms, thus are generalists capable of utilizing a wide range of physiological tolerances, while subdominant organisms are specialized in their environmental requirements and more limited in their physiological tolerances.

Lastly, the degree of dominance expressed by any one species appears to depend in part on the position the community occupies on a physical or chemical gradient. For example, at one particular point on a moisture gradient, species A and species B may be the dominants. As the gradient becomes drier, species B may assume a subdominant position in the community, its place as a dominant may be taken by a third species, C. Nutrient enrichment too, can change the structure of the community. Lakes receiving excessive sewage discharges shift from a diverse assemblage of nutrient-thrifty diatoms to a few blue-green algae that are able to exploit a nutrient-rich system (see **Smith**, 1977).

13-2·3 A. Synthetic Characters of Community

These include the following parameters :

13-2·3A. Presence and constance. The presence expresses the extent of occurrence of the individuals of a particular species in the community. The species on the basis of its percentage frequency may belong to any of the following five presence classes that were first proposed by **Braun Blanquet** :

I. **Rare**—present in 1 to 20% of the sampling units.

II. **Seldom present**—present in 21–40% of the sampling units.
III. **Often present**—present in 41–60% ,, ,,
IV. **Mostly present**—present in 61–80% ,, ,,
V. **Constantly present**—present in 81–100% ,,

A species with high frequency is considered constant. The term constancy is used when equal sample areas are taken in each stand, and presence is used when the area of the sample varies from stand to stand.

13-2·3B. Fidelity. Fidelity or "faithfulness" is the degree with which a species is restricted in distribution to one kind of community. Such species are sometimes known as **indicators.** The species have been grouped into the following fidelity classes which were first formulated by **Braun Blanquet** :

1. **Strangers.** Rarely occurring species which either have arrived from other communities or are relicts from earlier stages of succession.
2. **Indifferents** (or **ubiquitous**). Species which occur in any community without exhibiting any **preference** to any particular kind of community.
3. **Preferentials.** Species which occur in several kinds of communities but are predominant in one.
4. **Selectives.** Species occurring frequently in one kind of community and sometimes also in others. Sometimes preferential and selective species are considered together as **characteristic.**
5. **Exclusive or true.** Species which are completely restricted to one kind of community.

13-2·3C. Importance value. To determine dominance, ecologists have used several approaches like relative abundance, relative dominance and ratio of basal area. Often all the three measurements are combined to arrive at an **importance value** for each species. This index is based on the fact that most species do not normally reach a high level of importance in the community, but those that do serve as an index, or guiding species. Once importance values have been obtained for species within a stand, the stands can then be grouped by their leading dominants according to importance values. Such techniques are useful in the study and classification of communities on some environmental gradient.

13-3. ECOTONE AND EDGE EFFECT

The zone where two or more different communities meet and integrate is a **transition zone** or **ecotone.** This zone of integration may be narrow or wide, local (*e.g.*, a zone between a field and a forest) or regional (*e.g.*, the transition between forest and grassland. **Daubenmire,** (1968) have recognised three types of ecotones. One is an **abrupt transition,** the result of a sudden change in environmental conditions, such as soil type or soil drainage. The second is a **sharp transition** brought about by plant interaction, particularly competition. The third type is a blending of two or more adjacent vegetational types. In the second and third types the superior plant competitors on the side of the ecotone extend as far out as their ability to maintain themselves allows. Beyond this point, competitors of the adjacent community take over. In consequence, the ecotone or transition zone exhibits a shift in dominance of the conspicuous species of both sides; it may also include a number of highly adaptable species that tend to colonize such transitional areas (Fig. 13·5). Because of this, the variety (*i.e.*,

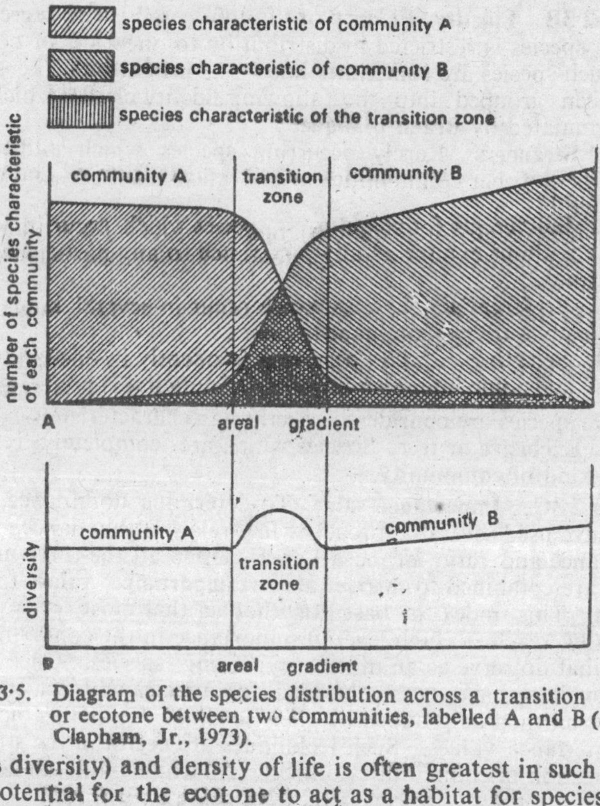

Fig. 13·5. Diagram of the species distribution across a transition zone or ecotone between two communities, labelled A and B (after Clapham, Jr., 1973).

species diversity) and density of life is often greatest in such areas. This potential for the ecotone to act as a habitat for species found in neither major community is termed the **edge effect**. A common example of the edge effect in action can be seen in those species of owl that live in or near ecotones between forests and grasslands. They depend on forest trees for nesting, and do their hunting in the grassland, where they depend on field rodents for food. In man-made communities such as agricultural fields, the ecotone between the field and the forest act as a refuge for species formerly found in the ploughed area, as well as for other plants such as weeds. Ecotones of this type are also the prime habitat of many species of insects, game birds, and mammals.

13-4. ECOLOGICAL NICHE

The diversity of species in a biotic community indicates that some communities have their resources more finely divided among individuals of many species than do others. This suggests that such communities have more positions to fill, more ways in which their resources can be utilized. These positions or functional roles in the community can be considered as **niches** and the community can be viewed as an aggregation of niches (**Smith**, 1977).

In fact, the niche is the **property** of the community and it represents the place of the species in the formal community structure. The term niche has been used in ecology with a variety of meanings and the following discussion will make it more clear.

13-4·1. Area, Habitat, Microhabitat and Ecological Niche

The **area** of a species or any other taxon refers to the total extent of its geographical range of dispersal ; this can be plotted on a map. The **habitat** of a species described in a single word or phrase the totality of abiotic factors to which the species is exposed in this area. Thus one can talk of marine habitats, coastal habitats, marsh habitats, forest habitats, disturbed habitats or even dry habitats.

In many cases, however, the habitat of the species is highly specialized. Certain species of leaf miners, for example, live only in the upper photosynthetic layer of the leaves of certain species of plants, while other species live in the lower cell layer. These patterns of location are so consistent that it is clear that the habitats of the species are different. Subdivisions of the environment on this scale are commonly called **microhabitats**. Thus the leaf constitutes a microhabitat for leaf miners within the total forest, and the different cell layers of leaf constitute different microhabitats within the leaf for different species of leaf miner. Conditions within the leaves are quite different from the general conditions in the forest. The specific environmental variables in the microhabitat of a population is called **microenvironment** or the **microclimate** (**Clapham, Jr.**, 1973).

Though habitats and microhabitats indicate the places where organisms of a community live, but, if one wishes to understand how a species fits into the fabric of the ecosystem, it is not enough to merely describe the creature's habitat or microhabitat, he must also understand its functional role within the community—"address" of the organism but its "profession" as well. This includes what it eats, what it does, where it lives–everything about it that influences the community. Thus, the ecological niche is a particular combination of physical factors (microhabitat) and biotic relations (role) required by a species for its life activities and continued existence in a community (**Kendeigh**, 1974).

13-4·2. Growth of Niche Concept and Types of Niches

Although **Steere** (1894) appreciated the niches in his explanation of the way bird species within a genus were segregated on different islands in the Philippines, the term and concept were really first developed by American ornithologist **Joseph Grinnell** (1917, 1924, 1928). In his study of California thrashers and other birds he suggested that the niche be regarded as the ultimate distributional or spatial unit occupied by just one species, or subspecies, to which that species is held by structural and instinctive limitations, such as climatic factors, kind and amount of food, suitable nesting sites and cover. Thus, he considered the niche as a subdivision of habitat (**Udvardy**, 1957) or microhabitat (**Odum**, 1971), or habitat (**Smith**, 1977). **Charles Elton** (1927), in his classic book *Animal Ecology*, considered the niche as the fundamental role of the organism in the community—what it does, its relations to its food and its enemies. Basically, this idea stresses the occupational status of the species in the community, its "profession". In 1957 **G.E. Hutchinson** proposed that the niche be considered the total range of environmental conditions under which an organism lives and reproduces. The Hutchinsonian concept initially was limited to environmental variables affecting a species, such as light, temperature, moisture, height of vegetation, and time. If biological variables are added to the environmental variables, then this, concept of the niche considers both the organism's place and its function in community.

In a way the community can be considered an aggregation of many environmental and functional variables, each of which can be thought of as a point in a volume of space of infinite dimensions. This **Hutchinson** termed the **multidimensional space** or **hypervolume**. He also recognised the following two niches : the **fundamental niche** which is the maximum abstractly inhabited hypervolume when the species is not constrained by competition with others ; and the **realized niche** which is a smaller hypervolume occupied by a species under biotic constraints. In other words, each species has a fundamental niche, the full range of variables within a community to which it is adapted through natural selection and evolution. However, because of competition few species occupy their fundamental niche ; rather they occupy a smaller realized niche. The following two examples will explain Hutchinsonian concept more clearly :

1. **Root** (1967) studied the exploitation of the niche by the bird blue-gray gnatcatcher (*Polioptila caerulea*) in California oak woodland. He characterized the niche of bird in part by the size of its food (prey-insect) and the height above the ground at which it was captured (Fig. 13·6). For the sake of example one could consider the maximum range of its food size, prey between 1 and 3 mm, and its foraging area, from ground level to 11 feet, as the gnatcatcher's fundamental niche. The gnatcatcher's niche centre, indicated by frequency of capture and stomach content analysis, consists of insects 3 to 5 mm in length taken 8 to 18 feet above the ground. Further the height and food dimensions are diverted from this centre, the more the gnatcatcher's niche overlaps those of other

species. Depending on the degree of competition, the boundaries of realized niche could be defined by any one of the contour lines.

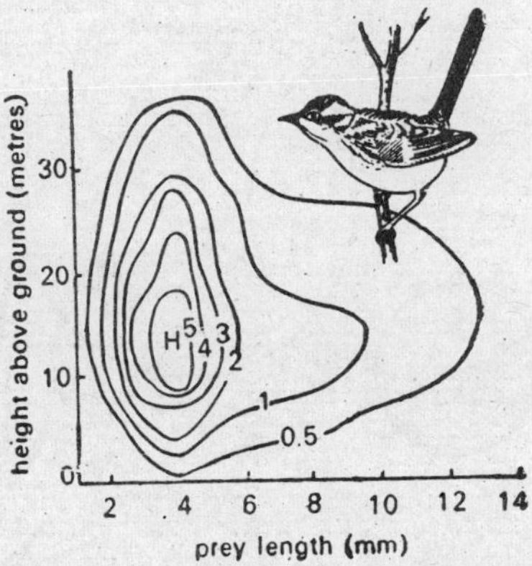

Fig. 13·6. Representation of a niche space for blue-gray **gnatcatcher** based on the combination of two variables : size of prey (food) and feeding height above the ground. The contour lines map the feeding frequency (in terms of total diet for the two niche axes) for adult birds during the incubation period in July and August in California oak woodlands. The maximum response level is at H. This represents optimum conditions. Contour lines spreading out from this optimum (*i.e.*, H) represent decreasing response levels. The outer contour line represents the outer boundary of fundamental niche for these two variables (after Smith, 1977).

2. **Putwain** and **Harper** (1971) studied the niches of two species of dock, *Rumex acetosa* and *Rumex acetosella*, each growing in two different grasslands. The two species must compete withgrass species and other forbs. To determine the relationship of these two species, they treated the flora with certain specific herbicides to selectively remove in different plots (1) grasses ; (2) forbs, except *Rumex* species ; (3) the *Rumex* species ; and (4) all species except *Rumex*. *R. acetosa* spread rapidly after the grasses were removed, but *R. acetosella* increased only when both grasses and forbs were removed. In this case, the fundamental niche of *R. acetosella* (R) overlaps the fundamental niche of both grasses (G) and forbs (D) (Fig. 13·7). Only when these competitors are eliminated does *Rumex* realise its **fundamental niche.** However, *R. acetosa* overlaps only with the grass species whose removal permits the expansion of *Rumex* throughout its fundamental niche. The niches for the seedlings differ from that of mature plants. This shows that the fundamental and realized niches of an organism can change with the growth and development of the organism. This fact is further proved by following examples: an insect with a complex life cycle can occupy one niche space as larva and an entirely different

niche space as an adult (*e.g.*, mosquitoes). In other organisms niche space can change as the organism matures, because food and cover (shelter) requirements change as the organism grows larger.

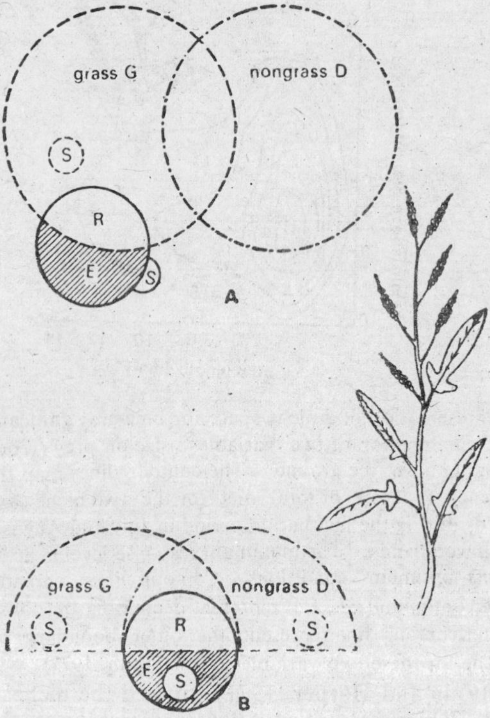

Fig. 13·7. A diagrammatic representation of niche relationships of (A) *Rumex acetosa* and (B) *Rumex acetosella* in mixed grassland. In each diagram the fundamental niches of grass species (G - - -) and forb species, excluding *Rumex* spp (D-.-.-), are indicated as overlapping areas. The fundamental niche of *Rumex* spp (R) is shown as a continuous outline and realized niche is shaded (E) is the part of the fundamental niche of *R. acetosa* which is expressed in the presence of forbs species alone and does not overlap the fundamental niches of (G) and (D). The fundamental niche of seedlings is shown the circles labelled "S" Note that the fundamental niches of seedlings are contained within the fundamental niches of grass and forb species (after Smith, 1977).

Odum (1971) has suggested that an ecological niche includes three aspects : the spatial or habitat niche, the trophic niche and hypervolume niche. We have already explained the concept of hypervolume niche. The other two types of niches can be discussed as follows :

1. Spatial or habitat niche. It represents the ultimate distributional or spatial unit occupied by a species. In a particular habitat shared by several species, each of the species may be confined to its own microhabitat or spatial niche because no two species in the same general territory can occupy for long identically the same ecological niche. In section 13-4·1, an example of spatial niches of leaf miners has been given. **O'Neill** (1967) has provided another interesting example of microhabitat segregation in seven species of millipedes. All these species live in the same general habitat, the forest floor of a maple-oak forest and all belong to the same basic trophic level, *i.e.*, all are detritus feeders. But each of the seven species predominates in a different spatial niche : *Euryurus erythropygus* (in heartwood at centre of logs) ; *Pseudopolydesmus serratus* (in superficial wood of logs) ; *Narceus americanus* (at outer surface of logs beneath bark) ; *Scytonotus granulatus* (in under log, but on log surface) ; *Fontaria virginiensis* (in under log, but on ground surface) ; *Cleidogonia caesioannularis* (within leaves of litter) ; and *Abacion lacterium* (beneath litter on ground surface).

2. Trophic niche. In this case two species live in the same habitat but they occupy different trophic niches, because of differences in food habits. For example, two aquatic bugs, *Notonecta* and *Corixa* live in the same pond but occupy different trophic niches. The former is an active predator **that** swims about grasping and eating other animals, but the latter feeds mainly on decaying vegetation.

13-4·3. Overlapping Niches and Niche Separation

Though two populations cannot coexist in a given habitat with precisely the same niche, it is highly probable that the niches of different populations will show varying degrees of overlap (Fig. 13·8). This occurs when two organisms use the same resource, such as food. Thus the concept of the niche is closely associated with the concept of competitive exclusion and competitive relationships among species (see chapter 16). Basically, Gause principle or competitive exclusion theory states that no two species can persist in the same niche. Either one becomes extinct or both, through natural selection, diverge into different niches. The latter phenomenon is called **niche separation.** Consider, for example, two species of plants that are able to grow in saline water,

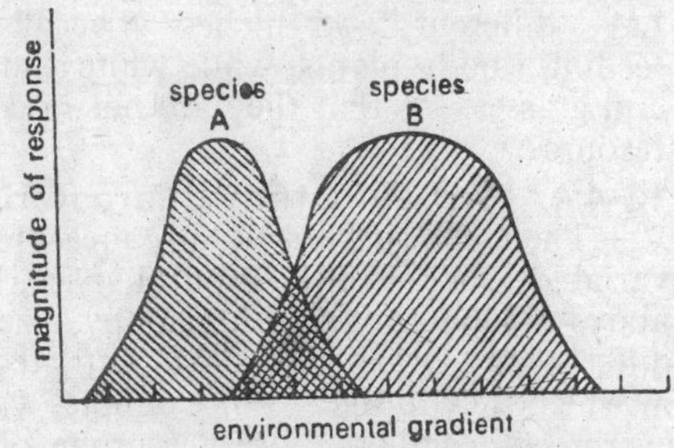

Fig. 13·8. Diagram showing overlapping of niches of two species A and B (after Clapham, Jr., 1973).

as in salt marsh. Species A can grow successfully if the salt content of the soil solution is 0–2 per cent, and species B can grow successfully if the salinity of the soil solution is 1–3 per cent. Individuals from both populations which are physiologically best adapted to survive under salinities of 1–2 per cent lie in the area of overlap and are in direct competition with one another. Those individuals of species A that favour a salinity lower than 1 per cent and those individuals of species B favouring a salinity greater than 2 per cent lie in the areas of non-overlap and are not in inter-specific competition. As a general rule, individuals from populations with overlapping niches but which themselves lie outside the area of overlap are likely to show greater survivourship and reproductive success, because they are not competing with individuals from the other species. Thus natural selection tends to favour these individuals, and the non-overlapping protions of the population's niche will tend to increase in relation to the overlapping portion. The reduction in the area of overlap is equivalent to separation of the niches.

The amount of niche overlap is usually proportional to the degree of competition for that resource (Note, competition involves only a resource that is in short supply). In extreme cases niches may be adjacent to one another with no overlap. In another extreme case the fundamental niche of one may be completely within the fundamental niche of another, as in the case of seedling *Rumex* and grass. In such instances the *Rumex* seedling is competitively superior. In most cases, however, fundamental niches may partially overlap. Some niche space is shared and some is exclusive, enabling the organisms to coexist. For example, in Natal (Africa), two species of rhinoceros, the black and white live in the same area but have different food niches. The black rhinoceros is a browser, feeding woody plants, while white rhinoceros is grazer, eating herbs and grasses, so that they are not in competition for the same food resource.

13-4·4. Certain Other Characteristics of Niche

The niche space is usually measured in terms of one to several variables, such as utilization of space, height of feeding or nesting above ground, position in canopy, size of food, and morphological differences. Organisms have restricted areas of the vertical profile in which they forage. For example, warblers are numerous in the evergreen-deciduous forest ecotone of Eastern North America because they nest and feed in so many diverse niches (**Kendeigh**, 1945; **MacArthur**, 1958) :

1. Blackburnian warbler : Top level of evergreen trees.
2. Black-throated green warbler : Middle level of evergreen trees.
3. Magnolia warbler : Low level of evergreen trees.
4. Redstart : Secondary deciduous growth.
5. Black and white warbler : Tree trunks.
6. Black-throated blue warbler : Shaded shrubs.
7. Chestnut-sided warbler : Sunlit shrubs.

8. Canada warbler : Wet shaded ground.
9. Yellowthroat : Wet sunlit area.
10. Ovenbird : Dry shaded ground.
11. Nash Ville warbler : Dry sunlit ground.
12. Louisiana waterthrush : Stream margin.
13. Northern waterthrush : Bog forest.

A similar division exists among the bird titmice, *Paridae*, of Europe (**Lack**, 1971). Such a division of feeding space may even exist between sexes of same species. The male red-eyed vireo (an American singing bird called greenlet), for example, picks up its insectivorous food in the upper canopy, the female in the lower canopy, with only about a 35 per cent overlap in the feeding area between the two (**Williamson**, 1971). Although similar foods may be utilized, each **secures** the insects from different levels. Similar sexual differences in foraging area exist between males and females of the woodpeckers. Even niches may be separated by the size of food, dictated in birds, perhaps by bill-size (**Lack,** 1971). Such morphological differences influencing food procurement may exist either between two species or between the sexes of a given species. For example, two African birds, the greater flamingo (*Phoenicopterus antiguorum*) and the lesser flamingo (*Phoeniconaias minor*) share same pond but exploit different niches for food. Here the fine platelets in the bill of the lesser flamingo restrict its diet largely to microscopic blue-green algae, wheras the coarser filter structure in the bill of the greater flamingo allows it to ingest larger food particles like small animals, as tiny crustaceans and mud-dwelling insect larvae. Similarly a pronounced sexual difference in bill-size exists between some male and female woodpeckers (*Dendrocopus arizonae*). Differences in bill-length are correlated with differences in foraging behaviour. The male seeks food on the trunk, the female on the branches (**Lignon,** 1968).

Further, most niche measurements involve plots along at least two axes, the resulting graph covers a certain area along an axis, called **niche width**. The wider the niche, the more generalized a species is considered to be. A species with a narrow niche is considered more specialized. This hypothesis has led to the suggestion that organisms with wide niches show more variability in environmental tolerance or food size, in behaviour or in some applicable morphological measurement such as bill-size.

Certain islands have a number of **empty niches** (**Boughey,** 1971). This term is used to show that the habitats of the island ecosystem include a number of resources not as yet being fully exploited by the island communities.

13-4·5. Ecological Equivalents

Different communities in ecosystems characterized by similar environments are often exceedingly similar in their structure, and they may contain one or more niches that are essentially identical. The adaptations of populations inhabiting these niches may also be exceedingly similar, even though they are totally unrelated. For example, the succulent (xerophytic) plants of the Southwestern United States are cacti ; in the deserts of Southern Africa, there are plants of the spurge family that are virtually identical. In two pairs of birds namely the little auk (a bird of order of gulls) and Magellan diving petrel (a bird of the order of albatross) ; and the eastern meadowlark (an American icterid bird) and yellow throated longclaw (an African motacillid bird), though each bird is totally unrelated to the other, but they all are virtually identical in appearance and basic adaptations. This phenomenon of similar but unrelated species in different geographical areas is called **ecological equivalence**. It gives much insight into the essence of the niche. The adaptations of an organism to its niche are the result of an extensive period of evolutionary change. In order, for two unrelated species to have evolved to a point of ecological equivalence, the niches to which they are adapted must be essentially identical, even though they may be separated by many thousands of miles. It would be impossible for different species to inhabit the same niche in widely separated areas if the niche and the interactions of which it is composed were a property of population alone. Clearly, the structure of communities set in similar environments is so similar that essentially identical niches can develop in them. This means that the principles by which biological communities are organized surpass the individual communities, and that the evolution of species may proceed in a fairly predictable manner, controlled by the structure of the community in which it exists, which itself is a function of the fundamental abiotic features of the environment.

13-4.6. Advantages of Niche Segregation (or differentiation)

The major advantage which organisms gain by occupying different niches is escape from continuous intense competition. In fact, the niche occupied is favourable to the species physically in furnishing suitable substratum and microclimate, although many species have the ability to live elsewhere where competition is not involved. Automatic segregation of a species into its niche through inherited behaviour patterns or imprinting avoids the great expenditure of energy and loss of time that would be required if this segregation had to be worked out anew each year or each generation. Segregation into niches also avoids confusion of activities between organisms in the community and permits a more orderly and efficient life cycle on the part of each species. Furthermore, the segregation of each species into different niches permits the occupancy of the area by a large number of species, since they will better divide the available resources between them. Similarly, the more distinct the niche of a species is, the more it can avoid conflict with its neighbours and lead a life that is orderly, productive and quite efficient (**Kendeigh**, 1974).

13-5. COMMUNITY STABILITY

One of the most important considerations about communities is their stability, or lack of variation in time. There are two discrete senses in which this term is used. The first, **community stability,** refers to the degree of fluctuation in population size of the populations comprising the community. The second, **environmental stability**, refers to the fluctuations in the abiotic factors of the ecosystem. The two types of stability are closely related, and stable communities are generally to be found in stable environments, while unstable environments are more often characterized by unstable communities.

The global range of variation in community stability is extraordinarily broad. Some communities like the tropical rainforest or the coral reef, appear exceedingly stable, and the densities of their populations are constant over long periods of time. Others, such as the tundra, are extremely variable, and they are made up of populations that fluctuate widely in abundance. Much of this fluctuation is due to variations in environmental factors which the community cannot avoid. However, genetically determined adaptations that have arisen through natural selection also play a part in determining the responses of a population to environmental fluctuations.

Further, the structure of community in which a population is found has a major role in determining the degree of environmental variability to which the species is subjected. There are at least two mechanisms by which community structure affects the stability or variability of a community. The first mechanism involves the relationship between community structure and microenvironment. Many organisms are protected from the full impact of environmental fluctuations because of the presence of other members of the community. For example, the floor of a conifer forest is protected from winds, evaporation, and temperature shifts, making it a much less harsh and variable environment for organisms than the forest canopy or an open field adjacent to the forest. The second mechanism by which community structure affects community stability is community diversity. But about the correlation of community stability and community diversity, many ecologists have expressed their doubts (**Hairston** *et al.*, 1968).

Ashby (1958) pointed out the relationship between the number of interspecific interactions and community stability in his **"law of requisite variety"**. This principle states that for maximum stability, a community must be characterized by a very high number of niche interactions (*i.e.*, if it is a complex community, Fig. 13·9A), the probability of there being a sufficient number of internal feedback mechanisms within the community to give it stability is good. Conversely, if a community is characterized by a low number of niche interactions (*i.e.*, if it is a simple community ; Fig. 13·9B), the likelihood that there will be sufficient internal feedback mechanisms to give it stability is poor.

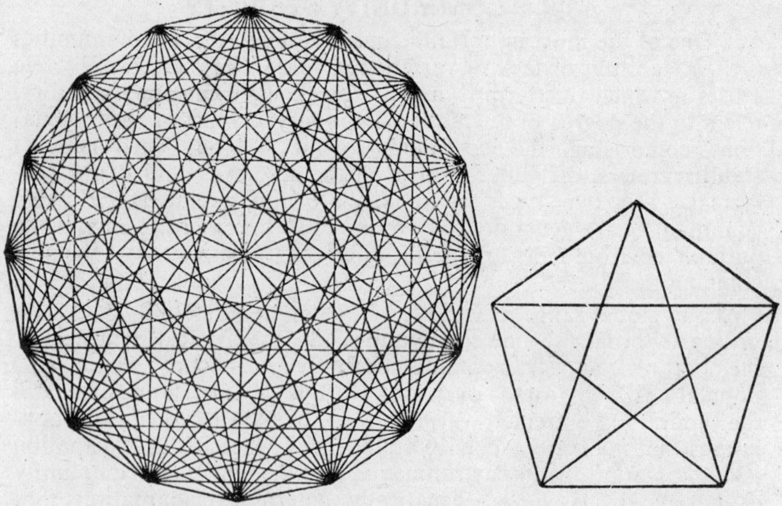

Fig. 13·9. Diagram showing the potential inter-relationships between the species in two communities. Each line represents two interactions—the effects of species A on species B and the converse effect of species B on species A : (A) a complex community, with a large number of stabilizing interactions ; (B) a simple community, with relatively few stabilizing interactions (after Clapham, Jr., 1973).

13·6. CORRELATION OF ENVIRONMENTAL GRADIENTS AND COMMUNITY GRADIENTS

Environmental units which afford uniform environmental conditions to the inhabiting species are called **biotopes**. The physical features of a biotope maintain the populations within the communities. In drastic environmental conditions the flora and fauna undergo abrupt changes and only those that are adapted to withstand the rapid variations are preferred by the environment. For example, in communities of littoral zone or of deserts, only those organisms that are typically adapted to withstand the environmental hazards can survive. Thus, different environmental factors (*e.g.*, temperature, rainfall, salinity, etc.) have a profound control over the populations so that habitats and communities vary along different environmental gradients or there is intergrading among them (Fig. 13·10 and Fig. 13·11). For example, the gradual slope of land into the sea exhibits various grades of environmental factors and animal communities also correspondingly show a gradation from the shore through the intertidal zone to the deep sea region. Characteristic of an environmental gradient is a distinct zonation, each zone harbouring its own assemblage of species adapted to one another and giving way at a sharp boundary to another assemblage

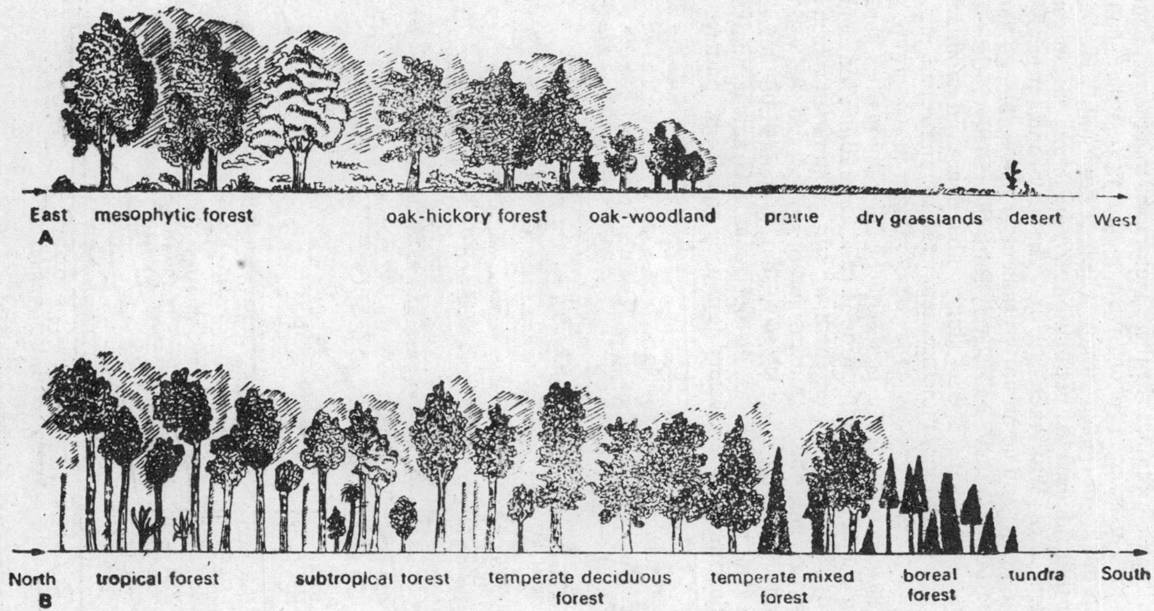

Fig. 13·10. Gradients of vegetation in North America from east to west (A) and north to south (B). The east to west gradient is largely a result of variation in precipitation. The north to south gradient reflects temperature (after Smith, 1974).

of species adapted to one another. For example the **elevation gradient** includes decreasing mean temperatures, decreasing lengths of growing seasons, increasing rainfall, increasing wind speeds, and so on, toward high elevation. All these factors act together on plants and animals. The assemblage of environmental factors that change together through the space along which a community gradient occurs and that influence its population may be termed as a **complex gradient** (see **Whittaker**, 1970). The environment gradient (complex-gradient) and community gradient or **coenocline,** go hand in hand and both gradients collectively form the **ecocline**, the gradient of ecosystem. Thus, species shift in abundance and dominance due to change in altitude, moisture, temperature, and other physical conditions. One species may be dominant in one group, an associated species in another. This sequence of communities showing a gradual change in composition is called a gradient or **continuum** (**Curtis**, 1959). However, there are some communities such as ponds, tidal beaches, grassy balds, island of spruce and fir within a hardwood forest, all of which have sharply defined boundaries. Here the vegetational pattern is **discontinuous**. Gradient analysis of different ecoclines have revealed following trends and patterns :

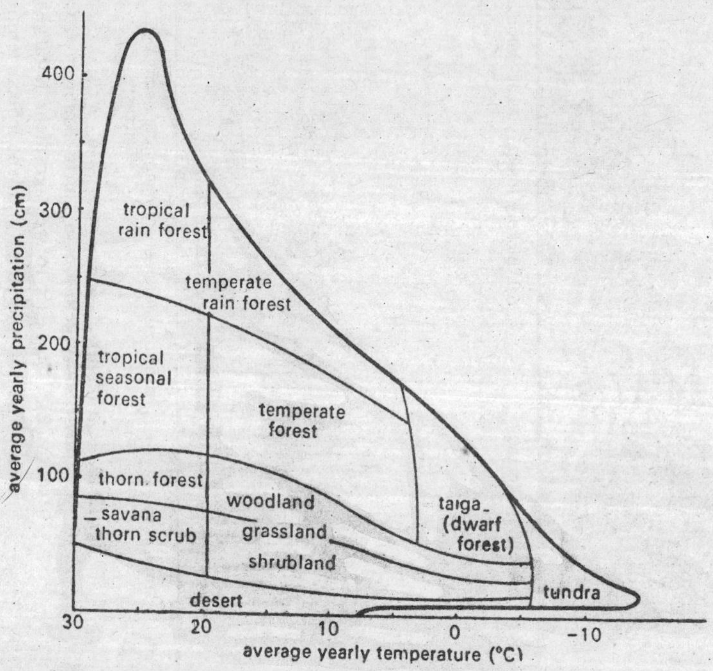

Fig. 13·11. A pattern of world plant formation types in relation to climatic variables of temperature and moisture (after Herreid II, 1977).

1. Along a gradient from a "favourable" environment to an "extreme" environment there is normally a decrease in the productivity and massiveness of communities. The decrease in amount of organic matter per unit area is expressed in decrease of height of dominant organisms and percentage of the ground surface covered.

2. Towards increasingly unfavourable environments there is a stepping-down of community structure and a reduction of strata differentiation, with generally smaller numbers of growth forms, (The different classes or kinds of forms of plants such as trees, lianas, shrubs, epiphytes, herbs and thallophytes are referred to as **growth-forms**) arranged in fewer and lower strata.

3. Some kinds of species-diversities of certain communities decreases from favourable to extreme environments, whether the latter are **extremes** of drought, or of cold, or of adverse soil chemistry or (for the sea cost) of tidal exposure.

4. Each growth-form has its characteristic place of maximum importance along the ecoclines—the rosette trees in some tropical forests, semishrubs in desert and adjacent semiarid communities, and so on. Some growth forms, for example, grasses and grass-like plants may have more than one area of importance along the major ecoclines.

5. Some growth-forms may be dominant in similar environments in widely different parts of the world. Thus, similar environments on different continents tends to have communities of similar physiognomy. This adaptive convergence at the level of the community is one of the major generalizations about the geography of life.

13-7. NATURE OF COMMUNITY

The nature of the biotic community has been the subject of study and dispute for years. There are following two opposing views concerning the nature of the community : the organismic school and the individualistic school.

13-7·1. The Organismic School

This approach is developed by **Clements** (1916) and supported by ecologists such as **Daubenmire** (1966, 1968) and **Langford** and **Buell** (1969). The organismic viewpoint regards the community as a sort of superior organism, the highest stage in the organization of the living world—rising from cell to tissue, organs, organ systems, organisms, population, community. The whole is more than the sum of its parts (see chapter 6). Just as tissues have certain characteristics and function above and beyond those of the cells that comprise them, so the biotic community has characteristics and functions above and beyond the various populations it embodies. The distribution and abundance of one species in the community are determined by the species interaction with others in the same community. Species making up the plant community are organised into discrete groups. Groups of stands similar to one another form associations. Stands of one association are clearly distinct from stands of other associations. Thus, the community acts as an unit in seasonal activity, in competition with other communities, in trophic functions, in succession, etc.

The organismic school also argues that stands studied by gradient analysis are disturbed stands or ones that are not in equilibrium. If undisturbed, all stands develop to an endpoint in a few hundred years ; and stand heading for an end point will naturally show a continuum of species. In section 13-6, we have already observed, that the technique of gradient analysis forces data into a continuum. These gradients are usually based on one variable such as moisture ; but it is impossible to restrict a continuum to one variable alone because many interacting variables influencing plant distribution are involved. The continuum approach assumes that all species are equal when in fact some are dominant.

13-7·2. The Individualistic School

This approach is advanced by **Gleason** (1926) and further developed by **Curtis** (1959), **McIntosh** (1958, 1967), and **Whittaker**

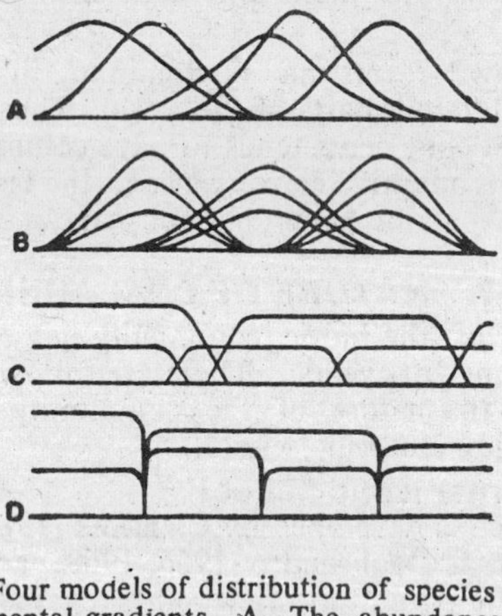

Fig. 13·12. Four models of distribution of species along environmental gradients. A—The abundance of one species on an environmental gradient is independent of the others. Thus the association of several species along the gradient changes with the response of the individual species to that gradient. B—The abundance of one species is associated with that of another. The two or more species are always found in association with each other. C—The distribution of one species is independent of another in an environmental gradient, but the abundance and distribution of each species is sharply restricted at some point on the gradient by interspecific competition. D—The distribution of a species is sharply restricted by a change in some environmental variable. This occurs in ecotone (after Smith, 1977).

(1962, 1970). The individualistic viewpoint emphasizes the species rather than the community as the essential unit in the analysis of interrelationships and distribution. Species respond independently to the physiological and biotic environment according to their own genetic characteristics. They are not bound together into groups of associates that must appear together. Instead, when species populations are plotted along an environmental gradient, long or short, the resulting graph suggests a normal or bell-shaped curve (Fig. 13·12). The curves of many species overlap in a heterogeneous manner. In this way, the vegetation and its associated animal life show a gradient or continuum from one extreme (*e.g.*, dry conditions) to another (*e.g.*, wet conditions). In this view the community is regarded as a collection of populations of species requiring the same environmental conditions. It is a continuous variable, not an integrated unit like the superior organism.

The supporters of individualistic approach suggest that community discontinuties do exist but that these reflect environmental discontinuties, such as breaks in soil types or sharp changes in moisture and salinity. If stands are as distinct as the organismic group considers them to be, then the boundaries between associations should be distinct. But they rarely are. Except where man or some environmental catastrophe such as fire has interfered, it is rare for the vegetation to be a mosaic of discontinuous units. Instead, boundaries between units are more or less diffuse. If separate stands of an association are similar, all such associations should have similar distribution, and the plants that comprise such associations should have distribution locally and over continental limits. If the association is a natural unit, then the species that comprise it should be bound together by obligate inter-relation ships. But most species are not obligatory.

R. L. Smith (1977) has suggested that both of these schools may be partly correct. As the species that collectively form the biotic community may respond individually to the environment ; yet, each community may also operate as a functional unit, especially in regard to the flow of energy and the cycling of nutrients.

13-8. CLASSIFICATION OF COMMUNITY

The current well-accepted view considers a community as a continuum, however, the plant and animal life of any large area is so complex that it must be separated into subdivisions. In fact the aggregation of organisms in any given locality or habitat must be regarded as a unit if the community is to be studied, described, or compared with similar community stands in other habitats. To give order to the study of communities, some system of classification is needed, even though the communities of a region often cannot be placed in discrete units. The most widely used classification systems are based on physiognomy, habitat, floristic characteristics like species composition and dominance and community dynamics.

13·8 1. Physiognomic Classifications

Physiognomy, or general appearance, is a highly useful method of naming and delineating communities, particularly in surveying large areas, and as basis for further subdivision of major types into their component communities. Because animal distribution is mainly depended on the structure of vegetation and not on the species composition, therefore, classification by physiognomy will relate both the animals and the vegetation of an area. Communities so classified are usually named after the dominant form of life, usually plant, such as the coniferous or deciduous forest, sagebrush, short-grass prairie, and tundra. A few are named after animals, such as the barnacle—blue mussel (*Balanus—Mytilus*) community of the tidal zone. One may grade into the other ; so even here the classification may be based on arbitrary, although specific, criteria. Some of the earlier attempts of physiognomic classification by different plant ecologists can be summarized as follows :

Warming (1909) recognised autotrophs and heterotrophs. The aquatic autotrophs, called **hydatophytes**, were grouped into seven classes, while, the aerial autotrophs were **divided** into **atmophytes**, **ombrophytes**, (plants growing in areas of abundant rainfall), **psychrophytes** (plants of cold soils), **halophytes** (plants of soils having high salt contents), **psammophytes** (plants of sandy soils), **agrophytes** (plants having natural habitat in crop-fields), **oxylophytes** (plants of acidic soils), **chasmophytes** (plants growing in rock crevices), etc. **Raunkiaer** (1934) classified plants on the basis of adaptations of life forms to their climates. He selected about

Table 13·3. Biological spectrum of some communities of India.

Region	Phanerophytes (Ph)	Chamaephytes (Ch)	Hemicryptophytes (H)	Geophytes or cryptophytes (G or Cr)	Hydrophytes (HH)	Therophytes (Th)	Plant climate
1. Indian desert	14	19	16	—	3	40	Therophytic
2. Semi-arid region of North India	23	18	11	5	3	33	Thero-phanerophytic
3. Semi-arid region of South India	30	13	1	6	4	28	—do—
4. Andhra	51	12	14	1	3	16	Phanerophytic
5. Mysore	38	31	2	—	—	12	Phanero-chamaephytic
6. Madras	33	23	9	1	4	18	—do—
7. Mount Abu	41	2	2	—	—	47	Therophytic
8. Allahabad	32	9	4	—	8	42	—do—
9. Varanasi	40	6	1	—	10	43	—do—

1000 plant species from all over the world and assigned them to their life forms. Certain plant ecologists have applied Raunkiaer's classification method to **classify** Indian communities (Table 13·3). **Iversen** (1936) based his classification on the water requirements of plants. He recognised the following classes **terriphytes** (land plants), **telmatophytes** (aerial plants with aerenchyma), **amphiphytes** (marsh plants) and **limnophytes** (water plants). The physiognomic classification also includes symbolic classification in which different characteristics of a species, such as life form, plant size, leaf shape and size, function, leaf texture, etc., are represented by symbols. **Dansereau** (1957) has used graphic symbols, while **Kuchler** (1967) has used several alphabets.

13-8·2. Habitat Classification

In areas where the habitat is well defined, physiography is used to classify and name communities. Examples of such are sand dunes, cliffs, tidal mud flats, lakes, ponds, and streams.

13-8·3. Floristic Classification

The floristic classification requires detailed study of the individual community and also involves many concepts like frequency, dominance, constancy, presence and fidelity. The Scandinavian ecologist **Du Rietz** (1921) classified communities into units called **sociation**, which was defined *as a recurring plant community of essentially homogeneous species composition with at least certain dominant species in each layer*. Another floristic unit **consociation** was proposed for the community if only the upper stratum of a multi-layered community is dominated by one species. Thus, the consociation can be defined as *a class composed of individual concrete sociations whose upper strata are dominated by the same species.*

European ecologists group communities into classes, orders, alliances, and associations. **Association** is defined as *a community of definite floristic composition, uniform physiognomy and occurring in uniform habitat conditions.* The associations are recognised on the basis of common dominant species in several stands and they are used in naming the association, for example, *Anogeissus-Boswellia* association, *Shorea-Terminalia* association, or *Balanus-Mytilus* association. The floristic system of community classification works much better with plants alone than with animals or with both.

The floristic classification system is modified when the stands are treated as a continuum. The community complex of a major physiognomy is subdivided by species composition and correlated with an environmental gradient arbitrarily divided into five segments : **wet, wet mesic, mesic, dry mesic**, and **dry (Curtis**, 1959). For example, the deciduous forest in Wisconsin (USA) has been divided into southern and northern hardwoods and northern forest. These are further divided in a moisture gradient. The southern hardwoods include the dry southern hardwoods with bur or black or white oak as the dominants; the dry mesic with red oak or basswcod; the mesic with sugar maple and beech; the wet mesic with silver maple, elm, and ash; and the wet woods with willow or cottonwood. Such a system recognises the influences of habitat on community composition. Detailed studies on animal distribution also reveal similar influences of animal composition.

13-8·4. Dynamic System

Clements (1928) modified the floristic system by laying great emphasis on the community dynamics. He called the climax vegetation in a macroclimatic region as "**formation**" (whenever the vegetation reaches highest development becoming more or less stable for more or **less** definite period, under the existing climate, it is called a **plant formation**). Each formation was divided into a few **associations** which were considered units of climax communities in which a few species were dominant. An association was named after the two or three dominant species. The association was further subdivided in a hierarchical series as follows :

1. **Consociation**. Unit with a single dominant species. The developmental or seral communities of a climax consociation are known as **consocies**.

2. **Faciation**. An association developed under different microclimatic conditions within the same general climate. It contains two or more dominant species and is characterized by specific precipitation, evaporation and temperature. The seral communities are accordingly known as **facies**.

3. **Lociation**. This is a localised variant of an association, differing from it in the composition of some main subdominant and chief secondary sepcies. Seral communities are accordingly called **locies**.

4. **Society**. A community with one or more sub-dominants. The seral communities are called **socies**.

13-8·5 Biome System

Usually animal and plant communities are studied separately, which unfortunately obscures the wholeness of the community and limits our understanding of its function. To escape this dilemma in part, the distribution of animals can be related to the life form of plants and types of vegetation. This results in a more inclusive classification, which embraces several plant communities but includes all animal life associated with them; this classification is called a **biome**. The biome is a broad ecological unit characterized by uniformity and distinctive life forms of the climax species, plant or animal. It is subdivided into smaller units distinguished by uniformity and distinctness in the species composition of the climax and its successional stages. Thus the life forms of plants is emphasized, rather than the taxonomic composition.

13-9. REVISION QUESTIONS

1. What is biotic community ? Discribe the main characteristics of the biotic community.
2. What is life form? Discuss the important use of this concept in the studies of biotic community.
3. Write an essay on the phenomenon of stratification.
4. Describe with suitable examples, the ecotone.
5. Discuss the concept of niche with suitable examples.
6. What are the two views about the nature of community ? Describe in detail.
7. Give an account of different methods of classification of biotic communities.
8. Write short notes on the following :
   (i) Physiognomy;
   (ii) Raunkiaer's growth forms;
   (iii) Vertical stratification;
   (iv) Species dominance;
   (v) Fidelity;
   (vi) Inportance value;
   (vii) Edge effect;
   (viii) Microhabitat;
   (ix) Ecological equivalents;
   (x) Continuum.

14
Ecological Succession and Community Evolution

Biotic communities are not static, they change through time. This change can be understood on several levels. The simplest is the growth, interaction and death of individual organisms as they pass through their life-cycles, affected by the cycles of seasons and other natural phenomena. But there are other levels of community change that act over longer time spans and that account for much larger changes in community composition and structure. These include ecological succession and community evolution.

14-1. ECOLOGICAL SUCCESSION

As a lake fills with silt it changes gradually from a deep to a shallow lake or pond, then to a marsh, and beyond this, in some cases, to a dry-land forest. When a cropfield is deserted or a forest is severely burned over, it is just like a plot of bare ground and a series of plant communities grow up there and replace one another—first annual weeds, then perennial weeds and grasses, then shrubs, and trees—until a forest ends the development. Likewise, if a landslide exposes a surface of rock in the mountains, the surface may be successively occupied by a sparse cover of lichens; a spreading moss mat; grasses, which enter and become a meadow; a shrub thicket, which overtops and suppresses the grasses; a first forest stage of smaller trees, which seed into the shrub thicket, grow through it, and replace it; and a final stage of larger trees, which take dominance from the first trees and may form a larger and potentially permanent forest community. The changes in the structure and composition of the community are rapid at first, slowing gradually until a point of dynamic equilibrium is reached, and the community is more or less stable.

Such an orderly and progressive replacement of one community by another until a relatively stable community, called the **climax community**, occupies the area, is called **ecosystem development (Odum,** 1971) or **ecological succession (Smith,** 1977). In the first example the principal cause of the change in the community was a physical process-the filling in of the lake with silt. In the second example, a principal cause was the growth of plants on an existing soil. In the third example the succession proceeded by a back-and-forth interplay between organisms and environment : as one dominant species modified the soil and microclimate in ways that made possible the entry of a third dominant, which in turn altered its environment. Causes of successional changes are, to varying degrees, external to the community or internal to the community, many successions involve both kinds of causes and reciprocal influences (**Whittaker**, 1970).

Succession is the "birth" of an ecosystem and subsequent "aging" process of its abiotic and biotic features. **Odum** (1971) has rightly included the following three parameters in his definition of ecological succession : (1) It is an orderly process of community development that involves changes in species structure and community processes with time; it is reasonably directional and, therefore, predictable. (2) It results from modification of the physical environment by the community; that is, succession is community-controlled even though the physical environment determines the pattern, the rate of change and often sets limits as to how far development can go. (3) It culminates in a stabilized ecosystem in which maximum biomass (or high information content) and symbiotic function between organisms are maintained per unit of available energy flow.

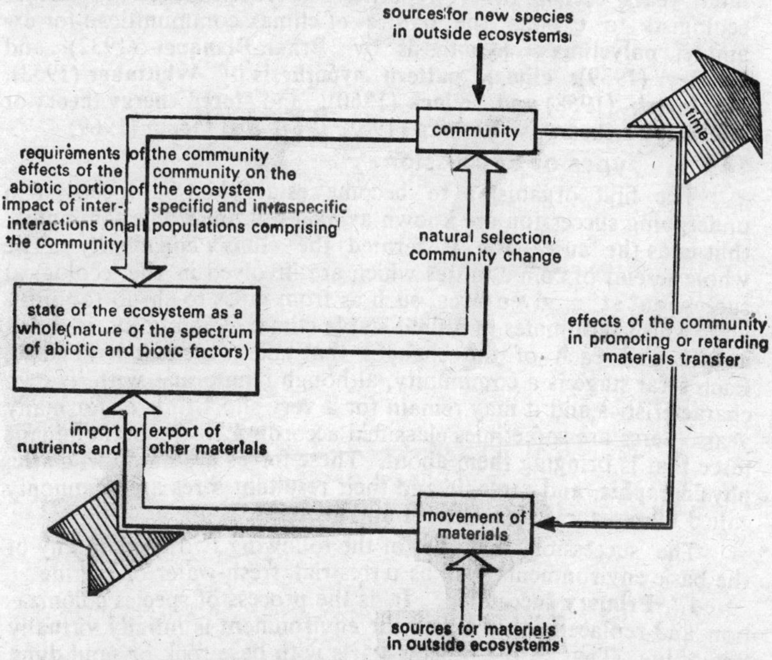

Fig. 14·1. Diagrammatic representation of the interactions between various aspects of the ecosystem as the mechanism for ecological succession (after Clapham, Jr., 1973).

In fact, ecological succession is a complex dynamic interaction between the community inhabiting the ecosystem, the geological factors determining the patterns of materials transfer, the current state of the ecosystem as a whole, and time. Insofar as changes in the general state of the ecosystem are related to the community inhabiting it, and alteration of the community is a function of the evolution of the general state of the ecosystem, succession proceeds under feedback control (Fig. 14·1).

The basic idea of succession was first of all forwarded by **Anon Kerner** (1863) in his book "*Plant Life of the Danube Basin*" during the description of the regeneration of a swamp forest. However, the term ecological succession was first of all used by **Hult** (1885) in the study of communities of Southern Sweden. **H.C. Cowles** (1899) laid the foundation of the succession studies by publishing his classic paper dealing with vegetation of sand dunes of Lake Michigan. The succession of animals on these dunes was studied by **Shelford** (1913). Later on, **Olson** (1958) restudied the ecosystem development on these dunes and has given us an updated information about it. **Federick Clements** (1907-1936) is credited with elaborating the principles and theory of succession. He proposed the **monoclimax hypothesis** of succession. During the later years certain other hypotheses were proposed by various ecologists to explain the nature of climax communities : for example, **polyclimax hypothesis** by **Braun-Blanquet** (1932), and **Tansley** (1939); **climax pattern hypothesis** by **Whittaker** (1953), **MacIntosh** (1958) and **Sellack** (1960); and **stored energy theory** or **information theory** by **Fosberg** (1965, 1967) and **Odum** (1969).

14·1·1. Types of Succession.

The first organisms to become established in an ecosystem undergoing succession are known as **pioneers**; the stable community that ends the succession is termed the **climax community**. The whole sereies of communities which are involved in the ecological succession at a given area, such as from grass to shrub to forest, and which terminates in a final stable climax community, is called a **sere** and each of the changes that take place is a **seral stage**, Each seral stage is a community, although temporary, with its own characteristics and it may remain for a very short time or for many years. Seres are sometimes classified according to the predominant force that is bringing them about. These forces are **biotic, climatic, physiographic, and geologic** and their resultant seres are commonly called **bioseres, cliseres, eoseres** and **geoseres.**

The succession may be of the following two types in any of the basic environments such as terrestrial, fresh-water or marine :

1. **Primary succession.** It is the process of species colonization and replacement in which the environment is intially virtually free of life. That is, the process starts with base rock or sand dune or river delta or glacial debris and it ends when climax is reached. The sere involved in primary succession is called **presere**.

2. **Secondary succession.** It is the process of change that occurs after an ecosystem is disrupted but not totally obliterated.

In this situation, organic matter and some organisms from the original community will remain; thus the successional process does not start from scratch. Consequently, secondary succession is more rapid than primary. It is seen in areas burned by fire or cut by farmers for cultivation. The sere involved in secondary succession is called **subsere.**

The primary and secondary successions may be of the following types depending on the moisture contents :

(a) Hydrach or hydrosere. The succession when starts in the aquatic environment such as ponds, lakes, streams, swamps, bogs, etc., it is called **hydrach** or **hydrosere.**

(b) Mesarch. The succession when begins in an area, where adequate moisture is present, it is called **mesarch.**

(c) Xerach or xerosere. The succession when starts in xeric or dry habitat having minimum amounts of moisture, such as dry deserts, rocks, etc., it is called **xerach.** A temporary community in an ecological succession on dry and sterile habitat is called **xerosere.** It may be of three types: **lithosere**-succession initiating on rocks, **psammosere**-succession initiating on sand; and **halosere**-succession starting on saline water or soil.

Further, a succession progressing entirely as a result of interactions of the organisms and their environment (*i.e.*, "driving force" is internal to the community) is known as **autogenic community** (*e.g.*, succession on sand dunes); and the one moving under the influence of external factors, as input of nutrients, is called **allogenic community** (*e.g.*, succession in a small pond or bog).

Sometimes, succession is classified as autotrophic and heterotrophic on the basis of community metabolism. **Autotrophic succession** is characterized by early and continued dominance of autotrophic organisms like green plants. It begins in a predominantly inorganic environment and the energy flow is maintained indefinitely. **Heterotrophic succession** is characterized by early dominance of heterotrophs, such as bacteria, actinomycetes, fungi and animals. It begins in a predominantly organic enviroment and there is a progressive decline in the energy content. The miniature succession of micro-organisms such as bacteria and different types of fungi on the fallen logs of the decaying wood, tree bark, etc., is called **serule.**

14-1·2. General Process of Succession

The complete process of a primary autotrophic ecological succession involves the following sequential steps, which follow one another :

1. Nudation. The process of succession begins with the formation of a bare area or nudation by several reasons, such as volcanic eruption, landslide, flooding, erosion, deposition, fire, disease, or other catastrophic agency. New lifeless bare areas are also created by man, for example, walls, stone quarrying, burning, digging, flooding large land areas under reservoirs, etc.

2. Invasion. The invasion is the arrival of the reproductive bodies or propagules of various organisms and their settlement in the new or bare area. Plants are the first invaders (pioneers) in any area because the animals depend on them for food. The invasion includes the following three steps: (1) **Dispersal** or **migration**: The seeds, spores or other propagules of the species reach the bare area through the agency of air, water or animals. The process starting from the time a propagule leaves the parent plant to the time it arrives the bare area is called migration. (2) **Ecesis.** This is the successful establishment of migrated plant species into the new area. It includes germination of seeds or propagules, growth of seedlings and starting of reproduction by adult plants Only a few immigrant propagules are capable of doing this under primitive hard conditions, and thus most of them disappear. (3) **Aggregation**: In this final stage of invasion, the successful immigrant individuals of a species increase their number by reproduction and aggregate in a large population in the area and in consequence individuals of the species come close to each other.

3. Competition and reaction. As the numbers of individuals of a species increase due to multiplication and all aggregate at the limited place, the competition for space and nutrition is started among them (intraspecific competition). They also compete with individuals of other species that may enter the area (interspecific competition). All individuals also interact with the environment. Consequently due to increased intra-and interspecific competitions and other type of biotic and abiotic interactions, the environment is modified and progressively becoming unsuitable for the existing community which sooner or later is replaced by new invaders or another community (seral community). Finding the modified environment more suitable more species enter the area and compete with the previous occupants. This results in a balance among the species in which the former species is brought down to a subdominant status or is completely eliminated. The addition of organic matter, nutrients and more moisture in the substratum by small plants make it suitable for larger ones. Increase availability of food allows various kinds of animals to join the community and the resulting interactions further modify the environment and pave the way for fresh invasion by other species of plants and animals to move on the process of succession.

4. Stabilization or climax. Eventually a stage is reached when the final terminal community becomes more or less stabilised for a longer period of time and it can maintain itself in the equilibrium or steady state with the climate of that area. Theoretically at least, this last seral stage is mature, self-maintaining, self-reproducing through development stages and relatively permanent. The vegetation is tolerant of the environmental conditions it has imposed upon **itself**. This terminal community is characterized by an **equilibrium** between gross primary production and total respiration, between the energy captured from sunlight and energy released by decomposition, between the uptake of nutrients and the return of nutrients by litter fall. It has a **wide** diversity of species, a well-developed spatial structure, and complex food chains; and its living biomass is in a steady state. This final stable community of the sere is the **climax community**, and the vegetation supporting it is the **climax vegetation.**

14-1·3. Concept of Climax

The concept of climax has since long been a subject of much controversy and discussion. There are following three theoretical approaches to the climax :

1. **Monoclimax theory**. This theory is developed largely by **Frederick Clements**. This theory recognizes only one climax, determined solely by climate, no matter how great the variety of environmental conditions is at the start. All seral communities in a given region, if allowed sufficient time, would ultimately converge to a single climax. The whole landscape would be clothed with a uniform plant and animal community. All other communities than the climax are related to the climax by successional development and are recognized as subclimax, disclimax, preclimax, post-climax and so on. A **subclimax** is a stage in succession of forests **just** preceding the climate climax community. The **disclimax** is the particular type of vegetation maintained in an area as a result of recurrent disturbance, chiefly the biotic, thus preventing a successful establishment of climate climax community. The **preclimax** is a vegetation of lower-life forms than the one adjacent to it and results from different edaphic conditions. The **postclimax** is a strip of vegetation of higher life-forms occurring within a climate climax, for example, a forest along a stream in a grassland community constitutes postclimax community. Later on numerous other terms such as coclimax, superclimax, quasiclimax, anticlimax, pseudoclimax, etc., were coined by some post-Clements ecologists to describe specific situations.

The monoclimax theory is supported by **Cowles**, **Ranganathan**, and **Puri**, but strongly objected by **Daubenmire** (1968).

2. **Polyclimax theory**. This theory was developed by **Tansley**. This theory considers that the climax vegetation of region consists of not just one type but a mosaic of vegetational climaxes controlled by soil moisture, soil nutrients, topography, slope exposure, fire and animal activity. The advocates of polyclimax theory preferred to call each stable community as a climax and described them with a prefix as edaphic climax, topographic climax, biotic (zootic, grazing or anthropogenic) climax and fire climax. For example, grassland communities of central India, Sri Lanka and parts of California have developed under the influence of fire, grazing and other biotic factors and so have been considered as biotic climaxes by **Misra**, **Pandeya** and **Holmes**.

3. Climax pattern hypothesis. This theory was developed by **Whittaker, MacIntosh** and **Sellack.** According to this theory, the composition, species structure and balance of a climax community are determined by the total environment of the ecosystem and not by one aspect, such as climate alone. Involved are the characteristics of each species population, their biotic interrelationships, availability of flora and fauna to colonize the area, the chance dispersal of seeds and animals, and the soils and climate. The pattern of climax vegetation will change as the environment changes. Thus the climax community presents a pattern of populations that corresponds with and changes with the pattern of environmental gradients, intergrading to form ecoclines The central and most widespread community in the pattern is the prevailing or climatic climax. It is the community that most clearly expresses the climate of the area.

4. Information theory. This theory is proposed by **Leith, Odum** and **Golley.** It considered succession and climax in terms of ecosystem development. In autotrophic succession (ecosystem development), diversity of species tends to increase with an increase in organic matter content and biomass supported by the available energy. Thus, in a climax community, the available energy and biomass, which is called **information content**, increase. In contrast to it, in a heterotrophic succession occurs a gradual depletion of energy, because the rates of respiration always exceed production rates. However, in an ecosystem, both the autotrophic and heterotrophic successions operates in a co-ordinate manner. The autotrophic individuals drives mineral elements from the soil and atmosphere, while the heterotrophic individuals carry on the return of the nutrients to soil and atmosphere, through decomposition of complex dead organic matter. Thus, succession reaches a stage, the climax stage, when the amount of energy and nutrients received from the environment by the plants is again returned in more or less similar amount to the environment by decomposition through heterotrophs.

14·1·4. Examples of Succession

One of the classical studies of ecological succession was made on the shores of Lake Michigan. As the lake has become smaller it has left successively younger sand dunes, and one can study the stages in ecological succession as one goes away from the lake. The youngest dunes nearest the lake, have only grasses and insects; the next older ones have shrubs such as cottonwoods, then evergreens and finally there is a beech-maple climax community, with deep rich soil full of earthworms and snails.

As the lake retreated it also left a series of ponds. The youngest of these ponds contain little rooted vegetation and lots of bass and blue-gills. Latter the ponds become chocked with vegetation and smaller in size as the basins fill. Finally the ponds become marshes and the marshes get transformed into a meadow with further addition of soil. Soon grasses grow, shrubs develop and later trees appear. Eventually, an aquatic plant community of the pond becomes transformed into a terrestrial forest community. Further, in this case of ecological succession, the aquatic animal community gets replaced in an orderly manner from purely fresh-water forms through marsh inhabitants and meadow grazers to terrestrial dwelling forms.

Another dramatic example of community succession began in August 7, 1883. when a volcanic explosion occurred on the Indonesian island Krakatoa, causing part of the island to disappear. The remainder was covered with hot volcanic debris to a depth of 60 metres and all life was obliterated. A year later some grass and a single spider were found. By 1908, 202 species of animals had taken up abode on the island. This increased to 621 species by 1919, and to 880 species by 1934, when there was a young forest on one part of the island.

The process of ecological succession can be best followed by studying the following types of successions in different habitats :

14-1·4 A. Hydrosere (Aquatic succession). The development of a fresh-water body—lake, pool, or pond into a mesic forest can be observed in the limited area, often a freshly-built deep reservoir or a man-made pond (Fig. 14·2). As the soil has been dug out from such a pond, there are very little or no nutrients in the substratum below the water and the water itself does not contain any nutrients. Due to this fact, this stage is characterized by a bottom barren of plant life or animal life. In such a new and virgin pond, aquatic succession or hydrosere starts with the colonisation of some phytoplanktons which forms the pioneer plant community, and finally terminates into a forest (the climax community). The complete process of aquatic succession includes the following stages :

1. **Phytoplankton stage**. During the pioneer stage of succession algal spores may be brought by wind along with soil particles and deposited on the water. The unicellular and colonial phytoplanktonic forms such as bacteria, diatoms and microscopic unicellular algae are the first invaders and therefore this stage is called **phytoplankton stage**. In the presence of traces of phosphorus in the medium, large blooms of blue-green algae appear as they could utilized atmospheric nitrogen. Later on filamentous algae like *Spirogyra* and *Oedogonium* appear.

Simultaneously, certain pioneer **zooplanktons** including protozoons like *Amoeba*, *Paramecium*, *Euglena*, etc., make their appearance. If the plankton growth becomes rich enough, the pond support other forms of animal life as blue-gill fish, sunfish, large-mouth bass and small caddishflies which build cases of sand and feed on micro-organisms living on the bottom.

All these organisms add large amount of organic matter and nutrients due to their various life activities and upon their death, they settle to the pond-bottom to form a layer of muck.

2. **Submerged stage**. The developing layer of loose, oozy material (*i.e.*, mud) on the pond bottom creates a substrate for

rooted submerged hydrophytes such as the branching green algae, *Chara*, *Hydrilla*, *Myriophyllum*, *Elodea* (water weed), *Potamogeton*

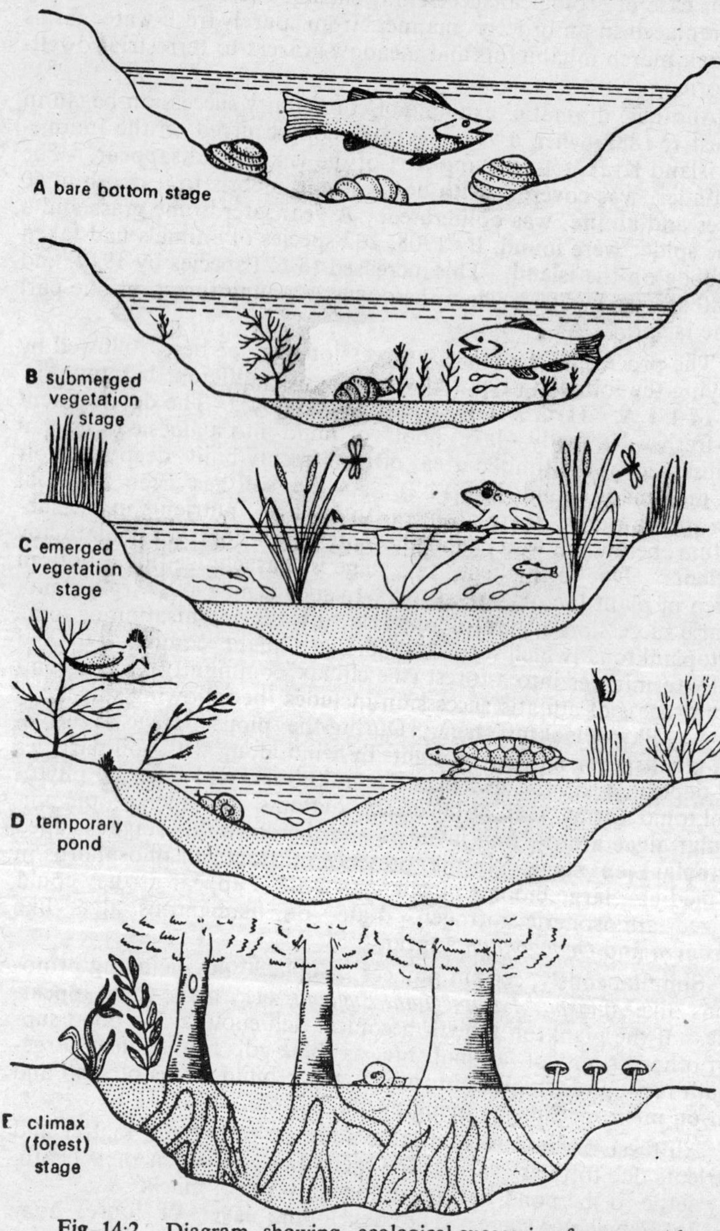

Fig. 14·2. Diagram showing ecological succession from a pond to a climax forest (after cockrum and Mc Cauley, 1965).

(pond weed), *Vallisneria* (eel grass), *Ceratophyllum* (hornwort), *Utricularia* (bladder wort), etc. Their seeds and propagules are brought by birds and other animals which visit the water body frequently for food or other activities. These plants reproduce, increase in number and bind the loose bottom sediment into a firmer matrix and add materially to the deposition of the bottom organic matter. Organisms common to the barren pond bottom cannot exist in the changed conditions of the submerged vegetation stage. The caddisflies of the pioneering stage are replaced by other species able to creep over submerged vegetation and build cases from plant material. Dragonflies, mayflies and some small crustaceans like *Asellas*, *Gammarus*, *Daphnia*, *Cypris*, *Cyclops*, etc. inhabit the pond at this stage.

3. **Floating stage**. Accumulation of sediment washed into the pond from the surrounding watershed supplemented by the organic matter from plant and animal remains reduce water depth and provide nutrients for more demanding plants. Floating hydrophytes like *Nelumbo*, *Nymphaea*, *Marsilea*, *Limnanthemum*, *Aponogeton*, *Trapa*, *Monochoria*, etc., with roots embedded in bottom sediments and leaves floating on the water's surface invade the pond. Some species of free-floating plants such as *Azolla*, *Lemna*, *Wolffia*, *Pistia*, *Spirodella*, *Salvinia*, *Eichhornia*, etc., also invade and colonize the water body. Since these plants shut out the light from the pond's depths, they tend to eliminate most of the submerged aquatic plants. However, certain shade-tolerant submerged hydrophytes like *Vallisneria*, *Najas*, *Nitella*, *Chara*, *Potomogeton*, *Nechamadra*, etc., thrive well at this stage.

In floating stage faunal living space is increased and diversified. Hydras, frogs, salamanders, gill-breathing snails, diving beetles (*Dysticus*) and host of new insects capable of utilizing the under surfaces of floating leaves appear. Some turtles and snakes also invade the pond.

By now, the water level becomes very much decreased due to progressive build up of pond substratum and the pond becomes more shallower. As the water body is reduced in area, the marsh vegetation encroaches upon newly exposed shallow water areas and a reed-swamp stage is reached.

4. **Reed-swamp stage**. During this stage the hydrophilous emergent marshy plants such a *Ranunculus*, *Saggittaria* (grow head), *Monochoria*, *Cyperus*, *Scirpus* (bulrushes), *Typha* (cattail), *Phragmites* (reed grass), *Rumex*, etc., are firmly anchored in the bottom muck by spreading their fibrous roots and rhizomes. Shoots of these plants are partly or completely exposed to air. Lacking the buoyancy and protection of water, the weak and soft-tissued floating plants fail to exist in the changing environment.

Animals of the floating stage are also replaced by those that inhabit the vegetation of reed-swamp stage. Gill-breathing snails give way to lung-breathers like *Lymnea*, *Physa*, and *Gyraulus*. Different species of mayflies (*Ephemera*) and dragonflies spend their nymphal stages on submerged stems and climb to the surface when they are ready to emerge as adults. Certain other insects such as water scorpion (*Nepa*), giant-water bug (*Lithocercus*, *Belostoma*) scavanger beetles (*Hydrophilus*), etc., are present at this stage. Red winged black birds, ducks, king fisher, great blue heron (*Ardea*), swamp sparrow, muskrats (*Ondatra*), beavers, etc., become common in the area. As the oxygen supply of the water decreases because of the increasing quantities removed through respiration by organisms of decay (reducers), breaking down the accumulated organic matter, only animals of low oxygen requirements can exist. Bullheads replace sunfish and annelid worms colonise the bottom muck.

5. **Sedge-meadow stage**. Since the root system and the annual deposition of leaf growth add great quantities of organic matter to the bottom and entrap sediments, the substrate build up rapidly after the emergents have appeared. Much of the old open water area is covered by species of Cyperaceae and Graminae, such as numerous species of sedges as *Juncus* (rushes), *Carex* (carices), *Eleocharis* (spike rushes), *Polygonum*, etc., and many species of forbs like *Mentha* (mints), *Colha* (marsh marigold), *Campanula* (bell flower), etc., to form **marsh** or **swamp**.

Further, as the bottom rises above the ground water level, the remnant of the open pond dries up in summer. It has now become a **temporary pond** drying in summer and freezing in winter. In a gradual manner as land builds higher, drainage improves, emergents disappear and the soil lies above the water table and organic matter exposed to the air decompose very rapidly. Meadow grasses accompanied by land animals invade to form a **marsh meadow** in forest regions and a **prairie** in grass country. With the approach of mesic conditions the herbs, and woody plants invade the area.

6. **Woodland stage**. This stage is characterized by the plants that can tolerate water-logged soil around their roots. Some common plant species of this stage include *Salix* (shrubby willows), *Cornus* (dog woods), *Cephalanthus* (button bush), *Alnus* (alders), *Populus* (cotton woods), tree willows, etc. The root systems of these plants spread horizontally instead of vertically in the soil due to high water table. By this time there is much accumulation of humus with rich flora-fauna of micro-organisms. Thus, mineralization of soil favours the arrival of new tree species.

7. **Forest stage**. This is the climax community. The woodland community is rapidly invaded by several trees. In tropical climates with heavy rainfall, there develop tropical rain forest, whereas in temperate regions, there develops mixed forests of aspen, elm (*Ulmus*), red and silver maples, ash (*Fraxinus*), oak (*Quercus*), white pine, etc. In regions of moderate rainfall, there develops tropical deciduous forests.

Moreover, as the forest floor becomes drier and the crown closes, seedlings of intolerant forest trees are unable to develop; but seedings of sugar **maple,** beech, hemlock, spruce and cedar able to grow in low light intensities of a temperate forest and dominate the understory and subsequently replace the intolerant trees. Since these trees tolerate the environmental conditions they create, the forest cover becomes stabilized.

14-1·4 B. Lithosere (Terrestrial succession). Lithosere is a typical xerosere and it occurs on bare rocks. Lithosere includes following seral stages :

1. Crustose-lichens stage. The rocks are dry and hard substrata which are characterized by full exposure to the sun, violent fluctuations in temperature and rapid changes in moisture conditions. The soil is absent for the penetration of roots and supply of nutrients. Thus, such a habitat is unsuitable for most **plants** but for blue-green algae and lichens, the pioneer species. The blue green algae as *Scytonema* are found to adhere to the rock by their mucilaginous cell walls. These algae tolerate extremes of temperature and moisture contents and can utilise atmospheric nitrogen. However, in cooler climates (high altitudes, and temperate regions) crustose lichens like *Rhizocarpon*, *Rinodina*, and *Lecanora* are common pioneers. They produce some acids which bring about weathering of rocks. The **dead** organic matter of algae and lichens become mixed with the small particles of rocks to form the very thin layer of moist soil on the rocks. The crustose lichens are then replaced by foliose type lichens.

2. Foliose-lichen stage. On the little soil which is accumulated on the rocks due to weathering effect of algae and crustose lichens, there appear foliose and fruticose lichens like *Dermatocarpon*, *Umbilicaria*, and *Parmelia*, etc., which have large leaf-like thalli. They can absorb and retain more moisture, are able to accumulate dust particles and produce more organic matter which help in the further build up of the soil on the rock.

3. Moss stage. The development of thin humus-rich soil layer on rock surface and in crevices and depressions in the rock, favours the growth of certain xerophytic mosses such as *Grimmia* (black moss), *Tortula* (twisted moss), *Polytrichum* (hair moss), *Bryum*, *Barbula*, *Funaria*, *Weissia*, *Hypnum* and *Fissidens*. These mosses grow, compete with the lichens and add more organic matter to the soil after their death and decay. The accumulated thick layer of soil retains great amount of water and provides suitable habitat for the invasion of herbs, grasses and ferns.

4. Herb stage. The herb stage is characterized by plant species like *Aristida*, *Festuca*, *Poa*, *Sporobolus*, *Lindenbergia*, *Solidago*, *Adiantum*, *Asplenium*, *Cheilanthes*, *Actiniopteris*, *Justicia*, *Tridax*, etc., Certain animals such as nematodes, insects and microarthropods, and other animals also invade the area. The weathering of the rocks by physical, chemical and biological means provides more soil and nutrients. Various bacteria and fungi decompose the accumulated organic matter. At this stage certain larger herbs and grasses such as *Themeda*, *Heteropogon*, *Cymbopogon*, *Adathoda*, etc., enter the community. Herb stage is followed by shrub stage.

5. **Shrub stage**. The shrub stage is characterized by small and large shrubs such as *Zizyphus*, *Capparis*, *Zygophyllum*, *Rhus*, etc., in moderate climates and *Rubus*, *Fragaria*, *Physocarpus*, *Symphoricarpos*, etc., in cooler climates The shrubs overshadow the herbaceous vegetation and produce more organic matter and thus increase the amount of humus and humidity in the soil. The species diversity of flora and fauna is increased. The shrubs are finally replaced by trees which form the climax community.

6. **Forest stage**. Forest climax community commence by invasion of certain xerophytic plants and this is determined by climate of the region. In dry climates, due to slow weathering of rocks, comparatively thin layer of soil is formed which supports small trees like *Acacia*, *Prosopis*, *Balanites*, *Anogeissus*, *Boswellia*, etc. All these plants require high rainfall to reach upto climax stage. In moist and wet climates and also in temperate climates dense climax forest is developed.

Like the hydrosere, the lithosere involves successive changes in animal life. The pioneer animals of the lichen stages are few species of mites, ants and spiders. These animals are exposed to harsh environment such as extreme fluctuations in temperature. During the moss stage, many new species of mites, small spiders, tardigrades and springtails invade the community. The herb stage is characterized by nematodes, mites, collembola, ants and various insect larvae. During the shrub and forest stages great qualitative and quantitative modifications occur in the fauna. Thus, there occur numerous kinds of animals such as slugs, snails, wire worms, millipedes, centipedes, mites, ants, sow bugs, springtails, amphibians like salamanders, frogs, etc., reptiles such as turtles, skinks and other lizards, snakes, birds like flycatcher and grouse, and mammals like mole, mouse, shrews, squirrels, chipmunk and fox.

14-1·4 C. Psammosere (Terrestrial succession). This kind of xerosere occurs on sand deposit (or sand dunes) existing on the banks of rivers, lakes and sea shores and also in the deserts. A product of pulverized rocks, sand is deposited by wind (**eolian sand**) and water (**alluvial sand**). The sand dunes are rigorous environments for life to colonise them due to their various characteristics such as very unstable existence, extreme **dryness** due to poor holding capacity of moisture and high surface temperatures by day and cold temperatures by night. Certain algae and grasses like *Cenchrus*, *Ammophila* and *Saccharum* are the most successful pioneers and binding plants. When these and such associates as beach pea (*Lathyrus japonicus*) have stabilized the dunes various deep-rooted shrubs like *Leptadenia*, *Citrulus*, *Calligonum*, *Tephrosia*, *Balanites*, *Sericostoma*, *Zizyphus*, *Salvadora*, *Capparis* and *Calotropis* invade the area and enrich the soil with moisture, nutrients and organic matter. Once the sand dunes become so stable, climax forest of deep-rooted perennial trees such as *Acacia*, *Albizzia*, *Anogeissus*, *Casuarina*, *Diospyros* and *Prosopis* is established in dry climates. Species of other trees and tall grasses like *Calamus*, *Arundo*, *Phragmites*, etc., are found to occur along river banks. In temperate regions, the shrub stage passes to become climax forest of oak.

14-1·5. Heterotrophic Succession

Within each major community and dependent of it for an energy source, is a number of heterotrophic micro-communities. Dead trees, animal carcasses and droppings, plant galls, tree holes, all furnish a substrate (microhabitat) on which groups of plants and animals live, succeed each other, and eventually disappear, becoming in the final stages a part of nutrient base of the major community itself. In these instances succession (called **microsuccession** or **heterotrophic succession**) is characterized by early dominance of heterotrophic organisms, maximum energy available at the start, and a steady decline in energy as the succession progresses.

Winston (1956) observed that an acorn supports a tiny parade of life from the time it drops from the tree until it becomes part of the humus. During the pioneer stage of the heterotrophic succession certain insects like the acorn weevil, *Curculio rectus*, invade the acorn by burrowing through the pericarp and its female deposits the eggs in the embryo of acorn. Upon hatching, the larvae tunnel through to the embryo and consume about half of it. Some fungi like *Penicillium* and *Fusarium* also invade the acorn, either simultaneously with the weevil or alone, and utilize the material. The embryo of acorn then turns brown and leathery and the weevil larvae become stunted and fail to develop.

When the acorn's embryo is destroyed, partially or completely, by the pioneering organisms, other animals and fungi enter the acorn. Weevil larvae leave the acorn through an exit hole, which they cut through the outer shell. Through this hole fungi feeders and scavengers enter. Most important is the moth *Valentinia glandenella*, which lays its eggs on or in the exit hole, mostly during the fall. Upon hatching, the larvae enter the acorn, spin a tough web over the opening and proceed to feed on the remainder of the embryo and the faeces of the previous occupant. At the same time numerous species of fungi enter and grow inside the acorn, only to be utilized by another set of occupants, the cheese mites, *Tryophagus* and *Rhyzozhyphus*. By the time the remaining acorn's embryo tissues are reduced to faeces and the acorn is occupied by cellulose-consuming fungi. The fruiting bodies of these fungi, as well as the surface of acorn, are eaten by mites and Collembola and if moist, by cheese mites too. At this time predaceous mites enter the acorn. Outside on the acorn, cellulose and lignin-consuming fungi soften the outer shell and bind the acorn to twigs and leaves on the forest floor. As the acorn shell becomes more fragile, new holes appear in it. Through these holes enter large animals such as centipedes, millipedes, ants, collembola and earthworms. Due to their activities the amount of soil in the **cavity** increases and greatly softened shell eventually collapses into a mound and gradually becomes incorporated into the humus.

Similarly, the heterotrophic succession of fungi on plant remains in the soil (*i.e.*, dead roots) involves the primary colonisers or pioneers in the form of weak parasites and saprophytes of sugar fungi group (mucoraceous phycomycetes) which feed on sugars and other carbon compounds. The ascomycetes and their imperfect forms (cellulose decomposers) are the **secondary colonisers**. Finally, the basidiomycetes (lignin decomposers) appear in the succession. Thus, chemical make-up of the substrate determines the qualitative characteristics of fungal flora appearing at a particular stage of decomposition.

Likewise, the primary colonisers of heterotrophic succession on dungs and similar excreta of herbivorous animals include phycomycete (mucorales). These are followed by secondary colonisers, the ascomycetes and deuteromycetes and in turn, being succeeded finally by the basidiomycetes.

14·1·6 Nature of Succession (Trends of Succession)

The studies of successional communities in the artificial and natural ecosystems have led some ecologists such as **H T. Odum** and **R.C. Pinkerton** (1955), **Cooke** (1967), **Margalef** (1968), **Mc Cormick** (1968), **E.P. Odum** (1969) and **Whittaker** (1970) to develop a model of succession that suggests the following widely applicable trend.

At the outset, this model makes clear that succession is not directed by an outside agency and that there is no utimate goal toward which an ecosystem heads. The forces that drive succession from the pioneer stages to a climax are natural selection and the normal organism--environment interactions which are present in all ecosystems regardless of whether they are undergoing succession or not. Populations that are efficient enough to withstand competition and that have a sufficiently broad niche to insure their survival can be found in the community at any point in the succession, where as those without these qualities either become locally extinct or never enter the community in the first place. The progreesive alteration of the community comes about because the array of niches in the ecosystem changes with the maturation of the system.

As a general rule, the environment that presents itself to a pioneer community is a harsh place. Pioneer species must be generalists with wide niches, able to withstand fluctuations of abiotic factors **unmitigated** by intracommunity forces. Consequently, early stages of succession are characterized by relatively few species, low biomass, and dependence on an abiotic source of nutrients. The ratio between gross primary production and biomass is high and net community production is greater than respiration, resulting in an increase of biomass. This ratio is a measure of the amount of photosynthetic energy fixed by the community that is needed to support a given level of biomass, and is thus a measure of the inefficiency of energy transfer. Its high level during pioneer stages in succession indicates either that it takes a great deal of **energy** to support a given amount of biomass, or that generalist pioneer populations are not very efficient in their disposition of energy throughout the community. Diversity of the pioneer community is low, so that food webs are poorly developed; food chains are short, linear and largely grazing and interspecific interactions are minimal. Ecosystems in early successional stages tend to be fairly open with respect to nutrients, that is, the organic phases of nutrient cycles are poorly developed, and nutrient movement into or out of the system may proceed readily. Thus, the pioneer community is not a highly organised and stable entity.

If there were no alteration in the characteristics of the ecosystem, the pioneer community would last indefinitely. For example, **Tilly** (1968) found that the community of a cold spring supported largely by detritus was young or immature yet was a stable ecosystem in the sense that it was not subject to rapid change. It was maintained in that condition by restricted space, a heavy flow of water and a low but constant temperature. In such physically controlled environments succession may never occur. However, in most biologically regulated ecosystems, as succession progresses, the ecosystem becomes less harsh, either because the environment is moderated by the organisms themselves or because materials that act to moderate the environment are added from outside. The former involves phenomena such as the weathering of rocks or stabilization of shifting sand by plants or the addition of organic material to the soil ; the latter might involve the importation of material such as nutrients or organic detritus from an adjacent ecosystem.

The mature stages in succession are characterized by a greater diversity of species, great physical stratification, complex food webs, high biomass and a nutrient source largely organic in nature, high net production and a low ratio between gross production and biomass and by gross community production that about equals respiration. In these advanced stages of succession energy is channelled down many diverse pathways and shared by many units. Food chains are complex and largely detrial. Inorganic nutrients accumulate in soil and vegetation. The soil, too reflects the progressive development of the ecosytem, exhibiting increasing depth, increasing organic content, and increasing differentiation of horizons as the mature soils of the final community are approached. Finally mature systems or climaxes are more stable and their population of long-lived species resistent to external disturbances.

Connell and **Orias** (1964) have proposed a useful model to show the mechanisms of community change during succession (Fig. 14·3). Succession proceeds through a classic positive feedback

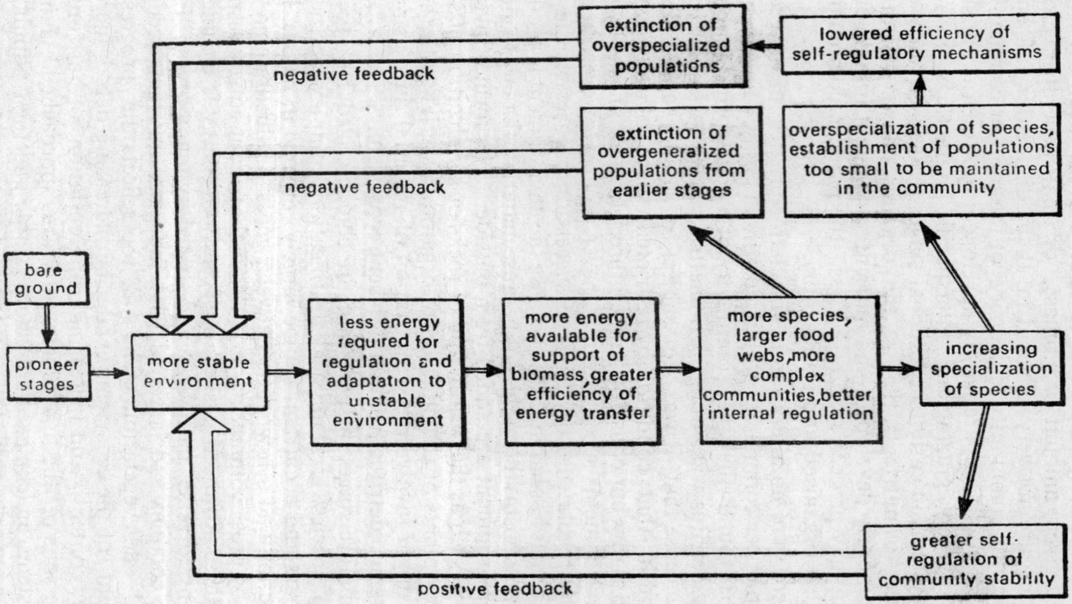

Fig. 14·3. Schematic representation of the progressive changes in communities during ecological succession as a series of feedback loops (after Clapham, Jr., 1973).

circuit, as changes in the total environment which make it more habitable lead to the successful introduction of increasingly specialized populations which further stabilize the ecosystem and allow the establishment of even more specialized forms. At the same time, generalist species which were present during the earlier stages of the succession are forced out, as they become competitively inferior to the more specialized species of the later stages. The climax is reached when the ecosystem reaches its maximum stability and has been modified as far as possible. Additional species at this point tend to stabilize it further, and are likely to be overspecialized forms which cannot remain in the community. Thus, the negative feedback loop through which superfluous populations are lost to the community becomes more important than the positive feedback loop that drives the progression, and succession change ceases.

Recently, certain serious anomalies have been recorded which contradict these trends of succession. Diversity of species does not necessarily increase with advancing successional changes. For example, in some old fields, the very early stages may have greater diversity than the later stages (**Tramer**, 1975). Some of the later stages may be dominated by plants with strong alleopathic chemical or other method of effective interference that reduce species diversity (**Bazzaz**, 1975).

The net productivity of the community does not necessarily increase through succession. A recent study suggests that the average net productivity decreases as relative dominance declines and diversity increases. Further, vegetation types do not necessarily follow one another sequentially. Several stages may be skipped, telescoped or extended. River bottom farmland, for example, may grow up quickly in yellow poplar and other forest trees, missing earlier successional stages completely.

14·1·7. Community Retrogression

It is in some cases possible to drive the successional process backwards by disturbing a community. A grassland under heavy grazing may show first a reduction in coverage of the plants most palatable to the cattle. With continued grazing the total grass coverage is reduced, the more palatable plant species disappear, and weeds that may be almost absent from the undisturbed grassland appear and spread with continued grazing and tampling, most of the grass may be destroyed and, despite the unpalatable weeds that survive, **the** soil is exposed to erosion. The end result of such community retrogression may be depending on the topographic position and soil, a mud field with erosion channels or a rocky mountain slope stripped of much of its soil. Thus by severe and **continued** disturbance all the trends of succession—soil development, community stature, productivity, diversity, stability, and extent **of** modification of environment by community—have been reversed toward zero.

14-1·8. Influences on Succession

Man has greatly modified natural communities the world over. **Fire**, more often man-caused than not, sets back succession to an earlier stage and changes the composition and type of community. By **lumbering** or cutting the forest for timber, man often has tended either to remove the trees completely or to leave the poorest-quality and genetically inferior individuals to reproduce the forest. **Overgrazing** of forests and grasslands by both domestic stock and mismanaged wild grazing animals has resulted in the denudation of grasslands and serious disturbances and even destruction of forest. To provide food for himself, man has cleared away natural vegetation and replaced it with simple, highly artificial communities of cultivated species adapted to grow on disturbed sites. This has brought about an explosion of insect pests, accelerated the erosion of unprotected soil and changed the nature of the successional communities that follow abandonment. Further, in the extraction of minerals, particularly coal, vast acreages of mountain and prairie land are overturned, their natural vegetation destroyed, and their water regimes altered.

14-2. COMMUNITY EVOLUTION

Like the responses of communities to changing abiotic conditions, community evolution involves progressive changes in climax communities. Because the evolution is exceedingly slow, it cannot be observed in operation, and few instances from the fossil record are sufficiently complete to show the process in action.

The example that best demonstrates evolution of the basic structure of the community is that of the development of a terrestrial community of a modern type by early reptiles some 250 million years ago. Between the time when vertebrates (amphibians) first became able to lead a predominant terrestrial existence some 350 million years ago and the establishment of an essentially modern type food web some 100 million years later, the structure of terrestrial community was decidedly different from what it is now. Development of the modern type of community structure required not only a complete rearrangement of the niche structure of the community but also the evolution of new species that could fill the new niches (**Olson**, 1961, 1966).

Attainment of the adaptations needed for terrestrial life by the first amphibians did not in itself establish a land-based vertebrate community. These early amphibians were carnivores, and the only animals inhabiting the land environment were insects. It is inconceivable that the clumsy locomotor system of early amphibians would have allowed them to prey effectively on animals such as insects. Thus the first communities inhabited by terrestrial vertebrates are best regarded as extensions of aquatic communities, with the land habit as an adaptation to improve the capabilities of organisms whose prime food supply was aquatic invertebrates and fish (Fig 14·4A).

By some 300 million years ago, reptiles had evolved that could feed effectively on terrestrial invertebrates (Fig. 14·4B). An entirely land-based community was theoretically possible in which all herbivore niches were assumed by invertebrates and some of the carnivore niches by vertebrates. However, the palaeoecological

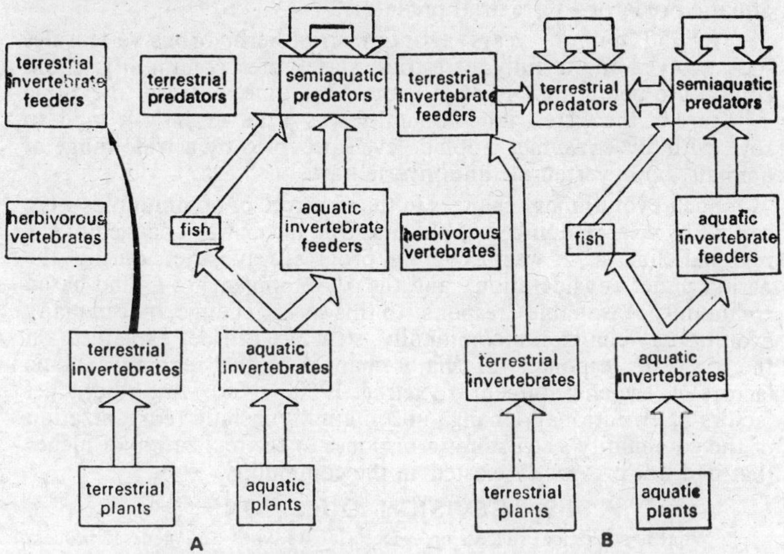

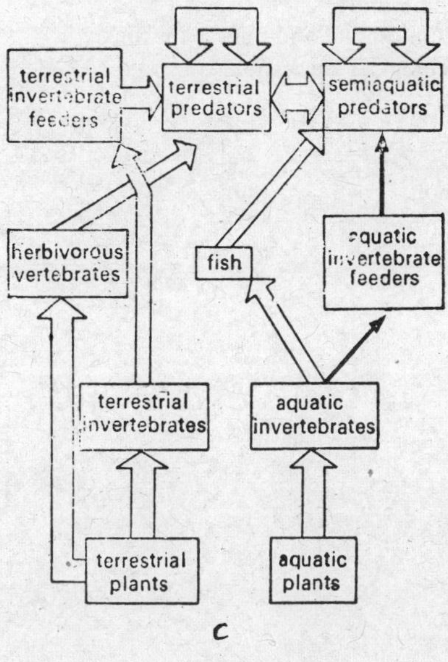

Fig. 14·4. Stylized food webs at several stages of the development of the modern type of terrstrial vertebrate community : (A) immediately after the realization of the land habit by early amphibians; vertebrate food chains entirely water based; 350 million years ago; (B) intermediate stage. 300 million years ago; vertebrate food chains are largely water based; (C) modern type of land vertebrate community, realized by the development of terrestrial herbivorous vertebrates; food chains largely land-based; 250 million years ago (after Clapham, Jr., 1973).

evidences suggest that most contemporary carnivorous vertebrates were unable as yet to realize an entirely terrestrial carnivore niche, so that the great majority of the energy flow through the community continued to pass through the aquatic route. The typical food chain to the highest terrestrial vertebrate carnivore was plant→aquatic invertebrate→aquatic-invertebrate-feeding vertebrate→semi-aquatic predator→terrestrial predator.

By 250 million years ago terrestrial herbivorous vertebrates had evolved and a fully terrestrial vertebrates community could come into being(Fig. 14·4C). From this time onward the basic structure of the terrestrial community was of an essentially modern sort, with all consumer trophic levels occupied by a wide range of animals, both vertebrate and invertebrate.

Such evolutionary changes in the structure of communities were caused by a large number of factors. One factor was changed in the regional climate. It was becoming progressively drier during the period under consideration, and the development of a land-based community reasonable response to this sort of change. Indeed many evolutionary changes in community structure can be explained on the basis of responses of major changes in the regional abiotic factors of the environment (**Axelrod**, 1950, 1958). But other chief factors of evolutionary change in community include reorganization of the community's structure in response to the realization of niches that had not previously existed in the community.

14-3. REVISION QUESTIONS

1. What is ecological succession ? Describe its types and general process
2. Describe the concept of climax community.
3. Describe the ecological succession in a pond and on a bare rock,
4. Write an essay on the nature of ecological succession.
5. Write short notes on the following :
   (i) Heterotrophic succession;
   (ii) Psammosere;
   (iii) Human influence on succession;
   (iv) Community retrogression.

15

Population

All living organisms exist in groups of the same species, or **populations**. The word population (*L. populus*, people) was originally referred to a group of people occupying a particular space Thus, a population can be defined as the total assemblage of individuals of the same species occupying a particular space at a given time. Population is a unit of the ecosystem through which energy flows and nutrients are cycled and it is a self-regulating system that helps in maintaining the stability in the ecosystem. Population has a density and age structure, a birth rate and a death rate, a growth form, and numerical dispersion in time and space. The study of ecology of individuals of same species is called **population ecology**, *i.e.*, it studies the number of organisms and causes of their abundance and distribution.

15-1. POPULATION DENSITY

The density of a population refers to its size in relation to some unit of space on observed time. The size of the population can be measured in several ways, including **abundance** (absolute number in population), **numerical density** (number of individuals per unit area or volume), and **biomass density** (biomass per unit area or volume). The density of a species population can be expressed either with reference to total area (**crude density**) or with reference to the actual area of habitat available to the species (**ecological density**). Thus, we might have a crude density of 500 rabbits per square mile but if only half of the square mile had suitable habitat for rabbits, the ecological density would be 1000 per square mile of rabbit habitat. When the size of individuals in the population is relatively uniform, as mammals, birds or insects, then density is expressed in terms of number of individuals (**numerical density**). But, when the size of individuals is variable, such as true of fishes, trees or mixed populations (**biomass density**). To measure density of a population in terms of measurement of biomass will be satisfactory as a measure of density biomass one can take wet weight, dry weight, volume and carbon and nitrogen weight.

Thus, **population density is the total number of the species within some natural habitat. It can be easily calculated by using density formula as follows :**

$$D=\frac{n/a}{t}$$

where, D is population density ; n is a number of individuals ; a is area and t is time unit.

The density of organisms on any area varies. It may change with the seasons, with weather conditions, with food supply, and with many other influences. **There is, however, an upper limit to** the density of a population within a unit area, imposed **by size** and trophic level. Generally the smaller the organism, the greater its abundance per unit area. A 100 acre forest will support a greater number woodland mice than deer. A forest stand will contain many more trees 5 to 8 cm dbh (diameter breast height) than trees 30 to 35 cm dbh. The larger its size or the higher its position on the trophic levels, the less is the numerical density of an organism. Further, the existence of a population in a particular area for a given time depends upon its rate of reproduction (**natality**) and mortality or death rate, besides the rates of immigration or ingress and emigration or egress. These phenomena exhibit structural attribute of a population and are intimately correlated with the functional aspects such as maintenance, growth and expansion of population.

15-2. NATALITY OF POPULATION

Populations increase because of natality. Natality is equivalent to birth rate and is an expression of the production of **new individual** (*viz.*, reproduction) in the population by birth, hatching, germination, or fission. At this stage, two aspects of reproduction namely fertility and fecundity must be distinguished. The fundamental notion of **fertility** is an actual level of performance in the population based on the numbers born. While **fecundity** is the potential level of performance (or physical capacity) of **the** population (**Krebs**, 1972). For example, the **fertility rate** for an actual human population may be only one birth per 8 years per female in the child-bearing ages, whereas the **fecundity rate** for humans is one birth per 9 to 11 months per female in the child bearing age.

Natality rate may be expressed as the number of organisms born per female per unit time. The **crude** natality rate of population can be expressed as follows :

$$\text{Birth (Natality) rate } (b)=\frac{\text{number of births per unit time}}{\text{average population}}$$

or

$$b=\frac{\text{dNn}}{\text{Ndt}}$$

where b=natality rate per unit time, d (or Δ)=entity that is changing, N=initial number of individuals in the population, Nn =number of new individuals added to the population by natality, t=time.

The birth rate is the function of the density of the population. As a general rule, if the population density is unusually low, the birth rate is also low, because the probability of males and females finding each other to mate is small. If population density is unusually high, the birth rate may also be relatively low, because of inadequate nutrition or the physiological or psychological aberrations associated with crowding. There is often an optimum density at which the birth rate is maximized. This is generally a relatively low density. Thus the **potential, physiological, maximum** or **absolute natality** is the theoretical maximum production of new individuals under ideal conditions. To attain this maximum, a species would exist under ideally optional ecological and genetic conditions. **Ecological** or **realised natality** refers to increase of population under given set of environmental conditions. It is not constant for a population but may vary with the size, composition and fecundity of a population.

15-2·1. Fecundity

Species vary greatly in the characteristic number of generations, broods or litters produced per year, and in the size of them. Protozoans often divide so rapidly that they produce a new generation every few hours. Plankton organisms, less fecund, may produce a new generation every few days. Many vertebrates breed but once a year, some large animals only once every two or three years. Several species of small birds and mammals have two or more broods per year. Rodents may continue to breed throughout the winter under favourable environmental conditions, so that their reproductive potential is enormous.

The maximum size of a litter is determined by the physiological and morphological characteristics of the species. In mammals, which produce viviparous young, the size of uterus and body cavity as well as the number of mammary glands for suckling the young after birth acts as limiting factors for the size of the litter. In birds there is a limit on the number of eggs that one individual can cover and successfully incubate. However in species that do not take care of their aggs after laying, the number produced may be limited only by the energy resources of the parent.

The number of eggs or young produced per litter is correlated inversely with the amount of attention (*viz.*, parental care) that they require. When parental care is altogether lacking, invertebrates may lay 1,000 to 500,000,000 eggs at one maturation ; where there is some protection afforded by brood pouches, 100 to 1,000 eggs may be laid ; with a high degree of brood protection, 1 to 10 or more eggs may be laid. Mammals seldom have more than a dozen young in a single litter and in larger species usually only one. Characteristic clutch size among birds varies from 1 to 15 ; rarely, 20.

Further, there is a limit on the size of the brood or litter **that** adult warm-blooded animals can successfully feed and raise to maturity. There is no advantage, for example, for starlings to have broods larger than five. In larger broods, each individual receives less food, and hence has less vigour and weight on leaving the nest. In such weak individuals, mortality increases either before fledging or in subsequent months (**Perrins**, 1965). In those species that feed their young in the nest, the clutch size has evolved through natural selection to the greatest number that can be hatched and raised successfully through efforts of the adults (**Lack**, 1967). In those species whose young leave the nest and feed themselves at hatching, the clutch size depends in large part on the capability of the female to mobilize energy in her body to produce eggs of a particular size (**Ryder**, 1970). The variability in clutch and litter size for most species allows them to take advantage of temporarily improved conditions.

The size of brood, litter or clutch is also regulated by weather. For example, the clutches laid by birds during periods of hot weather are usually smaller than those laid when temperature is moderate (**Kendeigh**, 1941). Clutches laid by related species in temperate latitudes tends to be larger than those laid in tropics. Likewise, the fecundity of white-tailed deer is higher with good forage than with poor forage. The reproduction is generally more successful after periods of high mortality than during years of abundance.

Among plants the emphasis is generally laid on the **reproductive capacity** which is the number of individuals that can be raised from each parent individual after successful germination of the seeds.

15-3. MORTALITY OF POPULATION

Mortality is the population-decline factor and thus, in terms of its effect on the group, is antithetic to natality. As natality can be expressed as a rate of gain or birth of individuals in unit time mortality can be expressed as a loss of individuals in unit time—the **death rate**. Generally mortality is expressed as **specific mortality**, that is, the number of members of an original population dying after the lapse of a given time. The crude rate of death of a population can be calculated by the following equation :

$$\text{Death rate (d)} = \frac{\text{number of deaths per unit time}}{\text{average population}}$$

The rate of mortality (death) is related with density like natality. Mortality is almost highest at very high density because the hazards of over crowding and the greater likelihood of concentrated predation or spread of disease, but it is often quite high at very low densities as well, because a number of individuals of a given species are often better able to survive during a period of stress than is a single individual. For instance, heat generated by a cluster of bees is sufficient to allow survival of the cluster in temperatures low enough to kill the bees were they not clumped together.

The death rates vary among species and are correlated with rates of reproduction (natality). The death rate of a species is influenced by a number of factors such as destruction of nests, eggs or young by storms, wind, floods, predators. accidents and desertation of parents, but of fundamental importance is the number of young that are born in relation to the carrying capacity of the habitat. When more young are born than the habitat can support, the surplus must either die or leave the area. Because, the number of survivours is more important than the number dying, so mortality is better expressed as survival or as **life expectancy**, the average number of years the members of a population have left to live.

Mortality rate is usually expressed as a percentage of individuals dying within a given time. The mortality rate in many species varies from one age level to another ; thus, a mean death rate has only general significance. In birds, however, the death rate is nearly constant once they become adult, and it is then apparent that it varies inversely with adult longevity. In adult penguins, pelicans, shore birds, gulls, and swifts, the annual mortality rate is commonly between 12 and 30 per cent ; in herons, hawks, and owls it is about 30 per cent ; in ducks, doves, and song birds it is between 40 and 68 per cent, while in gallinaceous birds it is the highest, 60 to 80 per cent. As mortality varies positively with age in the majority of organisms, the specific mortalities at particular ages can be illustrated in the form of life table.

15-3·1. Life Tables

Species differ widely in the number of young produced each year, in the average age to which they live and in their average rate of mortality. When sufficient facts about a species are known, a **life table** that tabulates the vital statistics of mortality and life expectancy for each age group in the population may be formulated. In such tables, age is usually represented by the subscript index x and is some convenient fraction of a species mean life span, such as year or stage of development. The life table is set up on the basis of an initial cohort of 100, 1000, or 100,000 individuals ; and the number living to the beginning of each successive age interval is symbolized as lx. Plotting these data gives a **survivourship curve** for the species and a number of survivourship curves have been worked out for different species. The number dying within each age interval is designed as dx and gives a mortality rate. The rate of mortality during each age interval is commonly expressed as the percentage of the number at the beginning of the interval 100 (dx/lx) and is indicated as qx. Survival rate is the difference between the mortality rate and 100 per cent $(100-qx)$ and is expressed as sx. Life expectancy (ex) is the mean time that elapses between any specified age and the time of death of all animals in the age group. For example we can consider following life table (15·1) for budworms of Canada.

15-3·1A. Uses of life tables. Life tables are useful for computing the average longevity of a population, for showing the age composition of a population, for indicating critical stages in the life cycle at which mortality is high, for showing differences between species, for showing the success of the same species in different biotopes, for furnishing information of value in game and fish exploitation (yield) and in control of pests.

Table 15·1. Life table for the 1952-53 generations of the spruce budworm in New Brunswick, Canada (after Morris and Miller, 1954, source Kendeigh, 1974).

x	lx	Factors responsible for dx	dx	qx
Eggs	1000	Parasites	17	2(—)
		Predators	86	9(—)
		Others	6	1(—)
			Total=109	12
Instar I	891	Dispersal, etc.	428	48
Hibernacula	463	Winter	79	17
Instar II	384	Dispersal	242	63
Instars III–VI	142	Parasites	51	36
		Disease	3	2
		Birds	20	14
		Others	61	43
			Total=135	95
Pupae	7	Parasites	0·6	8
		Predators	0·7	10
		Others	1·3	18
			Total=2·6	36
Moths	4·3	Total for generation	=995·7	99·5

15·3·1B. Types of survival curve. What is really vital for the population is not which members die, but which members survive ? Consequently, specific mortality rate of a population is expressed by **survivorship curve**. If it could be assumed that all members of an original population had the same capacity for survival (environmental effects for the moment are ignored), plotting the number of surviving individuals against time would produce a survivorship curve in the form of a right angle. The curves of survival plotted from life table may be of three types. In type 1 a cohort finds environmental conditions ideal ; all members, born at the same time, live out the full physioloical life span characteristic of the species, and all die at about the same time. The survivorship curve of modern industrialized man approches type 1. In type 2, the rate of mortality is constant at all age groups, so that an individual's chance of living another year is just as good at one age as another. This is approximately true for some adult birds. Type 3 shows extremely heavy mortality early in life, as will such forms as the oyster, where most of the millions of eggs spawned never hatch, but those few individuals that survive have a high life expectation there after.

Most survivorship curves observed under natural conditions are intermediate between these three types (*e.g.*, deer, annelid worms, insects, and isopods).

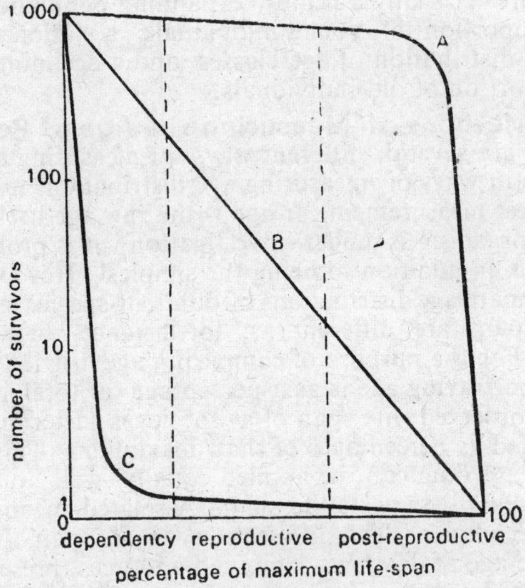

Fig. 15·1. To construct a survivorship curve, a total population of individuals, such as one thousand, is considered at age 0 (birth). At even increments of time, the total number of survivors from this thousand is plotted, and the curve is drawn. The slopes of the three basic types of survivorship curves show the following rates of changes : A–Curve for organisms out the full physiological life span of the species (Type 1). B–Curve for organisms in which the rate of mortality is fairly constant at all age levels – a more or less uniform percentage decrease in the number that survive (Type 2)· C–Curve for organisms with high mortality early in life (Type 3) (after Clapham, Jr., 1973).

15-4. AGE DISTRIBUTION OF POPULATION

It is obvious that populations that differ in the relative numbers of young and old will have different characteristics, different birth and death rates, and different prospects. In other words, age distribution is an important population characteristic which influences both natality and mortality. Mortality usually varies with age, as chances of death are more towards the earlier and later periods of life span. Likewise, reproduction (natality) is restricted to certain age groups, for example the middle age groups in the higher animals. The distribution of ages in a population may remain fairly constant, or it may change throughout the generation time. The population age distribution is related to the growth rate of the population, and it can be used to estimate whether a population is expanding or contracting. Further, the ratio of the various age groups in a population determines the current reproductive status of the population and indicates what may be expected in the future. Usually a rapidly expanding population will contain a large proportion of young individuals, a stationary population more even distribution of age classes and a declining population a large proportion of old individuals.

15-4·1. Methods of Measuring the Age of Population

There are several different ways of measuring age, and there are two main ways of measuring age distribution in a population. The simplest measurement for age is the **raw age** itself. If only one species of organism is under investigation, it is probably the best scale to use, in addition to being the simplest. However, it is difficult to compare age distributions of different species using raw ages, if the life spans are different, as, for instance, between elephants and mice. For the purpose of comparing age distributions, a useful means of portraying age is as a percentage of total life span. The maximum observed life span of a species is determined, and ages are presented as percentages of that maximum. This presentation format for age data can be applied equally well to any species of organism, regardless of its life span. A related means of presentation is percentage of deviation from mean life span. In this format, the average age at death is computed and ages presented in terms of deviation from this mean age. Both of these formats require that the maximum or mean life span, respectively, be shown for each species, because this information is not presented in data converted to percentage form.

A somewhat different type of age portrayal is the **ecological** or **functional age**, where the life span of a species can be divided up into various functional units, such as **prereproductive, reproductive,** and **postreproductive**, or egg, larva, nymph, and adult. The appearance of individuals of the various functional ages may be quite similar, as in man, or quite different, as in animals that undergo metamorphosis. Functional age breakdowns can be extremely useful in comparing certain features of different populations, such as the relative importance of specific environmental factors on the population as a whole. If we know both the percentage of time spent by an organism in specific functional ages and the environmental factors that have the gratest effect on each age, we have gained considerable insight into the characteristics of the species in question. This knowledge is essential if we are trying to design a procedure to control a best species on an economically important crop, and it is necessary to ascertain which stage in the life cycle of the pest is most vulnerable to attack by which means.

15-4·2 Age Pyramids

The first means of portraying age distribution is the **age pyramid**. An age pyramid is a vertical bar graph in which the number or proportion of individuals in varions age ranges at any given time is shown from youngest at the bottom of the graph to oldest

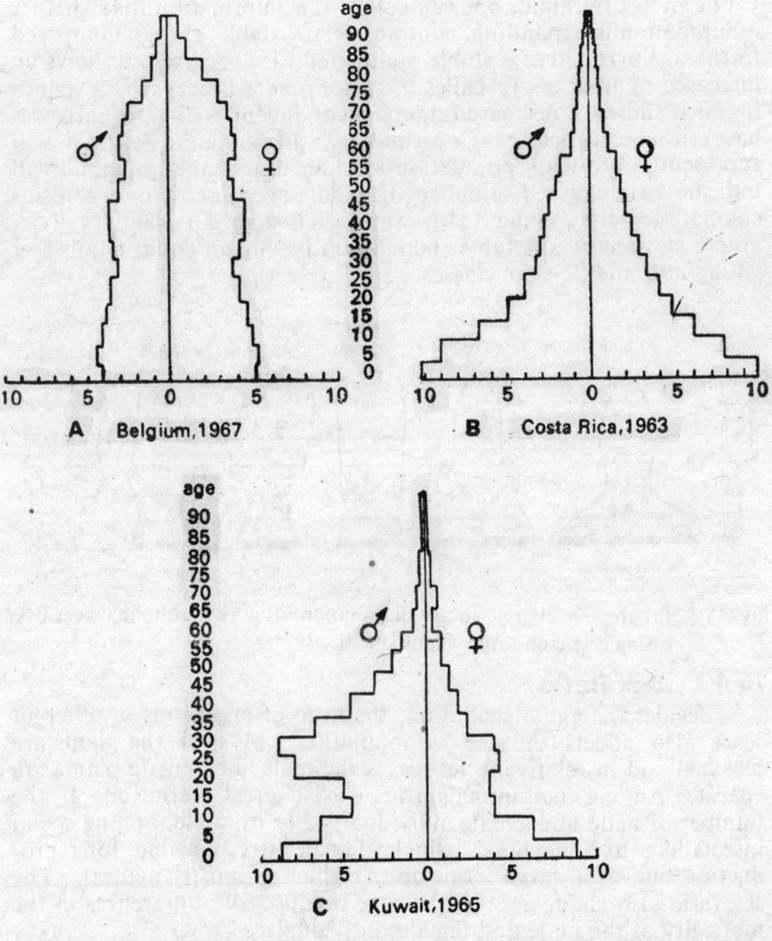

Fig. 15·2. Age pyramids for various human populations, drawn with males and females shown separately as percentages of the total population : A–Belgium, a stable population; B–Costa Rica, a rapidly expanding population; C–Kuwait, a population with extensive immigration. showing a highly skewed sex ratio (after Clapham, Jr., 1973).

at the top. (Fig. 15·2). It is very useful in monitoring the commerical exploitation of food species by man among other things. Ages may be shown in any of the formats we outlined. An age pyramid represents the age distribution of a living population at a specific moment; thus, as the population age distribution changes over a period of time, the form of the age pyramid does too (Fig. 15·3). From an age pyramid, one can collect the information that whether a population is expanding, contracting, or stable. When compared to the age pyramid of a stable population, if age pyramid shows an increased number of juveniles (*i.e.*, a broader base), it is expanding, if it shows a decreased number of juveniles (*i.e.*, a narrower base), it is contracting. Consequently, a triangular pyramid will **represent** a growing population and an urn-shaped pyramid will indicate an increased number of middle aged and old organisms compared to the young. However a bell-shaped pyramid will denote a stationary or stable population having an equal number of young and middle-aged classes.

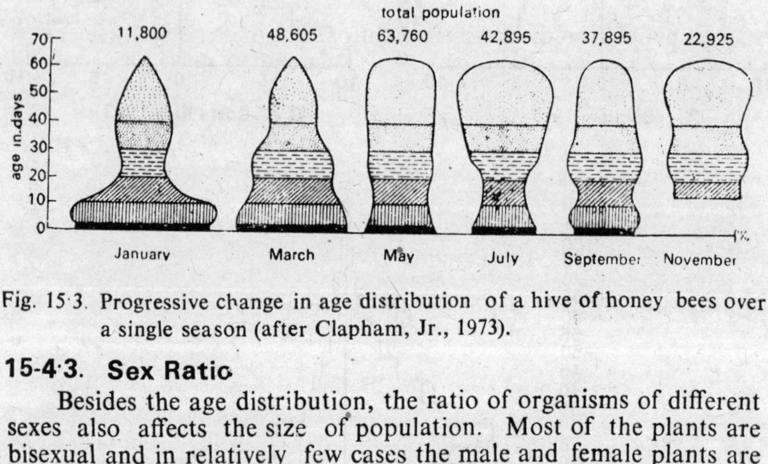

Fig. 15·3. Progressive change in age distribution of a hive of honey bees over a single season (after Clapham, Jr., 1973).

15·4·3. Sex Ratio

Besides the age distribution, the ratio of organisms of different sexes also affects the size of population. Most of the plants are bisexual and in relatively few cases the male and female plants are sparate. Among the animals, there exist great variations in the number of male and female individuals. For example, among social insects like in honeybees, a single female is responsible for reproduction out of a large population which is mostly neutral. The sex ratio also changes with the age because of differences in the mortality of the male and female individuals.

15·4·4. Biotic Potential and Environmental Resistance

The term **biotic potential** or reproductive potential, refers to the inherent power of a population to increase in numbers when the age ratio is stable and all environmental conditions are optimal. The biotic potential is defined mathematically as the slope of the population growth curve during the logarithmic phase of growth. When environmental conditions are less than optimal, the rate of population growth is less. The difference between the potential ability of a population to increase and the actual change in size of the population is a measure of environmental resistance.

The sum of the physical and biologic factors which prevent a species from reproducing at its maximum rate is termed the **environmental resistance**. Environmental resistance is often low when a species is first introduced into a new territory so that the species increases in number at a fantastic rate as when the rabbit was introduced into Australia and English sparrow and Japanese bettle were brought into the United States. But as a species increases in number the environmental resistance to it also increases, in the form of organisms that prey upon it or parasitize it, and the competition between the members of the species for food and living space, and consequently, population decreases. An equilibrium will be reached either by decreasing the birth rate (natality) or by increasing the mortality rate.

15-5. GROWTH FORM OF POPULATION

The growth is the most fundamental dynamic feature that a species population displays. Populations characteristically increase in size in a sigmoid, S-shaped or logistic fashion. When a few organisms are introduced into an unoccupied area, the growth of the population is at first slow (**positive acceleration phase**), then becomes very rapid (**logarithmic phase**) and finally slows down as the environmental resistance increases (**the negative acceleration phase**) until an equilibrium level is reached around which the population size fluctuates more or less irregularly according to the constancy or variability of given environment. The level beyond which no major increase can occur represents the saturation level or **carrying capacity**. The carrying capacity or equilibrium density is repre-

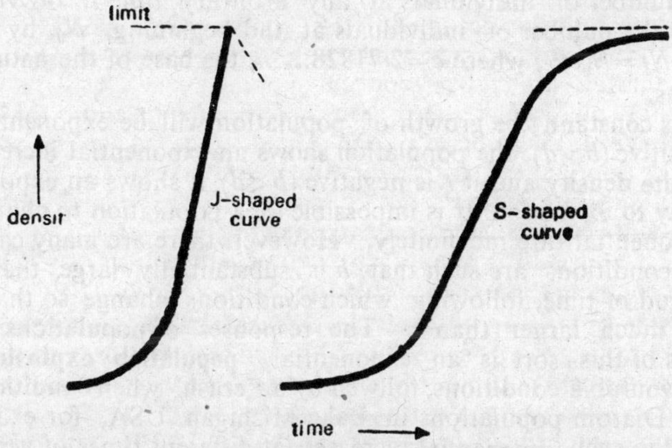

Fig. 15·4. J-shaped and S-shaped population growth curves.

sented by the letter **K**. It is often useful to define the maximum rate of growth of the population. This parameter, generally termed the **intrinsic rate of natural increase**, is symbolized r_0 and represents the growth rate of a population that is infinitely small. Accordingly such type of population growth can be described by following **logistic equation :**

$$dN/dt = r_0\, N\, (K-N)/K$$

If one desires to represent sigmoid growth of the population graphically, he has to plot the time on the X-axis and the number of organisms on the Y-axis, and consequently, will get a characteristic S-shaped sigmoid curve. Human population and growth of yeast, *Drosophila* or rabbit in laboratory conditions, follow such S-shaped population growth.

Another kind of population growth curve, called **J-shaped growth curve**, is obtained when the density of the organisms increases rapidly, in compound interest fashion, and then stops abruptly as environmental resistance or other limits become effective more or less suddenly.

15-6. GROWTH RATE OF POPULATION

The rate of **growth** of a population is expressed as the number of individuals by which the population increases divided by the amount of time that elapses while this population increase is taking place :

$$\text{Growth rate } (r) = \frac{\text{number of births } (b) - \text{number of deaths } (d)}{\text{average population in time interval}}$$

The actual change in population number ($\triangle N$) over any span of time ($\triangle t$) is equal to rN ($\triangle$ is the entity that is changing). This can be written $\triangle N/\triangle t = rN$ or, using the symbology of the calculs, the rate of change of the population at any instant time (dN/dt) can be expressed $dN/dt = rN$. This is equivalent to saying that the number of individuals at any arbitrary time t, or Nt, is related to the number of individuals at the beginning, N_0, by the equation $Nt = N_0 e^{rt}$, where $e = 2{\cdot}71828......$, the base of the natural logarithms.

If r is constant, the growth of population will be exponential. If r is positive ($b > d$), the population shows an exponential increase to **indefinite** density and if r is negative ($b < d$), it shows an exponential decay to extinction. It is impossible for a population to change at an exponential rate indefinitely. However, there are many cases in which conditions are such that b is substantially large than d for a period of time, following which conditions change so that d becomes much larger than b. The responses of populations to variations of this sort is an exponential "**population explosion**" during fovourable conditions, follwed by a "**crash**" when conditions change. Diatom populations in Lake Michigan, USA, for example, undergo such **exponential** increases at different times of years, triggered by variations in abiotic factors within the lake, followed by equally rapid declines.

15-7. POPULATION DISPERSION

Populations have a tendency to disperse, or spread out in all directions, until some barrier is reached. Accordingly, population dispersion is the movement of individuals into or out of the population area. It takes three forms : **emigration**—one way outward movement ; **immigration**—one-way inward movement ; and **migration**—periodic departure and return. Dispersion supplements natality and mortality in shaping population, growth form and density, and also it plays a significant role in the distribution of plants and animals even to the areas previously unoccupied by the members of the population. Most types of population dispersion occur due to a number of reasons such as for obtaining food, avoiding predators, preventing overcrowding, result of action of wind and water, environmental factors as light and temperature, breeding behaviour, physiological reasons as secretion of some hormone or for interchange of genetic material between populations.

15·7·1. Emigration

Emigration under natural conditions occurs when there is overcrowding in the migratory locust, lemming, grouse, snowy owl, snowshoe rabbit, Arctic fox, gray squirrel and occasionally in other species. This is generally regarded as an adaptive behaviour that regulates the population on a particular site and prevents overexploitation of the habitat. Further, it leads to occupation of new areas elsewhere. By dispersing into new localities, there is opportunity gained for interbreeding with other populations leading to more genetic heterozygosity and adaptability.

It is, of course, population pressure that is responsible in large part for the dispersal of the young and extension of ranges into new areas. Under normal conditions adult animals especially among the higher vertebrates, are well established on their territories and the youngs are forced to seek homes elsewhere. Among insects, there is a relation between emigration and inherited behaviour tendencies. Individual tent caterpillars, both larvae and adults, differ innately in the extent to which they show activity even within same colony. In the development of populations of excessive size, spread of infestations of the insect into new regions is largely by the more active individuals. The outbreak finally terminates when the proportion of sluggish individuals comes to predominate in the population (**Wellington**, 1966). Continuous emigrations are rare and when they occur, result in **depopulation**. Equilibrium of populations is maintained in such circumstances by enhancing the reproductive ability as well as by decreased mortality among the populations.

15·7·2 Immigration

Immigration leads to a rise in population level, causing an **overpopulation** which may lead to an increase beyond the carrying capacity. These immigrations result in increased mortality among the immigrants or decreased reproductive capacity of the individuals. Both emigration and immigration are initiated by weather and other abiotic and biotic environmental factors.

15-7 3 Migration

Migration is a peculiar kind of population dispersion which involves the mass movement of entire population. This can occur only in mobile organisms and best developed in insects like the butterfly *Danaus plexippus* and in the migratory dragonflies *Libellula quadrimaculata* and *Pantala flavescens* ; in fishes like eels, in birds and in certain mammals. Most two-way migratory movements are rhythmic processes of population and regular periodicity is a common feature. Very often, environmental periodicities control these migratory movements, as for example day and night rhythms, lunar periods, tides and changing seasons. The monarch butterflies-*Danaus plexippus*, travel very long distances and their migrations are found to be pathed every year through their same conventional routes, and their migratory movements are initiated by the oncoming winter and the return trip being influenced by spring.

In most cases, migration of population may occur for food, shelter, or reproduction. Better utilization of uninhabited or hitherto untouched habitats and their resources are the greatest benefits derived from the migratory movements. However, during migration of population mortality of numerous individuals may occur due to different ecological hazards such as temperature fluctuations, scarcity of food, predation, etc. Anyhow, migration has certain benefits for populations—as it enables wider dispersion of populations; it avoids intrasepecific competition for food, shelter or any other means.

At last we can say that population dispersion is brought about by population increase, improvement of habitats or deterioration of habitats, sparseness of population and by changes in or extremes of weather.

15-8. REGULATION OF POPULATION SIZE

The inherent tendency of all animal populations is to increase in number. But this increase in number is however, not infinite since the carrying capacity of the environment always imposes a restriction upon it. Thus, after reaching the carrying capacity level, the population density tends to fluctuate above and below this level and such fluctuation in population between upper and lower limits tend to give some stability to the population. The nature of the processes that regulate the numbers of plants and animals is a major problem of population ecology.

Many groups of animals and plants are provided with unique and intrinsic self-regulatory mechanism such as failure of reproduction and self-inflicted mortality for controlling the size of the population. Populations grow when natality exceeds mortality, and they decline when mortality exceeds natality. However, limitation of animal number in a population is brought about by the action and interaction of two basic regulatory processes namely **density independent** and **density dependent factors**. Density **independent** factors are the extrinsic factors which tend to regulate the density of a population under different conditions, appearing to act on the population and inflict loss of individuals irrespective of the population density. Variations in space or cover, favourable weather and food occur independently of population densities and may cause drastic changes in the abundance of animals. Such ecological or environmental factors influence negatively or positively all the individuals of a population irrespective of density.

The density dependent factors are intrinsic or biotic factors and they depend on coaction between individuals within the same population or between populations of different species. Density-dependent factors may stabilize populations at an asymptote, the level of which is determined by the carrying capacity of the environment. Some of the important density dependent factors are competition reproductivity, predation, emigration and disease. The combination of factors or any specific factor involved in the density dependent action may vary from species to species. Further these species have been known to display the property of **intercompensation**. According to intercompensation, if there is a change in the environment to relieve the population of pressure from an existing effect, then the population increases till it reaches a level when the second effect takes over. Thus, if the predators that normally keep herbivorous animals down are removed, the population of herbivores may increase to become overcrowded and result is starvation. A supply of abundant food at this junction would make the individuals susceptible to diseases due to the intensity of crowding (**Wilson** and **Bossert**, 1971). Generally, the population fluctuations controlled by extrinsic factors tend to be irregular and correlated with the variation in one or more major physical limiting factors such as temperature, food, water, etc., while fluctuations of populations controlled by intrinsic factors exhibit regularity and population cycles.

15·8·1. Population Cycles

Populations are said to be cyclic when they alternatively errupt and subside in a more or less uniform manner between high and low levels of density. Different animals exhibit population cycles at different times. The best established cycles of population density of fluctuations are those of periodicities of 3–4 years and 9–10 years. The 3–4 year cycles are most commonly observed in many birds such as snowy owl, willow ptarmigan, capercaillie. Blackgame, Hazel grouse etc.; mammals such as lemmings, voles, arctic foxes, etc., and fishes. 9–10 years cycles of population density fluctuations have been observed in birds such as ruffed-grouse, sharp-tailed grouse, willow ptarmigan, etc., and mammals such as snowshoe rabbit, muskrats. Canada lynx, etc. Among invertebrates, insect pests of coniferous forests in Germany fluctuate in periods variously from 6 through 18 years. *Asterias forkesi* is found to contain periodicity of 14 years.

Attempts to explain these vast oscillations in numbers on the basis of climatic changes have been unsuccessful. At one time it was believed that these were caused by sunspots, and the sunspots and lynx cycles do appear to correspond during the early part of the nineteenth century. However, the cycles are of slightly different lengths and by 1920 were completely out of phase, sunspot maxima corresponding to lynx minima. Attempts to correlate these cycles with other periodic weather changes and with cycles of disease organisms have been unsuccessful.

The snowshoe hares die off cyclically even in the absence of predators and in the absence of known disease organisms. The animals apparently die of "shock" characterized by low blood sugar (hypoglycemia), exhaustion, convulsions and death, symptoms which resemble the "alarm responses" induced in laboratory animals subjected to physiologic stress. Visualizing this similarity, **J. J. Christian** (1950) proposed that their death, like the alarm response is the result of some upset in the adrenal-pituitary system. As population density increases ; there is increasing physiological strees on individual hares owing to crowding and competition for food. Some individuals are forced into poorer habitats where food is less abundant and predators more abundant. The physiologic stresses stimulate the adrenal medulla to secrete **epinephrine**, which stimulates pituitary via the hypothalamus to secrete more **ACTH** (adrenocorticotropic hormone). This, in turn, stimulates the adrenal cortex to produce corticosteroids, an excess or imbalance of which produces the alarm response or physiological shock.

In the latter part of the winter of a year of peak abundance, with the stress of cold weather, lack of food and the onset of the new reproductive season putting additional demands on the pituitary to secrete gonadotropins, the adrenal-pituitary system fails, becomes unable to maintain its normal control on carbohydrate metabolism, and low blood sugar (hypoglycemia), convulsion and death ensue.

According to **Kendeigh** (1974) several instrinsic factors such as disease, predation, food factor and natural selection affect animal cycles in addition to several extrinsic factors such as weather, solar radiaton, and so on.

15-9. REVISION QUESTIONS

1. What is population ? Describe its different characteristics with suitable examples.
2. Discuss that in what way a population is a self-regulating system.
3. Write short notes on the following :
   (i) Fecundity;
   (ii) Natality;
   (iii) Mortality of population;
   (iv) Survivorship curve;
   (v) Age pyramids;
   (vi) Population dispersion.

16
Intraspecific and Interspecific Interactions

(Biotic Environmental Factors and Social Behaviour)

No organism under natural conditions is divorced from all other living things. Generally a number of different species will exert a direct or indirect effect on a particular individual because the vital processes such as growth, nutrition and reproduction depend upon the coactions or interactions of other members within the species (**intraspecific interactions**) or between individuals of different species (or populations) of the community of an ecosystem (**interspecific interactions**). Many of these interactions act to regulate density or, along with climate cause fluctuations or even more drastic changes in a number of individuals of the populations from time to time and so constitute **biotic environmental factors.**

16-1. INTRASPECIFIC INTERACTIONS

It is only under extremely rare and unusual circumstances that an individual is the only representative of his species existing in a given region. The individual generally has many types of intraspecific interactions with old and young males and females of the same species and this interaction involves associations between individuals including social organisations (societies), territorial segregations and communication between individuals.

16-1·1. Types of Association

Organisms vary in their ability to move about in their environment. Many animals such as cephalochordates, vertebrates, most arthropods, many invertebrates and few plants and microbes are **free-living** in that they can crawl, swim, walk, fly, or otherwise move. Others are **sessile**, being attached permanently to a rock, or other structure in their environment. Among the free-living forms, there are many different grades of mobility. Many marine invertebrates are **sedentary**, remaining in one place for most of their lives. Other animals and micro-organisms are **planktonic** in that they float in the water and are carried about by the action of currents and winds. Many, however, are **motile**, actively moving about in their environment, often with relative rapidity.

The type of locomotive ability often influences the types of association existing between individuals. Some kinds of animals are **solitary**, living alone except perhaps during the reproductive seasons. Others are **gregarious**, living together in various groups such as schools (of fish), flocks (of birds), packs (of wolves), herds (of ungulates), swarm (of locusts), colonies (of various invertebrates) and similar other associations. Different kinds of associations of organisms can be studied as follows :

16·1·2. Colonisation

An early manifestation of intraspecific cooperation in the evolution of animals is the grouping of free-living protozoans to form **colonies**, and the further development of such colonies into multicellular metazoans that thereafter behave and respond as unit organisms. Whether the first gathering of protozoan cells to form colonies developed for better protection from some predator or environmental condition, improved utilization of food supplies, or more efficient reproduction, it is impossible to say (**Kendeigh**, 1974). The colonial form, however, must have had some survival value.

Further, there are various sessile metazoan animals which frequently occur in colonies often with structural connections between the different members of the colony. Such colonial forms include many coelenterates (Cnidarians) such as *Ceratella*, *Obelia*, *Physalia*, *Porpita*, *Millepora*, *Tubipora*, *Alcyonium*, *Gorgonia*, *Corallium*, etc.; ectoprocts (*Bugula*, etc.) and ascidians. Among the plants also, grasses, sedges and several herbs form dense stands.

Outcomes of colonisation. During the course of evolution the colonisation quickly led to **division of labour** between somatic and reproductive cells, as occurs in *Volvox*, and later to division of labour between **somatic** cells themselves, so that different cells or organs become specialized to serve particular functions of digestion, respiration, circulation, excretion, and so on. Cooperation among cells, tissues, and organs gave greater metabolic efficiency to the whole individuals and resulted in evolution to the highest types of animals.

The colonial hydroids, also exhibit polymorphism and division of labour in different individuals, called **zooids** or **polyps** of the colony. For examples *Physalia* is a floating colony of polyps or zooids specialized into four major polyp types—for floating (pneumatophore), feeding (gastrozooid), fighting (dactylozooid) and reproduction (gonozooid). Such specialization of polyps is called **polymorphism.**

16·1·3. Aggregations

Many associations of individuals appear to be the result of random groupings that furnish some survival advantage for the group. The term **aggregation** or **temporary aggregation** is usually applied to a group of animals which come together for some external reason. For example, a group of insects is more apt to survive a dry environment than is a single insect, since, in a limited space, each would lose less water before the relative humidity of the surrounding environment would be raised to a less dangerous level. Other similar examples of so-called "unconscious cooperation" include migration of locusts in swarms and of birds in flocks. Many species of membracid insects (Homoptera) exhibit good example of the gregarious habit; sometime the same host plant (*e.g.*, *Prosopis spicigera*) lodges two different gregarious species (*e.g.*, *Oxyrhachis tarandus* and *Otinotus oneratus*), the individuals of both crowding together and mingling with each other very closely. Such associations may be asexually gregarious, combining individuals without respect their age or sex, or may be sexually gregarious, often bringing together males and females during the reproductive season. For example, mayflies, midges, and mosquitoes swarm for mating purposes.

16·1·4. Social Organisations and Behaviour

Individuals often actively associate for mutual gain. The term **permanent aggregation, society** or **social organisation** is usually applied to a group of individuals belonging to the same species and organised in a cooperative manner. In a social organisation, success is often measured in terms of survival of the colony, with individual survival having only secondary importance. The social organisation has evolved separately in several different groups of insects including termites, ants, bees and wasps. According to **Eisenberg** (1965), the organised societies have the following defining attributes :

1. **A complex system of communication**. All organised societies have some form of complex communication system.
2. **A division of labour based on specialisation**. In organised societies, animals of different castes, sexes, or age groups have different functions in maintaining the society.
3. **Cohesion, a tendency for members to remain together**. The individuals constituting a society tend to remain in close proximity to one another.
4. **A permanence of individual composition**. The individuals making up a society tend to be the same from day to day; there is little migration from the group.
5. **A tendency to be impermeable to conspecifics that are not members of the group**. Most organised societies resist immigration by "outsiders".

Both kinds of aggregations (*viz.*, temporary and permanent) have following advantages :

1. It is very often the environment that derives animals into collective groups. Scarcity of space, light, utilization of the same food and shelter, collective breeding and various other conditions induce the animals to aggregate or congregate. These aggregations are capable of tolerating different ecological hazards more efficiently than they have been living solitarily. Insects and vertebrates survive the acute cold of winter by forming groups which help to generate, retain and conserve heat. This is one of the basic advantages of aggregations and is termed as the **group survival value**. Coveys of bobwhite quail roost in close circles, at night.

Perhaps this enables detection of predators approaching from any direction but by that behaviour the birds can tolerate lower air temperatures and for a longer time than isolated birds can. Similarly, mice huddle in low air temperatures, a behaviour that reduces heat radiation and consequent need for frequent feeding.

2. Aggregation may have a protective function against predators. The benefits may occur because of mutual protection of the members of the group. Sheep, when threatened, will cluster together, with the rams encircling the young sheep and ewes. The benefits of aggregation may result because they warn one another of impending danger or beeause of the confusion caused by the prey scattering in several directions as the predator strikes.

Further, a single musk ox or bison may succumb to a pack of wolves. When in a group, the males form a circle facing outward with the females and young inside, whereby they are usually able to ward off the attack.

3. Aggregation facilitate food gathering. Large numbers of animals seeking food can shorten the time to discovery. Also they may aid each other in capturing and subduing the prey. Cooperative hunting by wolves is a well known example. A single wolf has difficulty in killing a deer; a single coyote in killing a pronghorn antelope. But in packs the wolves can overpower a deer and a pack of coyote can chase a pronghorn to exhaustion.

4. Aggregation of two or more organisms is necessary for sexual reproduction. Aside from the obvious need for two animals to participate in sexual reproduction, group activities, sometimes facilitate breeding by mutually stimulating one another or, as in the case of male frogs that sing together in ponds, it increases the likelihood that a female will find the pond and be fertilised.

5. Transfer of learning may occur in aggregations. Mammals in general teach their young routes to food, techniques of food capture, places of danger, and so on.

6. Groups of animals can cooperate in building structures for various functions that would be difficult or impossible for one animal alone. The social insects (ants, bees, termites, wasps) generally erect rather substantial nests by cooperative building.

7. Aggregations permit division of labour and much specialisation to occur in structure, function and behaviour of the participating members of the group, so the individuals cannot survive outside the colony. Such specialisation allows a higher degree of efficiency when performing a task. For example, ants and termites have several castes and are called **eusocial insects**. There are winged fertile males (kings) and females (queens) which are specialised for reproduction ; the wingless sterile soldiers that possess large mandibles and irritating glandular secretions and are specialised for fighting and defending the colony against the predacious enemies ; smaller, wingless sterile workers which are specialised for carrying out the daily normal activities of the nest and so on.

In cases, where such specialisations of individuals like the eusocial insects do not occur, individuals may take turns with a workload. For example, in some primate groups, the care of the young is partly a community responsibility, Certainly this is the case in the social insects like bees and wasps, where workers or "nurses" care for the larvae. For example, the colony of honeybee has a single fertile female that is specialised for egg-laying and numerous sterile workers specialized for various activities for the benefit of the colony. In fact, a beehive is a marvellous society. Just as a mammalian body has elaborate homeostatic mechanisms that maintain a constant temperature, water balance, and nutrient level, so a beehive has elaborate mechanisms that maintain hive homeostasis. For example, when the hive is hot, worker bees fan air throughout it and cool it off. When the nurse workers attending the larvae are short of water, they turn to the nearest workers and signal a need for water. The latter give what water they have and then turn to their neighbours ; thus, the shortage of water moves from bee to bee until it reaches a worker who leaves the hive and returns with more water. When a predator or parasite enters the hive, workers rush forward and defend the colony, with the result that many die. It is almost as if the hive itself, and not the individual bees, were the organism. **Martin Lindauer** (1961) found that in a beehive, a horde of workers, as they go through their life cycles, perform one major task after one another.

After emergence an adult worker bee first cleans the hive, then tends the brood, build more comb, guards the nest entrance and finally forages for pollen and nectar. This behavioural sequence of events is **correlated** with physiological changes in the nurse glands (which produce nutritive materials for bee larvae) in the head and the wax glands (which produce comb-building material) in the insects abdomen (**Alcock**, 1979). Thus, the majority of females, (*i.e.*, workers) of the bee society do not reproduce, acting instead to care for their sisters. This pattern of behaviour is called **altruistic behaviour** and it is required for the following three evolutionary processes : **inclusive fitness** (*i.e.*, the sum of an **individual's** own fitness plus its influence on the fitness of its relatives other than direct descendants; **Hamilton**, 1964), **kin selection** (*i.e.* differential representation of genes resulting from favouring the survival and reproduction of relatives who possess the same genes by common descent), and **reciprocal altruism** (*i.e.*, the trading of altruistic acts by two or more individuals at two different times ; in effect, an individual engages in an altruistic act, such as jumping into the water to save a drowning person, in exchange for the promise that the other individual will do the same in like circumstances ; **Trivers** 1971).

There exist several more examples of the altruistic behaviour in animals : In the colony of paper wasp (*Polistes*), a diploid fertile and dominant female performs the work of reproduction, while a horde of subdominant sterile female workers increase the size of the nest, add cells, feed the larvae, detect and drive off parasites and descend *en masse* on potential predators, stinging them into retreat. In **lions,** females of a given social group (called a **pride**) care for, defend and even nurse all youngesters in the pride. The nest of a social bird, Florida scrub jay has numerous helper birds other than the parents that help in feeding the nestlings and fledg**lings.** On spotting a predator, group living birds (Blue **titmouse** black bird), Belding's ground squirrel and primates emit a "alarm call" or "warning cry" which alerts other group members to the presence of a predator but may make it easier for the predator to locate the signaler (**Dewsbury**, 1978).

Further, the social animals also exhibit the following behavioural phenomena.

16-1·4A. Dominance hierarchies. Most social animals recognise pattern of authority or dominance-subordination, the most basic being parent-child or provider-dependent and the relationship of the stronger to the weaker. **Van Kreveld** (1970) has defined **dominance** as a priority of access to an approach situation or of leaving an avoidance situation that one individual has over other. Approach situations include access to food, mating partners, opportunities for aggression and space. Avoidance situations include remaining in undesirable areas as well as having to tolerate threats, attacks and punishments. The essential feature of dominance is that there is some resource that is of limited availability and is con**tested** One animal has gained **privileged** access to the limited resource relative to another animal. Such a behaviour pattern is simply adaptive, ensuring that the strongest will be best fed and so remain strong and successfully reproduce, while the weaker are less likely to survive (**Alcock**, 1979).

The classical examples of dominance hierarchies are from the so called "**peck orders**" of chickens and pigeons. The organisation of an established flock of chickens is based on a set of dominance relationships. If food is given in a restricted location, it becomes apparent that one animal has priority of access to the food. It will exclude other chickens from the area of the food, perhaps by delivering a peck at them. The other animals never return the peck. Such pecking is particularly prominent in newly established flocks ; hence **Schjelderup-Ebbe**, an early student of dominance hierarchies, referred to dominance hierarchies as "peck orders". Typically, in a group of a dozen or so chickens, there is but a single despot or **alpha animal** who has priority of access over all other animals in the flock and who literally can peck any other. A second animal, the **beta individual**, has priority over all animals in the flock except the alpha animal. In an ideal **linear hierarchy**, there is a perfect straight-line ordering of all animals in the flock, so that each is dominant to all animals below in the hierarchy and subordinate to all animals above it in the hierarchy. The hieararchy gives dominant animals priority of access to food, water, roosting places, mates, and any other appropriate resource. The peck order, thus, can be an exceedingly effective means of pegging the population size to the current state of the environment, generally the availability of food, water, or nesting space, while insuring that a number of individuals in the population both survive and reproduce.

Sometimes, as in rabbits, there are two linear hierarchies, one for males and one for females. In other cases, for example in certain monkeys and rats, the hierarchy for food may be different from that for access to reproductive females. Finally, not all hierarchies are linear but rather a complex web. For example, in complex monkey troops, triangular relationships may occur, in which animal A dominates B, B dominates C, but C dominates A. Dominance hierarchies are found to exist in a variety of animals such as crabs, crayfish, cockroaches, lizards, rodents, wolves, hyaena, dairy cattle, reindeer and various non-human primates (**Dewsbury**, 1978).

16-1·4B. Social hierarchies or leaderships. The term "leadership" generally is used to refer to the ability of a given individual to influence the movement pattern of the group as it goes from place to place. Thus, the leader determines the time, rate and direction of the groups movements. The most effective "leader", in this case, is not necessarily the most dominant individual. Experience often counts more than physical prowess in determining leadership. In many primate troops, it is the females that play an important role in controlling group movements. Among the red deer protection of the herd from danger falls to the most experienced female, rather than to a dominant stag. Leaders may be of any age and of either sex.

16-1·4C. Territoriality. Many species of animals establish "ownership" and defend specific areas of land called **territories**. This is called **territorial behaviour** or **territoriality**. **Kaufmann** (1971) has defined a territory as an area in which the resident enjoys priority of access to limited resources that he or she does not enjoy in other areas. The male stickleback fish provides an excellent example of territoriality. During the breeding season, the male leaves the school and stakes out a distinct territory, defending its borders by displaying to intruders. Examples of such territorial defense are found in a wide variety of insects, fish, amphibians reptiles, birds and mammals.

In birds, the territorial behaviour is generally exhibited by males early in the breeding season. When the females arrive in the area, they are courted by various males and eventually settle down in one territory to build a nest. The territory that is established is sufficiently large to provide food for the family of birds for the season.

The fighting between contending males seldom leads to bloodshed; it is largely ritualized and stylized behaviour filled with bluff, threat and display. In this way, the entire suitable habitat is parceled out among the aggressive healthy males. If too many birds are present in an area, all the suitable sites are taken and many animals (especially the young) are forced to try to establish territories in less hospitable regions or will not take up residence at all. The chances that these subordinate animals will successfully raise a brood is very low. Territorial behaviour assures that the natural resources of the bird population are not overexploited by over-population. A certain area of habitat is required for a pair of birds to raise a brood to maturity. In the years of low population density, all birds will find suitable real estate, and breeding success will be high for all, tending to drive the population up. In years when population is dense, the extra birds (without territories) will fail to raise young and will probably have a low survival rate themselves ; this will drive the population down.

Territorial behaviour has similar regulatory function in controlling population size in most of the animals. For example, seabirds, such as gannets, nest on rock cliffs but feed from the ocean. The adults defend nesting territories on the cliffs symbolically might be said to establish fishing rights. Only birds with territories will raise young, thus limiting the population in that area. Likewise, mature **males of fur seals which breed on the Pribilof Island, establish territories** along the beach, collecting females into their "harems". Young males (bachelors) cannot hold territories, so they do not breed their first year. Only so many seals can occupy the beaches, thus population is limited. Certain colonial species of the primates often set up **communal territories**, which they defend jointly. Group size is roughly limited, and "tribes" will compete with each other for dominance over selected feeding areas.

Forms of territoriality. Wilson, (1975) has recognised the following five forms of territorial organisation :

Type A. A large, defended area within which sheltering, courtship, mating, nesting and most food gathering occur (*e.g.*, benthic fishes, arboreal lizards, insectivorous birds, some small mammals).

Type B. A large defended area within which all breeding activities take place but which is not the site for most food gathering (*e.g.*, nightjars and reed warblers).

Type C. A small defined area around a nest (*e.g.*, many colonial birds, such as ibises and herons, and sphecid wasps).

Type D. Pairing and/or mating territories (*e.g.*, insects such as damselflies and dragon flies ; birds such as sage grouse and ungulates such an Uganda Kob).

Type E. Roosting positions and shelters (*e.g.*, many species of bats, starlings, domestic pigeon).

16-1·4D. Home range. The home range of an individual is the area that it habitually travels in the course of its normal activities. Many species, such as most rodents, occupy relatively small home ranges for their entire lives. Home ranges may be completely undefended and convey no privilaged access to resources, thev should not be confused with territories.

16-1·4E. Individual distance and flight distance. Many species act so as to keep a certain distance between themselves and other species (such as a distance between birds and primates). Such a distance is referred to as "**individual distance**".

The **flight distance** refers to the distance to which an individual will permit a predator to approach before it flees.

16-1 4F. Communication. Communication between members of a species is almost universal. It generally occurs during reproduction in sexual species ; however, it plays a daily role in the life of many social animals. Communication is said to have occurred when an animal performs an act that alters the behaviour of another organism. Such interactions are often quite specialised and usually are adaptive for one or both organisms. A communication system involves the following seven essential components (**Dewsbury,** 1978) : 1. **Sender** : an individual which emits a signal. 2. **Receiver** : an individual whose behaviour is altered by signal. 3. **Message** or **signal** : the behaviour emitted by the sender. The variety of signals animals use to communicate a behavioural state such as aggressiveness or an intention to take flight or attack includes odours (chemicals), postures, colours, sounds, shapes and motions. These are meaningful to friend and foe of the same species as are raised eyebrows denoting surprise among humans. 4. **Channel** : a pathway through which the signal travels (*e.g.*, vocal-auditory channel). Several different sensory receptors are involved in animal communication. 5. **Noise** : background activity in the channel that is unrelated to the signal. 6. **Context** : the setting in which a signal is emitted and received. 7. **Code** : the complete set of possible signals and contexts. The accurate transmission and reception of social signals can mean life or death to an individual or its offspring.

Some important methods of animal communication can be studied as follows :

1. Chemical signaling. Odours may be the most primitive communication techniques. Chemicals produced by animals and serving either as **gustatory** or **olfactory signals** that convey information between different members of a species are called **pheromones**. Pheromones may be released into the air or water or deposited onto or near the ground by animals when they are in certain behavioural states. Pheromones then generate complementary behaviour when they are detected by other members of the species. These chemical signals play a part in behaviour comparable to that played by hormones as chemical messengers between different groups of cells within the body.

Pheromones may be of the following two basic types : **signaling pheromones** that result in a more or less immediate effect on the behaviour of the recipient animal ; **priming pheromones** that trigger hormonal activities that may become manifest in overt behaviour only at a latter time. The sources of pheromones in different species include faeces and urine as well as the secretions of a bewildering array of glands located in many regions of the bodies of different species. Pheromones may be deposited by means of object marking, marking of a social partner, self-marking, substrate marking, or release of the pheromone into the air. Pheromones convey a variety of different kinds of information in different species, including species and racial identity, sexual identification and reproductive state, individual identification, age and mood. Pheromones are functional in affecting reproductive behaviour (sexual and maternal) and other forms of social behaviour (withdrawal and submission, aggression and dominance and scent marking).

Best known pheromones are the chemical markers serving to bound the territories of individual animals. Various mammals produce a "personalized" secretion from anal or facial glands that is rubbed into trees or rocks. For example, a male dik-dik deer marks its territory by depositing secretions from an eye gland onto twig ends. Cat and dog species generally urinate to mark the edges of their territory. Any incursion into the territory evokes a response. In prairie dogs the emission of strong, musky odours is often associated with the defense of territory from other individuals of the same species.

Various female mammals such as cats and dogs are known to release a species-specific sex pheromone, when they are in "heat", which attracts males from considerable distances. Such a pheromone signals the sexual state of the female and acts to excite the males. Other striking examples of potency of sex pheromones are the **sex attractants** released by virgin female moths of many species, including the silkworm moth and the gypsy moth. These pheromones are secreted by special glands on the female abdomen and are detected by receptors on the male antennae. A "calling" female releases minute amounts of the sex attractant, which is then carried downwind. When it is detected by males, often kilometres away, it induces flight activity and a tendency to travel upwind in a search pattern for as long as they can detect the odour. These built-in instructions brings the males close to the calling female. When we realize that the abdomen of a silkworm moth (*Bombyx mori*) contains a total amount of only $1{\cdot}5 \times 10^{-6}$ g of sex pheromone called **bombykol** and that only a fraction of this amount is diluted in an air stream that may be several kilometres in length, the extreme sensitivity of the male moth's olfactory sense is not surprising. **Dietrich Schneider**, (1974) has estimated that a male moth will react when 40 out of 40,000 receptors on the antennae receive one strike of a pheromone molecule per second.

The best known and most elaborate systems of chemical communication are found in social insects such as termites, ants, bees and wasps. Honeybees and ants are found to have upto a dozen different exocrine (externally secreting) glands, each producing chemicals that elicit specific forms of behaviour in nest mates. For example, queen production in a honeybee colony is carried out by the workers, who feed larvae suitably so that they develop into future queens. To do this they build rather larger cells than usual—the queen cells. In each of these the queen deposits an egg ; the future of the egg is upto the workers. But they do not make queen cells so long there is a healthy queen in the hive and the hive population is not too large. Such a peculiar behaviour of worker bees is found to be regulated by a pheromone which is chemically known as 9–oxodec–2–enoic acid and is secreted by the mandibular glands of the healthy queen. She spreads this pheromone over the surface of her body as she cleans herself. The workers lick it off her as they tend her. This pheromone then enters the food-exchange stream and is passed from bee to bee round the hive. It produces a direct physiological effect on the workers making them sterile (*i.e.*, this pheromone inhibits ovarian development in worker bees). The effects of this pheromone disappears a few hours after the queen's death, releasing queen-raising behaviour in the worker bees. This pheromone also acts as sex attractant, for drones are drawn to virgin queens by it.

Beekeepers are well aware of the fact that if one bee stings, many others will attempt to. An **alarm pheromone** (isoamyl acetate) is left with the sting of the bee in the wound. It diffuses in the air and excites other bees to aggregate and sting. Likewise, when their colony is threatened ; ants produce various kinds of alarm pheromones that cause aggregation, excited running and readiness to attack. An injured fish also gives off a chemical that causes other fish to leave the vicinity and hide.

Further, an ant returning from a rich food source touches her stinger to the ground every few steps, leaving a chemical trail that is followed by other ants who in turn reinforce the trail on their return to the nest. Oleic acid is formed in the decomposing corpses of dead ants. This chemical causes a live ant to pick up the source and carry it out of the nest to the nearby refuse pile. The unity and stability of these insect colonies are reinforced by mutual licking and a continuous interchange of material regurgitated from the crop. This is called **trophallaxis** and serves as social bond by distributing a common "nest flavor" among nest mates.

2. Tactile communication. In prairie dogs and numerous other mammals and certain insects tactile signals such as a 'kiss', nose-rubbing or rubbing of antennae (in bees) are used to identify members of a group and perhaps, to communicate other information.

3. Visual communication. Visual communication is particularly important among certain fish, lizards, birds, mammals and arthropods including insects. Visual messages may be communicated by a variety of means, such as through colour, posture or shape, movement, or timing. The octopus is capable of rapid colour changes that are involved with sexual display. Plumage colouration in many male birds (*e.g.*, peacock) serve the same purpose. The colour of eye ring in certain sea gulls is the major key in maintaining species separation. Some male butterflies are attracted to the female by her performance of a specific flight pattern. Fireflies are attracted to each other on the basis of their flash intervals, each species has its own frequency. Some spiders have males which raise their appendages in a specific ordered sequence much like a semaphore signal : these are required for proper species identification and they prevent the female spider's attack. Different postures of the trunk, ears and head of the African elephant display different messages. Trunk in a forward position indicates threat. Trunk, curled inward indicates fear and submissive or defensive postures. Raising of ears or head indicates increased aggressiveness. Similarly, body positions such as are used by dogs **and** wolves transmit such signals as anger, fear and joy. Male baboon exposes his long canine teeth to signal the threat.

Visual alarm signals are common in flocking birds which flash their brilliant tail and wing feathers when danger is near. Mammals, too, use alarm signals. Species of deer have a white patch under their tail which they flash during danger. This behaviour alerts the other deer in the area and they bound off together. Visual signals sometimes have temporary nature such as the reddened rump areas in female chimpanzees and baboons during estrus.

4. Auditory communication. Auditory signaling (acoustic and vocal) is the most highly developed means of communication. Sounds are produced in a wide variety of ways. For example, some insects (crickets, cicadas, grasshoppers, etc.) rub parts of the body together (wings, legs), while many fish vibrate their swim bladders by muscle contraction. The modified skin forming the vibrating rattle of a rattlesnake is well known and various mammals vibrate their tails or stamp their feet on the ground, thus producing sounds. Vocal cords are well developed in most frogs, salamanders, some lizards, birds and mammals.

For example, sound plays a significant role in reproductive, food-gathering and defensive behaviour in various animals. Most birds, insects and amphibians use sounds to attract mates The sound patterns are generally specific. In **cases** where several species look similar and live in the same area (**sympatric**), their calls are rather distinct. In areas where they live apart (**allopatric**), the differences in their calls may be small. Herring gulls and gibbons have **food-finding calls** which announce the discovery of food to other individuals. Buzzards signal the location of dead animals to other buzzards by circling above the site where the carrion is located. **Alarm calls** are common in birds. Small European finches have two types of alarm calls. One is used when they sight hawks overhead ; it is a high frequency sound, starting and stopping gradually. The other call is used when they see danger on the ground, it is loud, sharp, staccato sound. Some birds (such as U. S. eastern crows) have **mobbing calls** or **assembly calls** to draw members of the group together to deal with predators like owl or a cat. Some animals have developed **departing calls**. This call is used if no danger is present, so their sudden departure from a group will not cause alarm. Eastern crows produce a departing call before they fly off. If this is not given, which is the case when danger is near, the flock is alerted and they all fly off when one abruptly leaves.

Communication in honeybees. Honeybees of species *Apis mellifera* have developed a complex language that utilizes chemoreception as well as tactile, visual and auditory stimuli. The German ethologist **Karl Von Frisch** and his colleagues have deciphered the language of bees. They found that honeybees leave the hive and forage for food. When they return, other bees gather around the scout to get the information about the food source. Scout bees inform other bees of a source of nectar by performing the following two types of dances on the wall of the hive : a **round dance** for nearby food and a **tailwagging dance** for distant food. Tailwagging dance also gives information of direction of food source. Both dances are performed in the dark hive, so that the kind of food is determined by chemoreceptors on the antennae of the other bees identifying the pollen or nectar on the dancing bee, and the location by placing the antennae on the dancing bee and following the pattern of the dance. Sometimes a low humming noise is also associated with the dance, apparent also indicating distance. After repeating the dance a few times the bee then returns to the food source for more nectar. As long as the supply remains large, the dance is repeated after each trip, but when the supply is limited the bee continues to make trips until the supply is exhausted but does not perform the dance in the hive.

16-1·5. Psychological Factors

There are certain reactions by populations to unusual conditions, most commonly crowding, which do not come into play under normal conditions, and which seem like breakdowns in the normal psychological balances of the organisms. These include **cannibalism, inability to breed** or **care for young, increase in rate of spontaneous abortion** and similar problems. In many species, they act to control the population size when the more normal controls have failed. Certain evidences have suggested that the social stresses act on the individual through a physiological feedback involving the endocrine system (Fig. 16·1). In vertebrates, this feedback is most closely associated with the pituitary and adrenal glands (**Christian**, 1963). Increasing populations of mice held in the laboratory resulted in the suppression of somatic growth and curtailment of reproductive functions in both sexes. Sexual maturation was delayed or totally inhibited at high population densities, so that in some populations female reached normal sexual maturity. Intrauterine mortality of foetuses increased. Increasing population density resulted in an inadequate lactation and subsequent stunting of nurslings.

An example of a relatively simple psychological factor acting to control population size is shown in the fish guppy, *Lebistes reticulatus* (**Silliman** and **Gutsell**, 1958). When stable experimental populations are maintained by regularly removing a percentage of the fish, significant numbers of the young survive, although some may be cannibalized by their parents. When no artificial predation is practiced the guppies breed continuously, but they remove the entire breeding surplus through cannibalism. Evidently, the normal control mechanism for guppy populations is predation from other fish higher in the food chain. But if, for any reason this mechanism is lost the population can regulate its own number through cannibalism.

Such psychological regulation of population has also been studied in albino rat (**Calhoun**, 1962). When rats were raised in densities higher than normal, they showed psychological abnormalities such as hypersexuality, homosexuality, asexuality, cannibalism, inability of a mother to raise her pups successfully to weaning, increased rate of spontaneous abortion, breakdown in the premating rituals and abnormal crowding around certain places such as food hopper and water fountains. Different biotic factors of the environment thus tend to modify the activities of animals like the abiotic factors.

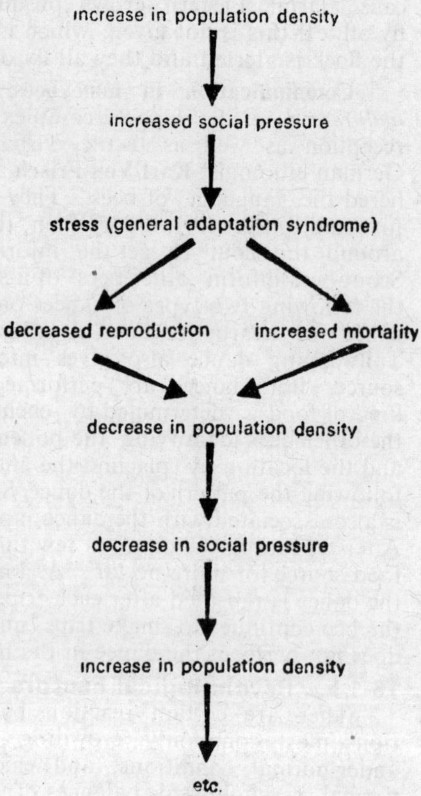

Fig. 16·1. Christian's stress hypothesis of population regulation, which operates through the general adaptive syndrome (after Smith, 1974).

16-2. INTERSPECIFIC INTERACTIONS

An individual in nature interacts not only with others of the same species but with individuals of other species. Many different types of interactions occur (Table 16·1). When, as in few cases, the presence of one species appears to have no measurable effect on a second species (*i.e.*, no interaction) a state of **neutralism** is said to exist. When the interactions result in a benefit for one or both of the species and harm to neither, the results are termed **positive interactions**. Several degrees of such interactions can be recognised : **commensalism**, wherein one species gains and the second is neither helped nor harmed ; **protocooperation**, wherein both gain, though each is also able to survive separately ; and **mutualism**, wherein both gain but neither can survive without the other. **Negative interactions** occur when one of the species is harmed by the relationship. Various levels of negative interactions are also recognised : **amensalism**, wherein one species is harmed by a second, but the latter neither gains nor is harmed by the association; **parasitism**, wherein one species gains (the parasite) while the second (the host) is harmed but not killed by the association ; **predation**, wherein one species (the predator) kills and feeds on the second (the prey): and **competition**, wherein each is adversely affected by the presence of the other in their search for food, shelter, living space, or other requiremnets for existence.

Table 16 1. Basic types of interspecific interactions (Odum. 1971).

Types of interaction	Effect on population growth and survivial of two populations, A and B — when not interacting A	when not interacting B	when interacting A	when interacting B	General results of interaction
1. Neutralism (A and B) independent)	0	0	0	0	Neither population affect the other
2. Commensalism (A commensal; B Host)	—	0	+	0	Obligatory for A; B not affected
3. Protocooperation (A and B cooperates)	0	0	+	+	Interaction favourable to both, but not obligatory
4. Mutualism (A and B partners or symbionts)	—	—	+	+	Interaction obligatory for both
5. Ammensalism (A amensal, B inhibitor or antibiotic	0	0	—	0	A inhibited ; B not affected
6. Parasitism (A parasite; B host) 7. Predation (A predator: B prey)	—	0	+	—	Obligatory for A ; B inhibited
8. Competition (A and B competitors)	0	0	—	—	Population most affected and eliminated from niche

\+ Population growth increased
— Population growth decreased
0 Population growth not affected

16-2·1. Neutralism

Neutralism is the most common type of interspecific interaction. Neither population directly affects the other. If there are interactions, they are of a more subtle and indirect type. The mere presence of the two species should not directly affect the population level of either. For example, robins (birds) and squirrels living in a forest present a good example of neutralism, as neither serves as food for the other neither has any direct interactions say that both may inhabit the same tree yet both are clearly part of the forest community.

16-2·2. Commensalism

Commensalism defines the coaction in which two or more species are mutually associated in activities centering on food and one species at least, derives benefit from the association, while the other associates are neither benefitted nor harmed. The concept of commensalism has been broadened in recent years, to apply to coactions other than those centering on food ; cover, support, protection, and locomotion are now frequently included (**Baer**, 1951; **Kendeigh**, 1974).

Lianas (vascular plants) are rooted in the ground and maintain erectness of their stems by making use of other objects and trees for support. The **epiphytes** such as orchids, "spanish moss" "old man's beard", hanging moss, *Usnea* and *Alectoria* grow in the branches of trees, where they are near light; their roots draw nourishment from the humid air. All these epiphytes depend upon the trees for support only. They manufacture their own food by photosynthesis. Likewise, ectocommensals or **epizoite** animals are associated with another animal for the purpose of anchorage and protection. Many small animals, for example, become attached to the outside of larger ones, such as the protozoans *Trichodina* and *Kerona* on *Hydra*, vorticellids (*Vorticella*, *Epistylis*) or various other aquatic organisms, branchiobdellid annelids on crayfish, the barnacles which attach themselves to the backs of whales and shells of horse-shoe crabs and so on. The remora fishes attach themselves to the bellies of sharks, swordfish, tunny, barracuda, or sea turtle by means of a dorsal fin highly modified into a suction disc on the top of the head. They are of small size and are not burdensome to the host. The host benefits the remora, however, for when the host feeds, the scraps of food floating back are swept up by the remora.

Commensals may also be internal. For example, many harmless protozoans occur in the intestinal tract of mammals, including man. The pea crab, *Pinnotheres*, lives as a commensal in the mantle cavity of certain sea mussels; the crab steals food collected by the host mollusc but does little if any other known injury. The hermit crab normally lives inside gastropod shell. The pitcher of the pitcher plant found in bogs furnishes a breeding site or home for certain specie of midge flies, mosquitoes, and tree toads. Many kinds of micro-organisms, plants and animals(such as small crabs, shrimps, polychaets, ophiuroids, ascidians, etc.) live in the canal system of sponges.

The nest of one species often furnishes shelter and protection for other species as well. Ant nests may contain guest species of various other insects ; some birds place thair nests close to wasps, bees or ants for the protection offered by these insects. Woodchuck burrows are used also by rabbits, skunks and racoons, especially in winter.

16-2·3. Protocooperation

Protocooperation is less extreme sort of interaction than mutualism in which the interaction is clearly beneficial to both species, allowing the equilibrium population levels of both to be higher than they otherwise would be. However, it is not obligatory for either species. Examples of protocooperation are many, including the relationship between water moccasins and large birds such as herons and ibises on several islands off the west coast of Florida. The birds nest in the lower branches of relatively unprotected trees, while the snakes congregate around the bases. This protects the bird from tree climbing predators such as racoons. In turn, the snake feed, in part, on fish dropped by the birds and the occasional baby birds that fall out of the nest (**Ehrenfeld**, 1970)

There is no clear demarcation between protocooperation and mutualism, as in many cases where a mutually beneficial relationship is obligatory for one species but not for the other. For instance, the nitrogen-fixing bacteria in the roots of legume plants could not survive without the host plant. However the host plant could probably survive without the bacteria.

16-2·4 Mutualism

Mutualism is an obligatory interaction that is strongly beneficial to both species. The term **symbiosis** has often been applied to this relationship (*i.e.*, mutualism), but symbiosis properly refers to intimate association of two or more disimilar organisms, regardless of benefits or the lack of them and hence includes mutualism, commensalism and parasitism (see **Smith**, 1974). Mutualism may be **facultative**, when the species involved are capable of existence independent of one another, or **obligate**, when the relationship is imperative to the existence of one or both species.

Mutualism in plants is demonstrated in the association of fungi and algae to form lichens, of nitrogen-fixing bacteria with the roots of legumes, and of fungal mycorrhizae with the roots of many flowering plants. Further, there are many intimate relations between plants and animals. Mutualism is evident in the presence of photosynthetic algal cells in the protective ectoderm of green hydra and those associated with turbellarians, molluscs, annelids, bryozoans rotifers, protozoans and the egg capsules of salamanders. The algae give off oxygen, benefiting the animals, which in turn supply CO_2 and N_2 to the plants. Likewise, certain beetles, ants and termites cultivate fungi for food. Bacteria in the caeca and intestine of herbivorous birds and mammals aid in the digestion of cellulose. The cross pollination of flowers by the agency of insects and birds seeking nectar and pollen is of such great importance that many structural adaptations in both plants and animals fit the one to the other to ensure the success of the function. Moreover, birds and mammals are of great importance as agents of plant distribution. Seeds, fruits, even entire plants become attached to feathers or fur or ingested seeds are eaten and eliminated unharmed with the faeces.

Interspecific mutualism is nicely demonstrated by the flagellate protozoan, *Trichonympha* an obligate anaerobe in the gut of several species of wood-eating termites where it digest cellulose. *Trichonympha* also occurs in the alimentary canal of wood-eating roach *Cryptocerus*. The termite and roach reduce the wood to small fragments, passing them through the alimentary canal to hind gut, where the protozoans digest the cellulose, changing it into sugar. The host benefits the protozoa by removing harmful metabolic waste products and maintaining anaerobic conditions in the intestine.

16-2·5. Amensalism

Contrasting with commensalism is **amensalism**, a situation in which one population definitely inhibits the other while remaining unaffected itself. By so modifying the environment, the organism improves its own chance of survival. Ammensalism commonly involves some type of chemical interaction, by which the organisms of one species affects the well-being and growth of individuals or the population biology of other, is called **allelochemic** (see, **Whittaker**, 1970, **Smith**, 1974, **Kendeigh**, 1974). Thus, chemicals released by one species, which act on the growth, health, and behaviour of another species, are called **allelochemicals**. Allelochemical effects are of three general sorts : allomones, kairomones and depressants. **Allomones** are chemicals which give adaptive advantage to the organism that produces the chemical. Examples include repellents, which provide a defense against attack, such as the odour of a skunk; escape substances, such as the ink that is used by an octopus to confuse predators; venoms and toxins, as would be found in snakes and scorpions; and attractants, which are used by a predator to attract a prey, such as those used by carnivorous plants and fungi or flower scents used to attract pollen carriers.

Kairomones are chemicals that give adaptive advantage to the receiving organism. These include chemicals given off by food sources, such as chemical cues used by parasites or predators to find their victims; the chemical released by nematode worms, which stimulates certain fungi to develop traps for these worms; signals that warn of danger or toxicity to the receiver, such as the odour of a cat smelled by a mouse.

Depressants are a class of allelochemicals that inhibit or poision the receiver without benefit to the releasing organism, such as some bacterial and fungal toxins. Depressants may be inorganic chemicals such as acids or bases which are produced by pioneering organisms and which reduce the competition for nutrients, light, and space, between the amensal species and inhibitor species. Depressants may be simple organic toxins which inhibit growth of seedlings of other species in the vicinity. Finally, depressants may be **antibiotics**, toxic substances produced by an organism which in low concentrations, can inhibit or kill the growth of another organism.

The majority of inhibiting chemicals are produced as secondary substances by plants and released into the soil through the roots or leaf wash. The suppression of growth through the release of chemicals by a higher plant is known as **allelopathy**. For example, walnut tree produces a non-toxic substance, juglone (**Bode**, 1958) which is found in its leaves, fruits and other tissues. When the leaves or fruits fall on the ground, juglone is released to soil, where it is oxidized to a substance that inhibits the growth of certain under story species and garden plants such as heaths and broad-leaf herbs, and favours others such as bluegrass and blackberries.

Antibiotics produced by bacteria, fungi, actinomycetes, and lichens are widespread in nature and may be one of the reasons why bacteria pathogenic to man cannot multiply well in soils. A number of antibiotics such as penicillin, have been used extensively in human medicine.

16-2·6. Parasitism

Parasitism is a kind of harmful coaction (disoperation) between two species. It is the relation between two individuals wherein one individual called **parasite** receives benefit at the expense of other individual called **host**. Parasitism is mainly a food coaction, but the parasite derives shelter and protection from the host, as well. A parasite usually parasitizes a host which is larger in body size than it. Further, a parasite does not ordinarily kill its host, at least not until the parasite has completed its reproductive cycle. However, a host may die due to some secondary infection or suffer from stunted growth, emaciation, or sterility. The balance between parasite and host is upset if the host produces antibodies or other substances which hamper normal development of the parasite. In general the parasite derives benefit from the relation while the host suffers harm. But this does not mean that all parasites are harmful. The former category constitutes the **pathogenic parasites**.

16-2·6A. Classification of parasites. Parasites exhibit a tremendous diversity in ways and adaptations to exploit their hosts. The parasites may be **viral parasites** (*e.g.*, bacterial, plant and animal viruses), **microbial parasites** (*e.g.*, bacteria, protozoa, fungi, etc.), **phytoparasites** (*e.g.*, plant parasites) and **zooparasites** (*e.g.*, animal parasites such as platyhelminthes, nematodes, arthropods, etc.). They may parasitize micro-organisms, plants and animals. They may occur on the outside of the host (**ecto-parasites**) or live within the body of the host (**endoparasites**). The endoparasites usually live in the alimentary tract, body cavities, various organs or blood or other tissues of host. Ectoparasites may be parasitic only in the immature stages—the hairworm larvae, parasitic in aquatic insects; only the **adults** may be parasitic—fleas on birds and mammals; or both larvae and adults may be parasitic—the blood sucking lice and flies, biting lice, mites, and ticks that occur on birds, mammals, and sometimes reptiles, and the monogenetic trematodes on fish. Similar relation occurs in endoparasites but in them usually all the stages of life are parasitic—entozoic amoebae, trichomonad flagellates, opalinid ciliates, sporozoans, pentastomids, nematodes, digenetic trematodes, acanthocephalans, cestodes, and some copepodes.

Further, endoparasites rely on various means of transport from one host to another so that their survival, range of **distribution** and life cycle are unaffected. For the endoparasite; the interior of its host is its **microhabitat** and **microenvironment**, while the outside environment is the **macroenvironment**. The host acts as a buffer between the parasite and the outside environment—changes in the microenvironment directly influence the parasite while alterations in the macroenvironment influence it indirectly. Thus, mutual adaptations and tolerance between the host and parasite are important factors facilitating successful transmission of endoparasites, and climatic fluctuations tend to alter the physiology of the parasite, host or vector. Mutual adaptations are the parasite's **abilities** to establish and maintain itself in a favourable location within the host and exist in a proper form for sufficient length of time enabling **subsequent** transport to another host and continuation of species.

Animals may also be parasitic to plants. Nematodes infest the roots of plants. Wasps or gnats form galls on plants such as oaks, hickories, willows, roses, goldenrods, and asters. Mites stimulate formation of witches brooms in hackberry. A variety of insects the larvae of which are leaf miners, wood borers, cambium feeders, and fruit eaters should be included here.

Lastly, parasites may be **full-time** (or **permanent**) **parasites** or **part-time** (or **temporary**) **parasites.** Mosquitoes and bugs that **suck** the blood of their hosts are temporary parasites. Some temporary parasites spend only a part of their life cycle as parasites. For example, glochidium larva of *Anodonta* (fresh-water mussel) attaches itself to the body of the fish by means of its hooks and penetrates inside the fish integument to remain burried there for several weeks and finally emerges out as the young mussel to lead an independent existence. Permanent parasites, however, spend their life completely on other organisms. The common examples of permanent parasites are—*Plasmodium*, *Entamoeba histolytica*, and other protozoan pathogens, different platyhelminthes, nematodes, arthropods, etc.

16-2·6B. Parasitic adaptations. While studying parasitism as a biotic factor influencing the life and activities of parasites and hosts we often encounter manifold adaptations, both offensive and defensive, being developed by the host, as well as by the parasite. An endoparasite entering a host often meets with the antibodies or phagocytic cells produced by the host. The ectoparasites, and endoparasites have following parasitic adaptations :

1. In parasitic animals, a reduction of organs of special sense of nervous system, and of locomotory organs occur.

2. Most ectoparasites develop some clinging organs such as hooks, suckers, etc., to get attached with the body of their hosts. They also develop special piercing and sucking organs to suck the blood of animals or sap of plants. Certain blood-sucking ectoparasites such as leeches, blood-sucking mosquitoes, etc., contain certain anticoagulant enzymes in their salivary secretions.

3. Most endoparasites exhibit anaerobic respiration, high rate of reproduction, parthenogenesis, hermaphroditism, polyembryony, intermediate hosts and a complicated life-cycle. Some endoparasites, such as tapeworm, have become so adapted to the host that they no longer require a digestive system. They simply absorb their food directly through their body wall. Further, parasites that live within the bodies of plants and animals possess cuticles or develop cysts resistant to the digestive enzymatic action of the host.

4. Many parasites pass their entire existence in a single host; others require one, two, even three intermediate host. It is of ecological significance that both primary and intermediate hosts of a parasite occur in the same habitat or community. Even then the hazards to successful passage from one host to another are so great and mortality so high that prodigious quantities of offrpring are produced to ensure that at least a few indivinuals will complete the cycle.

5. Parasites are transferred from one host to another by active locomotion of the parasite itself; by ingestion, as one animal sucks the blood of or eats another; by ingestion as an animal takes in eggs, spores, or encysted stages of the parasite along with its food or drinking water; as a result of bodily contact between hosts; or by transportation from host by way of vectors. For example, the bacteria that cause tularemia in man are carried from rabbit to rabbit by ticks. Man contacts the disease when he handles infected rabbits but the incidence of infection is greatly reduced in the autum when cold wheather forces the ticks (*i.e.*, the vectors of bacteria) to leave the rabbits and go into hibernation.

16-2·6C. Host specificity of parasites. Certain parasites, such as copepodes, contain a wide range of hosts and they are ubiquitously present in various invertebrates and fishes. Most parasitic genera, however, are adapted to hosts of one phylum only. For example, each order of birds possesses its own particular species of tapeworms. This is true even when several order of birds live in the same habitats as do, for instance, grebes, loons, herons, ducks, waders, flamingoes and cormorants (**Baer** 1951). Species of flagellate protozoans that occur in termite alimentary canals are largely host-specific. Some species of gall wasps attack only one species of oak. Where a single species parasitizes two or more host **species**, the shape and structure of the gall formed around the egg and larva on both hosts is essentially similar. When several insects are found on the same oak, each kind of parasite produces its own characteristic gall form. The parasite-specific gall formation is found to be related with parasite-specific enzyme (**Kinsey**, 1930).

Certain parasites remain restricted to special habitats within the host. The roundworm lives near the duodenum of the digestive tract and soil nematodes live in the rootlets of plants. Biting lice remain restricted to the head or body regions of birds. Some zooparasitic nematode species are found throughout the body in connective tissue, but not in the gut ; some occur only in the digestive tract and associated organs ; certain species occur in the **glandular** crop of birds, but others only in the caecum; many species occur exclusively in the lungs or in the frontal sinuses. Such fine restrictions of parasites to particular microenvironments—hosts or organs is a consequence of precise physiological and morphological adaptations that permit the parasite to survive and complete the life cycle only under very special condions. Likewise, the life cycle of the parasite is often closely synchronized in time with that of its host (**Foster**, 1969).

16-2·6D. Effects of parasite on the host : disease. Parasites may not cause immediate mortality but they cause damage to body structures, should it become excessive, may cause death. Because of these parasitically caused anatomical damages, the finely adjusted balance of different vital processes of host's body become disturbed and host is said to be diseased. There are a number of causes of diseases. Parasites are one; physiological stress, nutritional deficiency and poisoning are others. Some of common parasitic agents of disease and consequent mortality of animals are following:

1. Viruses are the potent agents of several disastrous diseases of plants and animals including man. For instance, some viruses are the agents of hoof and mouth disease in deer, spotted fever in rabbit, encephalitis and distemper in foxes.

2. Bacteria may produce localized inflamatory changes in tissue, enter the blood stream, or produce powerful poisons known as toxins. They cause a variety of disease, notably tularemia paratyphoid, and tuberculosis among birds and mammals, as well as other diseases in lower **types** of organisms.

3. Fungus spores of ***Aspergillus*** may be drawn into the lungs of ground-feeding birds, where they germinate and grow, causing

aspergillosis disease. Fungus may also develop on the external surface of animals.

4. Protozoan parasites are especially important in the alimentary tract and in the blood. A sporozoan species of *Eimeria* damages the walls of the intestine in upland game birds and causes **coccidiosis disease** ; *Toxoplasma* becomes encysted in the brain of rodents ; *Leucocytozon* is a common blood parasite of waterfowl and game birds.

5. Worm parasites, such as tapeworms, nematodes, and acanthocephalans may wander through the hosts body doing mechanical injury as well as destroying and consuming tissues. The host may respond by forming a fibrous capsule or cyst around an embedded parasite.

6. External parasites such as ticks, fleas, lice, mites, and flies do not commonly produce serious mortality by themselves, but they are often vectors, transmitting protozoa, bacteria and viruses from one animal to another. Heavy infestations of external parasites however lower the vitality of an animal and cause diseases of fur or feathers.

7. Nutritional deficiency in vitamins or minerals, or improper balance among carbohydrates, proteins, and fats may produce malformations, lack of vigour, even death.

8. Food poisoning, **botulism**, occurs when certain food become contaminated with the toxins released by the bacterium *Clostridium botulinum.*

9. Physiological stress (**Selye**, 1955) is a term that has come to be applied to changes produced in the body non-specifically by many different agencies which may accompany any disease. Effects of stress include loss of appetite and vigour, ache and pains, and loss of weight. Internally, physiological stress causes following abnormalities, such as, acute involution of the lymphatic organs, diminution of the blood eosinophils, enlargement and increased secretory activity of the adrenal cortex, and a variety of changes in the chemical constitution of the blood and tissues.

10. Accidents, aging, starvation, and so on, must also be included as important causes of mortality.

Organisms that produce disease generally fall into one or two categories. They are either present in the body at all times but not normally virulent, or they are normally absent but are virulent from the moment the host is infected by them. Further, a single attack, even a mild one of some diseases often confers a partial or complete **immunity** from further attacks of same disease. Like parasites, disease can reduce populations, exterminate them locally, or restrict the distribution of host. For example, smallpox, intentionally or unintentionally introduced nearly exterminated some tribes of North American Indians (*viz.*, Human populations). Malaria in Sourthern Asia has depressed population growth.

16-2·6E. Social parasitism. Social parasitism describes the exploitation of one species by another, for various advantages. It is a kind of parasitism in which the parasite foists the rearing of its young into the host. Social parasitism in various stages of development is found among some higher vertebrates and insects. For example, there occurs a **egg parasitism** in two species of birds—old world cuckoos and the brown-headed cowbirds of North America, both of which do not build nests of their own, rather they deposit their eggs in two nests of other species, abandoning eggs and young to the care of foster parents. Similary, there occurs a **brood parasitism** in between two Indian birds—Indian koel and crow. The Indian koel (*Endynamys scolopacea scolopacea*), like many of its cuckoo cousins (family **Cuculidae**), neither incubates its eggs nor feed its young. But these functions are secretly forced upon the house crow (*Corvus splendens*) and jungle crow (*Corvus macrorhynchus* (**Davis, De** and **Pal**, 1977). One species of ant waylays foraging workers of another species and snatches away the food they are transporting or the robber species may deliberately rob another nest of food. Some species of ants make slaves of the workers of other species. Such kind of dependency of one species on another also occur in other social insects such as termites, waps and bees. Social insects are apparently the only animals other than man to have succeeded in domesticating other species and of cultivating plants particularly fungi for food.

16-2·6F. Parasitodism. Some Diptera and Hymenoptera deposit their eggs in the immature stages of other insects ; the larvae on hatching feed on the host until they are fully grown. In such type of parasitism in which host generally dies of the larval depredations before the larva emerges and the parasitoid larva generally lives in spite of the hosts death, is called **parasitodism**. It stands in between parasitism and predation. The parasitoid may in turn be infested with **hyperparasitoids**. In the Chicago, *Samia cecropia*, a saturniid moth, suffers the destruction of nearly 23% of its cocoon by an ichneumonid parasitoid, *Spilocryptus extrematis*, which deposits an average of 33 eggs on the inside of each cocoon or on the surface of larva. The host larva dies in a few hours after the parasitoid hatches and the ichneumonid larva moves about freely, feeding on the cuticle by burrowing into the tissue to drink the body fluids. *S. extrematis* is infested by another hyperparasitoid, *Aenoplex smithii* which causes about 13% destruction of *S. extrematis* cocoons. A chalcidid, *Dibrachys boucheanus*, fed both upon *S. extrematis* and as a tertiary parasitoid, upoh *A. smithii.*

16-2·7 Predation

Predation is a form of disoperation, at least in point of immediate effects, since one animal kills another animal or plant for food. Like parasitism, predation is important in community dynamics but both differ in the point that a predator tends to be larger than its prey, and it catches its prey from without, while a parasite is smaller than its host and consumes it from within (see **Clapham**, Jr., 1973).

16-2·7A. Components of predation. The main components of predation are predators and prey. A successful **predator** has following characteristics :

1. The hunting ability of a predator remians well developed.

2. By their hunting activities predators can be regarded as specialized or generalized. **Specialized predators** are those adapted to hunt only a few species. They are forced to move where the vulnerability of a staple prey item drops to a point where the predator population cannot support itself. For example, Peale's falcon shows a marked preference for ducks and pheasants. Deer exhibits a pronounced preference for certain species of browse (**Klein**, 1970). **Generalized predators,** not so restricted in diet, adjust to other food sources. The horned owl and buted hawk have a large range of collective prey available. Foxes can shift to vegetable and carrion diet, should conditions require it.

3 Hunting ability and success of predator involve the development of a searching image on the part of the predator. Once it has secured a palatable item of prey, the predator finds it progressively easier to find others of the same kind.

4. Though a predator may have a strong preference for a particular prey, it can turn in the time of relative scarcity to an alternate, more abundant species that provides more profitable hunting. For example, if rodents are more abundant than rabbits and quail, foxes and hawks will concentrate on them insetead of game animal.

5. Habitat preference or overlapping territories can bring predator and prey into close contact, increasing prey risks. For example, predatory rainbow trout in Paul Lake British Columbia, moves into the shoals when their prey, the red side shiner, are most heavily concentrated there.

6. Age, size, and strength of prey influence the direction that predation takes. Predators select food on the basis of size. Mountain lions, for example, avoid attacking large healthy elk, which they cannot successfully handle, and concentrate instead on deer and young or feeble elk (**Hornocker**, 1970).

7. Predators hunt only when it is necessary for them to procure food. The searching rate of predator is influenced by the speed of the predator relative to the speed and escape reactions of the prey, to the distance at which predators first notice and attack the prey and to the proportion of attacks that result in successful capture.

Like the predators, the prey has its certain defensive specializations. The prey risk is determined by density of prey population, availability of food and protective cover (concealment place), movement, activity, habits, size, age, strenght and escape reactions of prey.

16-2·7B. Regulating effect of predation. It is commonly concluded that the predator—prey interaction causes a reduction in the prey population, it is detrimental in the same fashion to the prey This belief has led to extensive "predator control" efforts conducted in the name of wildlife conservation. But the coevolution of species within natural ecosystems has in fact led to a dynamic balance between the populations in a community, so that the population sizes of predator and prey species are inter-regulated by feedback mechanisms that effectively control the population of both (Fig. 16·2). As the models of **Lotka-Volterra** and **Nicholson-Bailey**

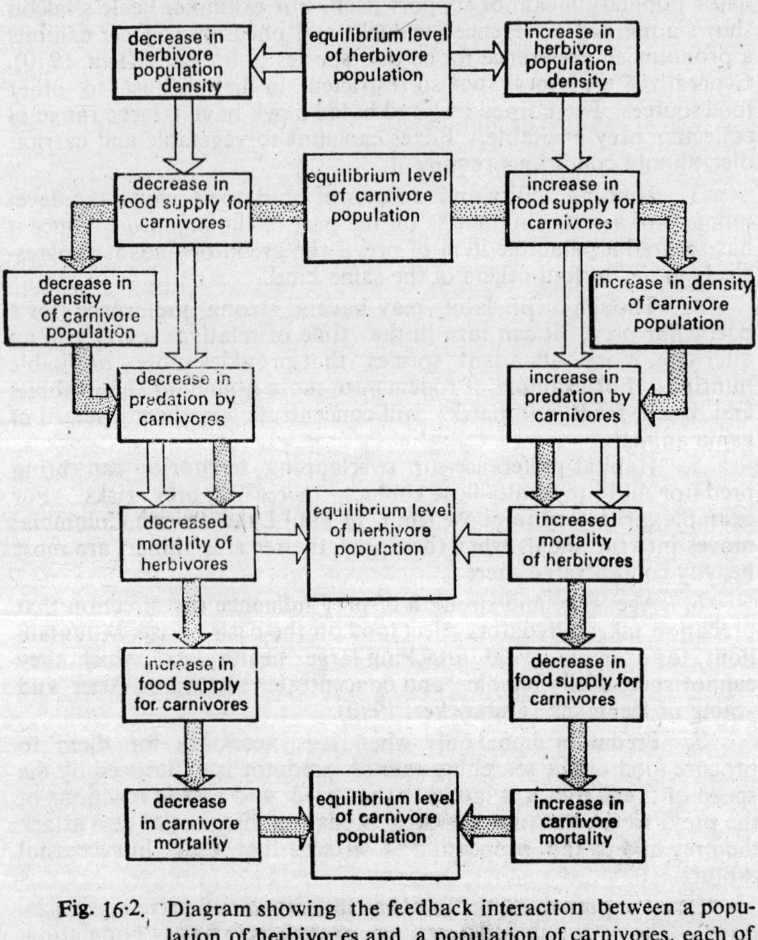

Fig. 16·2. Diagram showing the feedback interaction between a population of herbivores and a population of carnivores. each of which acts to regulate the population size of the other. Herbivores shown with unshaded arrows, carnivores shown with shaded arrows (after Clapham, Jr., 1973).

stated a predator that consistently destroyed more prey organisms than the prey population could stand would drive itself, as well as the prey, into extinction, because it would cut off its own food supply. Likewise, if the level of predation is too low, a prey population—whose evolutionary history has, after all, conditioned it to a normal level of losses to predation—may become too large, where upon it can destroy its food supply.

Ricker (1954) has recognized two kinds of predation—A and B. In type A predation, predators of any given abundance take a fixed number of prey species during the time they are in contact, usually enough to satiate themselves. The surplus prey escape Trout feeding on an evening hatch of mayflies would come under this category. There is a functional response but no numerical response. Type B predation exists if predators of any given abundance take a fixed fraction of a prey species, as though the prey were captured at random encounters. In other words, the amount of prey eaten is proportional to the abundance of predator and the abundance of the prey. There is both a functional and numerical response. Many vertebrates, however, do not follow the same functional and numerical response in relation to prey and predator. As predators take most or all the individuals of the prey species that are in excess of a certain minimum number, as determined by the carrying capacity of the habitat and social behaviour. The prey species compensates for its losses through increased litter and brood size and greater survival of young. For this reason this type of predation, **Ricker's** type C, is called **compensatory**. The population level at which predators no longer find it profitable to hunt the prey species has been called the "**threshold of security**" by **Errington** (1946). As prey numbers increase above this threshold, the surplus animals are no longer tolerated in the area, and become vulnerable to predation. Below the threshold of security, functional response of the predator is very low and numerical response is non-existent. Above the threshold, functional response is marked and numerical response could occur.

16-2·7C. Herbivores and plants. Herbivores come in many sizes ranging from elephants to insects. Their effects on plants can be devastating or subtle. A dramatic example of a herbivore regulating a plant involves the prickly pear cactus (*Opuntia*) which was introduced into Queensland, Australia, from south America. It spreads rapidly over millions of hectares, turning pasturel and for cattle into areas where grazing was difficult at best. When the natural enemy of the cactus, a caterpillar, was introduced into Australia, it virtually destroyed most of the cactus

Why don't herbivores eat up all their food supply ? Some biologists believe that this is because the herbivores are held below the carring capacity of the environment either by predators, diseases or by intraspecific competition. An alternative explanation is that plants have evolved mechanisms of protection. Protective devices such as spines, thorns, toxins, etc., of cacti, thistles, roses, buttercups, conifers, etc., are obvious.

16-2·8. Competition

Competition refers to the type of interaction in which two individuals or species compete for a limited resource. It may occur between the individuals of the same species (called **intraspecific competition**), or between the individuals of two or more species of a community (called **interspecific competition**). Both types of competitions have population-regulating and evolutionary roles.

16-2·8A. Intraspecific competition. In intraspecific competition organisms compete for some resource often in short supply–food, space or mate. In a long-term experiment involving the sheep blowfly, *Lucilia cuprina*, **Nicholson** demonstrated the influence of intraspecific competition in a population. He recognised the following two basic types of intraspecific competitions :

1. **Scramble competition,** in which the resource is subdivided into many small parts to which all have access. Individuals "scramble" for the resource, which results in wastage. Each individual obtains such a small amount of resource that it is unable to survive.
2. **Contest competition,** in which each successful individual claims a part of the resource and the unsuccessful are denied any access to it. Contest competition shows up in territoriality, social dominance and hierarchy, migration and psychological stresses.

Because an energy source is essential for the maintenance of a population, the food supply, for example, directly or indirectly, independently or associated with other limited resources, is found to regulate the population size. **Lack, Pitelka** and **Schultz** developed a hypothesis called rodent-vegetation supply hypothesis (Fig. 16·3) to show the role of overutilization of food in population regulation. The hypothesis relates the 4-year cycle of the Arctic lemming to ecosystem function. According to the hypothesis the **forage** of the lemming is rich in calcium and phosphorus, but because of high production of forage much of these and other mineral elements are tied up in organic matter. Grazing, burrowing and tunneling by lemmings reduce the organic insulation of the soil, and the soil, thaws deeper. The next year the forage production is low, there is poor cycling of nutrients and the lemming population is now under stress from a low and nutrient-poor food supply. Nutrient depletion affects reproduction and lactation, and the lack of adequate protective cover exposes the population to excessive predation. A combination of the two causes a rapid decline of the population. The lower population permits the vegetation to recover and organic matter to accumulate, but the forage quality is still low. By the fourth year the plants have fully recovered, protective cover has turned, nutrient cycling has speeded up and nutrients accumulate in the plants. The nutrient content of the forage is high enough to stimulate reproduction and support a large population of lemmings again.

16-2·8B. Interspecific competition. When two species in the same community seek the same resources, such as food and space, which are in short supply (in relation to the number seeking it), or interact in a way that affects their growth and survival, they are said to compete with one another. The idea of competition, or the "struggle between species" was emphasized by **Darwin.** Later **Lotka** and **Volterra** separately developed mathematical formulae describing relationships between two species utilising the same food source. Essentially these formulae indicate that eventually only one species, the least susceptible to food shortage or most adaptable to changed environmental conditions, will survive.

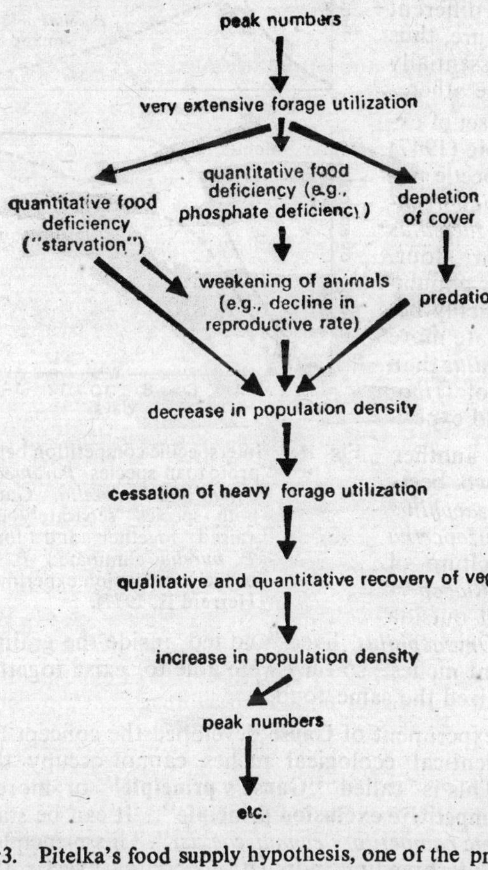

Fig. 16·3. Pitelka's food supply hypothesis, one of the proposed regulatory mechanisms in an animal population (after Smith, 1974).

American ecologist **G. F. Gause** (1934) set out to test these formulae experimentally. He introduced two species of *Paramecium,*

P. aurelia and *P. caudatum*, in one tube containing a fixed amount of bacterial food. *P. caudatum* died out. The success of *P. aurelia* resulted from its higher rate of inerease (Fig. 16·4). But in another experiment *P. caudatum* occupied the same solution with *P. bursaria*. Here both species were able to reach stability because *P. bursaria* confined its feeding to bacteria on the bottom, whereas the other fed on that suspended in the solution. Although the two used the same food supply, they occupied different parts of the culture, thus utilising food essentially unavilable to the other.

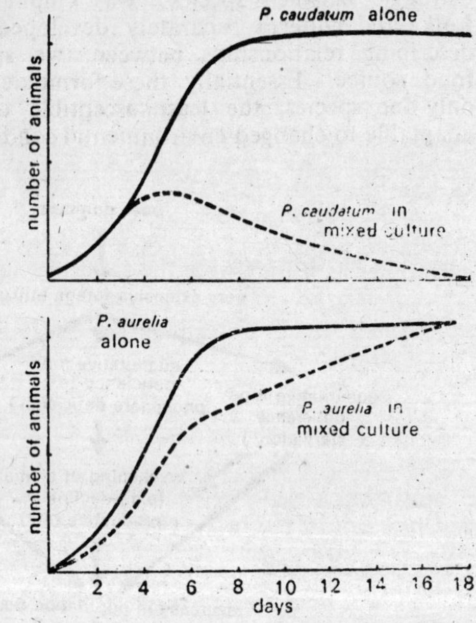

Fig. 16·4. Interspecific competition between two protozoan species. *Paramecium caudatum* and *P. aurelia*. Gause raised both species separately and then raised together and found that *P. aurelia* eliminated *P. caudatum* in the competition experiment (after Herreid II, 1977).

In another set of experiment **Crombie** (1947) confined two beetle populations, *Tribolium confusum* and *Oryzaephilus surinamensis*, in flour. The *Oryzaephilus* population died out, cheifly because *Tribolium* ate more eggs of *Oryzaephilus* than the latter did of *Tribolium*. In a second experiment he reared another population of two beetles, one *Oryzaephilus* and the other *Rhizopertha dominica*, in a culture of cracked wheat. *Rhizopertha* lived and fed outside the grain and *Oryzaephilus* lived and fed inside the grain. They occupied different niches, so they were able to exist together even though they utilized the same food.

From this experiment of **Gause** developed the concept that two species with identical ecological niches cannot occupy the same environment. This is called **"Gause's principle"** or more appropriately as **"competitive exclusion principle"**. It can be stated briefly as: "*complete competitors cannot coexist*". This principle means that if two non-interbreeding populations occupy exactly the same ecological niche, if they occupy the same geographic territory, and if population A multiplies even the least bit faster than population B, then A eventually will occupy the area completely and B will become extinct.

1. Strategy of competition. Interspecific competition involves the following two strategies, exploitation and interference. **Exploitation** takes place when two or more **species** have access to a limited resource. The outcome is determined by how effectively each of the competitors utilizes the resource. For example, two species of butterflies, the monarch and the queen monarch in Florida are attracted to the flowers of the same milkweed plant, upon which they both feed and lay eggs and on which the larvae feed. The non-migratory queen monarchs, however, avoid some species of milkweed that the migratory monarch utilise. But after the monarch have flown northward, the queen monarchs oviposit on this milkweed. Thus the disappearance of the monarchs coincides with a population increase in the queen monarchs.

In **interference**, one competitor is denied access to the resource, usually by some form of aggressive behaviour. For example, when brook trout and rainbow trout occupy the same stream, they treat each other as if they were the same species. The largest fish occupy the deepest and darkest parts of the stream and the smaller ones are forced to live in less favourable parts.

2. Character displacement. Interspecific competition acts as a means of natural selection (*i.e.*, speciation) in the following way : if two species are competing for a limited resource (say, food) over a period of time, both species would benefit if their preferences would shift away from the type of food for which there was the greatest competition. Suppose that two bird species had slightly different bill sizes, with certain overlap in the food habits. Even though they both ate the same seeds, the larger species could take large seeds as well as middle-sized seeds, while the smaller species could take the smallest seeds plus middle-sized seeds. Competition would be most severe for middle-sized seeds, especially during times of food shortage. Therefore, individuals that could best exploit that part of the seed supply for which competition was reduced would be at an advantage. Over generations, one would expect the bill-size and feeding habits of the two species to become more distinct, the small ones getting smaller and the large ones getting larger. This divergence of character is called **character displacement**, and there are many examples of it.

For example, the I4 species of Darwin's finches belonging to two genera *Camarhynchus* (insect-eating) and *Geospiza* (seed-eating) of Galapagos Islands, differ primarily in size and bill shape. Three species of ground finches (*G. magnirostris*, *G. fuliginosa* and *G. fortis*) have partially overlapping food preferences. When they exist together on an island, their bill-sizes show almost no overlap in size range yet when they live on different islands, considerable duplication of bill measurements occur. Thus, competition for limited resources, if it is continued over a long period of time without extinction of one or another species leads to divergence in characteristics and the reduction of competition. It can lead to the filling of otherwise unfilled niches.

3. Results of interspecifiic competition. Interspecific competition thus have several alternative outcomes depending upon the intensity of the **interaction** and the heterogeneity of the environment. The following categories are a matter of scale. At one extreme, there occurs total dominance by one species; at the other extreme, virtual coexistence :

1. Complete competitive exclusion. One species eliminates the other completely from a region, perhaps leading to extinction.

2. Reciprocal competitive exclusion. Because of species and environmental differences, one species is completely dominant in one geographical area and the other species is completely dominant in another, with a sharp boundary line between them. In some cases, removal of one species immediately allows the other to invade the vacant area. In other cases, no invasion would occur because specialisation of anatomy, behaviour or physiology for one environment might preclude an immediate shift into the empty niche.

3. Sympatric species with different habitats. In a given biome both species occur, but each is dominant in his own habitat. one species might live on muddy bottoms, for example, and the other on sandy bottoms.

4. Coexistence in a given habitat with other niche differences. Species may live in the same habitat but use different foods, or exploit them at different times, which avoids competition.

5. Coexistence with shifting advantages. Species can coexist, even if they compete strongly, if the environment first favours one, then another, species. This could occur in a changing environment (seasonal variations) or if stages of the life history compete differently.

6. Coexistence with niche similarity. Coexistence of similar species is possible if the populations are kept below levels at which competition for common resources occur. In Africa dozens of similar antelope species exist together. If their population levels are controlled well below the carrying capacity of the environment by predators or disease, their competition may not be important. Several species have shown that removal of predators increased competition among the prey for common resources, resulting in the elimination of several species. Predation can increase species diversity in a habitat.

16-3. REVISION QUESTIONS

1. Describe different kinds of intraspecific interactions.
2. What is aggregation ? Discuss its advantages.
3. Describe the role of communication in the social animals.
4. What is interspecific interaction ? Describe different kinds of interspecific interactions with suitable examples.
5. Discuss how competitiion regulates the population-size and causes speciation ?

6. Describe different kinds of allelochemicals and their effects in interspecific interactions.
7. Write short notes on the following :
   (i) Colonisation;
   (ii) Social organisation;
   (iii) Altruistic behaviour;
   (iv) Dominance hierarchy;
   (v) Territoriality;
   (vi) Communication;
   (vii) Pheromone;
   (viii) Ammensalism;
   (ix) Parasitism;
   (x) Predation;
   (xi) Character displacement;
   (xii) Scramble competition.

17
Ecological Genetics

In the previous chapter, we have discussed the role of competition along with several other intraspecific and interspecific interactions in population dynamics. In this chapter, a discussion will be made of the ways by which genetic make-up (*i.e.*, genotype) of the population is changed, species become adapted to changed environmental conditions and new species are formed.

17-1. POPULATION GENETICS AND NATURAL SELECTION

The theory of evolution developed independently by **Charles Darwin** and **Alfred Wallace** states that natural selection favours those individuals whose traits render them best able to survive and reproduce ; thus, their offspring will be represented in greater numbers in future generations. This theory posited the concept of constant and gradual change as characteristic of all life. To this, geneticists contributed the finding that characteristics favoured by selection are passed on from generation to generation according to Mendelian laws of particulate inheritance, thus preserving genetic variation in a population. Further, individuals do not evolve, populations do. That is, the genetic make-up (genotype) of an individual remains constant throughout life, except for special situations, while the types and numbers of gene flow in a population of creatures can vary. The field of **population genetics** deals with questions of gene flow among individuals in interbreeding group.

Populations comprise groups of individuals in one locale or within defined geographic boundaries capable of freely interbreeding. All of the genes of these individuals form a single **gene pool** and the way in which each allele is represented in the gene pool determines a population's variation or **gene frequencies** (A frequency is the ratio of actual number of individuals falling in a single class to the total number of individuals).

Variation is both phenotypic and genotypic, the range of both ultimately determined by the individual genotype. Phenotypic variation represents adaptation by the individual, whereas genotypic variation represents adaptation by the population, enabling it to change over time. Genotypic variations is both **continuous** or **regular** (*e.g.*, quantitative or polygenic traits such as body size and skin pigmentation), and **discontinuous** or **sharp** (*e.g.*, qualitative traits such as form, structure, etc.) (see **Verma** and **Agarwal**, 1982 for more details). When individuals within a population differ strikingly from each other in one or more qualitative traits the population is considered **polymorphic**.

Variation originates in mutation and recombination, and is affected by natural selection, migration (introgression), and genetic drift (random changes in gene frequencies). In the absence of these factors it will remain constant from one generation to the next, provided the population is randomly mating. This principle is known as the **Hardy-Weinberg principle** because it was developed independently in 1908 by a British mathematician, **G.H. Hardy** and a German physician, **W. Weinberg**. In mathematical terms, this principle is stated as follows :

$$(p+q)^2=1$$

which can be written in expanded form as

$$(p+q)\times(p+q)=1$$

$$\text{or,}\quad p^2+2pq+q^2=1$$

where, p=frequency of the dominant gene ; q=frequency of the recessive gene, p^2=**frequency** of the homozygous dominant genotype, 2pq=frequency of heterozygous genotype, q^2=frequency of the homozygous recessive genotype, and 1=a population in genetic equilibrium.

The Hardy-Weinberg principle, thus, enables us to determine equilibrium genotypic frequencies in future generations from any given gene frequencies. The degree to which actual genotypic frequencies depart from equilibrium frequencies indicates the rate of evolution, for evolution can be considered a change in a population's gene frequencies.

Sexual reproduction permits **recombination** of genes in instances of crossing-over at meiosis in gamete formation and in the pairing of parental chromosomes at zygote formation. **Mutations** take the form of changes in the gene's DNA (thereby called **point mutations**) and changes in chromosome number and changes in chromosome structure (both are called **gross** or **chromosomal mutations**). Recombination and mutation are regarded basic source of evolutionary change.

Migration operates as a force of evolution when it occurs between relatively isolated populations with distinct gene pools, as immigrants from one population to another can contribute totally new alleles. This exchange of genes is known as **gene flow**, and functions to increase variation in a population.

Operating upon genetic variation are the forces of selection and genetic drift. **Natural selection** occurs in nature, whereas **artificial selection** is engineered in laboratory conditions ; both, however, result in differential reproduction of the genotypes of a population. Selection acts upon the individual phenotype to increase or decrease the representation of the corresponding genotype in a population. The following three types of selection have been categorized by population geneticists : (i) **Directional selection** describes the change that occurs when a population shows a particular trend through time. An example of this change is seen in the evolution of the horse, in which the average size has increased through time, presumably because there are selective advantages for being large. The case of industrial melanism in peppered moth (*Biston betularia*) represents another popular example of directional selection. (ii) **Stabilising selection** describes the change when extreme individuals are eliminated from the population. The result of this process is a reduced variability in the population. Most selection that occurs in populations is stabilising and homeostatic because it tends to maintain the *status quo*. (iii) **Disruptive selection** describes the process where certain types of a character have a high survival value, whereas intermediate types are not advantageous. This would be the case if big horses and small horses of a population had a high survival rate, while middle-sized horses were not favoured. Directional selection, combind with stabilising selection, is probably most significant in terms of evolution.

Genetic drift is most likely to become an evolutionary factor in a small population, especially one characterised by a high degree of inbreeding. Due to more frequent pairing, homozygous recessives may appear in relatively high frequencies. On the other hand, some genes may be carried by so few members that they become lost by chance. When such changes occur as a result of a non-representative original population, they are ascribed to **founder effect**. Founder effect describes the presence of one or more deleterious allele in relatively high frequencies, resulting from a non-representive sample in the seed stock, or founders of a population. For example, founder effect may apply to the surviving members of a population considerably reduced by plague or famine or some other catastrophe. If as this small population expands by inbreeding and it remains isolated, due to geographical inaccessibility or general environmental hardship, a harmfull recessive gene which may have been insignificant in the parent population (before the disaster) because of the small chances of its combining with another such gene, may reappear more frequently (**Goldsby, 1976**).

Variation is protected or maintained in any population by the combination of a deleterious recessive gene in heterozygous form and selection favours modifying genes that render the heterozygote phenotypically similar to the dominant homozygotes. Moreover, standard patterns or canalisation of development are imposed upon genetic variation for traits essential to survival. A demonstrable influence on continued variation is **heterosis** or **heterozygote superiority**. This adaptive superiority may be due to the fact that the heterozygote has the benefit of two kinds of metabolites, one produced by each allele. Since it has the capabilities of both homozygotes, it is better adapted, in certain conditions, than either. The most frequently quoted example of heterosis is the human variant known as **sickle-cell anemia**, in which the haemoglobin structure is modified so that a homozygote for the condition usually dies of anemia in childhood. However, the heterozygotes of this abnormality are found to have considerable tolerance for attacks of malaria (**Boughey**, 1968). Greater fitness of the heterozygotes ensures the maintenance of both alleles in populations operated upon by opposing selective forces. This results in **balanced polymorphism**, a stable proportion of two or more forms; any imbalance in either direction will be disadvantageous for the population as a whole, but selection against the homozygote will tend to restore equilibrium. Polymorphism is particularly useful to species living in a variable environment, for it enables one or another form to thrive under various local conditions. Finally, variation is assured by sexual reproduction, and selection has favoured the evolution of some form of outbreeding even in self-fertilizing or asexual organisms.

17-2. ADAPTATIONS AND ECOTYPES

Organisms can respond to environmental stress in such a way that their tolerance zones may change. The genetic changes that occur during the evolution of the species because of mutation and natural selection are called **adaptations**. Presumably, each species is fairly well adapted or fit, if it has been in its particular environment for many generations. The range of tolerance of a species may be narrow (*i e.*, species has a low **ecological amplitude**) or broad (*i.e.*, species has a high ecological amplitude). Thus, organisms are "slaves" to the physical environment; they adapt themselves and modify the physical environment so as to reduce the limiting effect of temperature, light, water and other physical conditions of existence. Such **factor compensation** is particularly effective at the community level of organisation but also occurs within the species. Species with wide geographical distribution almost always develop locally adapted populations called **ecotypes** that have optima and limits of tolerances adjusted to local conditions. Compensation along gradients of temperature, light or other factors may involve **genetic races** (with or without morphological manifestation) or merely physiological acclimatisation.

Acclimatisation describes all the non-genetic modifications of the organisms in response to an environmental stress during their own life-time. For example, nude man exposed to moderately cold temperatures shiver violently. However, after several weeks of exposure, this response decreases and they become better able to withstand the cold. Likewise, within a few days after a move from low to high altitudes, people experience an increase in the number of red cells in their blood. Red blood cells carry oxygen, so these people can then tolerate the lower oxygen level in the air at high altitudes without signs of stress. Acclimatisation is an important weapon in the arsenal of the organism trying to maintain homeostasis.

The concept of ecotype was developed by the Swedish botanist, **Gote Turesson** as a result of a series of observations and experiments conducted in the 1920s and 1930s on variation on Swedish plant species such as *Plantago maritima.* **Turesson** collected live plants from different parts of the country and grew them together in the same garden. **Turesson** described the study of ecotypes as a new research field, called **genecology** (**Krebs**, 1972). In fact, he started a procedure for **transplant experiments** that has since been followed by many genecologists or biosytematists, as they are now more commonly called.

Turesson's basic conclusions includes the following aspects : (1) Widely distributed species exhibit a variation in morphological and physiological characters ; these vary from place to place within the habitat. (2) This variation is largely correlated with observable habitat differences. (3) This correlated variation, where it is not simply phenotypic variation, is the result of the natural selection of particular genotypes from the pool of genetic variability available within the species.

The population sampling and subsequent procedures that Turesson followed led him to conclude that species were composed of a mosaic of populations—the ecotypes—each adapted by natural selection to a particular habitat and each being discontinuously distinct from the others. Thus populations of ecotypes were seen to differ from clinal populations or **ecoclines**, in which variation is both continuous and directional in response to a habitat gradient. Ecotypes can also be distinguished from the **ecads** or **ecophenes**, which are the morphological variants of the same population in relation to some habitat characteristic but not fixed by genetic composition. One of the earliest studies in India showed that the two closely related species of *Lindenbergia* namely, *L. polyantha* and *L. urticaefolia* were not two distinct species but only two ecotypes of the same species (**Mira** and **Sivarao**, 1948). In another study, **Ramakrishnam** showed that the two populations, red and green in colour, of ***Euphorbia** thymifolia* were two ecotypes which had their distribution related to the exchangeable calcium content of the soil. The red ecotype grows in calcium rich soil but may also occur in calcium deficient soil. On the other hand, the green ecotype is an obligate calcifuge, *i e.*, it cannot grow in calcium rich soil. These ecotypes were found to be caused by paired dominant alleles (red) and paired recessive alleles (green) in simple Mendelian way. Ecotypes have been reported in several plant species such as *Euphorbia hirta*, *Cassia tora*, *Xanthium strumarium*, *Ageratum conizoides*, *Cenchrus ciliaris*, *Portulaca oleracea*, etc.

17·2·1. Types of Ecotypes

Ecotypes have been classified into the following three types :

1. Climatic ecotypes. **McMillan** found that in fifteen out of seventeen species in eight of nine American tree genera examined (*Acer*, *Betula*, *Fraxinus*, *Larix*, *Picea*, *Prunus*, *Pseudotsuga*, and *Ulmus*), photoperiod ecotypes could be correlated with the latitude of seed origin.

2. Edaphic ecotypes. Edaphic ecotypes occur especially in saline soils and serpentile soils, but also in soil-moisture stress.

3. Biotic ecotypes. Biotic ecotypes have been selected by response to some biological feature of the environment such as animal grazing and browsing. In areas of grassland subjected to particularly high grazing pressure (for example, around a drinking hole in a game reserve), grasses and other herbaceous plants with normally an upright height will be selected for those ecotypes that are prostrate and thus less liable to have their vegetation and reproductive portions removed.

17-3. SPECIATION

Through mutation and introgression followed by selection, modification of gene frequencies (genetic drift) and recombination, many species populations come to be composed of a number of freely interbreeding **biotypes** or **races**, each adapted to a particular combination of parameters in the habitat. Such a species is called **polytypic species**. When the component populations of a polytypic species are readily distinguishable by morphological characters, they are **frequently** called **subspecies**. This term is especially applied to geographical races of animal species. In plants, as already discussed, races of a species that have distinct habitat requirements are commonly described as ecotype. From time to time a mutation occurs in a segment of a polytypic species, such as an ecotype or a subspecies, that limits its ability to exchange genes with other segments of the parent species. The population that develops from the mutant is **reproductively isolated** and is commonly recognised as a separate species population, distinct from the original species. Before further mutations completely restricts all gene exchange between the parent population and the new one, there may be for one reason or another some further crossing between them which is known as **hybridisation**. Hybrids frequently fail because of the weakness of their phenotypes or through failure to reproduce normally. In plants the doubling of the chromosome content or **polyploidy** of hybrid individuals sometimes overcome this sterility.

Species populations that occupy different geographical areas are said to be **allopatric**, where as those that occupy the same area are said to be **sympatric**. These terms are now usually applied only to related populations ; allopatric populations are defined as segments of an original population that have become separated by **spatial isolation** (**Boughey**, 1968).

Lastly, **ecospecies** is an unit of classification which contains one or more ecotypes, which although interfertile but do not cross or, at least do not produce viable offspring, if crossed with ecotypes of other ecospecies.

17-4. ECOLOGICAL INDICATORS

Since, specific factors often determine rather precisely what kinds of organisms will be present in a particular habitat, we can turn the situation around and judge the kind of physical environment from the organism present. It is found that certain species of micro-organisms, plants and animals have one or more specific requirements and they become very much limited in their distribution. Thus, the occurrence of such species in a particular area indicates special habitat conditions, and such species are called **biological** or **ecological indicators,** since they indicate some very specific condition of the environment. Some of the Indian plant species which serve as ecological indicators have been tabulated in (Table : 17·1) :

Table : 17·1. Certain ecological indicators of plant species.

Name of species	Indicator (s) of
1. *Utricularia, Chara, Wolffia, Ottelia alismoides*	Water pollution
2. *Petridium* sp.	Burnt and highly disturbed coniferous forest
3. *Argemone maxicana*	Recently disturbed or flooded soil
4. *Rumex acetosella*	Acid grassland soil
5. *Ccrissa spinarum* and *Capp..ris spinosa*	Intense soil erosion
6. *Enicostema littorale*	Well drained soil
7. *Saccharum spontaneum*	Poorly drained soil
8. *Zizyphus rotundifola*	Soil deposition
9. *Andropogon scoparium*	Sandy loam type soil
10. *Lippia nodiflora* and *Rumex* sp.	Nitrate rich soil
11. *Woodfordio foribunda* and *Chloris virgata*	High lime content of soil
12. *Atriplen*, *Salsola*, and *Saueda*	Saline water condition
13. *Hydrilla vertic llcta* and *Ceratophyllum demersum*	Hard water
14. *Waltheria indica ste culiasceae*	Copper, lead and zinc rich soil
15. *Equisetum plebejum* and *E. arvense*	Gold rich soil
16. *Silene cobalticola* and *Crotalaria cobalticola*	Cobalt rich soil
17. *Spermacoce st·icta*	Iron rich soil
18. Pines and junipers	Uranium rich soil

In fact, Central Arid Zone Research Institute, Jodhpur has made the use of certain plant communities of Indian deserts to indicate the salinity and depth of groundwater (Table 17 2).

Table 17·2. Certain ecological indicators for groundwater of Indian deserts (Chatterjii and Bhaskar, 1977).

	Plant Community	Indicated depth to water in metres (m)	Indicated total soluble salt contents in pph) in groundwater
1.	*Euphorbia caducifolia*	12 to 18 m	1500 to 7000
2	***Acacia** senegal—Anogeissus pendula*	12 to 18 m	500 to 1500
3.	*Salvadora persica—Tamarix* sp.	6 m	1500 to 3200
4.	*Salvadora oleoides—Prosopis cineraria*	10 to 20 m	500 to 2000
5.	*Prosopis cineraria—Zizyphus nummularia, Capparis decidua*	6 to 18 m	5000 to 10,000
6.	*Salvadora oleoides, Capparis **deci**dua*	6 to 12 m	—do—
7.	*Salvadora oleoides—Zizyphus nummularia*	18 to 28 m	—do—
8.	***Panicum** turgidum—Calligonium polygonoides*	6 to 18 m	1500 to 3200
9.	*Panicum turgidum—Zizyphus complex*	6 to 18 m	5000 to 10,000
10.	*Crotalaria burhia—Leptadenia pyrotechnica*	6 to 20 m	5000 to 3200
11.	*Suaeda fruticosa—**Aelurophus** lagopides*	6 m	11,000 to 12.000
12.	*Capparis decidua*	12 to 20 m	180 to 15000
13.	*Acacia indica—Nilothica* sp.—*Prosopis cineraria—Salvadora oleoides*	12 to 20 m	400 to 1500

Besides plants, certain micro-organisms serve as ecological indicators. For example, *Escherichia coli*, a bacterium, is an index organism for aquatic pollution. Some species of diatoms (Arphidinae) indicate the degree of sewage pollution of lakes. The presence of *Fusilinids* (protozoans) in the soil indicated the petroleum deposits.

Similarly, certain animal species serve as ecological indicators. Chironomid larvae and maggots increase in number as the pollution increases. Burrowing may fly *Hexagenia* serves as indicator for well **oxygenated** aquatic environments. Emigration and consequent disappearance of certain Indian fish species as *Catla catla Labeo gonius*,

Labeo bata, *Labeo rohita*, *Notopterus notopterus*, etc., from Kalinadi river near Mansurpur Sugar Factory, indicate industrial pollution of water, as these fishes are found to be very sensitive to aquatic pollution (**Verma**, 1971).

17-5. REVISION QUESTIONS

1. Describe the process of natural selection in correlation to population genetics.
2. What is ecotype ? Describe the phenomenon of **polymorphism** and different types of ecotypes.
3. Write short notes on the following :
   (i) Variation ;
   (ii Hardy-Weinberg principle ;
   (iii) Directional selection;
   (iv) Genetic drift ;
   (v) Founder effect ;
   (vi) Heterosis ;
   (vii) Adaptation ;
   (viii) Acclimatisation ;
   (ix) Genecology ;
   (x) Ecological indicators.

Part 4

DYNAMICS OF ECOSYSTEM

The existence of the living world depends on the flow of energy and the circulation of nutrients or materials through the ecosystem. Both influence the abundance of organisms, the rate at which they live, and the complexity of the community. Energy and nutrients flow through the community together, each of them cannot be segregated from the other. The flow of energy is one way ; once used by the ecosystem, it is lost. Nutrients, on the other hand, recirculate. An atom of carbon, oxygen or calcium may pass through the living and the non-living many times, or it may even be exchanged between ecosystems. This one-way passage of energy and the round trip of materials or nutrients (called biogeochemical cycle) are the cogs on which the living world turns **(Smith, 1977).**

18

Energy Flow in Ecosystems

Energy can be defined as the capacity to do work, whether that work be on a gross scale as raising mountains and moving air masses over continents, or on a small scale such as transmitting a nerve impulse from one cell to another. There are two kinds of energy, **potential** and **kinetic**. Potential energy is energy at rest. It is capable of and available for work. Kinetic energy is due to motion, and results in work. Work that results from the expenditure of energy can both store energy (as potential energy) and arrange or order matter without storing energy.

The expenditure and storage of energy is described by two laws of thermodynamics. The first law of thermodynamics called **law of conservation of energy** states that *energy is neither created nor destroyed.* It may change forms, pass from one place to another, or act upon matter in various ways, but regardless of what transfers and transformations take place, no gain or loss in total energy occurs. Energy is simply transferred from one form or place to another. When wood is burned the potential energy present in the molecules of wood equals the kinetic energy released, and heat is evolved to the surroundings. This is an **exothermic reaction.** In an **endothermic reaction**, energy from the surrounding may be paid into a reaction. For example, in photosynthesis, the molecules of the products store more energy than the reactants. The extra energy is acquired from the sunlight, but even than there is no gain or loss in total energy. Further, though the total amount of energy involved in any chemical reaction, such as burning wood, does not increase or decrease, much of the potential energy stored in the substance undergoing reaction is degraded during the reaction into a form incapable of doing any further work. This energy ends up as heat serving to disorganize or randomly disperse the molecules involved, thus making them useless for further work. The measure of this relative disorder is named **entropy**.

The second law of thermodynamics states that whenever energy is transformed from one kind to another, there is an increase in entropy and a decrease in the amount of useful energy. Thus, when coal is burned in a boiler to produce steam, some of the energy creates steam that performs work, but part of the energy is dispersed as heat to the surrounding air. The same thing happens to energy in the ecosystem. As energy is transformed from one organism to another in the form of food, a large part of that energy is degraded as heat and as a net increase in the disorder of energy. The remainder is stored in living tissue.

Three sources of energy account for virtually all the work of the ecosystem : **gravitation, internal forces within the earth** and **solar radiation.** The last is very significant for ecosystem. The solar radiation which originates from sun is the source of energy for life and life is what sets the ecosystem apart from other natural systems.

18-1. ENERGY IN AN ECOSYSTEM

Living organisms can use energy in several forms, but all can be grouped under one of two headings : **radiant** and **fixed**. Radiant energy is in the form of electromagnetic waves such as light. Fixed energy is potential chemical energy bound up in various organic substances which can be broken down or reacted with something else in order to release their energy content.

Only a small fraction of the light energy reaching the earth is trapped; considerable areas of the earth have no plants, and plants can utilize in photosynthesis only about 3 per cent of the incident energy. This radiant energy of sun is converted into chemical or potential energy by photosynthesis in plants, the **photosynthetic autotrophs** :

$$6CO_2 + 6H_2O + hv \rightarrow 6C_6H_{12}O_6 + 6O_2$$

CO_2 from atmosphere	water from soil	light energy	sugar in plant cell	Oxygen released to atmosphere

The reaction is catalyzed by certain pigments (*e.g.*, chlorophylls) in the cells. The product of this reaction is carbohydrate such as sugar (glucose) shown above. This sugar can have several destinations : it can be converted to a relatively inert energy-rich organic substance such as starch, and stored ; it can be combined with

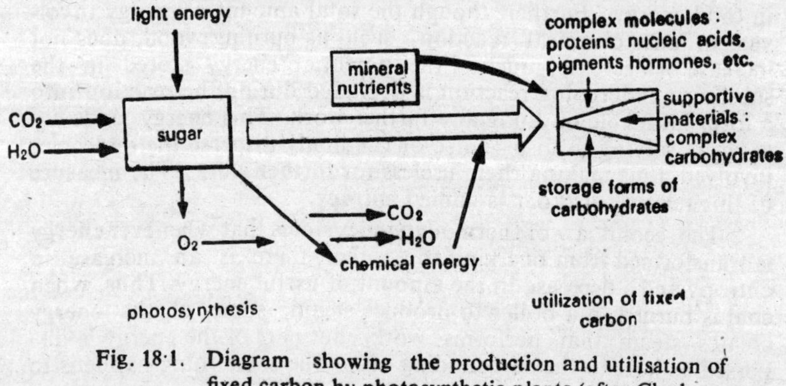

Fig. 18·1. Diagram showing the production and utilisation of fixed carbon by photosynthetic plants (after Clapham, Jr., 1973).

other sugar molecules to form specialized carbohydrates such as cellulose, which are used by the plants for specific purposes; and it can be combined with other substances—such as nutrient substances such as nitrogen, phosphorus, and sulphur, to build complex molecules like proteins, nucleic acids, pigments, and hormones. All these types of reactions are necessary for normal growth and maintenance of body tissues and functions of the plants (Fig. 18·1). All required energy which is provided by oxidation of some of the sugar produced by photosynthesis to give CO_2, H_2O and usable chemical energy :

$$C_6H_{12}O_6+6O_2 \rightarrow 6CO_2+6H_2O+\text{usable energy}$$

Oxidation of sugar or any other organic molecule to get usable energy by organism is called **respiration**. The energy released by respiration is lost permanently to the ecosystem.

18-1·1. Primary Production

All the energy that the plant actually fixes results in the formation of sugar. Conversely, all the sugar produced in the leaves of green plants is derived from carbon dioxide and water which have been combined by the solar energy. Thus, one can speak of the energy incorporated into living tissue either in terms of the light energy utilized or in terms of the sugar produced. Because all the energy used by the plant is converted into chemical energy, it should theoretically be possible to determine the entire energy uptake of plants by measuring the total amount of sugar produced. This quantity is termed **gross primary production** or **gross productivity**. Because, it is not easy to measure gross primary production in most ecosystems because some of the sugar produced by photosynthesis is lost immediately through respiration. So, if one measures the total organic material actually present in the plant (*i.e.*, biomass) then indirectly he will measure the gross primary production less the respiration, or **net primary production**. In addition, there are direct measurement techniques involving the rate of uptake of radioactive carbon ^{14}C. Productivity is generally expressed in terms of grams or kilo-calories per sq metre per day or per year. It can also be expressed as $P_g=P_n+R$, where P_g is the gross productivity or energy used to do work, P_n is net productivity, and R is respiration or energy to do work. Net production is an increase of stored energy with time, while biomass is total stored energy at any time. If gross production equals respiration ($P_g=R$) no change in energy content results. But when P_g is less than R biomass decreases, and when P_g is greater than R there is an accumulation of biomass.

The gross productivity is dependent on many things, including climatic conditions such as temperature, rainfall, and total solar radiation, and such other features of the abiotic environment as the availability of nutrient material essential for life, such as nitrogen, phosphorus and sulphur. Following table (Table 18·1) lists the gross primary production or gross productivity over one year for several different environments :

Table 18·1. Estimated annual gross primary production of a range of **ecosystem** (Odum, 1971 ; Clapham, Jr., 1973).

Ecosystems	Area (10^6 Km.^{2})	Gross primary productivity (ly./Yr.)
Marine		
1. Open ocean	326·0	100
2. Coastal zones	34·0	200
3. Up welling zones	0·4	600
4. Estuaries and reefs	2·0	2,000
Terrestrial		
1. Deserts and Tundras	40·0	20
2. Grasslands	42·0	250
3. Dry forests	9·4	250
4. Boreal coniferous forests	10·0	300
5. Cultivated land	10·0	300
6. Moist temperate forests	4·9	800
7. Mechanized agriculture	4·0	1,200
8. Wet tropical, Subtropical forests	14·7	2,000

18-1·2. Secondary Production

The potential energy resulting from primary production furnishes the energy required by other trophic levels in an ecosystem. Some energy, in the form of food, is consumed by herbivores or omnivores who may be eaten by carnivores, who in turn may be eaten by other carnivores. However, much of this ingested food is not assimilated ; herbivores may assimilate only 10 per cent of the ingested food, where as the assimilation coefficiency of carnivores is typically higher. For example, various species of fish absorbed between 86 and 96 per cent of the ingested food, but the efficiency absorption of different classes of food substances varied : about 92—98 per cent of the ingested protein was absorbed, while only 80 per cent of the fat and about 63—79 per cent of carbohydrates was absorbed by the fishes (**Pandian**, 1967). The unassimilated material leaves the animal's body as waste material to serve as an energy source for other organisms (detritus feeders and saprotrophs). Assimilated energy is used by these consumer species for various metabolic processes, such as respiration, excretion, and secretion. The resultant amount of energy stored in the tissues of heterotrophs is termed **net secondary production**. The **gross secondary production** is equivalent to the total plant material ingested by the herbivore, less the materials lost as faeces. That is, it is the food that actually passes through the walls of the gut. Unlike gross primary production, gross secondary production can be measured directly by determining the amount of material defaecated.

Respiration of the population. In previous sections it has been discussed that gross production and net production and the total energy involve respiration or the cost of maintenance of the population. This is expressed in terms of oxygen utilized for vital processes or of carbon dioxide eliminated and so termed **respiration of the population.**

18-2. PATTERNS OF FLOW OF ENERGY THROUGH THE ECOSYSTEMS

In an ecosystem one can observe the transfer or flow of energy from one trophic level to other in succession. A **trophic level** can be defined as the number of links by which it is separated from the producer, or as the *n*th position of the organism in the food chain. The patterns of eating and being eaten forms a linear chain called **food chain** which can always be traced back to the

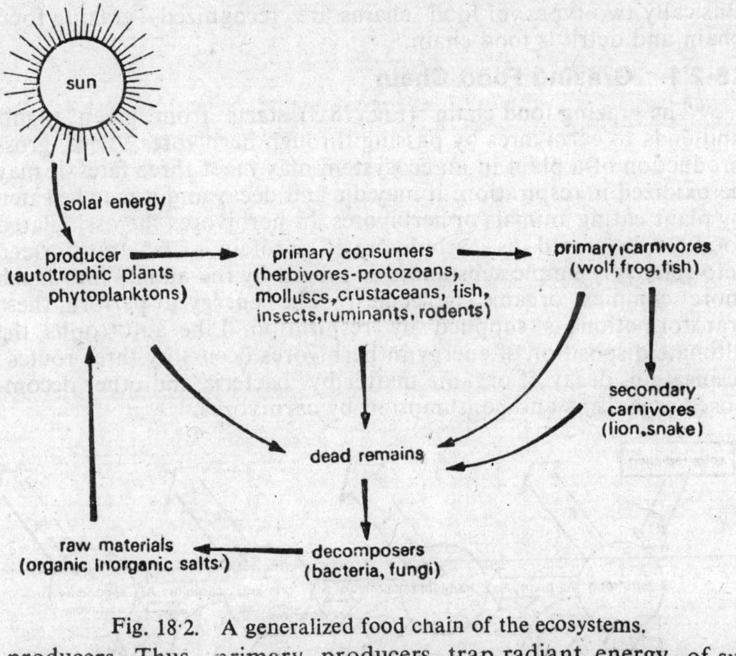

Fig. 18·2. A generalized food chain of the ecosystems.

producers. Thus, primary producers trap radiant energy of sun and transfer that to chemical or potential energy of organic compounds such as carbohydrates, proteins and fats. When a herbivore animal eats a plant (or when bacteria decompose it) and these organic compounds are oxidized, the energy liberated is just equal to the amount of energy used in synthesizing the substances (first law of thermodynamics), but some of the energy is heat and not useful energy (second law of thermodynamics). If this animal, in turn, is eaten by another one, along with transfer of energy from a herbivore to carnivore a further decrease in useful energy occurs as the second animal (carnivore) oxidizes the organic substances of the first (herbivore or omnivore) to liberate energy to synthesize its own cellular constituents. Such transfer of energy from organism to organism sustains the ecosystem and when energy is transferred from individual to individual in a particular community, as in a pond or a lake or a river, we come across the food chains (Fig. 18·2). The number of steps in a food chain are always restricted to four or five, since the energy available decreases with each step. In other words, the energy fixed in organisms such as in aquatic plants in a pond passes through the ecosystem in the form of plant→herbivore→carnivore, where only three steps are encountered in the chain. But very often, the chains are very complicated with several steps being involved. For example, the food chain of shore ecosystem includes plant→herbivore→carnivore 1→carnivore 2→carnivore *n*. Basically two types of food chains are recognized—grazing food chain and detritus food chain.

18-2·1. Grazing Food Chain

The grazing food chain (Fig. 18·3) starts from green plants and ends to carnivores by passing through herbivores. Thus, gross production of a plant in an ecosystem may meet three fates–it may be oxidized in respiration, it may die and decay and it may be eaten by plant eating animals or herbivores. In herbivores the assimilated food can be stored as carbohydrates, protein or fat, transformed into relatively simple substances or rebuilt by the animal into much more complex organic molecules. The energy to perform these transformations is supplied by respiration. Like autotrophs, the ultimate disposition of energy in herbivores occurs by three routes: respiration, decay of organic matter by bacteria and other decomposer organisms and consumption by carnivores.

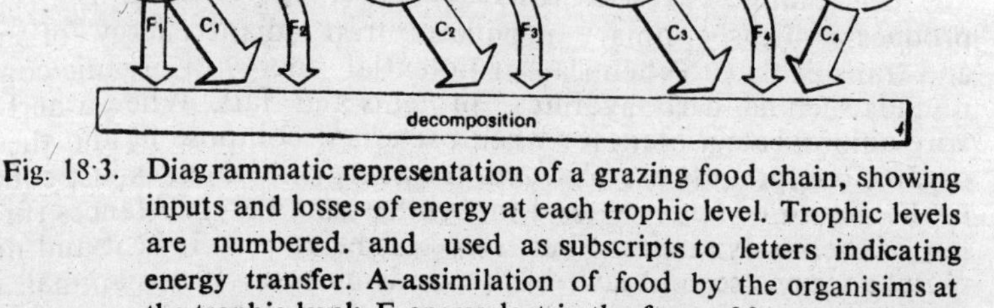

Fig. 18·3. Diagrammatic representation of a grazing food chain, showing inputs and losses of energy at each trophic level. Trophic levels are numbered. and used as subscripts to letters indicating energy transfer. A–assimilation of food by the organisims at the trophic level; F–energy lost in the form of faeces and other excretory products; C–energy lost through decay; and R–energy lost to respiration (after Clapham, Jr., 1973).

The **primary carnivores** or **secondary consumers** eat herbivores or primary consumers of the ecosystems. And likwise, **secondary carnivores** or **tertiary consumers** eat primary carnivores. The total energy assimilated by primary carnivores, or **gross tertiary production**, is derived entirely from the tissues of herbivores, and its disposition into respiration, decay and further consumption by other carnivores is entirely analogous with that of the herbivores. It seems clear that much of the energy flow through grazing food chain can be described in terms of the categories or **trophic levels**, outlined as follows :

{**Autotroph** primary producers} → {**Herbivore** primary consumer} → {**Primary carnivore** secondary consumer} → {**Secondary carnivore** tertiary consumer} → etc.

As a general model for the direction and extent of energy flow within communities, the food chain is very useful. It demonstrates the amount of energy found at any trophic level, the transfer of energy from one trophic level to the next, and the amount of energy lost from the grazing food chain, either through respiration or decay. Further the producer→herbivore→ carnivore chain is a **predator chain**. **Parasitic chains** also exist wherein smaller organisms consume larger ones without outright killing as the case of the predators.

18-2·2. Detritus Food Chain

The organic wastes, exudates and dead matter derived from the grazing food chain are generally termed **detritus**. The energy contained in this detritus is not lost to the ecosystem as a whole ; rather it serves as the source of energy for a group of organisms **(detritivores)** that are separate from the grazing food chain, and generally termed the **detritus food chain** (Fig. 18·4). The detritus food chain represents an exceedingly important component in the energy flow of an **ecosystem**. Indeed in some ecosystems, considerably more energy flows through the detritus food chain than through the grazing food chain. In the detritus food chain the energy flow remains as a continuous passage rather than as a stepwise flow between discrete entities. The organisms of the detritus food chain are many and include algae, bacteria, slime molds, actinomycetes, fungi, Protozoa, insects, mites, Crustacea, centipedes, molluses, rotifers, annelid worms, nematodes and some vertebrates. Some species are highly specific in their food requirements and some can eat almost anything. Many Protozoa, for instance, need certain specific organic acids, vitamins, and other nutrients before they can thrive ; on the other hand, the guts of small Collembola (a group of tiny soil insects) have been reported to contain decaying plant material, fungal fragments, spores, fly pupae, other Collembola, parts of decaying earthworms, and cuticle from their own faecal casting (**Hale**, 1967). In contrast to the grazing food chain, in which energy storage is entirely within the tissues of living organisms, energy storage for the detritus food chain may be largely external to the organisms, and is the detritus itself.

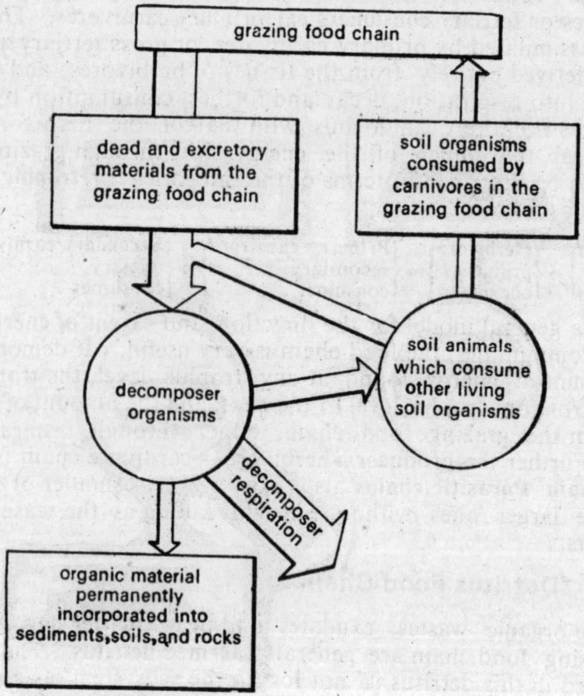

Fig. 18·4. Diagrammatic representation of the detritus food chain, showing energy transfers between it and the grazing food chain, as well as energy losses to the detritus food chain, (after Clapham, Jr., 1973).

Detritus organisms ingest pieces of partially decomposed organic matter, digest them partially, after extracting some of the chemical energy in the food to run their metabolism, excrete the remainder in the form of slightly simpler organic molecules. The wastes from one organism can immediately be utilized by a second which repeats the process. Gradually, the complex organic molecules present in the original wastes or dead tissue are broken down to much simpler compounds, sometimes all the way to carbon dioxide and water. In most instances, however, the organic material is broken down until all the easily biodegradable material has been degraded, and all that is left are some refractory organic substances termed **humic acids**, or simply, humus. In a normal environment the humus is quite stable and will form an essential part of the soil.

18·2·3. Links Between Grazing and Detritus Food Chains

Just as energy enters the detritus food chain from the grazing food chain as leaf litter, or dead organisms, energy can re-enter the grazing food chain from the detritus food chain if organisms from the latter are consumed by grazers. Considering the diversity of the detritus community, it is not surprising that many of these serve as prey for carnivores in the grazing food chain. A common example is robins eating earthworms. In addition, many insects, among them beetles and flies, spend their larval period in the detritus food chain and their adulthood in the grazing food chain. But through such links through which energy passes from detritus food chain basic into grazing food chain, very

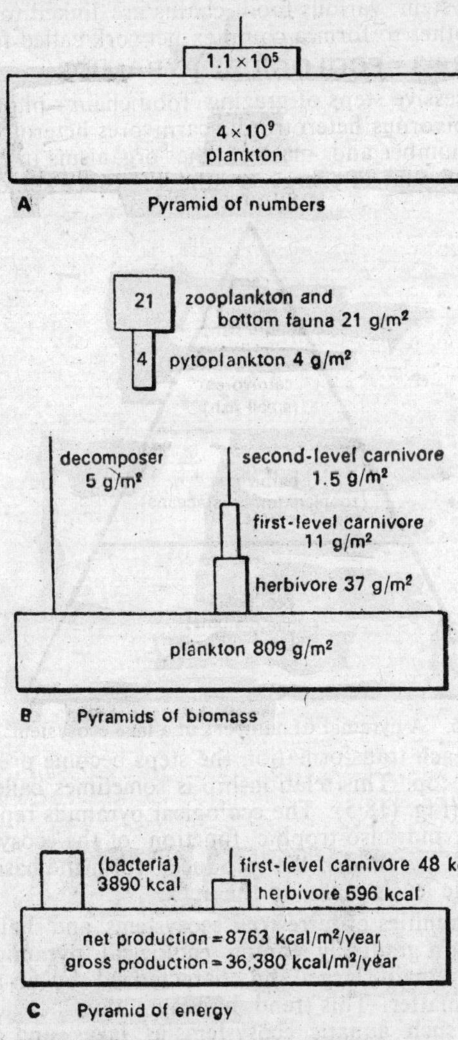

Fig. 18·5. Different kinds of ecological pyramids (after Smith, 1977).

little amount of energy is passed in comparison to the amount of energy which flow from grazing food chain to detritus food chain.

Food web. In nature simple food chains occur rarely. The same organism may operate in the ecosystem at more than one trophic level, *i.e.*, it may drive its food from more than one source. Even the same organism may be eaten by several organisms of a higher trophic level or an organism may feed upon several different organisms of a lower trophic level. Usually the kind of food changes with the age of the organism and the food availability. Thus, in a given ecosystem various food chains are linked together and intersect each other to form a complex network called **food web**.

18-3. ECOLOGICAL PYRAMIDS

In the successive steps of grazing food chain—photosynthetic autotroph, herbivorous heterotroph, carnivores heterotroph, decay bacteria—the number and mass of the organisms in each step is limited by the amount of energy available. Since some energy is

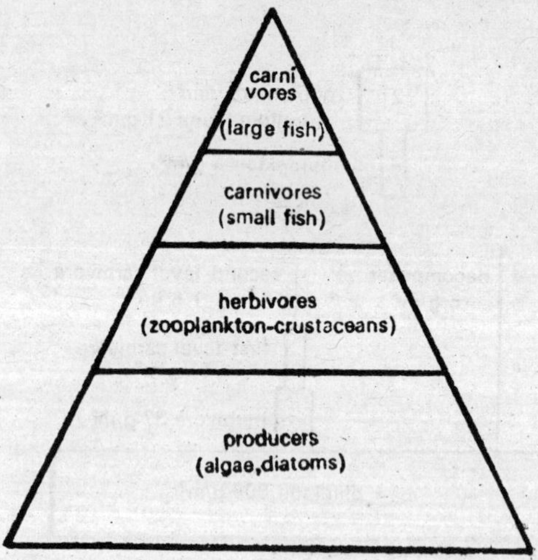

Fig. 18·6. A pyramid of numbers in a lake ecosystem.

lost as heat, in each transformation the steps become progressively smaller near the top. This relationship is sometimes called "**ecological pyramid**", (Fig. (18·5). The ecological pyramids represent the trophic structure and also trophic function of the eeosystem. In many ecological pyramids, the producer form the base and the successive trophic levels make up the apex.

Thus, communities of terrestrial ecosystems and shallow water ecosystems contain gradually sloping ecological pyramids because these producers remain large and characterized by an accumalation of organic matter. This trend, however, does not hold for all ecosystems. In such aquatic ecosystems as lakes and open sea, primary production is concentrated in the microscopic algae. These algae have a short-cycle, multiply rapidly, accumulate little organic matter and are heavily exploited by herbivorous zoo-plankton. At any one point in time the standing crop is low. As a result, the pyramid of biomass for these aquatic ecosystems is inverted: the base is much smaller than the structure it supports.

18·3·1. Types of Ecological Pyramids

The ecological pyramids may be of following three kinds :

1. **Pyramids of number.** It depicts the number of individual organisms at different trophic levels of food chain. This pyramid was advanced by **Charles Elton** (1927), who pointed out the great difference in the numbers of the organisms involved in each step of the food chain. The animals at the lower end (base of pyramid) of the chain are the most abundant. Successive links of carnivores decrease rapidly in number until there are very few carnivores at the top. The pyramid of number ignores the biomass of organisms and it also does not indicate the energy transferred or the use of energy by the groups involved. The lake ecosystem provides a typical example for pyramid of number (Fig. 18·5 and 18·6).

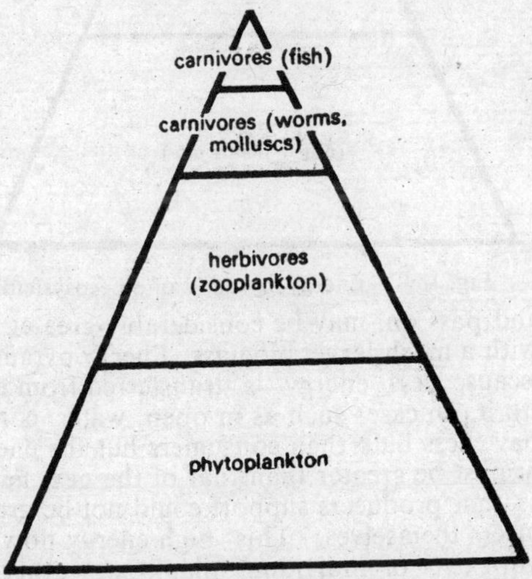

Fig. 18·7. Pyramid of biomass for a marine ecosystem.

2. **Pyramid of biomass.** The biomass of the members of the food chain present at any one time forms the pyramid of the biomass. Pyramid of biomass indicates decrease of biomass in each trophical level from base to apex (Fig. 18·5 and Fig. 18·7). For example, the total biomass of the producers ingested by herbivores is more than the total biomass of the herbivores in an ecosystem. Likewise, the total biomass of the primary carnivores (or secondary consumer) will be less than the herbivores and so on.

3. Pyramid of energy. When production is considered in terms of energy, the pyramid indicates not only the amount of energy flow at each level, but more important, the actual role the various organisms play in the transfer of energy (Fig. 18·5 and 18·8). The base upon which the pyramid of energy is constructed is the quantity of organisms produced per unit time, or in other words, the rate at which food material passes through the food chain. Some organisms may have a small biomass, but the total energy they

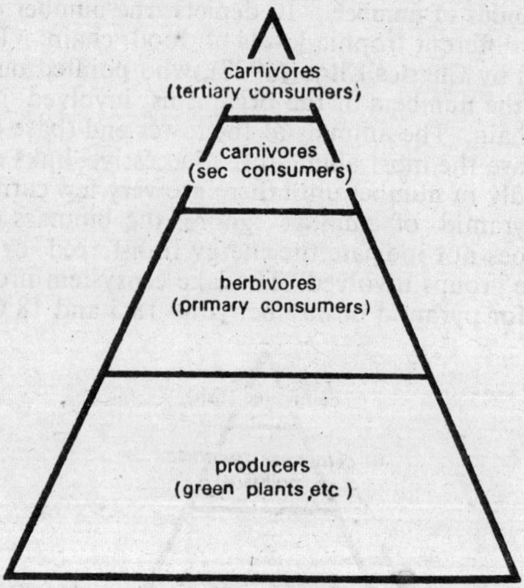

Fig. 18·8. Energy pyramid of an ecosystem.

assimilate and pass on, may be considerably greater than that of organisms with a much larger biomass. Energy pyramids are always slopping because less energy is transferred from each level than was paid into it. In cases such as in open water communities the producers have less bulk than consumers but the energy they store and pass on must be greater than that of the next level. Otherwise the biomass that producers support could not be greater than that of the producers themselves. This high energy flow is maintained by a rapid turn over of individual plankton, rather than an increase of total mass.

18-4. ECOLOGICAL EFFICIENCIES

The relationship between the amount of energy assimilated and the amount of energy utilised within one trophic level of a food chain has an important bearing on how much energy one trophic level in the food chain passes on to the next. **Efficiency** is the ratio of useful output of energy in relation to input. There exist various methods of measuring different kinds of efficiencies, all of which require the understanding of the following parameters ; **Ingestion** describes the quantity of food or energy taken ir by a consumer or the amount of light available or absorbed by a producer. **Assimilation** is the amount of food absorbed in the alimentary canal of a consumer, the absorption of extracellular products by a decomposer, the energy fixed by a plant in photosynthesis. **Respiration** is all of the energy lost in metabolism and activity in all forms, including energy (heat) lost in urine. **Net productivity** is the energy accumulated and represents that left over after respiration. **Production** is the portion of productivity a trophic level passes on to the next trophic level. It is the portion actually available and does not include losses to decomposers, losses to other systems, or increases or decreases in the standing crop. In this context production is considered as **yield.** Different kinds of efficiencies are calculated by the following methods :

1. Assimilation efficiency among producers would be :

$$\frac{\text{Energy fixed by plants}}{\text{Light available}}$$

2. Assimilation efficiency for consumer level would be :

$$\frac{\text{Food absorbed (assimilated)}}{\text{Food ingested}}$$

3. Ecological efficieney would be :

$$\frac{\text{Energy passed to trophic level } n+1}{\text{Energy taken in at trophic level } n}$$

4. Ecology growth efficiency would be :

$$\frac{\text{Net production}}{\text{Ingestion}}$$

5. Tissue growth efficiency would be :

$$\frac{\text{Net production}}{\text{Assimilation}}$$

6. Trophic level production efficiency would be :

$$\frac{\text{Assimilation at trophic level } n}{\text{Net production at lower trophic level } n-1}$$

These efficiencies vary among species, populations and trophic levels. Growth efficiencies among larger animals appear to be less than among smaller animals, and greater among younger animals than among older ones. Assimilation efficiencies appear to be higher among carnivores than among herbivores since the food of the carnivore more nearly matches its own chemical composition. Respiration in proportion to both ingestion and assimilation also increases at higher trophic levels. Consequently net productivity and production decrease in proportion to ingestion at the carnivore level (**Smith**, 1977).

18-5. REVISION QUESTIONS

1. How does energy flow in an ecosystem ?
2. Describe different kinds of ecological pyramids.
3. What is food chain ? Describe different food chains with examples.
4. What do you mean by ecological efficiency ? Discuss it briefly.

19
Biogeochemical Cycles

The total mass of all the organisms that have lived on the earth in the past 1·5 billion years is much greater than the mass of carbon and nitrogen atoms present. According to the **law of conservation of matter,** matter is neither created nor destroyed; obviously the carbon and nitrogen must have been used over and over again in the course of time. The earth neither receives any great amount of matter from other parts of the universe nor does it loss significant amount of matter to outer space. The atoms of each element as carbon, hydrogen, oxygen, nitrogen, phosphorus, calcium, and the rest are taken from the environment, made a part of some cellular component of an organism and finally, perhaps by a quite circuitous route involving several other organisms, are returned to the environment to be used over again. The cyclic movements of chemical elements of the biosphere between the organism and the environment are referred to as **biogeochemical cycles**, after **Vernadsky** (1934). "*Bio*" refers to living organisms and "geo" to the rocks, soil, air and water of the earth.

19-1. ORGANIC AND ABIOTIC PHASES OF GEOCHEMICAL CYCLES

In so far as element is necessary for the maintenance of life, its movement through biotic communities (organisms) can be viewed in the terms of food chain. The flow of a chemical element through the food chain can be viewed as the **organic phase** of the biogeochemical cycle. Further, the biogeochemical cycles also include **abiotic phases** which are the functions of the chemistry of the elements in question. These abiotic phases are of cirtical importance to the ecosystem, as the major reservoirs for all nutrient elements are external to the food chains, and flow in the abiotic phases tends to be much slower than in the organic phase. The rapidity and direction of nutrient cycling through the abiotic phases determine not only the distribution of the element in the total environment, but also its availability of living system. There are two classes of abiotic phases in biogeochemical cycles, a **sedimentary phase**, which is part of all cycles, and an **atmospheric phase**, which is possessed by some. In some cycles, such as nitrogen, the atmospheric phase is more important than the sedimentary. In others, such as phosphorus, the atmospheric phase is essentially non-existent. In still others, such as sulphur, both phases are present and their relative importance depends on other environmental factors. Biogeochemical cycles that have dominant atmospheric phases are often called **atmosphere-reservoir cycles** ; those whose sedimentary phase is dominant are termed **sediment-reservoir cycles.**

19-2. TYPES OF BIOGEOCHEMICAL CYCLES

There are two types of biogeochemical cycles, the **gaseous** and the **sedimentary**. In gaseous cycles the main reservoir of nutrients is the atmosphere and the ocean. In sedimentary cycles the main reservoir is the soil and the sedimentary and other rocks of the earth's crust. Both involve biotic and abiotic agents, both are driven by the flow of energy and both are tied to the **water cycle.**

19-2.1. Water Cycle

Living organisms, atmosphere and earth maintain between them a circulation of water and moisture, which is referred to as **water cycle** or **hydrologic cycle**. As we have already discussed in chapter 10, water forms a very significant factor of environment and without the cycling of water, biogeochemical cycles could not exist, ecosystems could not function, and life could not be maintained. Water is important for an ecosystem for several reasons—it is the medium by which nutrients are introduced into autotrophic plants; it is an important part of living tissue, either as liquid water or as part of essential organic molecules; it serves as a means of thermal regulation for both plants and animals ; it is the medium by which sediments—a prime source of mineral nutrients—are removed from or added to local ecosystems ; it covers the great majority of the earth's surface, and is the dominant feature of all aquatic ecosystems.

The hydrologic cycle is driven by solar energy and gravity. More than 80 per cent of the total insolation that is not lost immediately as electromagnetic radiation goes to **evaporate** water. The atmospheric water vapour produced by this means can then condense around particles of dust in the atmosphere, often called **nucleation particles**. The atmosphere possesses a limited capacity for holding water vapour, thus, the droplets formed by this means are heavy enough to fall as precipitation under the influence of gravity. Eventually, the hydrologic cycle can be defined as an alternation of evaporation and precipitation, with the energy used to evaporate the water being dissipated as heat in the atmosphere as the water condenses.

Distribution of water in earth's surface. Water is not evenly distributed throughout the earth. Almost 95 per cent of the total water on earth is chemically bound into rocks and does not cycle. Of the remainder about 97·3 per cent is in the ocean, about 2·1 per cent exists as ice in the polar caps and permanent glaciers, and the rest is fresh water, present in the form of atmospheric water vapour, ground water, soil water, or inland surface water (**Nace,** 1967).

Table : 19 1. Distribution of water in the earth's crust and surface (Clapham, Jr., 1973).

A. Chemically bound water of rocks : Does not cycle		
1.	Crystalline rocks	$250{,}000 \times 10^{17}$ Kg.
2	Sedimentary rocks	$21{,}00 \times 10^{17}$ Kg.
B. Free water : Moves via hydrologic cycle		
1.	Oceans	$13{,}200 \times 10^{17}$ Kg.
2.	Ice caps and glaciers	292×10^{17} Kg.
3.	Ground water to a depth of 4000 m.	$83{\cdot}5 \times 10^{17}$ Kg.
4.	Freshwater lakes	$1{\cdot}25 \times 10^{17}$ Kg.
5.	Saline lakes and inland seas	$1{\cdot}04 \times 10^{17}$ Kg.
6.	Soil moisture	$0{\cdot}67 \times 10^{17}$ Kg.
7.	Atmospheric water vapour	$0{\cdot}13 \times 10^{17}$ Kg.
8.	Rivers	$0{\cdot}013 \times 10^{17}$ Kg.

19-2·1A. The rate of cycling of water. The rate of cycling between surface and atmosphere is very rapid. The amount of water vapour in the atmosphere is sufficient, on the average, **turnover** the entire earth to a depth of 2·25 cm. But the average annual rainfall for the earth is about 81·1 cm (**Furon**, 1967) and in some places it ranges upto 1.200cm. This means that the average tornover time for atmospheric water is about 11·4 days, or that the equivalent of all the water vapour in the entire atmosphere fall as precipitation and is re-evaporated more than 32 times per year.

Further, the distribution of evaporation and rainfall is quite uneven. If we compare the annual evaporation and precipitation over land and sea, we find that relatively more water precipitates on land than evaporates from land. This is fortunate from the point of view of terrestrial organisms. Even so, the amount of rainfall on the open ocean is proportionately greater than that on land, taking into account the relative percentages of the earth's surface covered by land and sea.

19-2·1B. Nature of hydrologic cycle. The hydrologic cycle over the oceans is extremely simple–the water is evaporated from the surface of the ocean and water vapours form the clouds which when cool down precipitate the water as rain fall. But several routes are open to precipitation that falls on land–direct evaporation, transpiration, entry of water into ground water system and runoff (Fig. 19·1). Consequently, the routes of hydrologic cycles on land can be divided into following three main categories–the rapidly cycling portion or **evapotranspiration**, which includes the evaporation and transpiration, the less rapidly cycling water, or surface runoff, and very slowly cycling ground water that seeps into the soil can end up in any one of these three categories.

1. Evapotranspiration. Evapotranspiration includes evaporation and transpiration. **Evaporation** refers to water that is evaporated

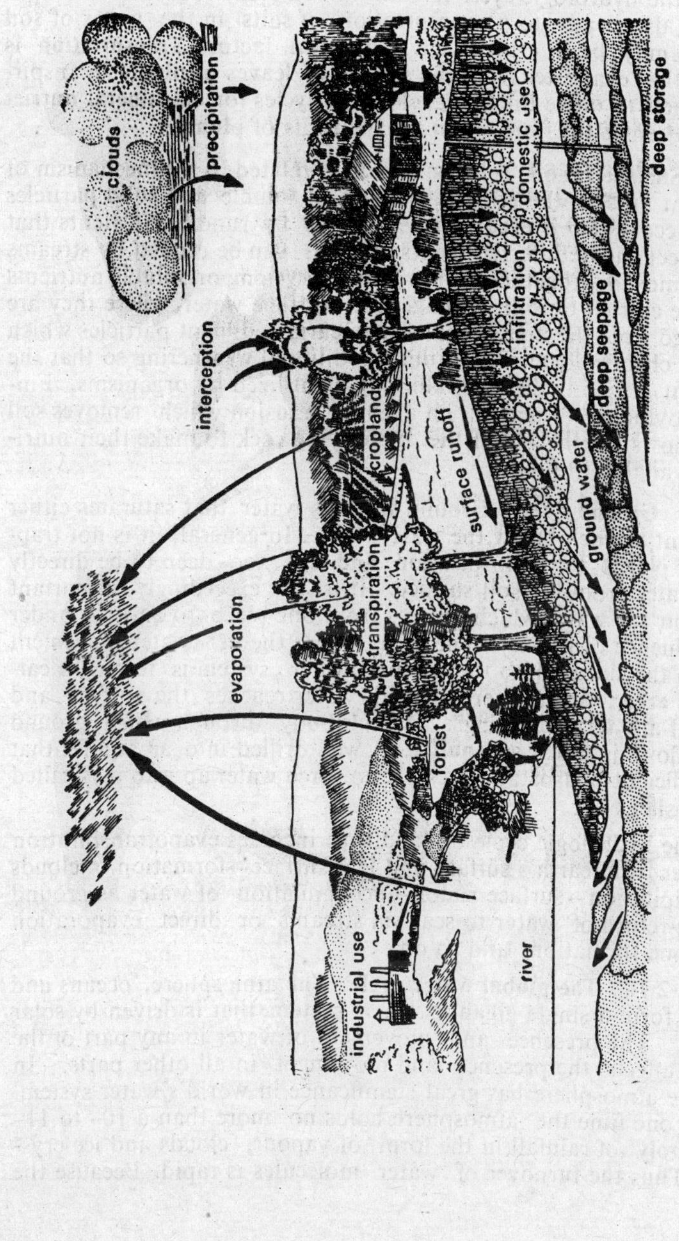

Fig. 19·1. The water cycle, showing the major pathways of water through the ecosystem (after Smith, 1974)

directly from any surface other than a plant, such as a .lake, soil surface, or animal skin. In most cases, the main effects of direct evaporation are to moderate the temperature of local area and to allow the hydrologic cycle to continue. In some ecosystems, evaporation also leads to a concentration of salts in the water of soil which may be a critical environmental factor. **Transpiration** is water that evaportes from the surface of leaves of plants. Transpiration acts to move the biogeochemical cycles for all mineral nutrients that enter the food chain via the roots of plants.

2. **Surface runoff.** If transpiration is related to the mechanism of nutrient uptake, the gross movement of soluble and solid particles in the ecosystem is accomplished largely by runoff. Nutrients that have accumulated in sediments or soils can be eroded by streams and removed altogether from a local ecosystem, or soluble nutrients may be carried by soil. Seepage into surface water, where they are removed from the area. Streams may carry sediment particles which can be chemically altered through additional weathering so that the nutrient elements they contain may be utilized by organisms. Finally moving water acts as an agent of **erosion** which removes soil and allows weathering of the underlying rock to make their nutrients available to plants.

3. **Ground water.** Ground water is water that saturates either sediment or rock below the water table. In general, it is not trapped by plants for transpiration and it is too deep to be directly evaporated from the soil surface. It is an exceedingly important reservoir for water which moves from one place to another under the influence of gravity. The area where the net water movement is from the surface into the ground water system is termed **a catchment area**; areas where ground water reaches the surface and runs off are termed **springs.** A rock body through which ground water flows is called an **aquifer.** A well drilled into an aquifer that has sufficient hydrostatic pressure to force water up into it is called an **artesian well.**

The hydrologic cycle on land thus includes evapotranspiration of water from earth's surface and leaf surface→formation of clouds →precipitation→surface runoff+accumulation of water as ground water→return of water to sea via streams or direct evaporation and cloud formation, and so on.

19-2·1C. The global water cycle. The atmosphere, oceans and masses form a single gigantic water system that is driven by solar energy. The presence and movement of water in any part of the system affects the presence and movement in all other parts. In fact, the atmosphere has great significance in world's water system. At any one time the atmosphere holds no more than a 10– to 11–day supply of rainfall in the form of vapour, clouds and ice crystals. Thus the turnover of water molecules is rapid. Because the source of water in the atmosphere is evaporation from land and sea, there are global differences in the amount of evaporation and the amount of moisture in the atmosphere at any given point. Evaporation at lower latitudes is considerebly greater than higher latitudes, showing the greater heat budgets produced by the direct rays of the sun. Evaporation is greater over ocean than over land.

Thus, evaporation, precipitation, detention, and transportation maintain a stable water balance on the earth. Consider the amount of water that falls on the earth in terms of 100 units (Fig. 19·2). On the average, 84 per cent of water is lost from the ocean by evaporation, while only 77 per cent of water is gained by it from precipitation. Land areas loss 16 per cent of water by evaporation and gain 23 per cent of it by precipitation. Water runoff through rivers from land to oceans makes up 7 per cent, which balance the evaporative deficit of the ocean. The remaining 7 per cent of water circulate as atmospheric moisture.

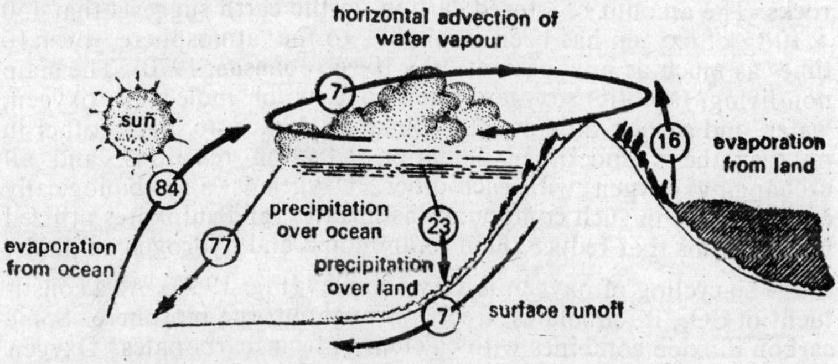

Fig. 19·2. **The global budget of water. The mean annual global precipitation of 83·6 cm. has been converted into 100 units (after Smith, 1977).**

19-2·2. Gaseous Cycles

The gaseous geochemical cycles are of following types :

19-2·2A. The oxygen cycle. Oxygen (O_2), the by-product of photosynthesis, is involved in the oxidation of carbohydrates with release of energy, carbon dioxide and water. Its primary role in biological oxidation is that of a hydrogen acceptor. The breakdown and decomposition of organic molecules proceeds primarily by dehydrogenation. Hydrogen is removed by enzymatic activity from organic molecules in a series of reactions and is finally accepted by the oxygen, forming water. Though oxygen is necessary for life, but being very active chemically, molecular O_2 may be toxic to living body cells. Therefore, for the protection from toxic effects of molecular O_2, cells possess the cellular organelles called

peroxisomes which mediate oxidative reactions resulting in the production of hydrogen peroxide which in turn is used through the mediation of other enzymes as an acceptor in oxidizing other compounds.

The major supply of free oxygen which supports life occurs in the atmosphere. There are two significant sources of atmospheric oxygen. One is the **photodissociation** of water vapour in which most of the hydrogen released escapes into outer space. The other source is **photosynthesis**, active only since life began on earth. Because photosynthesis and respiration are cyclic, involving both the release and utilization of oxygen, one would seem to balance the other, and no significant quantity of oxygen would accumulate in the atmosphere. However, at some time in the earth's history the amount of oxygen introduced into the atmosphere had to exceed the amount used in the decay of organic matter and that tied up in the oxidation of sedimentary rocks. Part of the atmospheric oxygen represents that portion remaining from the unoxidized reserves of photosynthesis—coal, oil, gas, and organic carbon in sedimentary rocks. The amount of stored carbon in the earth suggests that 150×10^{20}g of oxygen has been available to the atmosphere, over 10 times as much as now present, 10×10^{20}g (**Johnson**, 1970). The main non-living (abiotic) oxygen pool consists of molecular oxygen, water, and carbon dioxide, all intimately linked to each other in photosynthesis and other oxidation-reduction reactions, and all exchanging oxygen with each other. Oxygen is also biologically exchangeable in such compounds as nitrates and sulphates utilized by organisms that reduce them to ammonia and hydrogen sulphide.

The cycling of oxygen is very complex (Fig. 19·3). As a constituent of CO_2, it circulates freely throughout the biosphere. Some carbon dioxide combines with calcium to from carbonates. Oxygen combines with nitrogen compounds to from nitrates, with iron to ferric oxides, and with many other minerals to form various other oxides. In these states oxygen is temporarily withdrawn from circulation. In photosynthesis the oxygen freed is split from the water molecule. This oxygen is then reconstituted into water during plant and animal respiration. Some part of the atmospheric oxygen that reaches the higher levels of the troposphere is reduced to ozone (O_3) by high energy ultraviolet radiation.

19-2·2B. The carbon cycle. The carbon being a basic constituent of all organic compounds and a major element involved in the fixation of energy by photosynthesis, is so closely tied to energy flow that the two are inseparable. The source of all the fixed carbon both in living organisms and fossil deposits is carbon dioxide CO_2, found in the atmosphere and dissolved in the waters of the earth. During photosynthesis, carbon from atmospheric CO_2 is incorporated into the production of the carbohydrate, glucose, $C_6H_{12}O_6$, that subsequently may be converted to other organic compounds such as polysaccharides (sucrose, **starch**, cellulose, etc.),

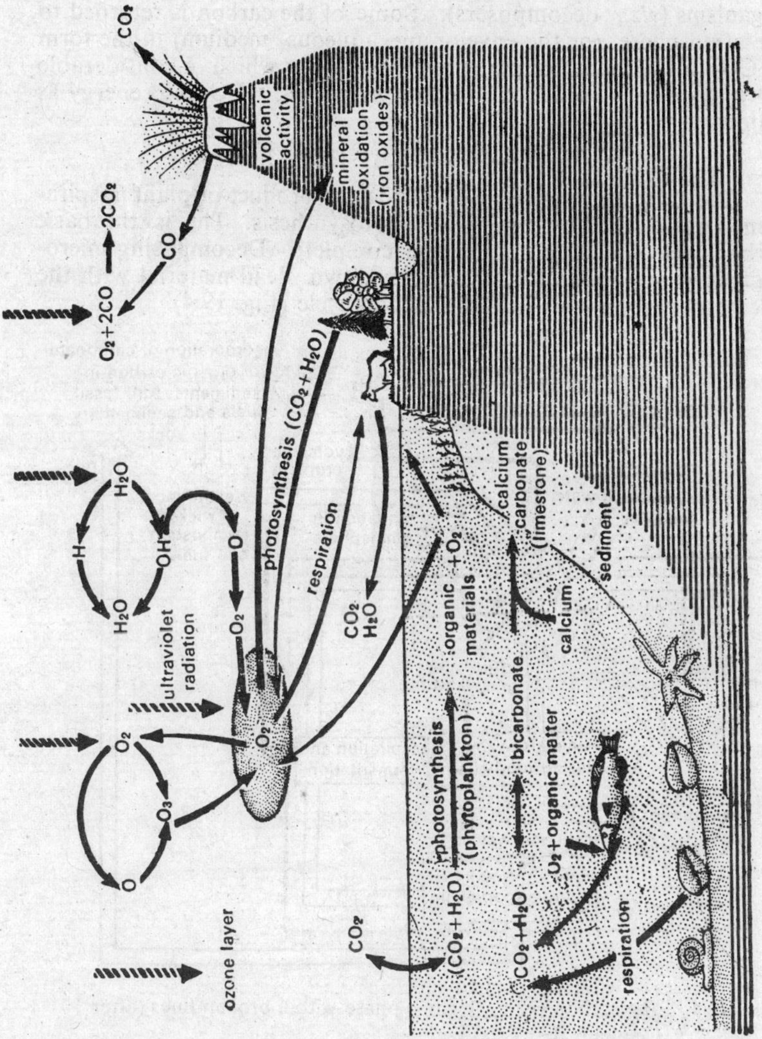

Fig. 19·3. Oxygen cycle (after Herreid II, 1977).

proteins and lipids. All the polymeric organic compounds containing carbon are stored in different plant-tissues as food and from them the carbon is passed on to the trophic levels of herbivores or phytoparasites, or retained by the plant until it serves as food for decay organisms (*viz.*, decomposers). Some of the carbon is returned to the atmosphere (or the enveloping aqueous medium) in the form of CO_2, a byproduct of plant respiration, in which, a considerable portion of glucose is oxidized to yield CO_2, H_2O and energy as follows :

$$1/6C_6H_{12}O_6 + O_2 \rightarrow CO_2 + H_2O + \text{Energy}$$

The CO_2 which is released as the by-product of plant respiration is again used by plants in photosynthesis. This is the basic carbon cycle which is simple and complete. Decomposing micro-organisms are important in breaking down dead material with the release of carbon back into the carbon cycle (Fig. 19·4).

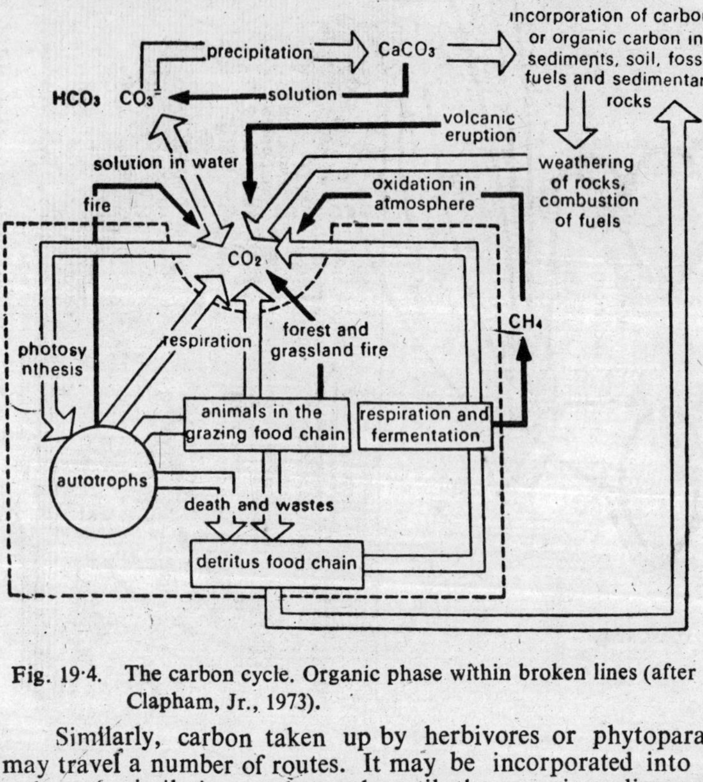

Fig. 19·4. The carbon cycle. Organic phase within broken lines (after Clapham, Jr., 1973).

Similarly, carbon taken up by herbivores or phytoparasites may travel a number of routes. It may be incorporated into protoplasm (assimilation) and stored until the organism dies, where upon it is utilized by decomposers ; it may be released through animal respiration ; it may serve as live food for other organisms ;

or finally it may be stored in the environment as CO_2. Similar fates await carbon at the carnivore trophic levels. In fact, all the carbon of plants, herbivores, carnivores and decomposers is not respired but some is fermented and some is stored. The carbon compounds that are lost to the food chain after fermentation, such as methane are readily oxidized to carbon dioxide by inorganic reactions in the atmosphere. As for the storage of carbon in sediments, just as deposition works to store materials, erosion may uncover them, and inorganic chemical weathering of rock can oxidize the carbon contained there. Some carbon is permanently stored in sediments and not uncovered by weathering ; it may be replaced by carbon dioxide released from volcanoes and other similar examples of intense geological activity. In modern age man has greatly increased the rate at which carbon is passing from sedimentary form to carbon dioxide. The combustion of fossil fuels is a significant means of recycling sedimentary carbon much faster than natural weathering.

Small portion of carbon, especially in the sea, is found not as organically fixed carbon but as carbonate ($CO_3^{=}$), especially calcium carbonate ($CaCO_3$). $CaCO_3$ is very commonly used for shell construction by such animals as clams, oysters, some protozoa, and some algae. Carbon dioxide reacts with water to from carbonate in the following three step reaction :

$$CO_2 + H_2O \rightleftharpoons \underset{\text{Carbonic acid}}{H_2CO_3} \rightleftharpoons H^+ + \underset{\text{Bicarbonate}}{HCO_3^-} \rightleftharpoons 2H^+ + \underset{\text{Carbonate}}{CO_3^{=}}$$

The precise amount of each of these constituents in the water depends on the pH of the water. Organisms such as clams can combine bicarbonate or carbonate with calcium dissolved in the water to produce calcium carbonate. After the death of the animals, this calcium carbonate may either dissolve or remain in sedimentary form.

Certain control mechanisms are inherent in the carbon cycle. The rate of carbon utilization is dependent on its availability. If excessive amounts of carbon are taken up in any one phase of the cycle, other phases of activity may be inhibited or slowed down. For example, if the pH of water is alkaline, more carbon is tied up in a carbonate and less is in solution. This removal of carbon in solution would upset the equilibrium established between the atmospheric and the dissolved CO_2 and the new effect would be a movement of CO_2 into solution until equilibrium was reached.

Peculiarities of carbon-cycle. Though carbon-cycle exhibits basic similarity with other biogeochemical cycles, yet it is unusual in that the organic phase is not essentially a complete cycle within itself. The organic (biotic) and atmospheric (abiotic) phases, however, are so closely intertwined that the rapid cycling typical of the organic phase is present. The multiplicity of paths along which carbon can flow is typical of biogeochemical cycles in general, and provides a well-buffered system with adequate feedback mechanisms to insure an adequate supply of the carbon. It is significant that all phases of the cycle yield carbon dioxide at some time, and carbon dioxide is the raw material for them Thus, despite its relative low concentration in the atmosphere (0·03 per cent), carbon in a form in which it can be used by living orgnisms is virtually always present.

19-2·2C. The nitrogen cycle. Nitrogen is an essential constituent of different biologically significant organic molecules such as amino acids and proteins, pigments, nucleic acids and vitamins. It is also the major constituent of the atmosphere, comprising about 79 per cent of it. The paradox is that in its gaseous state, N_2, abundant though it is, is unavailable to most life. Before it can be utilized it must be converted to some chemically usable form.

To be used biologically, the free molecular nitrogen has to be fixed and fixation requires an input of energy. In the first step molecular nitrogen, N_2, has to be split into two atoms : $N_2 \rightarrow 2N$. The free nitrogen atoms then must be combined with hydrogen to form ammonia, with the release of some energy :

$$2N + 3H_2 \rightarrow 2NH_3$$

This fixation comes about in two ways. One is by high-energy fixation such as cosmic radiation, meteorite trails, and lightning that provide the high energy needed to combine nitrogen with oxygen and hydrogen of water. The resulting ammonia and nitrates are carried to the earth in rain water. The second method of nitrogen-fixation which contributes about 90 per cent of fixed nitrogen of earth, is biological. Some bacteria, fungi, and blue-green algae can extract molecular nitrogen from the atmosphere and combine it with hydrogen to form ammonia. Some of this ammonia is excreted by the nitrogen-fixing organism, and thus becomes directly available to other autotrophs. Some of these nitrogen-fixing organisms may be free-living, either in the soil (*e.g.*, bacteria-*Azobacter* and *Clostridium*) or in water (*e.g.*, blue-green algae-*Nostoc*, *Calothrix* and *Anabaena*) and produce vast quantities of fixed nitrogen. In other cases, certain symbiotic bacteria of genus *Rhizobium*, although unable to fix atmospheric nitrogen themselves can do this when in combination with cells either from the roots of legumes (*e.g.*, peas, beans, clover and alfalfa) and of other angiosperms such as *Alnus*, *Ceanothus*, *Shepherdia*, *Elaeagnus* and *Myrica*, or from the leaves of African genera of *Rubiaceae* and *Pavetta* (Fig. 19·5). The bacteria invade the roots or leaves and stimulate the formation of root-nodules or leaf-nodules, a sort of harmless tumor. The combination of symbiotic bacteria and host cells remains able to fix atmospheric nitrogen and for this reason legumes are often planted to restore soil fertility by increasing the content of fixed nitrogen. Nodule bacteria may fix as much as 50 to 100 kilograms of nitrogen per acre per year, and free soil bacteria as much as 12 kilograms per acre per year. Further both free soil bacteria (*Azobacter*, and *Clostri-*

dium) produce ammonia as the first stable product and like the symbiotic bacteria, they require molybdenum as an activator and are inhibited by an accumulation of nitrates and ammonia in soil.

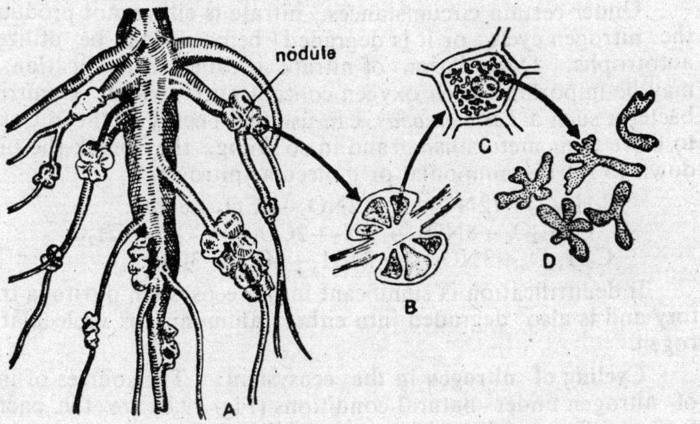

Fig. 19·5. Root-nodules of a legume plant. A—Legume root with root-nodules, B—T.S. of root-nodule. C—Single cell of nodule. D—Nitrogen-fixing bacterium, *Rhizobium*.

Recently, certain lichens (*Collema tunaeforme* and *Peltigera rufescens*) were also implicated in nitrogen fixation (**Henriksson,** 1971). Lichens with nitrogen-fixing ability possess nitrogen-fixing blue green species as their algal component.

Nitrogen fixed by symbiotic and non-symbiotic micro-organisms in soil and water is one source of nitrogen. Another source is organic matter. The nitrogenous wastes and carrion of animals are degraded by the detritus organisms, nitrogen is converted to the amino form (*e.g.*, L-Alanine). The amino group ($-NH_2$) is liberated from organic molecules to form ammonia; this process is called **deamination.** Certain specific bacteria, most notably of the genus *Nitrosomonas*, can oxidize ammonia to nitrite (NO_2) by the reaction.

$$2NH_3 + 3O_2 \rightarrow 2NO_2^- + 2H_2O + 2H^+$$

This reaction takes place in the soil, in lake or sea water or sediments, and whenever ammonia is being released and oxygen is present. As fast as nitrite is produced, other bacteria, such as *Nitrobacter*, can combine nitrite with oxygen to form nitrate (NO_3) by the reaction :

$$2NO_2^- + O_2 \rightarrow 2NO_3^-$$

Both of these ractions which are performed by two nitrifying bacteria—*Nitrosomonas* and *Nitrobacter* are the parts of a single biological process called **nitrification**. In nitrification process, thus ammonia is oxidized to nitrate and nitrite yielding energy. This, energy is used by the bacteria to make their organic materials directly from carbon dioxide and water. Nitrate can be taken up by autotrophs at the beginning of food chain.

Under certain circumstances, nitrate is either not produced in the nitrogen cycle or it is degraded before it can be utilized by autotrophs. Degradation of nitrate is called **denitrification**, and may be important when oxygen concentration is low. Denitrifying bacteria such a *Pseudomonas* can use the energy of the nitrate ion to drive their metabolism, and in so doing, they break the nitrate down to nitrite, ammonia, or molecular nitrogen :

$$C_6H_{12}O_6+12NO_3^- \rightarrow 12NO_2+6CO_2+6H_2O$$

$$C_6H_{12}O_6+8NO_2 \rightarrow 4N_2+2CO_2+4CO_3^-+6H_2O$$

$$C_6H_{12}O_6+3NO_3^- \rightarrow 3NH_3+6CO_2+3OH^-$$

If denitrification is significant in an ecosystem nitrite is transitory and is also degraded into either ammonia or molecular nitrogen.

Cycling of nitrogen in the ecosystem: The sources of inputs of nitrogen under natural conditions (Fig. 19·6) are the bacterial fixation of atmospheric nitrogen, addition of inorganic nitrogen in rain from such sources as lightning fixation and fixed "juvenile" nitrogen from volcanic activities, ammonia absorption from the atmosphere by plants and soil, and nitrogen accretion from wind-blown aerosols, which contain both organic and inorganic forms of nitrogen. In terrestrial ecosystems, nitrogen, largely in the from of ammonia and nitrates is taken up by plants, which convert it into amino acids and proteins. Animals (primary macro-consumers) may eat the plants and utilize the amino acids from the plant proteins in the synthesis of their own proteins and other cellular constituents. When animals and plants die, the decay bacteria convert the nitrogen of their proteins and other compounds into ammonia. Animals excrete several kinds of nitrogen-containing wastes–urea, uric acid, creatinine, and ammonia and the decay bacteria converts these wastes to ammonia. Ammonia may be lost as gas to the atmosphere, may be acted upon by nitrifying bacteria, or may be taken up directly by plants. The nitrates may be utilized by plants, immobilized by microbes, stored in decomposing humus, or leached away. This material is carried to streams, lakes, and eventually the sea, where it is available for use in aquatic ecosystems. There **nitrogen** is cycled in a similar manner, except that the large reserves contained in the soil humus are largely lacking. Life in the water contributes organic matter and dead organisms that undergo deposition and subsequent release of ammonia and ultimately nitrates. In aquatic ecosystem atmospheric nitrogen is fixed by numerous blue-green algae.

Under natural conditions nitrogen lost from ecosystems by denitrification, volatilization, leaching, erosion, wind blown aerosols, and transportation out of the system is balanced by biological fixation and other sources. Both chemically and biologically, terestrial and aquatic ecosystems constitute a dynamic equilibrium system in which a change in one phase affects the other.

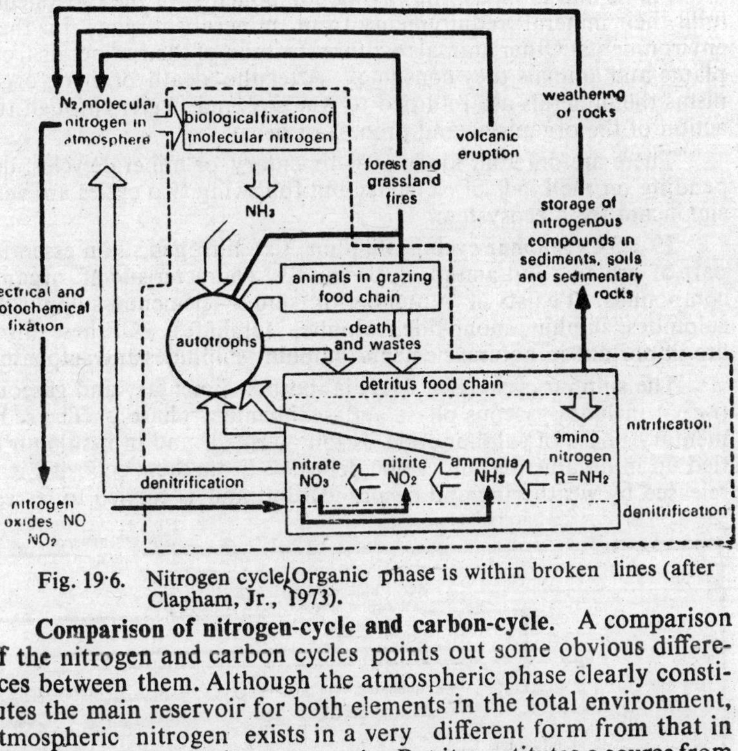

Fig. 19·6. Nitrogen cycle. Organic phase is within broken lines (after Clapham, Jr., 1973).

Comparison of nitrogen-cycle and carbon-cycle. A comparison of the nitrogen and carbon cycles points out some obvious differences between them. Although the atmospheric phase clearly constitutes the main reservoir for both elements in the total environment, atmospheric nitrogen exists in a very different form from that in which it is taken up by autotrophs. But it constitutes a source from which losses sustained by the organic phase can be made up. Also, there are more inorganic chemical reactions external to the food chain in the nitrogen cycle than in the carbon cycle. Despite these differences, both the nitrogen and carbon cycles are examples of finely tuned circuits which meet demands for the needed elements **through** several pathways, each controlled by effective self-regulating feedback loop.

19·2·3. Sedimentary Cycles

Mineral elements required by living organisms are obtained initially from inorganic sources. Available forms occur as **salts** dissolved in soil water or lakes, streams, and seas. The **mineral cycle** varies from one element to another, but essentially it consists of two phases; the **salt-solution phase** and the **rock phase**. Mineral salts come directly from the earth's crust by weathering. The soluble salts then enter the water-cycle. With water they move through the soil to streams and lakes and eventually reach the sea, where they **remain** indefinitely. Other salts are returned to the earth's crust through sedimentation. They become incorporated into salt beds, silts and limestones ; after weathering they again enter the cycle

Plants and animals, the living components of the ecosystems, fulfil their mineral requirements from mineral solutions in their environments. Other animals acquire the bulk of their minerals from plants and animals they consume. After the death of living organisms the minerals are returned to the soil and water through the action of the organisms and process of decay.

There are different kinds of sedimentary or mineral cycles, depending on the kinds of elements, but following two cycles are very significant for a ecosystem.

19-2·3A. Sulphur cycle. Sulphur, like nitrogen, is an essential part of protein and amino acids and is characteristic of organic compounds. It exists in a number of states—elemental sulphur, S, sulphides, sulphur monoxide, sulphite, sulphates. Of these three are important in nature: elemental sulphur, sulphides and sulphates.

The sulphur cycle (Fig. 19·7) is both sedimentary and gaseous (*i.e.*, it includes gaseous phase and sedimentary phase). The sedimentary phase of sulphur cycle is long-termed and in it sulphur is tied up in organic and inorganic deposits. From these deposits, it is released by weathering and decomposition and is carried to terres-

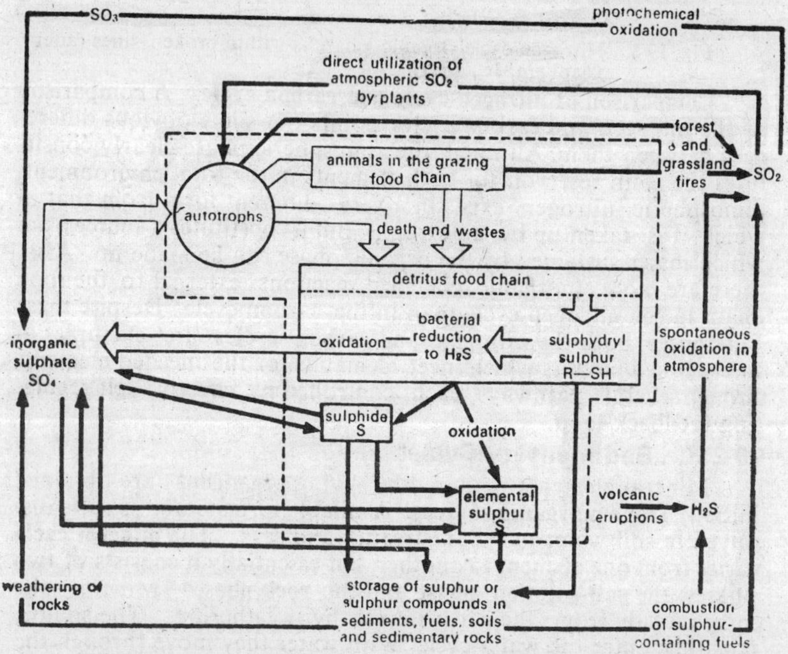

Fig. 19·7. Sulphur cycle. Organic phase is shown within broken lines (after clapham, Jr., 1973).

trial and aquatic ecosystems in a salt solution. Atmospheric (gaseous) phase of sulphur-cycle is less pronounced and it permits circulation on a global scale.

Sulphur enters the atmosphere from several sources—the combustion of fossil fuels, volcanic eruption, the surface of the oceans and gases released by decomposition. Initially sulphur enters the atmosphere as hydrogen sulphide, H_2S, which quickly oxidizes into another volatile form, sulphur dioxide, SO_2. Atmospheric sulphur dioxide, soluble in water, is carried back to earth in rainwater as weak sulphuric acid, H_2SO_4. Whatever its source, sulphur in a soluble form, mostly as sulphate ($SO_4^{=}$) is absorbed through plant roots, where it is incorporated into certain organic molecules, such as some amino acids (*e.g.*, cystine) and proteins. From the producers the sulphur in amino acids is transferred to the consumer animals, with excess being excreted in the faeces.

Excretion and death carry sulphur in living material back to the soil and to the bottoms of ponds, lakes, and seas where the organic material is acted upon by bacteria of detritus food chain. Within the detritus food chain, the sulphydryl group (—SH) of amino acids (*e.g.*, L-cysteine) is separated from the rest of the molecule as hydrogen sulphide (H_2S) by most decomposing bacteria as a normal part of the degradation of proteins. In an aerobic environment, the hydrogen sulphide is oxidized to sulphate by bacteria specially adapted to perform this conversion :

$$H_2S+2O_2 \rightarrow SO_4^{=}+2H^+$$

The sulphate produced then can be reused by the autotrophs. In anaerobic environments, such as bottom of certain lakes, it is impossible to oxidize sulphide by this means, because the process of oxidation requires oxygen. But if infra-red **radiation** is present in these environments, there are photosynthetic bacteria that can use it to manufacture carbohydrates and oxidize sulphide either to elemental sulphur or to sulphate :

$$6CO_2+12H_2S+hv \rightarrow C_6H_{12}O_6+6H_2O+12S$$

$$6CO_2+12H_2O+3H_2S+hv \rightarrow C_6H_{12}O_6+6H_2O+3SO_4^{=}+6H^+$$

Elemental sulphur can also be utilized by other bacteria to form sulphate. If oxygen is present, the reaction is quite rapid.

$$2S+3O_2+2H_2O \rightarrow 2SO_4^{=}+4H^+$$

Under anaerobic conditions, elemental sulphur can still be oxidized to sulphate by certain bacteria if nitrate is present :

$$6NO_3 +5S+2CaCO_3 \rightarrow 3SO_4^{=}+2CaSO_4+2CO_2+3N_2$$

None of these bacterial reactions is unidirectional ; under certain conditions, sulphate can also be reduced either to sulphide or to elemental sulphur by bacteria. This series of reactions operating within the organic phase of the sulphur cycle provides a rather finely tuned mechanism for **regulating** the availability of sulphur to autotrophs.

The sulphur is removed from the organic phase in the form of elemental sulphur which is insoluble and accumulates in sediments.

If iron is present in the sediment, it can combine with sulphide to form iron sulphides, all of which are highly insoluble :

$$Fe^{++} + S^{=} \rightarrow FeS$$

$$Fe(ionic) + 2S(ionic) \rightarrow FeS_2 \text{ (Ferrous sulphide or pyrite)}$$

FeS_2 is highly insoluble under neutral and alkaline conditions and is firmly held in mud and wet soil. Some ferrous sulphide is contained in sedimentary rocks overlying coal deposits. Exposed to the air in deep and surface mining, the ferrous sulphate oxidizes and in the presence of water produces ferrous sulphate and sulphuric acid :

$$2FeS_2 + 7O_2 + 2H_2O \rightarrow 2FeSO_4 + 2H_2SO_4$$

$$12FeSO_4 + 3O_2 + 6H_2O \rightarrow 4Fe_2(SO_4)_3 + 5Fe(OH_3) \downarrow$$

In this manner sulphur in pyrite rocks, suddenly exposed to weathering by man, discharges heavy slugs of sulphur, sulphuric acid, ferric sulphate and ferrous hydroxide into aquatic ecosystems. These compounds destroy aquatic life and cause acidic water.

19-2·3B. Phosphorus cycle. Phosphorus cycle (Fig. 19·8) has no atmospheric phase. It occurs naturally in environment as phosphate (PO_4^-, or one of its analogues, HPO_4^- or H_2PO_4), either as soluble inorganic phosphate ions, as soluble organic phosphate (*i.e.*, as a part of a soluble organic molecule), as particulate phosphate (*i.e.*, as part of an insoluble organic or inorganic molecules) or as mineral phosphate (*i.e.*, as part of a mineral grain as found

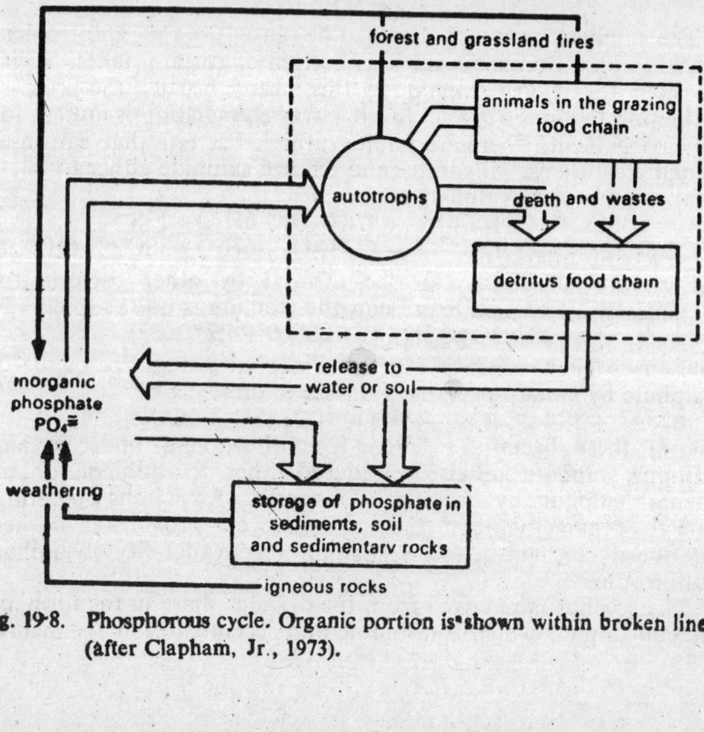

Fig. 19·8. Phosphorous cycle. Organic portion is shown within broken lines (after Clapham, Jr., 1973).

in a rock or sediment). The ultimate source of phosphate in the ecosystem is crystalline rocks. As these are eroded and weathered, phosphate is made available to living organisms, generally as ionic phosphate. This is introduced into autotrophic plants through their roots, where it is incorporated into living tissues. From autotrophs, it is passed along the grazing food chain in the same fashion as nitrogen and sulphur, with excess phosphate being excreted in the faeces. An extreme example of faecal phosphate is the tremendous guano deposits built up by birds on the desert west coast of South America. Phosphates can also be released as particulate matter from forest and grassland fires.

In the detritus food chain, as large organic molecules containing phosphate are degraded, the phosphate is liberated as inorganic ionic phosphate. In this form it can be immediately be taken up by autotrophs, or it can be incorporated into a sediment particle, either in the soil of a terrestrial ecosystem or in a sediment of an aquatic ecosystem. The sedimentary phase of phosphorus cycle remains comparatively slow than the organic phase.

Besides phosphorus, there are biogeochemical cycles for all the other nutrients (minerals) used by living organisms, as well as some that are not. Most of them has complete cycles in sedimentary phase. The availability depends on their solubility in water and availability of water as solvent.

Thus, the geochemical cycles of different chemical substances are closed : the atoms are used over and over again. To keep the cycles going does not require new matter but it does require **energy**, for the energy cycle is not a closed one. Further, the patterns of flow, both of energy and of chemical substances, are of great significance. The simpler patterns involves energy, as the sources of energy are external to the ecosystem, and flow is unidirectional through it. Chemical substances, on the other hand, are finite and have their orgin inside the ecosystem, thus they must continuously cycle within the system.

19-3. REVISION QUESTIONS

1. What are biogeochemical cycles ? What role they play in the ecosystem ? Discuss.
2. Describe nitrogen and phosphorus biogeochemical cycles with suitable **examples.**
3. Write short notes on the following :
   (a) Carbon cycle ;
   (b) Water cycle ;
   (c) Sedimentary biogeochemical cycle ;
   (d) Gaseous biochemical cycles.

Part 5

DIVERSITY OF ECOSYSTEMS

So far we have discussed various basic principles and concepts of ecology, as they apply to different levels of organisation, that is, the individual level (autecology), the population level (population ecology), the community level (community ecology) and the ecosystem level. During our discussion of these aspects, our approach has been mostly theoretical having strong emphasis on the functional aspect of ecology, that is, how sysytems of nature "work". In this part, however, our main emphasis would be on the structural aspect of nature, that is how nature "looks". This is usually called **habitat approach** of ecology. In fact, **by** studying a particular habitat we become acquainted with organisms and physical factors actually associated in a particular ecosystem (**Odum**, 1971). Thus, in habitat ecology, the approach is mainly descriptive, where we describe the environmental conditions and organisms present in different types of habitats, for example the four major habitats or ecosystems of the biosphere : freshwater, marine, estuarine and terrestrial, the following diagram illustrates some important types of habitats.

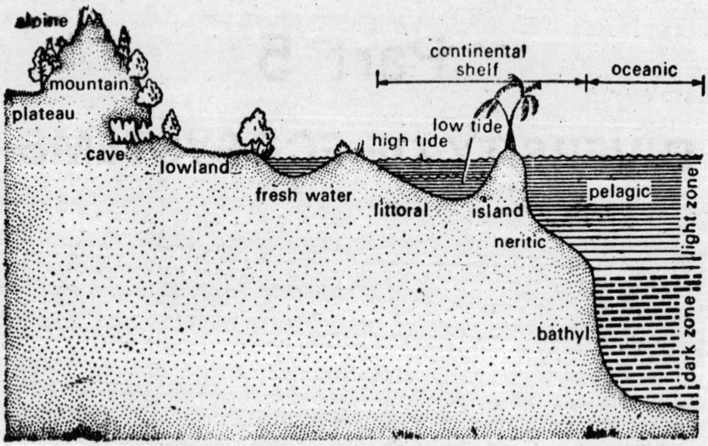

Figure. Some important terrestrial and aquatic habitats of earth (after Cockrum and McCauley, 1965).

20

Aquatic Ecosystems : Fresh-water Communities

As already discussed in chapter 10, water is a most important and essential abiotic factor of all kinds of ecosystems and it also forms the habitat for a enormous variety of organisms. In other words, water forms the biggest ecosystem, the **aquatic ecosystem** of the biosphere.

20-1. SUBDIVISIONS OF AQUATIC ECOSYSTEMS

Global aquatic system fall into two broad classes definable by salinity, or amount of material dissolved in water—the **fresh-water ecosystems** and **salt-water ecosystems**. The latter may include inland brackish water, as well as marine and estuarine habitats. In fact, estuary represents a transitional zone between a river and the sea and it contains dissolved solid content intermediate between those of fresh and marine waters. Fresh-water ecosystems, the study of which is known as **limnology**, are conventionally divided into two groups—**lentic**, **standing** or **still water habitats** and **lotic** or **running water habitats**. Both can be considered on an environmental gradient. The lotic follows a gradient from springs to mountain brooks to streams to rivers. The lentic involves a gradient from lakes to ponds to bogs, swamps and marshes.

20-2. FRESH-WATER ECOSYSTEMS : PHYSICO-CHEMICAL NATURE OF FRESH-WATER

The fresh-water of both kinds—lentic and lotic, has low percentage of dissolved salts and is subjected to the influence of a wide array of physical and chemical factors. The rise and fall of these factors very frequently affect the fauna, altering their number and diversity. Some of the important factors of fresh-water environment are following :

20-2·1. Pressure, Density and Buoyancy

The pressure imposed on a lake-dwelling organism is the weight of the column of water above it plus the weight of the atmosphere. In all fresh-water environments maximum pressure is much less than in the ocean, and organisms appear to adjust to them readily. The absence of animal life from deep water ordinarily a consequence of low oxygen supply, or low temperature, rather than pressure.

The density of water varies inversely with temperature and directly with the concentration of dissolved substances. Water is most dense at approximately 4°C and becomes progressively less dense as it cooled below +4°C. Ice also expands markedly the colder it gets. It is because the coldest water is at the surface in winter that ice forms there, rather than at the bottom of a lake. In summer, the coldest waters of deep lakes are at the bottom. Dissolved salts increase the density of water ; the density of most inland water-bodies is much less than that of the ocean. However, when great evaporation occurs in a lake having no outlet, the lake may come to contain a higher percentage of salts (*i.e.*, hypersalinity) than the ocean. The few species capable of living in these very salty lakes (*e.g.*, Great Basin of USA) include some algae and Protozoa, the brine shrimp *Artemia gracilis* and the immature stages of two brine flies, *Ephydra gracilis* and *E. hians*.

According to the **law of Archimedes** the buoyancy of an object is equal to the weight of the water it displaces. Buoyancy varies with the density of water, and is influenced by the factors that affect density. Viscosity, the measure of the internal friction of water, varies inversely with temperature and also influence buoyancy. Most aquatic organisms keep stations by swimming movements or have special adaptations to decrease the specific gravity of the body and take advantage of any turbulance in the water. For this purpose fresh-water aquatic organisms have some swimming adaptations, clinging organs (in case of animals of lotic habitat) or following adaptations—absorption of large amounts of water to form jelly-like tissues ; storage of gas or air bubbles within the body ; formation of light-weight fat deposits within the body or oil droplets within the cell ; increase of surface area in proportion to body mass, which increase frictional resistance (**David**, 1955). When an organism so equipped dies, the special mechanism quickly cease to function, and it sinks to the bottom.

20-2·2. Temperature

The unique thermal properties of water are best demonstrated by fresh-water environments. Diurnal and seasonal variations of temperatures are very much evident in these environments than in marine environments. A diurnal variation range of 4·8–5·0°C has been recorded by **Sreenivasan** (1964) in a tropical pond, with an average depth of 3·0 meters. In shallow water habitats, difference between day and night temperatures remain more conspicuous. For example, in a polluted moat with an average depth of 1·5 metres, the lowest night time temperature was 26·6°C, the highest day time temperature was 32·0°C with a variation of 5·4°C (**Sreenivasan**, 1964). However, the Kodaikanal lake in South India showed a diurnal variation of only 2·8°C (**Sreenivasan**, 1964). Flowing lotic waters of streams and rivers lack such wide fluctuations in temperature.

Further, the lentic water of lakes and ponds undergo thermal stratification phenomenon according to the seasons. Thermal stratification have been reported most frequently in the lakes of tropical countries like Java, Sumatra, etc. In fact, according to their temperature relations, lakes have been classified into three types : (1) **Tropical lakes** in which surface temperatures are always maintained above 4°C ; (2) **Temperate lakes** in which surface temperature vary above and below 4°C and (3) **Polar lakes** in which surface temperatures never go above 4°C. The seasonaly regulated thermal stratification of lentic habitats has a significant influence on their inhabitant biotic communities. Decreasing temperatures often cause a fall in metabolism, resulting in a lower **rate** of food consumption. The extremes of lower and higher temperature have lethal effects on the aquatic organisms. So fluctuations in temperature of aquatic media regulate the breeding periods, initiate hibernation, gonadial activitation and a number of other biological phenomena such as thermally oriented migration, etc., of fresh water biota. On the basis of their ability to tolerate thermal variations, most fresh water organisms are **stenothermic** with a narrow range temperature tolerance, but some **euthermic** with a wide range of temperature tolerance. For example, the stenothermic oligochaets includes **stenominimothermal forms** (narrow range of temperature, *e.g.*, *Aeolosoma*, *Megascolex mauritii*), **stenomaximothermal** (*e.g.*, *Dero limosa*) *or* **steno-optimothermal forms** (*e.g.*, *Branchiodrilus semperi and B. menoni*) (**Sitaramiah**, 1966).

20-2·3. Light

Light influences fresh-water ecosystems greatly. The fresh-waters often have a lot of suspended material. While affording protection to the light sensitive species, these substances more often obstruct the light that normally reaches the water. The degree of such obstruction of light influence the productivity of the fresh-water ecosystems. A shallow lake receive light to its very bottom resulting in an abundant growth of vegetation both phytoplankton and rooted vascular plants. These plants in living or dead states form nice food for consumers of grazing food chain or organisms of detritus food chain, respectively. The running water contain little plant or animal plankton not due to the lack of sunlight but because of the action of the currents in washing it away. Further, light controls the orientation and changes in position of attached species and their nature of growth and it also caused into diurnal migration of planktonic species of fresh-water.

20-2·4. Oxygen

Chemically pure water is biologically uninhabitable and all fresh-water contains an array of chemical substances. The oxygen, which is a most essential chemical component of life processes, remain dissolved in fresh-waters. The aquatic environments which remain in close proximity with atmosphere contain an abundance of oxygen, that reaches the water either by direct diffusion or by movements of water like wave action or water circulation. Lotic (moving) water of streams and rivers often have a high percentage of oxygen. Aquatic plants and phytoplanktons supply water with oxygen that is formed as a product of photosynthesis. The amount of photosynthetically produced oxygen remain high at warm temperatures and at greater light intensities.

The oxygen level in a tropical pond exhibits diurnal variation-it remains at peak between 14·00 and 17·00 hours of day (**Sreenivasan**, 1964). Oxygen contents of a fresh-water body are depleted in numerous ways. Primarily oxygen is utilised in the respiration of organisms and decomposition of dead organisms in the aquatic environment. While photosynthesis remains restricted to the surface layer of water containing phytoplankton and exposed regions of rooted vascular plants, respiration and decomposition occur at all levels. In stagnant pools with a lot of decaying vegetation oxygen content often reaches a stage of complete depletion. The reduction in dissolved oxygen is magnified by the release of many gases as end products of decomposition or by the mixing up of waters of low oxygen content reaching the habitat as an inflow. Aquatic animals with very few exceptions, *i.e.*, those that breath air, utilize the oxygen dissolved in water. Certain fresh-water inhabitants such as many anaerobic bacteria and insect larvae of chironomids perform anaerobiosis and requires no oxygen.

20-2·5. Carbon dioxide

Aquatic vegetation and phytoplankton require carbon dioxide for photosynthetic activity. The carbon dioxide of fresh-water environments is produced as the end product of respiration and of decomposition. Carbon dioxide also diffuses directly from the atmosphere and is readily dissolved in water to result in carbonic acid (H_2CO_3) which affects the pH of water. It is also present in the fresh-water as carbonates and bicarbonate of calcium, magnesium and other minerals. The growing plants and lime-depositing bacteria and other animals may cause a depletion in carbon dioxide resources. Photosynthesis is the major cause for its drain. The high saturation levels of O_2 and CO_2 have been found to have toxic effects on aquatic biota.

20-2·6. Other Gases

Streams and lakes contaminated by sewage and stagnant pools with decaying vegetation show an abundance of the gas hydrogen sulphide which is a decomposition product. This gas is highly toxic to living organisms and results into complete denudation of bottom fauna. Methane and carbon monoxide are other toxic gases which are the products of decomposition. Nitrogen, hydrogen, sulphur dioxide and ammonia are some of the other gases which are found dissolved in fresh-waters.

20-2.7. Dissolved Salts and Salinity

Fresh-water being efficient solvent contains many solutes in solution, but even then its salt contents remain under 1/5‰ than marine water which contains about thirty-five parts per thousand (%o) dissolved salts (see **Clapham, Jr.**, 1973). Different dissolved salts reach the water by erosion inflow and decay of aqatic forms. Dissolved substances have peculiar significance for floating aquatic vegetation and phytoplankton; since these organisms do not depend on the substratum for the supply of nutrients. Compounds of nitrogen, phosphorus and silicon are most important substances found dissolved in fresh-water. Nitrates, nitrites and ammonium salts are essential for the food of aquatic vegetation such as algae and water weeds. Nitrates always remain available due to nitrogencycles occurring between nitrogen fixing bacteria and nitrogen consuming plants. Ammonium salts in excess have a lethal effect on the fauna. Dissolved silicates of fresh-water are readily utilized by diatoms and sponges in constructing their body structures such as shell in case of diatoms and spicules in case of sponges.

All fresh-water environments also contain small amounts of phosphorus which more often acts as a limiting factor. Utilization of phosphorus by plankton during the periods of abundance may result in a total elimination of other plants that require the element (phosphorus). Many other elements like calcium, magnesium, manganese, iron, sodium, potassium, sulphur, and zinc are found dissolved in water and influence the fauna variously. Iron being a growth promoting element for plants exists as the compound of oxygen (ferrous oxide) or sulphur (ferrous sulphide) in different fresh-water bodies. Its influence is often modified by the pH of water. Calcium is an essential element for plants. The abundance and scarcity of carbonate of calcium determine the faunal composition. Deposition of calcium carbonate in water called **marl** is produced by the activity of plants. External coverings of arthropods and the shell of molluscs and tubes of some worms need calcium carbonate.

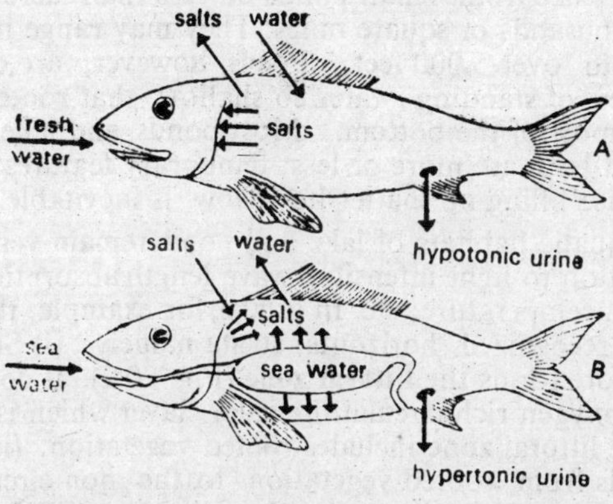

Fig. 20·1. Osmoregulation in (A) fresh-water and (B) marine fish (after Odum, 1971).

Snails are found to develop a heavy shell if the water in which they lived contained excess of calcium. Bryozoans, sponges, and cladocerans prefer an increased calcium content.

Due to low salinity of fresh-water animals face the problem of **osmoregulation**. Because the salt concentration of body fluids of animals remain higher than the fresh-water hence the water continuously tend to enter the body which should be readily removed. Most aquatic animals (*e.g.*, Protoza and fishes) have the means to excrete extra amount of water of body by osmoregulation. For this purpose Protozoa employ contractile vacuoles and other multicellular invertebrates and chordates use excretory organs, such as nephridia, kidney, etc.

20-2·8. pH or Hydrogen Ion Concentration

In fresh-water environments pH is a determining factor for the biota by becoming a limiting factor. The pH value of different fresh-water bodies may fluctuate seasonally and annually. The pH of surface waters and deeper waters exhibit marked differences. **Sreenivasan** (1968) has reported a marked pH variation of 2·2 units between surface water and deeper water in the Sandynulla reservoir in the Nilgiris (India). Though pH range is species specific, yet lower aquatic forms in general showed little reaction to alterations in pH, while higher aquatic organisms (*e.g.*, fishes) responded quickly to little pH variations.

20-3. LENTIC ECOSYSTEMS

Lentic ecosystems include all standing fresh water habitats such as lakes, ponds, marshes, swamps, bogs, meadows, etc. Some of the important lentic ecosystems or wetlands have been tabulated in table 20·1.

20-3·1. Lakes and Ponds

Lakes are inland depressions containing standing water. They may vary in size from small ponds of less than acre to large seas covering thousands of square miles. They may range in depth from a few feet to over 5,000 feet. **Ponds**, however, are considered as small bodies of standing water so shallow that rooted plants can grow over most of the bottom. Most ponds and lakes have outlet streams and both are more or less temporary features on the landscape because filling no matter how slow, is inevitable.

The aquatic habitats of lake and pond remain vertically stratified in relation to light intensity, wave length absorption, hydrostatic pressure, temperature, etc. In a lake, for example, there are three to five well recognized horizontal strata namely : (i) Shallow water near the shore forms the **littoral zone** (Fig. 20·2). It contains upper warm and oxygen rich circulating water layer which is called **epilimnion**. The littoral zone includes rooted vegetation. **(ii) Sublittoral zone**-extends from rooted vegetation to the non-circulating cold water with poor oxygen zone, *i.e.*, **hypolimnion**. **(iii) Limnetic zone** is the open water zone away from the shore. It is the zone upto

Table 20·1. Classification of wetland or lentic ecosystems (excluding lakes and ponds) (Smith, 1974).

A. Inland freshwater areas	
1. Seasonally flooded	Soil covered with water or waterlogged during variable periods but well-drained during much of the growing season. In upland depressions and bottomlands. Bottomland hardwoods to herbaceous growth.
2. **Freshwater meadows**	Without standing water during growing season ; water logged to within few inches of surface. Grasses, rushes, sedges, broad leaf plants.
3. Shallow fresh water marshes	Soil waterlogged during growing season; often covered with 6 inches or more of water. Grasses, bulrushes, spike rushes, cattails, arrowhead, smartweed, pickerelweed. A major water-fowl-production area.
4. Deep freshwater marshes	Soil covered with 6 inches to 3 feet of water. Cattails, reeds, bulrushes, spikes rushes, wild rice. Principal duck-breeding area.
5. Open freshwater marshes	Water less than 10 feet deep. Bordered by emergent vegetation : pondweed, naiads, wild celery, water lily. Brooding, feeding, nesting areas of ducks.
6. Shrub swamps	Soil waterlogged; often covered with 6 inches or more of water. Alder, willow, buttonbush, dogwoods. Ducks nesting and feeding to limited extent.
7. Wooded swamps	Soil waterlogged; often covered with 1 foot of water. Along sluggish streams, flat uplands shallow lakes, basins.. North of USA : tamarack, arborvitae, spruce, red maple, silver maple. South of USA ; water oak, overcup oak, tupelo, swamp black gum cypress.
8 Bogs	Soil waterlogged ; spongy covering of mosses. Heath shrubs, sphagnum, sedges.
B. Coastal freshwater areas	
9. Shallow freshwater marsh	Soil waterlogged during growing season ; at high tides as much as 6 inches of water on landward side, deep marshes along tidal rivers, sounds, deltas. Grasses and sedges. Important water-fowl areas
10. Deep freshwater marshes	At high tide covered with 6 inches to 6 feet of water. Along tidal rivers and bays. Cattails, wild rice, giant cutgrass.
11. Open freshwater	Shallow portions of open water along fresh tidal rivers and sounds. Vegetation scarce or absent. Important water-fowl areas.

the depth of effective light penetration where rate of photosynthesis is equal to the rate of respiration. (iv) **Profundal zone** is the deep-water area beneath limnetic zone and beyond the depth of effective light penetration. (v) **Abyssal zone** is found only in deep lakes, since it begins at about 2,000 metres from the surface. Ponds have little vertical stratification. In them littoral zone is larger than the limnetic zone and profundal zone. In a small pond the limnetic and profundal zones are not found. Lakes thus differ ponds in having relatively larger limnetic zone and profundal zone than littoral zone.

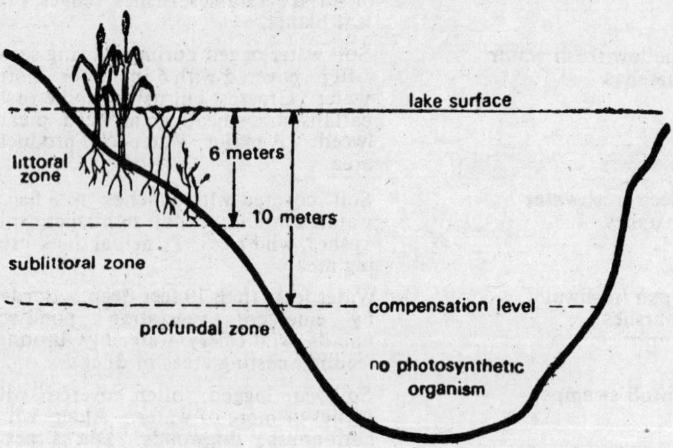

Fig. 20·2 Different zone of a deep freshwater lake.

In fact, the warm top layer, the epilimnion is heated by the sun and hamogenised by the wind and other currents, while, the deep cold layer, the hypolimnion is not heated by sun and not circulated by wind (Fig. 20·4). The transition between the two is the **metalimnion** or **thermocline.** The basis upon which the layers are maintained is strictly thermal and is caused by the fact that the warmer water is lighter than the colder water. Once a thermocline is formed, no exchange of water occurs between the epilmnion and hypolimnion.

Further, lentic water of a lake or pond is also classified on the basis of the depth of light penetration enabling photosynthesis into **trophogenic zone** (includes littoral plus sublittoral zones), and is **tropholytic zone** (upper part of profundal zone). The former is often distinguished by abundant plant growth and dependent fauna, while the latter denotes a general absence of vegetation and harbours mostly saprobes. In between the two zones is the **compensation level** which forms a boundary between two zones (Fig. 20·3). It exibits perfect equilibrium between respiration and photosynthesis.

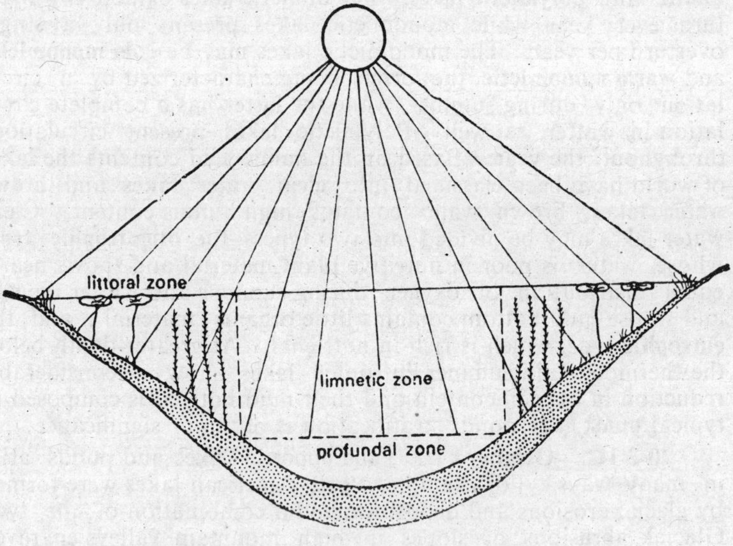

Fig. 20·3. Different zones of a fresh-water pond (after Odum, 1971).

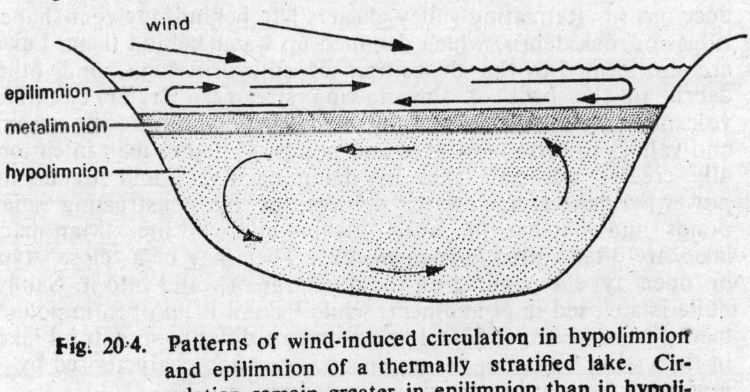

Fig. 20·4. Patterns of wind-induced circulation in hypolimnion and epilimnion of a thermally stratified lake. Circulation remain greater in epilimnion than in hypolimnion (after Clapham, Jr., 1973).

20-3·1A. Physico-chemical properties of lakes and ponds. Lakes have the tendency to become thermally stratified during summer and winter to undergo definite seasonal periodicity in depth distribution of heat and oxygen. Light too penetrates only to a certain depth, depending upon turbidity. These gradations of oxygen, light and temperature profoundly influence life in the lake, its distribution and adaptation.

20-3·1B. Kinds of lakes. Different classifications of lakes on the basis of physical factors, productivity, etc,, exist. Based on temperature **Hutchinson** (1957) classified lakes into **dimictic, mono-**

mictic and **polymictic** lakes. The dimictic lakes exhibit two overturns every year while monomictic lakes present only a single overturn per year. The monomictic lakes may be **cold monomictic** and **warm monomictic**, the former being characterized by a circulation only during summer, while the latter has a complete circulation in winter as well. Polymictic lakes present circulations throughout the year. Based on the humic acid contents the lakes of world have been classified into **clear water lakes** and **brown water lakes**, brown water contains high humus content. Clear water lakes may be divided into two types—the **oligotrophic type**, whose water is poor in nutritive plant material and shows nearly equal distribution of oxygen during summer and winter months and whose mud bottom contains little organic material ; and the **eutrophic type**, which is rich in nutrients. At greater depth below the thermocline in summer eutrophic lakes show a considerable reduction in oxygen content and their mud bottom is composed of typical muck. In ponds stratification is of minor significance.

20-3·1C. Origin of lakes and ponds. Lakes and ponds arise in many ways. For example, North American lakes were formed by glacial erosions and deposition and a combination of the two. Glacial abrasions of slopes in high mountain valleys engraved basins, which filled with water from rain and melting snow to produce tarns. Retreating valley glaciers left behind crescent-shaped ridges of rock debris, which dammed up water behind them. Lakes are also formed by the deposition of silt, driftwood, and other debris in the beds of slow-flowing streams. Craters of extinct volcanos may fill with water and landslides can block off streams and valleys to form new lakes and ponds. Further man intentionally creates artificial lakes by damming rivers and streams for power, irrigation, and water storage, or by constructing small ponds and marshes for water, fiishing and wild life. Man made lakes are often called **impoundments**. They may be a 'close' type or 'open' type depending on the discharge of water into it. Sandynulla is a closed impoundment, while Pykara is an open impoundment (**Sreenivasan**, 1957). Impoundments differ from natural lakes in thermal and oxygen variations and are often characterised by a low percentage of bottom dwellers or benthos.

Ponds occur in most regions of adequate rainfall. Temporary ponds, which are often shallow and quick drying contain a characteristic lentic environment. Permanent ponds contain water throughout the year, while flood-plain ponds are formed when a stream shifts its position, leaving the formed bed isolated as a body of standing water.

20-3·2. Biotic Communities of Lakes and Ponds

Different organisms of the lentic environment can be ecologically classified based on whether they are dependent on the substratum or free from it. Organisms depending on the substratum are called **pedonic forms** and those that are free from it are the **limnetic**

forms. Further, the aquatic organisms may also be classified into following groups depending upon their sizes and habits :

1. **Neuston**. These are forms which are found in all aquatic ecosystems at the air-water interface. They may include floating plants such as duckweed, as well as many types of animals. Animals that spend their lives on top of the air-water interface, such as water striders, are termed **epineuston**, while others, including insects such as diving beetles and back swimmers, which spend most of their time on the underside of the air-water interface and obtain much of their food from within water, are termed **hyponeuston**.

2. **Plankton.** These are forms which are found in all aquatic ecosystems except for fast-moving rivers. They are small plants and animals whose powers of self-locomotion are so limited that they cannot overcome currents in their ecosystems. Most planktons (phytoplanktons and zooplanktons) can move a bit, however, either, to contral their vertical distribution or to seize prey. Certain animal planktons or zooplanktons are exceedingly active and move relatively great distances considering their small size, but they are so small that their range is still controlled largely by currents and such planktons are also known as **nektoplankton.**

3. **Nekton.** Nektonic animals are swimmers and are found in all aquatic systems-except for the fastest-moving rivers. In order to overcome currents, these animals are relatively large and powerful ; they range in size from the swimming insects of quiet water, which may be only about 2 mm long, to the largest animal that has ever lived on earth, the blue whale.

4. **Benthos**. The benthos includes the organisms living at the bottom of the water mass. They occur virtually in all aquatic ecosystems. The benthos organisms living above the sediment-water interface are termed **benthic epifauna** and those living in the sediment itself are termed **infauna.**

20-3·2A. Neustons of lentic aquatic environment. The water surface of lake or pond contain certain free-floating hydrophytes such as *Wolffia*, *Lemna*, *Spirodella*, *Azolla*, *Salvinia*, *Pistia*, and *Eichhornia*. These plants remain in contact with water and air, but no soil.

20-3·2B. Biota of littoral zone. Lentic aquatic life is most prolific in the littoral zone. The littoral zone of a lake remains rich in pedonic flora especially upto the depth to which effective light penetration is possible facilitating the growth of rooted vegetation. At the shore proper are the **emergent vegetation** which remain firmly rooted in the shore substratum but their tops with thin chlorophyll bearing regions are exposed. Certain rooted emergent plant species of littoral zone are *Ranunculus*, *Monochoria*; *Cyperus* and *Rumex*. Interspersed with these plants are the cattails (*Typha*), bulrushes (*Scirpus*), arrowheads (*Sagitaria*) and pickerel-weeds. Out slightly deeper are the rooted plants with floating leaves such as the water lilies *Nymphaea*, *Nelumbo*, *Aponogeton*,

Trapa, *Marsilea*, and *Potamageton*. Still deeper in the littoral zone of lake are the fragile **thin**-stemmed water weeds, rooted but totally submerged. Such a submerged vegetation includes plants like *Elodea*, *Vallisneria*, *Myriophyllum*, *Isoetes*, *Hydrilla*, *Chara*, *Potamogeton*, etc.

The phytoplankton of littoral zone of lake is chiefly composed of diatoms (*e.g.*, *Navicula*, *Cyclotella*), blue green algae (*e.g.*, *Microcystis*, *Oscillatoria*), green algae (*e.g.*, *Cosmarium*, *Staurastrum*) and holophytic flagellates. **Sreenivasan** (1970) has observed the abundance of marine dinophycean flagellate *Ceratium hirudinella* in some impoundments of India.

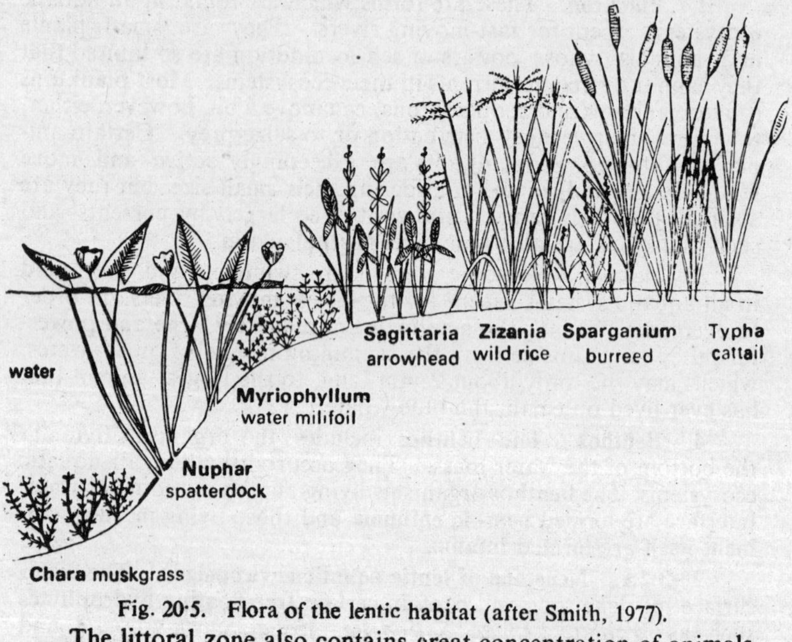

Fig. 20·5. Flora of the lentic habitat (after Smith, 1977).

The littoral zone also contains great concentration of animals, which remain distributed in recognizable communities. In or on the bottom are various dragonfly nymphs, crayfish, isopods, worms, snails and clams. Other animals live in or on plants and other objects projecting up from the bottom. These include protozoans like ***Vorticella***, *Stentor* ; larvae of *Dytiscus*, *Laccotrophes*, leech-like *Glossophonia*, climbing dragonfly, damsel fly nymphs, rotifers (*Brachionus*), flatworms, Bryozoa, hydra, snails (*Lymnaea*) and other. With the exception of a few rotifers like *Keratella*, *Brachionus*, *Diurella*, and *Trichocerca* and crustaceans like *Cyclops*, *Daphnia* and *Moina*, most freshwater animals of lentic habitat depend in one way or other on the aquatic vegetation. Larvae of *Chironomus* are found to inhabit the interior of rotting vegetation. Planarians are found underneath the leaves of floating plants. The zooplankton of littoral zone consists of water fleas such as *Daphnia*, rotifers and ostracods. The free swimming fauna (Nekton) includes *Paramecium*, *Euglena*, *Ranatra*, *Corixa*, *Dytiscus*, larvae of *Culex* and of *Chaoborus*, *Gerris*, *Gyrinus*, etc. Among the vertebrates are frogs, salamanders, snakes and turtles are the nektons of littoral zone. Floating members of the community (neuston) are whirliging beetles, water spiders and numerous Protozoa. Many pond fishes such as sun-fish, top minnows, bass, pike and gar spend much of their time in the littoral zone.

20-3·2C. Biota of limnetic zone. Limnetic zone is the region of rapid variation, with the water level, temperature and cxygen content varying from time to time. In certain Himalayan lakes which have a glacial origin, change in water level is a conspicuous phenomenon, the onset of summer bringing forth a rapid and

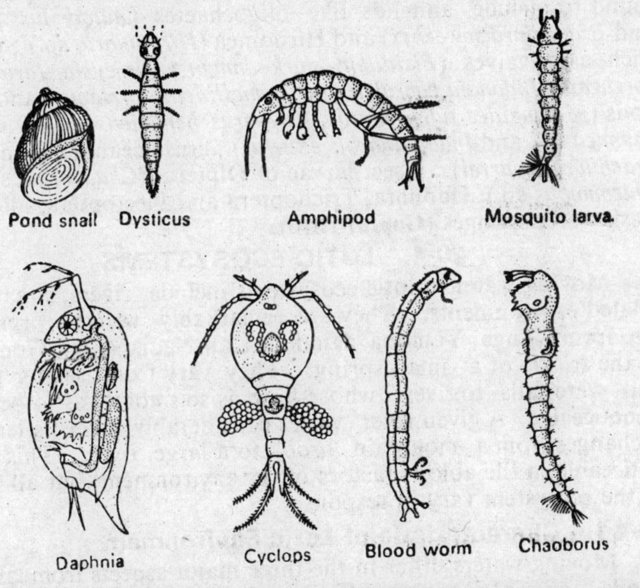

Fig. 20·6. Fauna of littoral zone of lentic habitat (after Odum, 1971).

sudden rise in water level. Various protozoans which are capable of encystment during adverse ecological conditions, tardigrades like *Macrobiotus*, rotifers like *Rotaria*, *Philodina*, copepods, snails and frogs occupy the limnetic zone. Sedentary and slow moving forms are excluded from this zone because of predators and lacking of permanent substratum for attachment. Many microscopic plants (*Volvox*, *Euglena*) and fishes also occupy this zone. The limnetic zone has autotrophs in abundance.

20-3·2D. Biota of profundal zone. The deep (profundal) life consists of bacteria, fungi, clams, blood worms (larvae of **midges**), annelids and other small animals capable of surviving in a region of little light and low oxygen. In the profundal zone, autotrophs cannot produce food, and the main source of energy is detritus that rains out of the limnetic zone. All the organisms of this zone are heterotrophs, either as detritus feeder or as carnivores. Most of our largest lake fishes inhabit the dark waters of hypolimnion most of their lives. In addition to detritus from the limnetic zone, food washed into lakes by rivers can settle out and serve as food for profundal animals.

20-3·2E. **Benthos of lake bottom.** The lake bottom in young lakes may be of the original rock; in older lakes, it would have been covered with sediment to form a uniform substrate of mud or sand. The benthos community includes several species of insect larvae, including those of small mosquito like midges, burrowing mayflies, clams snails, and tubeworms. Macrobenthic fauna of Loni reservoir of Rewa district of Madhya Pradesh (India) was found to include annelids like oligochaetes (*Aulodrilus pluriseta* and *Branchiura sowerbyi*) and Hirudinea (*Hirudinaria* sp.) ; molluscs such as bivalves (*Pisidium clarkeannum*, *Parreysia favidens*, *P. corrugata*, *Indonaia caerulea* and *Lamellidens corrianus*) and gastropods (*Melanoides tuberculatus*, *Viviparus bengalensis*, *Gyraulus* sp., *Lymnea* sp., and *Indoplanorbis exustus*) ; crustaceans (Prawn *Macrobrachium lamarrei*) ; insect larvae of Diptera (*Chaoborus* sp., and *Chironomus* sp.), Odonata, Trichoptera and Plectoptera, and a few fresh-water sponges (**Gupta**, 1976).

20-4. LOTIC ECOSYSTEMS

Moving-water or lotic ecosystems include rivers, streams, and related environments. They are remarkably variable, ranging in size from Ganga, Yamuna, Hindon, Kali Nadi, Sutlaj, Gomti, etc., to the trickle of a small spring. They vary from raging torrents and waterfalls to rivers whose flow is so smooth as to be almost unnoticeable. A given river varies considerably over its length, as it changes from a mountain brook to a large river. This is most noticeable in the abiotic factors of the environment, but all features of the ecosystem vary in response.

20-4·1. Characteristic of Lotic Environment

Moving waters differ in the three major aspects from lakes and ponds : current is a controlling and limiting factor ; land-water interchange is great because of the small size and depth of moving water systems as compared with lakes; oxygen is almost always in abundant supply except when there is pollution. Temperature extremes tend to be greater than in standing water. Besides these, the most distinctive features of moving water ecosystems are those related to their motion, *i.e.*, the **rate of flow** and the **stream velocity.** The rate of flow or discharge refers to the volume water passing a given observation point during a specific unit of time ; it is measured in units such as m.^{3}/sec., ft.^{3}/sec., or acre-feet/sec., (one acre-foot is equivalent to a volume of water 1 acre in area by 1 foot deep). It tends to increase steadily growing down stream, as tributaries join with the main river. The **velocity of flow** is the speed at which the water moves, and is measured in m./sec., ft./sec., or mi./hr. Velocity is variable, but it also tends to increase downstream with increasing discharge. A factor that may be even more important to the biotic community is the **turbulence**, or the irregularity of the motion of the water particles. Perfectly even flow, in which water particles move parallel to one ancther, is called **laminar flow** (Fig. 20·7). It is contrasted with **turbulent flow**, in which the movement of water particles is highly irregular. In highly turbulent water the erosive power is great, the sheer forces at the water-sediment interface at the bottom of the river are powerful, and the amount of oxygen incorporated into the water is very high. In streams with laminar flow the erosive power is lower, the sheer forces at the bottom of the stream are lower, and relatively less oxygen is incorporated into the water than turbulent waters.

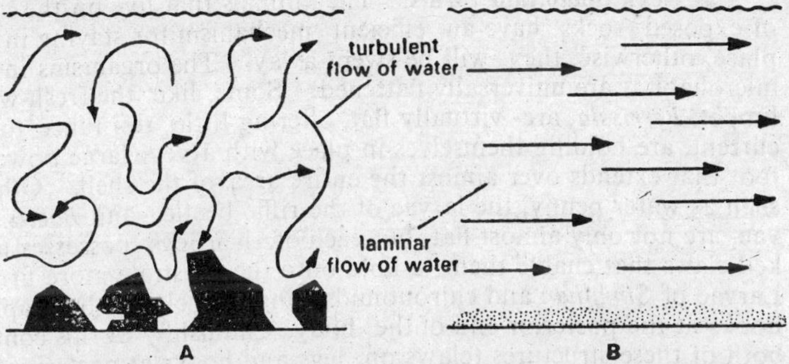

Fig. 20·7. Diagram showing the differences between turbulent (A) and laminar (B) flow of water in a stream (after Clapham, Jr., 1973).

The moving-water ecosystems can be divided into following two ecosystems depending up on the velocity and rate of flow of water body :

20-4·1A. Rapidly flowing water. Rapidly flowing water (of fast moving torrential streams, etc.) can be defined as the portion of the stream in which the flow is both rapid and turbulent. Everything that is not attached or weighty is swept away by the current; this includes organisms and sediment particles alike. The substrate tends to be rock or gravel, and the fragments are rounded and smoothed by the water. The habitat itself is an extremely diverse one. Physical parameters such as sheer force and rate of water movement tend to be quie different on top of a rock fragment, between rock fragments, or beneath rock fragments, and different species can exploit these differences in microhabitats. Aquatic plants, in addition, provide microhabitats for some torrential animals. The rarely occurring phytoplankton of **rapid** or **riffle zone** of stream includes diatoms, blue-green and green algae (*e.g.*, *Cladophora*, *Ulothrix*) and water moss (*e.g.*, *Fontinalis*) (Fig. 20·8).

1. Plant inhabiting torrential forms. Some animals of rapidly flowing water streams live among the mosses and flowering plants like *Eriocaulon miserum*, *Hydrolyum lichenoides*, and *Duroea wallichii.* Animals inhabiting these plants have a torpedo-like body which enables them to offer minimum resistance to the current. Some animals such as funnel mouthed tadpoles of *Megalophrys* and tipulid larvae live entangled in the roots of these plants. They possess devices for fixation and have hydrostatic organs.

Some animals such as *Cephalopteryx*, *Helodes*, *Phalacrocera*, *Gammarus*, etc., live among leaves and stems of aquatic plants. All these animals possess hook-like structures on their body helping them in anchoring on leaves and stems. The larvae of simulids and chironomids, however, live on the exposed surfaces of plants.

2. Rock inhabitant forms. The animals that live on the tops of exposed rocks have an efficient mechanism for staying in one place, otherwise, they will be swept away. The organisms in this microhabitat are universally flattened. Some, like the fresh-water limpet, *Ferrissia*, are virtually flat, offering little resistance to the current, are holding themselves in place with a very large powerful foot that extends over almost the entire area of the shell. Others, such as water penny, the larvae of the riffle beetle, and *Baetis* larvae, are not only almost flat, but each of their legs possesses hooked claws that enable them to hold onto the substrate more firmly. Larvae of *Simulium* and chironomids cling to rock top by grapling hooks at the posterior end of the body. Caddishfly worms contain both of these structures (claws on legs and hooks at posterior end of body). The mayfly nymphs *Iron* and *Psephenus* attach themselves to the rocks by means of functional pads. Some animals, including fresh-water sponges, actually cement themselves to the surface of the rocks. Others, such as caddish flies, build "houses" out of sediment or wood fragements, which are then cemented firmly to the rocks. Exposed surfaces of rocks also contain few sessile algae and few plants. A significant amount of organic detritus is washed into rapid-water ecosystems from up stream and adjacent terrestrial areas, however. This is the source of the most of energy in fast water ecosystems, and is much more important as an energy source then primary production within the stream itself. In consequence, the majority of primary consumers of such a microhabitat are detritus-eaters.

3. Inhabitant forms of spaces between rocks. Numerous different kinds of animals live in the spaces between the rocks. Many of them, such as mayfly and stonefly naiads, are flattened and have certain behavioral adaptations to hold them in place. These include **thigmotaxis,** by which they cling instinctively to any hard surface

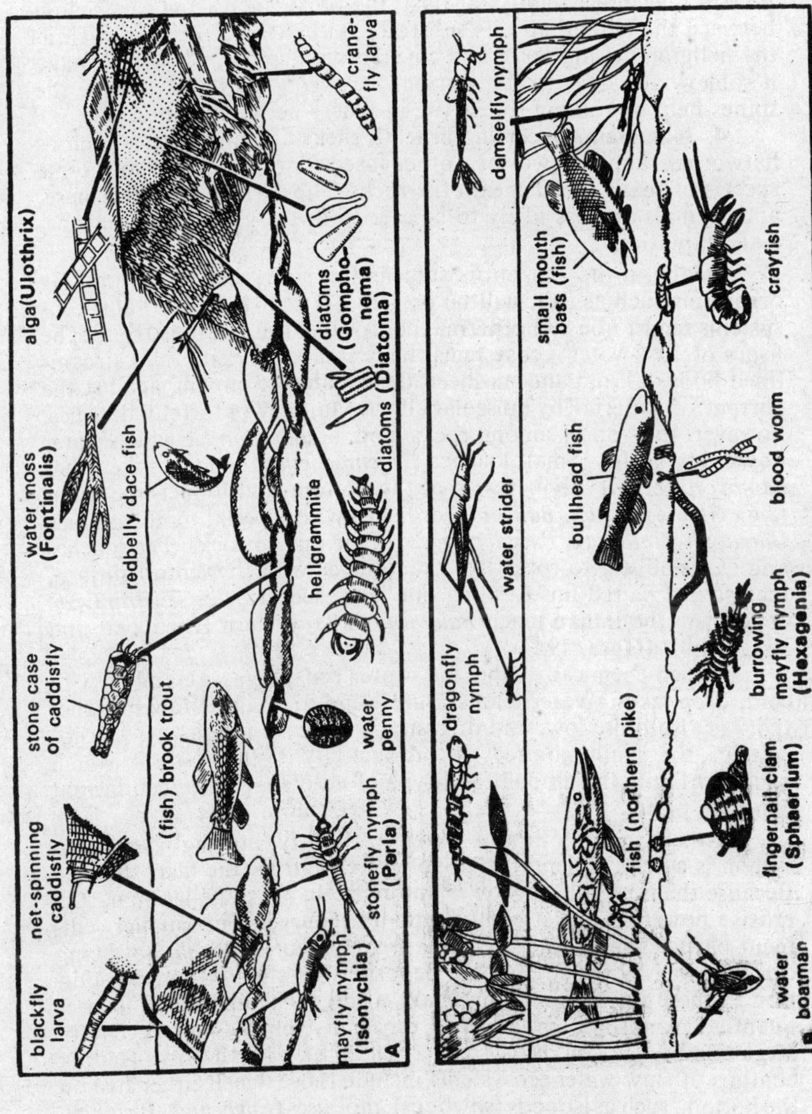

Fig. 20·8. Comparision of biota of rapidly-flowing stream (A) and slowly-flowing stream (after Smith, 1974).

such as rock or another insect larva. **Rheotaxis**, by which the animals orient themselves to face into current and move upstream is another adaptation of this sort. In many of these naiads, the combination of the organisms shape and behaviour is such that the current presses the insect tightly against the rock, increasing the friction between the animal and its substrate. Other insect larvae, such as the hellgramite, are large and covered with spines. Their size make it somewhat harder for the current to sweep them away, while the spines help in holding the larvae in places between stones.

4. Inhabitants occurring beneath rocks. Many species such as flatworms, annelid worms, other insect larvae, clams and some species of snails, live beneath the rocks. The current is weaker here, and animals are less likely to be carried away. They lack any special adaptation.

Finally, if the current is sufficiently slow, certain swimming organisms such as fish, will be present. The fish of fast-water ecosystems tend to be stenothermal cold-water fish, such as trout. The fishes of fast-water ecosystems have small body size and streamlined body. Trout and masheer are capable of moving against the current or waterfall by muscular effort. Majority of torrential fishes, however, live on or among rocks and boulders, over which water flows very swiftly. Small loaches (*Noemacheilus*) or loach-like fishes (*Ambyceps olyra*) are met with at the bottom and limpet-like fishes (*e.g.*, *Glyptosternum*, *Balitora*) or fishes with special modifications, *Garra*, *Glyptothorax*, *Pseudecheneis*) occur on the rocks. The loaches and carps adhere to rocks like the limpets, with the ventral surface, inclusive of paired fins forming one large sucker, *e.g.*, *Gastromyzon* of Borneo, the Indian loach *Baletora* of the eastern Himalayas and Assam hills (**Hora**, 1947).

Physico-chemically, the fast-water ecosystems resemble with cold, deep lakes. Water temperatures tend to be quite low, productivity is also quite low, and diversity is high. In the fast-water ecosystem, the main control of productivity is the current, which seriously limits the amount and type of autotrophs production that can take place.

20·4·1B. Slowly flowing water. A slowly flowing water ecosystem is a very different type of system from the fast streams. Because the flow is both slower and more likely to be laminar, the erosive power of the stream is greatly reduced, and smaller sediment particles (silt) and decaying organic debris, instead of being carried away by the stream, are deposited on the bottom. In addition to these the slow streams have a higher temperature, consequently, planktonic organisms, especially protozoans, occur in large numbers in this ecosystem (Fig. 20·8). The detritus-feeding benthos of slow water ecosystems include those which either live on the bottom, such as isopods (sowbugs), molluscs (*Sphaenius*, *Pisidium*, and *Anodonta dominata*), and mayfly and damselfly naiads, or which burrow into the sediment, such as tubeworms, naiads of the burrowing mayflies, *Sialis* (alderfly), midge *Chironomus*, and several other insect larvae, as well as, clams, nematodes, snails and rotifers. Swimming organisms are also abundant, including not only fishes, such as carps, cat fishes, suckers, stingers, spoonbills, etc., which tend to be different species from those of fast-water areas, but also larger Crustacea such as amphipods (fresh-water shrimps). Finally, several insects spend most of their time at the surface of the stream. These include forms such as water striders, water boatmen, back-swimmers, and predaceous diving beetles. Zooplankton is abundant, including a rich assemblage of Protozoa and smaller Crustacea such as Cladocera (water-fleas) and copepods. The richness of wood afforded by slow streams attract a large number of reptiles (water-snakes, crocodiles, turtles) and amphibians.

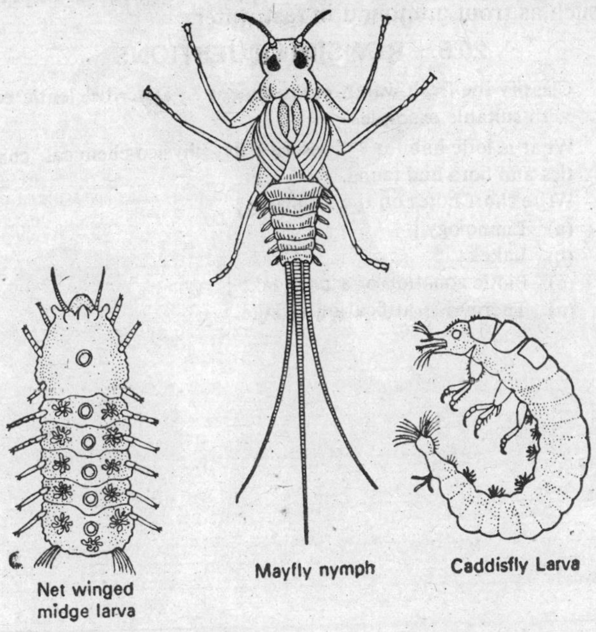

Fig. 20·9. Some inhabitants of lotic environment.

Plant life is also abundant in a slow-water ecosystem. It includes rooted vascular plants such as pond weeds and grasses, firmly attached aquatic mosses and large multicellular filamentous algae. Motile algae such as diatoms and flagellates may abound in the open water. In all, the primary productivity of slow-water ecosystem is higher than that of the rapid-water ecosystem, and the community is relatively less dependent on food materials from outside. Further, the detritus food chain consisting of organisms such as bacteria and fungi is much better developed in this community as the partially decomposed organic debris that comprises the main food source for these organisms accumulates in the mud bottom. In some slow-moving streams, in fact, the bottom muds contain more organic material than mineral fragments.

In slow-water streams oxygen concentration is main limiting factor. The high level of animal activity, coupled with an active detritus food chain, can withdraw a large amount of oxygen from water of a slowly moving stream. In addition, the low level of turbulence means that less oxygen is incorporated into the water at the surface. Thus the dissolved oxygen content of a slowly moving stream is likely to be much lower than that of a fast-moving stream. This is commonly reflected in the fauna of the stream. Fishes tolerant of low oxygen levels such as carps and catfishes are the most common fishes in slow water, while species with high oxygen demands, such as trout are found in fast water.

20·5. REVISION QUESTIONS

1. Classify the fresh water environment ? Describe lentic ecosystem with suitable examples.
2. What is lotic habitat ? Describe its physico-chemical characteristics and flora and fauna.
3. Write short notes on the following :
   (a) Limnology ;
   (b) Lakes ;
   (c) Biotic zonation of a deep lake ;
   (d) Thermal stratification of lake.

21
Aquatic Ecosystems : Estuaries and Marine Communities

21-1. ESTUARIES

Waters of all streams and rivers ultimately drain into the sea ; the place where this freshwater joins the salt water is called an **estuary**. Estuaries are the transitional zones between the sea and rivers and are the sites of unique ecological properties possessing a characteristic biological make up. They are semienclosed coastal bodies of water that have a free connection with the open sea and within which sea water is measurably diluted with fresh-water from rivers. Not all rivers open into estuaries : some simply discharge their runoff into the ocean. Estuaries differ in size, shape and volume of water flow, all influenced by the geology of the region in which they occur. As the river reaches the encroaching sea, the stream carried sediments are dropped in the quite water. These accumulate to form **deltas** in the upper reaches of the mouth and shorten the estuary. When silt and mud accumulations become high enough to be exposed at low tide, **tidal flats** develop, which divide and braid the original channel of the estuary. At the same time, ocean currents and tides erode the coast line and deposit material on the seaward side of the estuary, also shortening the mouth. If more material is deposited than is carried away, then **barrier beaches, islands**, and **brackish lagoons** appear.

21-1·1. Physico-Chemical Aspects of Estuaries

Current and **salinity**, both very complex and variable, shape life in the estuary, where the environment is neither fresh-water nor salt. Estuarine currents result from the interaction of a one-direction stream flow, which varies with the season and rain fall, with oscilation ocean tides, and with wind. The salinity of estuaries varies vertically and horizontally and fluctuates amazingly between 0·5—0·35 per cent. Due to low salinity fresh-water has a lowest density than sea water (±1·00, as opposed to 1·03). **Were there no tides in** an estuary to mix fresh and salt-water, the lighter **fresh-water** would simply flow over the heavier sea water and dissipate in the ocean (Fig. 21·1). However, the tidal action acts as a plunger to thoroughly flush the estuary and mix the fresh and salt water (Fig. 21·2). This has several results. The water level in the estuary fluctuates regularly unlike that of river. For this reason the habitat that is covered at high tide and uncovered at low tide is a prominent one in the estuary, and has no analogue in any purely fresh-water ecosystem. Further, the salinity is exceedingly variable and may change by a factor of ten over the course of a day at any location. At low tide, most of water passing through the estuary in fresh-water, and the salinity is correspondingly low. At high tide,

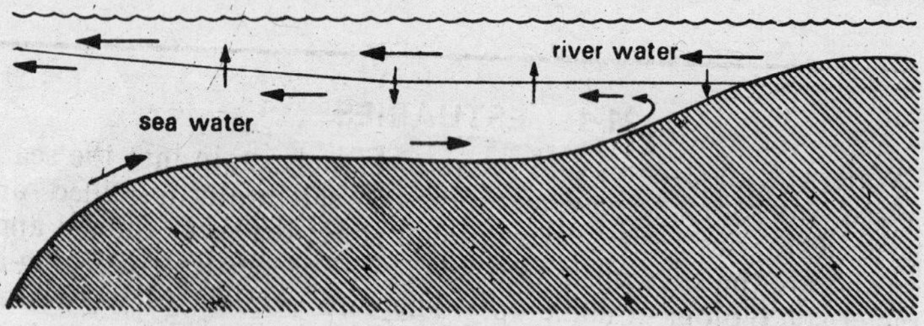

Fig. 21·1. **Theoretical** flow of freshwater over sea water in an are (*i.e.*, estuary) where a river enters the sea without tidal or other effects to mix fresh-water and sea-water (after **Clapham**, Jr, 1973)

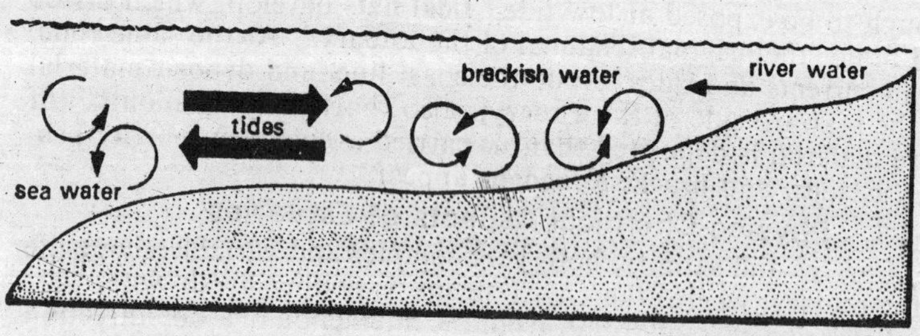

Fig. 21·2. Diagram showing the effect of tides on mixing fresh river water with sea-water (after Clapham, Jr., 1973).

most of the water may be of marine origin, and the salinity correspondingly high. Flow may be somewhat stratified, or mixing may be incomplete, and the salinity at any time may very considerably from one location to another. The intensity of current, and hence of the degree of mixing, is a function of the intensity of the tides and of the rivers flow rate. In other words, an estuary is an exceedingly variable environment. Further, salinity of estuaries remains highest during the summer and during periods of drought, when less fresh-water flows into the estuary. It is lowest during the winter and spring, when rivers and streams are discharging their peak loads. No stenohaline organism (such as Echinodermata, Cephalopoda, and **other** molluscs) could hope to survive in the estuaries. Likewise, no organism that could not tolerate strong currents and the turbid water that results from strong currents could live there.

The temperatures in estuaries fluctuate considerably diurnally and seasonally. Waters are heated by solar radiation and inflowing and tidal currents. High tide on the mud flats may heat or cool the water, depending on the season. The upper layer of estuarine water may be coolar in winter and warmer in summer than the bottom, a condition that, as in a lake, will result in a spring and autumn overturn.

In Cochin backwater, which is a tropical estuary, **Gopinathan** *et al.*, (1974) have found that its hydrological parameters fluctuate three times in the year. During the pre-monsoon season (February-May) a stable condition exists in the estuary without any vertical gradients (stratifications) in salinity or temperature. However, during monsoon season (June-September) there is considerable influx of fresh-water from precipitation and land run-off and during post-monsoon season (October-January), there is reduction in the discharge of fresh-water and the earlier brackish condition is restored by the incursion of sea water. The tidal cycles, upwelling, monsoon piling and sinking in the Arabian sea, also influence the seasonal pattern of the hydrological conditions of the back water. The seasonal and tidal cycles cause changes in nutrient concentration in the **estuary** which are found to be related with seasonal and spatial variation of phytoplanktons. Some common planktons of this estuary are diatoms (*Skeletonema costatum*), dinoflagellates (*Prorocentrum micans* and *Ceratium furca*), silicoflagellates (*Dictyocha fibula* and *Destephanus speculum*), chlorophyceans (*Spirogyra*), desmids (*Euastrum, Cosmarium, Closterium* and *Micrasterias*) and blue-green algae (*Trichodesmium theibautii*, *Oscillatoria* sp., and *Meresmopedia*), all of which exhibit seasonal fluctuations.

Anyhow, all estuaries have high productivity. As estuary's high productivity is exceeded among aquatic communities only by coral reefs. The reason is that although an estuary may be a harsh environment in some ways, it is not in others. Because the typical estuary is shallow and turbulant, the amount of dissolved oxygen tends to be fairly high although it may be low in bottom layers of those estuaries where water does not mix from top to bottom. More important, the tidal action acts to accumulate (concentrate) the nutrient and energy materials that wash in from upstream or, in some cases, that enter from the nutrient rich bottom waters of the sea. The mechanism of nutrient concentration is very simple–particulate nutrient material enters the estuary at its upper end, is carried seaward by the falling tide and is brought back through the estuary by the rising tide, and so on for several cycles. The length of time it takes from a nutrient particle to traverse the estuary is substantially greater than it would take for it to traverse a similar length of even the most slowly flowing river. Thus, the estuary acts as a nutrient trap, with an average nutrient level significantly higher than either the river or the sea that it connects. Likewise, the concentration of energy-rich organic materials remain high in estuaries.

The primary result of the concentration of nutrients and fixed carbon is a very high level of production within the detritus food chain. This has two aspects. First the nutrient material is broken down bacterially at a very high rate, and recycled back into soluble form. This allows a very high rate of gross primary production for estuarine plants. More important, perhaps, the amount of nutrient-rich organic detritus allows a level of productivity for detritus-eating animals much higher than could be maintained on the basis of the primary production by estuarine plants.

21-1·2. Biotic Communities of Estuaries

Carikker (1967) has classified the regions of estuaries into the **head**, where fresh-water enters the estuary, **upper**, **middle** and **lower reaches** with increasing range of salinities and the mouth with salinity nearly equal to the sea. He has also classified the animals inhabiting the estuarine region into **oligohaline** (0·5 to 5 per cent), **mesohaline** (5 to 18 per cent) in the upper reaches, **polyhaline** (18 to 25 per cent) in the lower reaches, 25 to 30 per cent in the middle and **euhaline** (30 to 50 per cent) in the mouth of the estuary. **Krishnamoorthy** (1963) in fact reported that the extent of penetration of the polychaets, *Marphysa gravelyi*, *Diopatra variabilis*, *Clymene insecta*, *Loimia medusa*, *Glycera embranchiata*, *Onuphis eremita* into the Adyar estuary in Madras from the Bay of Bengal vary with varying salinities at constant temperatures.

Furthermore an estuary being a transitional zone between the fresh-waters and seas is an ecotone and, therefore, typical estuarine forms are unique in their habitat. Thus, the estuarine community is a mixture of three components : the marine, the fresh-water and the brackish water. However, the diversity of both fresh-water and marine components reaches a maximum. Total diversity of the estuarine community is lower than that of either of the more normal environments. The number of interactions between species is not high and some estuarine populations may fluctuate greatly in size. The plants of the estuary are of four basic sorts : phytoplankton, marginal and marsh vegetation, mud-flat algae, and epiphytic plants-growing on the marginal marsh vegetation. Because of the turbid water found in estuaries, phytoplankton are normally uncommon. However, in Hooghly-Matla estuarine system in West Bengal, **Gopalakrishnan** (1971) has reported an abundance of phytoplanktonic forms—several species of diatoms, *Synedra*, *Navicula*, *Rhizostoma*, *Fragilaria*, *Asteriobella*, *Biddulphia*, *Planktoniella*, *Hemidiscus*, *Chaetoceros Cyclotella*, *Stephanodiscus*, *Triceratium* ; several species of green algae like *Pediastrum*, *Spirogyra*, *Eudorina*, *Tribonema*, *Closterium*, *Zygonema*, *Pandorina*, *Volvox*, *Chlorella* ; and blue-green algae like *Microcystis*, *Oscillatoria*, *Anabaena*, and *Trichodesmus*. Most estuarine algae are of marine origin. Further, the most significant estuarine plants are marsh grasses such as *Spartina*,

Salicornia, and *Scirpus*, as well as some submerged filamentous algae such as *Cladophora*, *Chara* and *Enteromorpha*. Very few animals feed on these plants directly. but a very large amount is consumed as detritus.

The estuarine animal communities include zooplanktons and other animals. For example, the planktonic animal forms of Hoogly-Matla estuary of India include flagellate protozoans such as *Euglena*, *Ceratium*, *Peridinium*, *Noctiluca*; other protozoans such as *Difflugia*, *Arcella*, *Vorticella* ; rotifer species such as *Brachionus*, *Keratella*, *Asplanchna*, etc., copepods like *Diaptomus*, *Pseudodiaptomus*, *Cyclops* and *Paracalanus* ; cladocerans of the genus *Bosmina*, *Bosminopsis*, *Ceriodaphnia*, *Moina*, besides a number of isopods (**Gopalakrishnan**, 1971). All these forms remain confined to water and lead a pelagic existence. The best known estuarine animals are detritus feeders like oysters (*Ostrea* sp. of Lake Chilka), clams, lobsters, and crabs. Several insect larvae, annelid worms, molluscs, enter the estuary from the fresh-water, while, most marine phytoplankton, Crustacea, annelid worms, anemones and Bryozoa enter a estuary from marine ecosystems. For example, out of 130 species of fishes and 30 species of prawns of Hoogly-Matla estuary, several fishes such as *Hilsa*, *Harpodon*, *Mugil*, *Trachyurus*, etc., migrate from sea in to the mouth of the estuary, to form important fish catches, while some like palaemonid prawns are fresh-water inhabitants and come to live in estuaries.

Adyar estuary of Madras coast is found rich in invertebrate and vertebrate fauna (**Aiyar** and **Panikkar**, 1937). The common invertebrates of this estuary are sea anemones (*e.g.*, *Phytocoetes gangeticus*, *Phytocoeteopsis ramunnii*. *Stephensonaetis ornata*, *Pelocoetes exul*, *Boloceractis gopalai*, etc.); hermaphrodite nereids (*e.g.*, *Lycastis indica*) ; the tubicolous polychaet *Diopatra variabilis* ; the burrowing polychaet *Marphysa gravelyi*, the oligochaet *Pontodrilus bermudensis* ; the mysids *Rhopalophthalmus egregius* and *Mesopodopsis orientalis* ; the crabs *Uca annulipes* ; *Uca triangularis*, *Neptunus pelagieus*, *Scylla serrata*, *Veruna litterata*, *Sesarma tetragonum* ; the hermit crabs *Clibanarius olivaceus* and *Clibanarius padavensis*, and several molluscs (*e.g.*, *Ostraea arkanensis*, *Modiolus undulatus*, *Meretrix casta*, *Cuspidaria cochinensis*, *Stenothyria blanfordiana*, *Amnicola stenothyroides*, *Potamides cingulatus*, *Styliger gopalai* and *Bursatella leachii*). The vertebrate fauna of Adyar estuary includes the fishes like *Etroplus maculatus*, *Etroplus suratensis*, *Acentrogobius viridipunctatus*, *Acentrogobius neilli*, *Panchax parvus*, *Aplocheilus melastigma*, and mud skipper *Periophthalmus koelreuteri* and snakes such as *Natrix piscator* and *Cerberus rhyncops*.

In short, the estuarine ecosystem is a complex and interesting one. It is also very vulnerable environment, because it has served as conduits for shipping and as sites for cities throughout the history of man. The organisms that inhabit an estuary are adapted to a vastly changing environment, but because estuaries are strategically located, man-made alterations to their ecosystems are substantially greater than in almost any other ecosystem.

21-2. MARINE ECOSYSTEMS

The marine environment of seas and oceans is large, occupying 70 per cent of the earth's surface. The volume of surface area of marine environment lighted by the sun is small in comparison to the total volume of water involved. This and the dilute solution of nutrients limits production. It is deep, in places nearly more than 6 kilometres. All the seas are interconnected by currents, dominated by waves, influenced by tides and characterized by saline waters. Not only the seashore and banks which are the homes of many organisms but the open ocean, many hundreds of kilometres away from land, supports plant and animal communities of great diversity and complexity.

21-2·1. Physico-Chemical Aspects of Marine Environment

In the marine environment, the most important physical factors which influence marine life are light, temperature, pressure, salinity, tides and currents.

21-2·1A. Light. Light is a very significant factor in regulating the pattern of distribution of marine animals and it contributes significantly to organic production. The autotrophic primary producers exploit the light energy in the photosynthetic production of food for primary macroconsumers of marine ecosystem. The amount of light exploited in photosynthesis depends on intensity of light and turbidity of water. For example, a ten metres deep turbid coastal zone receives an equal amount of light energy as a hundred metres deep clear oceanic zone, but the maximum intensity at a ten metres coastal water is greater. Light determines diurnal migrations of marine organisms, and it also regulates colour pattern of marine animals. The deep sea fauna which lives in total darkness exhibit either colourlessness or uniform colouration. It is also somehow related with the development of visual sense organs as shown by absence of functional eyes in deep-sea animals.

21-2·1B. Temperature. The range of temperature in sea is far less than that on land. Arctic waters at 27°F are much colder than tropical waters at 81°F, and currents are warmer or colder than the waters through which they flow. Seasonal and daily temperature changes are larger in coastal waters than in the open sea. The surface of coastal water is the coolest at dawn and the warmest at dusk. In general, sea water is never more than 2° to 3°F below the freezing point of fresh-water or higher than 81°F. At any given place the temperature of deep water is almost constant cold, below the freezing point of fresh-water. Unlike fresh-water, sea water does not have a density maximum at 4°C (39°F); rather it becomes continuously denser as it gets colder.

21-2·1C. Pressure. Pressure in the ocean varies from 1 atm at the surface to 1,000 atm at the greatest depth. Pressure changes are many times greater in the sea than in terrestrial environments and have a pronounced effect on the distribution of life. Certain organisms are restricted to surface waters when the pressure is not so great, where as other organisms are adapted to life at great depths. Some marine organisms, such as the sperm-whale and certain seals, can dive to great depths and return to the surface without difficulty.

21-2·1D. Zonation of marine environment. Just as lake exhibit stratification and zonation, so do the seas. The ocean itself is divided into two main divisions, the **pelagic,** or whole body of water, and the **benthic,** or bottom region. The pelagic region is further divided into two provinces ; the **neritic,** water that overlies the continental shelf, and the **oceanic provinces.** The **continental shelf** is the underwater extension of the continent and it generally extends to a depth of roughly 125-200 m. The edge of the continental shelf may be within a few kilometres of the shore, or it may be several hundred kilometres from shore. From the edge of the continental shelf there is a more rapid descent, the **continental slope** ; to the broad flat **abyssal plain** that underlies most of the ocean at a depth of 4,000—5,000 m. On the abyssal plain, there are extensive mountain ranges, or **midoceanic ridges**, some of which have tips projecting above sea level. In addition there are very deep troughs, which drop down below 11,000 m. below sea level. Life exists from the very top of sea to the bottom in virtually all areas, although its abundance is exceedingly variable.

The benthic life zones are defined in terms of these physical subdivisions. The **littoral**, or **intertidal** zone, is the zone between high tide and low tide levels. The **sublittoral** extends from the low tide mark to the edge of the continental shelf, with the **bathyal zone** comprising the continental slope. The **abyssal zone** includes the abyssal plains, and the **hadal zone** includes any life in the deep trenches below 5,000 m.

The pelagic zone is divided into three vertical layers or zones : from the surface to about 200 m is **the photic zone,** in which there is sharp gradients of illumination, temperature, and salinity. The region below this zone is called **aphotic zone.** From 200 to 1,000 m where very little light penetrates and where the temperature gradient is more even and gradual and without much seasonal variation, is the **mesopelagic zone** of aphotic zone. It contains an oxygen-minimum layer and often maximum concentrations of nitrate and phosphate. Below mesopelagic is the **bathypelagic zones**, where darkness is virtually complete except for bioluminescence, and where temperature is low and pressure is great.

21-2·1E. Stratification of marine environment. The upper layers of ocean water exhibit a stratification of temperature and salinity. Depths below 300 m are usually thermally stable. In high and low latitudes temperatures remain fairly constant throughout the year. In middle latitudes temperatures vary with the season, associated with climatic changes. In summer the surface waters become warmer and lighter, forming a temporary seasonal thermocline.

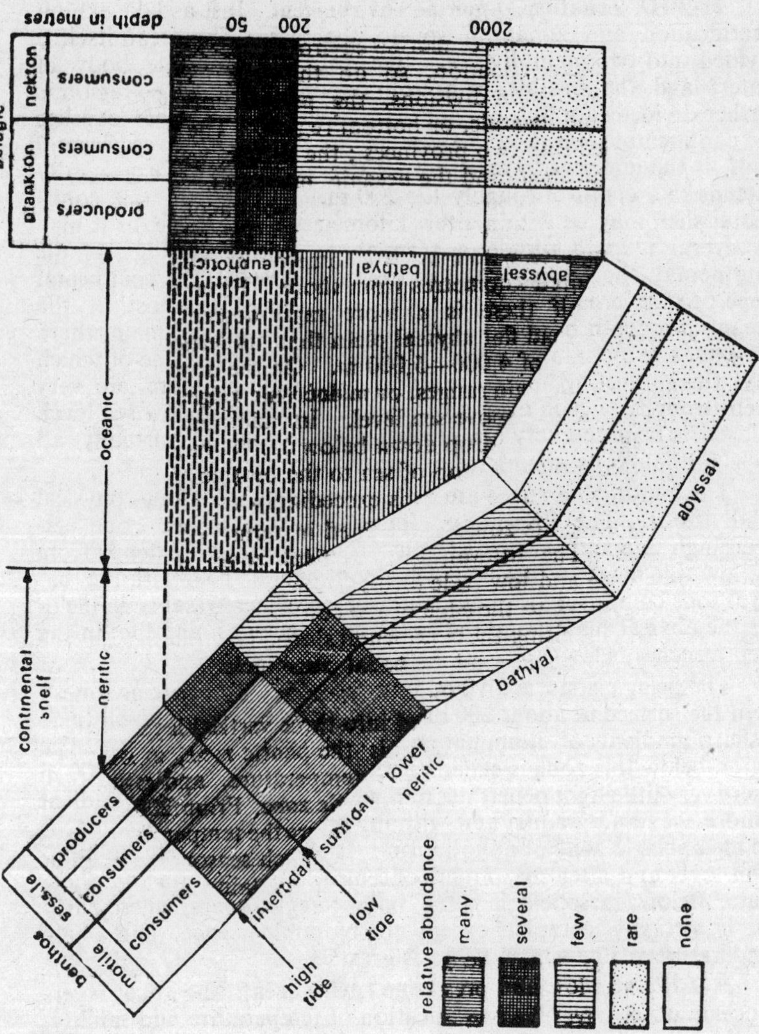

Fig. 21·3 Major ecological divisions of the marine habitat, (after Cockrum and Mc Cauley, 1965).

In subtropical regions the surface waters are constantly heated, developing a marked permanent thermocline. Between 500 and 1,500 m a permanent but relatively slight thermocline exists.

Associated with a temperature gradient is a salinity gradient or **halocline**, especially at the higher latitudes. There the abundant precipitation reduces surface salinity and causes a marked change in salinity with depth. Thus in the middle latitudes in particular, the two produce a marked gradient in density. Water masses form density layers with increasing depth. Because density of seawater does increase in depth and does not reach its greatest density at **4°C** as with freshwater, there is no seasonal overturn. This results in a normally stable stratification of density known as the **pycnocline**. Because there are marked changes in temperatures with depth in the open ocean and thus with density, the pycnocline often coincides with the thermocline. Below the pycnocline is the deep zone which is oxygen rich, cold and comprises 80 per cent of the ocean.

21-2·1F. Salinity. The marine animal life has specific osmoregulatory adaptation for high saline sea waters. The absence of many animal species in marine environment has been related to their inability to tolerate the high salt contents of sea water. Except few insects such as *Halobates* and shore collembolans like *Isotoma* and *Sminthurus*, most insects do not occur in marine ecosystems. The salinity of marine water fluctuates from place to place and it is caused due to the dissolved salts such as chlorides of sodium, potassium, calcium and magnesium, and also the sulphates of calcium and magnesium.

21-2·1G. Currents and tides. Sea water is never static and waves, currents, and tides are the regular features of sea water. All these phenomena are controlled by winds, cosmic forces and varying water densities. Most world-wide water currents occur at the surface and at great depths of sea water. These currents determine the interchange between the surficial and deep water masses, as well as horizontal movements. Both the horizontal and vertical movements of ocean water are significant, but for different reason. Within a given current, such as warm Gulf stream, for instance, the water mass retains its identity for great distances, and the warm-sea community ranges far north of what have been anticipated as its northernmost limits. In addition, the climate of terrestrial ecosystems is strongly affected by the nature of adjacent water masses. The prevailing winds blow across the ocean and are heated or cooled by the waters they traverse. In marine ecosystems, however, the oceanic current system controls the distribution of productivity of marine communities through vertical mixing of water masses. Because most of the ocean is permanently stratified, any essential nutrient that settles out of the upper layers, either as an inorganic mineral or as organic detritus, is not returned to the upper layers at the same location. This had led to such an **impoverishment** of the surficial waters of the oceans that productivity in most parts of the open oceans is about the same as that of desert, because of the very low concentrations of nutrients, especially phosphorus. At the same time, the lowest waters of the ocean are nutrient-rich. There are certain regions in the world, notably at the west coasts of continents and in subpolar latitudes in both hemispheres, where deep currents rise to the surface. These are called **zones of upwelling**, and even though their waters are very cold, they are sites of intense production because of their relatively high concentration of nutrients. Nutrients are then cycled back to the rest of the ocean via surface currents.

Tides are water movements which are caused by some astronomical factors. They represent a rhythmic rise and fall of water and often waves of long wave lengths characterize the process. The chief astronomic force behind the tidal rhythm is the attraction of the moon and sun which depends on the movement of earth in relation to the moon and the sun. Tides influence the sea-shore fauna variously, as low tides expose the shore and high tides flood the substratum.

21-3. MARINE COMMUNITIES

The sea biota is not abundant but contains well marked diversity. Every major group of algae and almost every major group of animals can be found some where in the oceans. The only striking omissions are the vascular plants and insects, which are abundant in estuaries but which have few or no marine representatives. The marine biotic communities can be studied separately for different life zones of the ocean as follows :

21-3·1. Biotic Communities of Oceanic Region

The oceanic region or pelagic zone is less rich in species and numbers than the coastal areas, but it has its characteristic species. Many of these are transparent or bluish and since the sediment-free water of the open sea is marvelously transparent, these animals are nearly invisible. Animals that are too thick to be transparent frequently have smooth shiny and silvery bodies which make them invisible by mirroring the water in which they swim. Further, the animals of the pelagic zone of ocean encounter stable and uniform environmental conditions due to the continuity of the waters of the sea. This continuity of sea waters causes the individuals of the open ocean to stay away temporarily or permanently from any solid objects, and they are provided with different locomotor organs. However, the planktonic forms lack a locomotor organ and float freely in the vast expanse of the sea.

21-3·1A. Biotic communities of pelagic zone. The pelagic plankton also called **epiplankton** are exceedingly diverse. **Phytoplankton** includes diatoms and dinoflagellates, which together produce most of the organic carbon in the sea (and most of the oxygen in the atmosphere), as well as other forms of golden-brown algae and flagellated green algae. Some seaweeds, such as the large brown algae *Sargassum*, may have a **floating** stage. The **zooplankton** of pelagic zone include representatives of every major phylum and most minor phyla, either as permanent members of the plankton community **(holoplanktonic forms)** or as transients during their larval stages **(meroplanktonic forms)**. Common among permanent planktonic forms are the Forminifera and Radiolaria, arrow worms (*Sagitta*), certain annelid worms, swimming snails, jelly fishes and most abundant

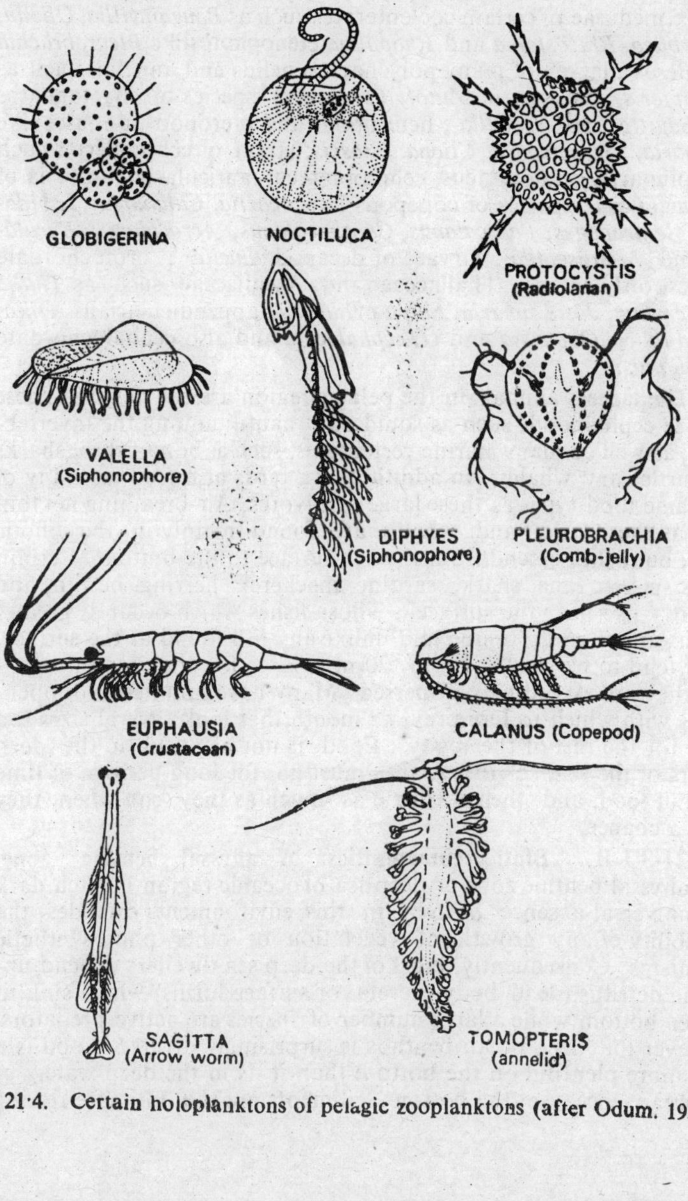

Fig. 21·4. Certain holoplanktons of pelagic zooplanktons (after Odum. 1971).

of all are the crustaceans such as shrimps, copepods and cladocerans. Among the temporary zooplanktons are the larva of larger animals from all marine environments and even some fresh-water environments. For example, in the zooplanktons of inshore waters of Mandapam (South India) **Prasad** (1956) reported following animal species—flagellate protozoan *Noctiluca miliaris*; larval forms of coelenterate like *Planula*, *Ephyra*, *Semper's* larva and *Cerianthus* larva ; medusae of certain coelenterates such as *Bougainvillia*, *Obelia*, *Charybdea*, *Rhizostoma* and *Rhopalina*;**ctenophores**like *Pleurobrachia* and *Beroe* ; larvae of many polychaet annelids and annelids such as *Tomopteris*, *Alciopa* and *Autolytus*, several species of chaetognaths like *Sagitta* and *Spadella* ; heteropod and pteropod molluscs like *Carinaria*, *Pterotrachea*, *Cliona*, *Creseis*; larvae of echinoderms such as bipinnaria, ophiopluteus, echinopluteus, auricularia; tornaria of *Balanoglossus* ; species of copepods like *Acartia*, *Calanopia*, *Labidocera*, *Centropages*, *Paracalanus*, *Canthocalanus*, *Acrocalanus*, *Eucalanus* and *Microsetella* ; larvae of decapod *Lucifer* ; protochordate species of Larvacea, Thaliaceae and Ascidiaceae such as *Thalia democratica*, *Jasis zonaria*, *Salpa cylindrica*; appendicularian—*Oikopleura diocia*,*O. parava* and *O. cephalocera* and also cephalochordate *Amphioxus*.

The largest animals in the pelagic region are the nekton. These include cephalopods such as squid and nautili among the invertebrates, as well as many marine vertebrates, such as bony fishes, sharks sea turtles and whales. In addition, sea birds also feed on many of the same food types as these large carnivores. Air-breathing nekton, such as the turtles, and whales, are found mainly in the photic zone, but fishes extend from the sea surface to the bottom. Certain fishes such as tuna, shark, sardine, mackerel, herring, bonito, and anchovy live near the surface. Those fishes which occur at greater depths are often grotesque and unlike any fish found at the surface. They tend to be small—15 to 20 m is large for the ocean deeps—and they are exceedingly dispersed. Many have luminescent appendages with which to lure prey, or mouth that look several sizes too large for the rest of the body. Food is not plentiful in the deep waters of the sea, so these fishes must go for long periods of time without food, and then consume as much as they can when they have a chance.

21-3·1 B. Biotic communities of abyssal benthic zone. The abyssal benthic zone or deep sea of oceanic region is pitch dark and universal absence of light in this environment excludes the possibility of any growth of vegetation or other photosynthetic organisms. Consequently, most of the deep sea dwellers depend upon the detritus (dead body excreta of surface forms) which sink to the sea bottom while a large number of species are active predators. However the deep ocean benthos is surprisingly diverse. Food is a little more plentiful on the bottom than it is in the deep waters of the ocean, because the botton sediments are the ultimate resting place for all detritus raining out of the upper layers (oceanic surface and pelagic zones). The bottom of the sea is a soft ooze, made of the organic remains and shells of Foraminifera, Radiolaria, and other animals and plants. Sea cucumbers, brittle stars, crinoids (sea lilies), sea urchins, certain benthic fishes, and several types of crustaceans, as well as sea anemones, clams, and similar animals are all

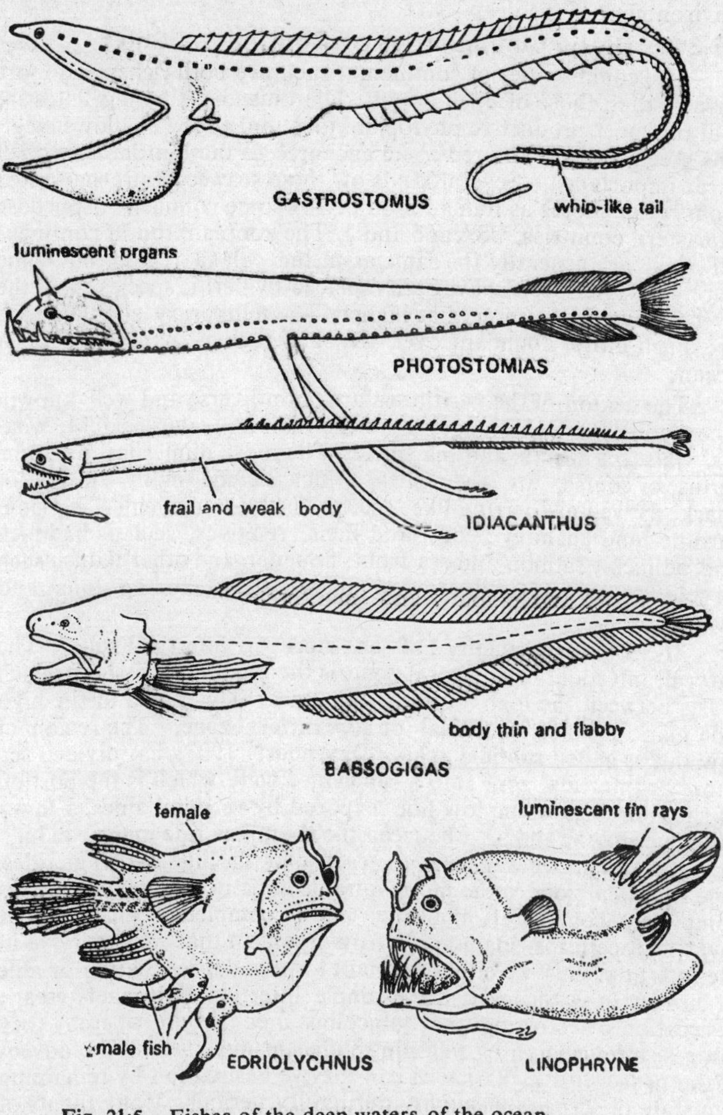

Fig. 21·5. Fishes of the deep waters of the ocean.

found on the bottom, not only on the abyssal plains, but some even at the bottoms of the deepest ocean trenches. Many of them are detritus feeders but a large number of them are carnivores. In fact, the biomass of carnivorous brittle stars is often higher than that of the detritus feeders that serve as their food source; however, their metabolic rate is much lower. The great diversity of sea bottom fauna has been related with great constancy in the physical environment of the deep sea.

21-3·2. Biotic Communities of Continental Shelf

The communities of continental shelf are both richer and more diverse than those of open ocean. Diatoms and dinoflagellates are still the most productive phytoplanktons, but in the shallower regions green, brown, and red algae anchored to the bottom may be of great importance. Several kinds of these seaweeds are harvested from rocky shores as human food or for some commerical purpose in eastern countries, USA and India. The zooplankton of continental shelf are generally the same as in the pelagic region, but some purely pelagic species have been replaced by neritic species, and the overall diversity is somewhat higher. The temporary zooplanktons are much more abundant over the shelf region than in the open ocean.

The nekton of the **neritic sea** are both diverse and well known. The significant nekton species of the region are large squid, whales, seals, sea **otters**, and sea snakes. The most numerous nektonic forms, of course, are the fishes, which include many species of shark as well as herring-like species such as menhaden, herring, sardine and anchovy ; cods and their relatives, such as haddock and pollack ; salmon and sea trout; flounder and other flatfish such as sole, plaice and halibut ; and mackerels, including tuna and bonito.

21-3·2A. Communities of sea shore or **intertidal zone**. The littoral, intertidal or eulittoral zone, is the region of sea shore which exists between the high and low tide lines. The region of the high tide mark is called **supratidal** or **supralittoral zone**. The region of low tide is called **subtidal zone**. **Davenport** (1903) has divided sea shore or intertidal zone into a **submerged zone** which is the portion of the shore below the low tide exposed by very low tides, **a lower beach** occupying the area between the mean low tide and mean high tide and an **upper beach** which is reached by the highest of high tides. The intertidal zone is the most variable zone in the entire sea. It is completely covered at high tide, and is completely uncovered at low tide except for tide pools. Any organism that is to survive in the intertidal must be either resistant to periodic desiccation or able to burrow to water level. For example intertidal polychaets escape desiccation by resorting to a tubicolous mode of life whereby they can resist desiccation by retaining sufficient moisture till the advent of the next wetting. Barnacles can survive desiccation by remaining in a state of suspended activity during dry periods. Many plants of sea shore contain jelly-like substances such as agar, which absorb large quantities of water and retain it while the tide is out. Further, the majority of shore animals utilize oxygen dissolved in water and when the low tide exposes them during the periods of drought, they utilize the oxygen stored in the blood plasma, *e.g.*, nereid like, *Arenicola*. Barnacles achieve this by enclosing an air bubble between the shells.

One of the outstanding characteristics of this region is the ever present action of the waves, and the organisms living on a sandy or rocky beach have had to evolve ways of resisting wave action. The many seaweeds have tough pliable bodies, able to blend with the waves without breaking, while the animals either are encased in hard calcareous shells, such as those of molluscs, Bryozoa, starfish, barnacles and crabs, or are covered by a strong lethery skin that can bend without breaking, like that of the sea anemone and octopus. Many sedentary shore animals (*e.g.*, sponges, tunicates, etc.) have special modes of adhesion which keep them firmly attached to the substratum. However, animals of sandy beaches escape wave action by taking burrowing mode of life, because, these lack a hard substratum for the attachment of the animals.

The biota of intertidal zone also have to overcome the wide fluctuations in salinity and temperature. Thus, like the estuary, this variable environment is a zone in which an organism must be adapted to a broad range of environmantal conditions. Like the estuary also, it is an area of very high productivity with a simple community many of whose members may be exceedingly abundant. The diversity of biota of intertidal zone is determined by nature of substratum whether loose sand or mud or rocky coast. Ecologists have recognized three major types of intertidal zones or seashores, namely, **rock, sandy** and **muddy. Eltringham** (1971) has included a fourth type of sea shore, the **pabble shore** in this classification. Each type of sea shore possess specific biota each of which remains specifically adapted for its peculiar habitats. Here we will discuss only rocky shore and sandy beach habitats.

1. Biotic communities of rocky shore. The rocky shore presents solid substratum for the attachment of many sessile animals which often remain abundant here. The animals which are sedentary and inhabit rocky shore are limpets like *Patella, Haliotis, Fissurella*, and oysters, barnacles, tunicates and bryozoans. Sessile organisms like the sponges, the colonial hydrozoans, anthozoans like *Gammaria* and *Zoanthus* attached to rocks. Certain animals such as some sponges, annelids and molluscs either bore the soft rocks like limestone or seek protection in the crevices of hard rocks. All these sedentary sessile animals are adopted for filter feeding.

Further, all these animals of rocky shore occurs in successive zones (Fig. 21·6). At the uppermost end is a zone of bare rock marking the transition between land and sea. Next is a spray zone with dark patches of algae on which the periwinkles (*Littoria*) graze.

Below this is the zone called barnacle zone regularly covered by the high **tide** ; rocks in this zone are encrusted with acorn barnacles, limpets and mussels (*e.g.*, *Mytilus*). This zone is a vulnerable zone being exposed to the brunt of wave action and active predation by starfishes and gastropods. The mussels act like hard substratum for tunicates, sponges and small mussels. The next one is the zone of oysters is less vulnerable and includes green algae (*Enteromorpha* and *Ulva*) and barnacles. The next zone includes mussels, *i.e.*, boring bivalves, which remain within crevices with exposed siphons only to be protected from waves and predation. The chief species of borer bivalves are *Pholas*, *Martesia*, *Hiatella*, *Tridacna*, etc. Besides bivalves a variety of filter feeder bryozoans,

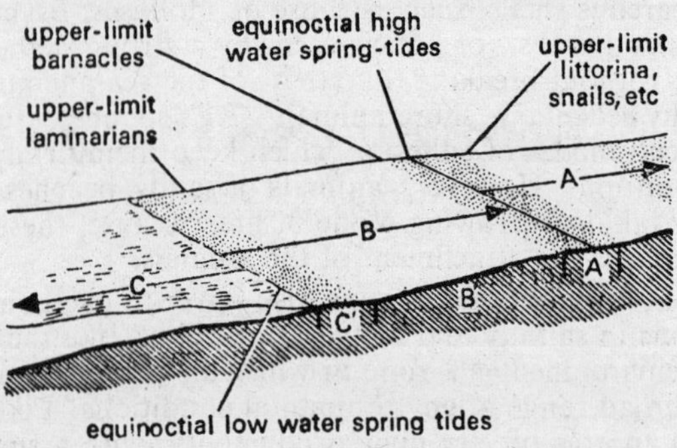

Fig 21·6. Basic zones on a rocky shore (after **Smith**, 1974).

and brachiopods and tubicolous polychaets like serpulids, terebellids, sabellids and cirratulids which live in tubes are the chief inhabitants of this zone. The subtidal zone include red algae, brown algae (Laminarians and fucoids), sea anemones, sea urchins, corals, etc. (Fig. 21·7 and Fig. 21·8).

For example, the Okha coast of Gujrat in India is a typical rocky sea coast which has well defined fauna and flora **(Gopalakrishnan**, 1970). Dominant weeds of subtidal zone of Okha coast are *Sargassum*, *Dictyopteris*, *Graciliaria*, *Padina*, *Ulva*, *Polysiphonia*, etc. This subtidal zone also contains sea urchins, serpulid tubes of species *Pomatoleius crosslandi*, and mollusc species like *Trochus radiatus Cypraea arabica*, *Thais alveolata*. The upper **limit** of the dense weed growth of subtidal zone marks the lower end of eulittoral or intertidal zone, where occurs a dense cluster of *Crassostrea cucullata* and *Littorina* sp. The intertidal zone includes three subzones, *i.e.*, 1. the subzone of patchy weed growth; 2. the subzone of limpets like *Sellana radiata* and barnacles and 3. the subzones of oysters and barnacles. The supratidal zone is inhabited by green algae and molluscs. Chlorophyceae like *Enteromorpha*, *Ulva*, *Bryopsis*, *Codium*, *Halimeda*, *Caulerpa*, *Dictyosphaeria* and *Cladophora*, Phaeophyceae like *Padina*, *Sargassum*, *Colpomenia*, *Iyengaria*, *Spathoglossum*, *Dictyopteris* and *Turbinaria*, ; Rhodophyceae like

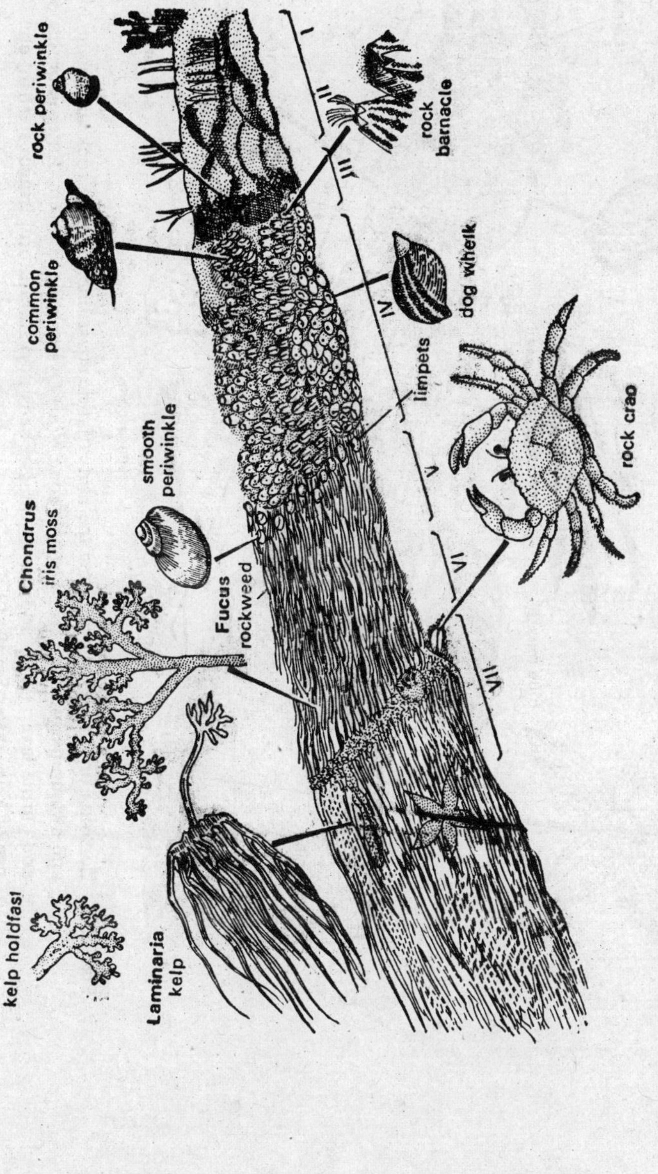

Fig. 21·7. Zonation on a rocky shore along the North Atlantic sea : (i) Land: lichens, herbs, grasses, etc. (ii) Bare rock; (iii) Zone of black algae and rock periwinkles; (iv) Barnacle zone : barnacles, dog whelks, common periwinkles, musseles ; (v) the Fucoid zone: rock weed and smooth periwinkles; (vi) Chondrus zone : Irish moss ; (vii) Laminarian zone : kelp (after Smith, 1974).

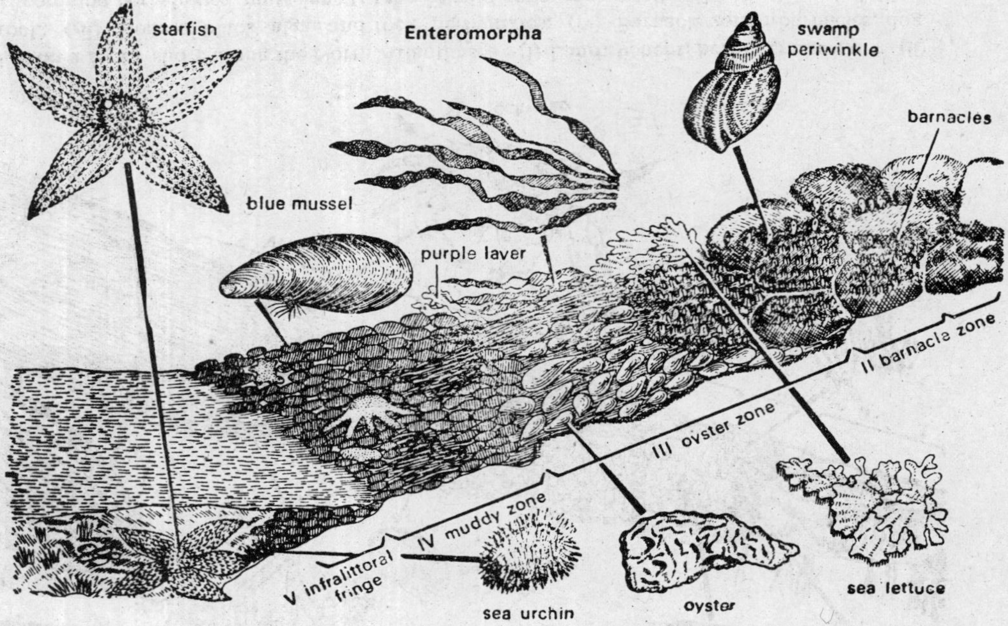

Fig. 21·8. Zonation on a rocky shore along the mid-Atlantic coast line: (i) Bare rock with some black algae and swamp periwinkle ; (ii) Barnacle zone ; (iii) Oyster zone : oysters, Enteromorpha, sea lettuce and purple lover; (iv) Muddy zone : mussel beds ; (v) Infralittoral fringe: starfish; etc. Note absence of kelps (after Smith, 1974).

Gracilaria, Hypnea, Acanthophora, Polysiphonia, Cryptonemia are the chief algae of the area. The common animals of this area are coelenterate species *Stoichactis, Ixalactis,* and *Gyrostama* ; Nereid like *Eunice, Tubifex*; molluscs like *Chiton, Nerita rumphii, Cypraea ocellata, Cypraea arabica* seaweed grazer (nudibranch) molluscs, *Aplysia benedicti, Onchidium, Elysia gradifolia, Eolis, Doris* and floating mollusc *Janthina*, etc. The midlittoral reef community of Okha coast consists of fishes like *Epinephelus, Petrachus,* the crab *Pilumnus vespertilio,* the annelid *Eurythoe complanata* and molluscs like *Murex, Bursa,* etc. Rock crevices of intertidal zone harbour within them crabs like *Atergatis, Grapsus, Pilumnus, Matuta, Charybdea, Gelasimus,* squillids like *Sesarma* and *Gonodactylus* ; coelenterates like *Stoichactis, Ixalactis, Zoanthus, Lobophytum* ; the hydroid species like *Lytocarpus, Sertularia* and *Plumularia* ; and annelids like *Serpula, Polynoe, Eurythoe, Sabella,* etc.

2. Biotic communities of sandy shore. The sandy shore may be even more harsh than the rocky shore. It is subjected to all the extremes of the latter (temperature, salinity, turbidity, wave action, etc.) plus inconvenience of a constantly shifting substratum. The last makes life on the surface almost impossible; life has retreated below the surface. Generally sandy beaches are characterized by gentler wave action. Because of the prolonged time taken for drying up, these beaches are suitable for animal life. Decaying seaweeds and dead remains of animals result in the addition of organic matter to the sand. The coastal subsoil water is the environment of a special interstitial brackish water fauna. Most of the interstitial organisms are small, elongate, vermiform with transparent bodies. They have no eye but possess well developed adhesive organs and sense organs. They are negatively phototactic and gregarious in habit. They may be herbivore, carnivore, omnivore or detritus feeder. Large animal forms which are few in number prefer coarse sand grains and smaller one which are very abundant, prefer finer sand. The principal animal groups of interstitial water are the Nematoda, Turbellaria, Annelida, Gastrotricha and Acrania. Certain animal groups such as Protozoa, occur sporadically. The top 20 cm of sand between low and mid-tide levels of the beach is occupied by the inhabitants such as Hydrozoa, Turbellaria, Nemertina, Rotifera, Archiannelida, Polychaeta, Ostracoda, Halacaridae and Nudibranchiata. Species of Gastrotricha, Kinorhyncha, and Isopoda occur at mid-tide levels preferably in deep layers of sand. Oligochaets are sporadic and are found at different levels. Harpactcoid copepods remain restricted to mid-tide level. Tardigrades occupy the region between mid-tide and high-tide. Thus, sandy beaches also possess characteristic successive zones or zonations of animals (Fig. 21·9).

For example, according to the studies of **Ganapathy and Rao** (1962) sandy beach fauna of Visakhapatnam coast though lacks or contains small number of tubicolous polychaets and Crustacea

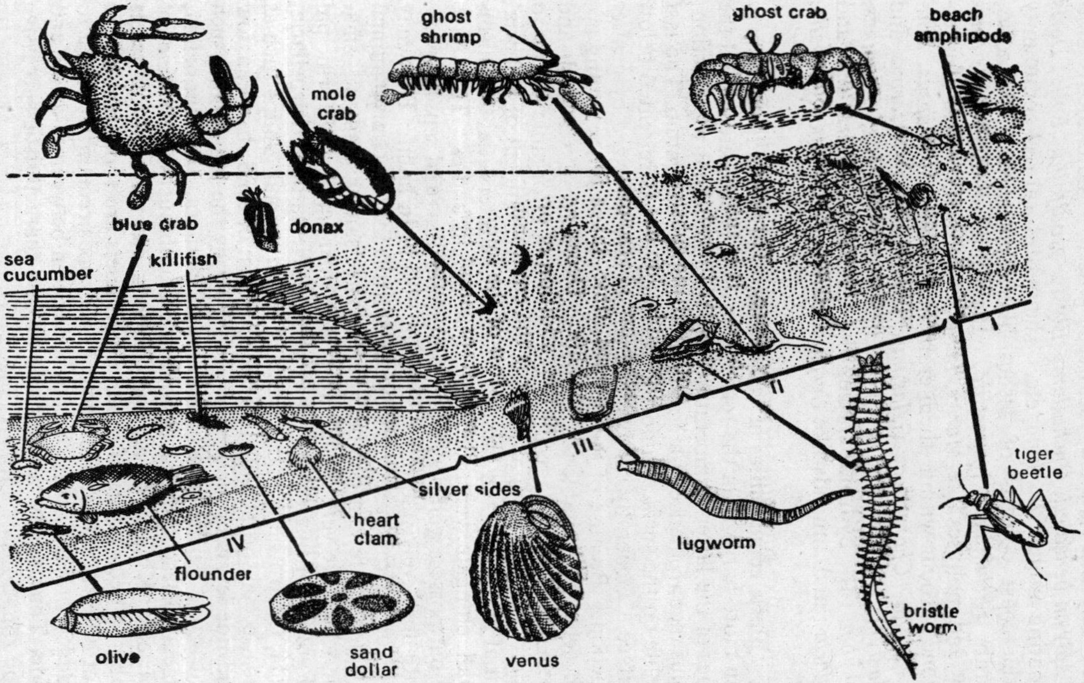

Fig. 21·9. Life on a sandy ocean beach along the Atlantic coast. (i) Supratidal zone : ghost crabs and sand fleas; (ii) flat beach zone : ghost shrimp, bristle worms, clams: (iii) Intertidal zone : clams, lugworms, mole crabs; (iv) Subtidal zone. The dotted line indicates high tide (after Smith, 1974).

due to very unstable substratum, but include following animal species—spinoid *Prionospio krusadiensis*, *Aricia* and *Lumbriconeries*, all of which possess the capacity to secrete mucus and formation of cover of sand grain over the body for protection; *Glycera lancadivae*, *G. alba*, *Nerina bonnieri*, *Pisionidens indica*, *Pisione complexa* and archiannelids *Saccocirrus minor* and *Protodrilus* also occur. The crustacean fauna of this zone includes *Emerita asiatica*, *Albunea symnista* (burrowing), amphipod species *Harpinia* sp., isopod *Sphaeroma walkeri*, mysid shrimp *Gastrosaccus spinifer*, hermit crabs *Clibanarius aretheustus* living within the shells of *Thais* and *Cerithium*, the eight-oared swimming crab *Matuta victor*, the spider crab, *Philyra scabruescula*, *Ocypoda platytarsis* and *O. macrocera*. Lamellibranchs like *Donax cuneatus* and *Pamphia textile*, burrowing snails like *Sinum neritoideum*, common gastropods like *Oliva gibbosa* and *Terebra* sp., *Bullia vittata* are also found along the coast of Visakhapatnam. This beach also contains eel *Ophichtys* and siphonophore coelenterates like *Physalia*, *Velella* and *Porpita* ; and medusae-like *Aequoria* and *Chiropsalmus*.

21-4. CORAL REEF AS A SPECIALIZED OCEANIC ECOSYSTEM

Coral reefs are the specialized ecosystems of ocean which are among the most productive of all ecosystems anywhere, with a diversity equalled only by tropical rainforests. The reef environment is very rugged one because it is constantly subject to the pounding of the waves, and parts of the reef are often above water at low tide. But reef is also a very rich environment for those species that can withstand the physical pounding of water. The amount of oxygen is very high. During the day, it may reach 250 per cent of saturation because of the production of oxygen by algae in the reef structure. Likewise, the tremendous amount of CO_2 produced by the metabolism can be carried off before it becomes too high.

Reefs can be found wherever a suitable substrate exists in the lighted waters of the tropics, away from areas of continental sediments which would tend to bury the reef in mud or cold upwellings which would tend to lower the temperature too much. They may form **barrier reefs** (*e.g.*, barrier reef in Australia which contains canal or lagoon between them and coast) adjacent to continents ; **atolls** or ring-shaped or horse-shoe-shaped islands on the top of submarine heights, or **fringing reefs**, as in Mandapam, Appa Islands and Hare Islands. In all modern reefs, the main reef structure is formed by madreporarian corals of following familes-Poritidae (*Porites minicoiensis*, *Porites mannarensis*), Acroporitidae (*Acropora intermedia*), Astraeidae and Milliporidae. Secretions of calcareous algae go into the building up of reefs. Equally important are calcareous Bryozoa, lamellibranchs and gastropod molluscs and Forminifera, which become significant elements in the total reef structure. Although the name 'coral-reef' implies that corals should be the most abundant organisms, algae of many sorts comprise over three times the biomass of the corals. In addition, the calcareous algae are not only producers; their skeletons act to cement the entire reef to make it more resistant to the pounding surf.

Coral reefs normally exist in water with mean temperature 20·5°C, but they can withstand very high temperatures. The growth of coral reef is influenced by salinity and reduced tides. Moderate water movement around the coral reefs help in the circulation of food.

21·4·1. Biotic Communities of Coral Reef

Coral reef ecosystems harbour an abundant and well diverse biota. The well known Krusadi Island is the coral island of gulf of manaar. It's dominant corals are *Porites*, *Montipora*, *Poecilopora*, *Acropora*, *Madrepora*, *Favia*, *Symphylia*, *Psammocora*, *Platygyra*, *Merulina* and *Leptastrea*. In the Krusadi area a luxuriant growth of sea weeds such as *Sargassum*, *Caulerpa*, *Turbinnria*, *Spatoglossum*, *Gracilaria*, *Ulva*, *Enteromorpha*, *Chaetomorpha* and *Polysiphonia*. The abundant sponges of this reef are *Spirastrella incorstans* and *Callyspongia fibrosa*. The other inhabitants of this reef are following—hydroid coelenterates like *Syncoryne*, *Sertularia*, *Eudendrium*, *Pennaria*, *Tubularia*, *Thyroscypus*, *Plumularia* ; alcyonarians like *Clavularia*, *Alcyonium*, *Lobophytum*, *Sclerophytum*, *Telesto*, zooantharians like *Zoanthus*, *Gemmaria* ; polychaete annelids like *Nereis*, *Polynoe*, *Eunice*, *Lepidonotus*, *Iphione*, *Eurythoe*, *Hesione* ; sipunculid species like *Dendrostomum*, *Aspidosiphon* and *Phascolosoma*; molluscs like *Doris*, *Dendradoris*, *Acanthochiton*, *Ischinochiton*, *Haliotis*, *Emarginula*, *Trochus*, *Turbo*, *Nerita*, *Natica*, *Cypraea*, and echinoderms like *Holothuria atra*, *H. scabra*, *H. edulis* and starfish *Proteaster lincki*.

Beautifully pigmented fishes like *Holocentrus*, *Lutianus* sp., *Pomacanthoides*, *Chaetodontopsis*, *Chaetodon*, *Linophora*, *Thallasoma* and *Zanclus* and the eels *Gymnothorax undulatus* and *G. punctatus* are commonly frequent in the coral reefs. These inhabitants of coral reef community, when exposed to lowtides are subjected to predation by large number of birds like the crested tern *Sterna bergii*, golden plover *Pluvialis dominica fulva*, etc.

According to **Clapham, Jr. (1973)** the great productivity of reefs is obviously caused by something more than the abundance of light and oxygen. He considers nutrient cycles which run very quickly and efficiently between the components of reef ecosystem to be main cause of great productivity of reef ecosystems. In fact, no organic matter is incorporated permanently into the sedimentary system portion of reef. The conditions of rapid nutrient cycling allow a mechanism by which the natural scarcity of nutrients in the ocean can be effectively circumvented. It may also allow a gradual increase in nutrient availability by incorporating the nutrients of the non-reef organisms that wash into the reef, while still continuing to prevent the reef's own nutrients from being lost to the deeper seas.

21·5. REVISION QUESTIONS

1. What are the physico-chemical characteristics of marine environment ? Classify the marine environment and describe in details the flora and fauna of the pelagic region.
2. Describe physico-chemical characteristics and biota of estuaries.
3. Describe the "coral reef as a specialized oceanic ecosystem."
4. Write short notes on the following :
   (a) Sublittoral zone ;
   (b) Deep sea fauna ;
   (c) Intertidal sandy shore.

22
Terrestrial Ecosystems

Only about a quarter of the earth's surface is dry land, yet the complexity and variegation of terrestrial ecosystems are much greater than those of aquatic ecosystems. There is no unifying theme in terrestrial ecosystems analogous to the physical properties of water. Rather, the variety of climates, the diversity of all lithosphere, and heterogeneity of terrestrial biotic communities all conspire to give rise to such a variety of themes that they are woven together in nature to give an incalculable diversity of biotic communities and terrestrial ecosystems.

22-1. PHYSICO-CHEMICAL NATURE OF TERRESTRIAL ECOSYSTEMS AND THEIR COMPARISON WITH AQUATIC ECOSYSTEMS

An aquatic system is essentially a **single-phase system**, where water sets the tone for entire habitat. A terrestrial ecosystem, on the contrary, is a **three-phase system**, where the characteristics of the habitat are a function of the atmosphere and climate, the soil, and the biotic community itself. The atmosphere is the source of oxygen for animals and carbon dioxide for plants. Both are sufficiently common and well-mixed in the atmosphere that they are essentially never limiting. Air as a medium of support is much less buoyant than water. Thus, an organism need relatively little support to survive in water, but it needs a fairly rigid skeleton to live and move on land. Climate in a terrestrial ecosystem is much more variable than it is in water. The most important aspects of climate are the temperature and water relations of ecosystems. These relations include mean annual temperature, amount of temperature fluctuation, annual rainfall, degree and time of fluctuation in rainfall and potential evapotranspiration. Rainfall tends to be somewhat irregular even in the most predictable terrestrial ecosystems, and in arid regions or during periods of drought, lack of water may be a source of extreme stress. The temperature is much more variable on land than in water. This is caused by the fact that aquatic ecosystems are heated by absorption of radiation by water, and water has a high specific heat ; terrestrial ecosystems, on the other hand, are heated by the absorption of heat by soil, rock, and vegetation, whose specific heat are much lower. A given quantity of heat energy absorbed by solid material can change its temperature two to five times as much as the same energy absorbed by water. Thus heat is gained and lost rapidly by terrestrial ecosystems, with a resulting wide fluctuations of temperature, both diurnally and seasonally.

Table 22·1. Specific heat of several common substances, (Clapham, Jr., 1973).

1.	Fresh water	1·00
2.	Sea water	0·93
3.	Wet mud	0·60
4.	Moist sandy clay	0·33
5.	Solid rock	0·20
6.	Copper	0·09

In terrestrial ecosystems, the soil serves two major functions—it provides support for living organisms, and it is the source for all essential nutrients except for carbon, oxygen, and hydrogen (for detail see chapter 9). Even its supportive role is different from the way in which a lake sediment supports rooted vegetation. Because the entire weight of all living organisms is borne by the soil, there is no partition of support between the soil and the air. The role of soil as a source of nutrients is unique to terrestrial environments. It is the site of the entire detritus food chain and thus is central to the biogeochemical cycling of nutrient materials. Different types of soils have different properties that affect the availability of nutrients to plants, so the productivity of the terrestrial ecosystems is very closely tied to the chemistry of the soil. Further, in terrestrial biotic communities interactions between species are real but they are simple, involving relationships like predation, mutualims, parasitism and the like. In contrast to aquatic ecosystems, living organisms of terrestrial ecosystems leave a permanent mark on the system. For example, the role of plants in breaking down and weathering rock and building soil is much more pronounced than it is in almost all aquatic ecosystems. In addition, the succession that takes place in a terrestrial community is almost entirely a function of the organisms within the ecosystem, unlike aquatic successions such as lake eutrophication, which are dependent in large part on materials being washed in from out side.

Lastly, in each of these two ecosystems, namely aquatic and terrestrial ecosystems, several types of major subdivisions can be recognized : thus one can distinguish fresh-water, estuarine and marine aquatic ecosystems, and several major types of terrestrial ecosystems such as prairies, forests, tundra, etc. The former are distinguished on the basis of a major chemical difference (*i.e.*, salt content), the latter generally on the basis of the predominant type of vegetation (grass, tree, etc.) (see **Kormondy**, 1976).

The earth surface—the continental land masses have been classified by biogeographers and ecologists into following regions :

(i) Biogeographic realms;

(ii) Biomes.

22-2. BIOGEOGRAPHIC REALMS OR REGIONS

Careful studies of the distribution of plants and animals over the earth have revealed the existence of six major biogeographic realms, namely **Palaearctic realm, Nearctic realm, Neotropical realm, Ethiopian realm, Oriental realm** and **Australian realm**, each characterized by the presence of certain unique organisms. Each of these realms embraces a major continental land mass and each remains separated by oceans, mountain ranges, or desert (For details see chapter 23).

22-3. BIOMES

Within these biogeographic realms and established by a complex interaction of climate, other physical factors and biotic factors, are large community units called **biomes.** A biome is a large community unit characterized by the kinds of plants and animals present. This may be contrasted to the ecosystem, which is a natural unit of living and non-living components that interact to form a stable system in which the exchange of materials follows a circular path. Thus, an ecosystem might be a small pond or a large area coextensive with a biome, but it includes the physical environment as well as populations of animals, plants and micro-organisms.

In each biome the **kind** of climax vegetation is uniform—grasses, conifers, deciduous trees—but the particular **species** of plant may vary in different parts of the biome. The kind of climax vegetation depends upon the physical environment and the two together determine the kind of animals present. The definition of biome includes not only the actual climax community of a region, but also the several intermediate communities that precede the climax community. Further, there is usually no sharp line of demarcation between adjacent biomes ; instead each blends with the next through a fairly broad transition region termed as **ecotone.** There is, for example, an extensive region in Northern Canada where the tundra and coniferous forest blend in the tundra-coniferous ecotone. Some of the biomes recognized by ecologists are **tundra biome, alpine biome, forest biome, grassland biome, desert biome.**

22-3·1. Tundra Biome

Tundra presents the most common example of "fragile ecosystem" (**Clapham, Jr.**, 1973). Tundra which means "marshy plain", lies largely north of latitude 60°N (*i.e.*, between the Arctic ocean and polar ice-caps and the forests to the south) and is characterized by the absence of trees, the presence of dwarfed plants, and an upper ground surface that is wet, spongy and uneven, or hummocky, as a result of freezing and thawing of this poorly drained land (**Kormondy**, 1976). Some five million acres of tundra stretch across Northern America, Northern Europe and Siberia. Although there is variation from place to place within the biome, temperature, precipitatian, and evaporation are characteristically low, the warmest months averaging below 10°C and the wettest with about 25 mm of precipitation. Despite the small amount of precipitation, water is usually not a limiting factor because the rate of evaporation is also very low.

The ground usually remains frozen except for the uppermost 10 to 20 cm, which thaw during the brief summer seasons. The permanently frozen deeper soil layer is called **permafrost**. The permafrost line which may exist at a depth of a few centimetres to several metres, is the ultimate limit of plant root growth, but the immediate control is the depth to which soil is thawed in summer. The rather thin carpet of vegetation of tundra biome includes few species : grasses and sedges are characteristic of numerous marshes and poorly drained areas, but large areas consist of ericaceous heat plants (bilberries and dwarf huckleberries), low flowering herbs, and lichens. Perhaps the most characteristic arctic tundra plant is the lichen known as "reindeer moss" (*Cladonia*). The animals that have adapted to survive in the tundra are caribou or reindeer, the arctic hare, arctic fox, polar bear, wolves, lemmings, snowy owls, ptarmigans and during the summer, swarms of flies, mosquitoes and a host of migratory birds.

Further, certain animals such as caribou and reindeer are highly migratory because there is not enough vegetation produced in any one local area to support them. Though tundras are barren areas yet a surprisingly large number of organisms have become adapted to survive the cold. During the long daylight hours of the brief summer, the rate of primary production is quite high. The production from the vegetation on the land, from the plants in the many shallow ponds that dot the landscape and from the phytoplankton in the adjacent Arctic Ocean provides enough food to support a variety of residant mammals and many kinds of migratory birds and insects. Diversity is very low, and despite the extreme hardiness of the tundra organisms, their growth rate tends to be very low, and the community as a structural unit is exceedingly vulnerable.

Moreover, Alpine tundra is quite similar to some of Arctic tundra, but in the absence of permafrost and in the growing season, mosses and lichens are less prominent, flowering plants more so.

22-3·2. High Altitude or the Alpine Biome

The region of mountain above the timber line contains a distinct flora and fauna and is referred to as the **alpine zone**. Alpine zone remains conspicuous in those mountains whose peaks reach up to the niveal or snow zone, as in the Himalayas. The alpine zone (zone which lies between timber line and snow zone) includes in the descending order, a **sub-snow zone** immediately below the snow zone a **meadow zone** in the centre and a **shrub zone** which gradually merges into the timber zone. According to **Mani** (1957) snow zone of Himalayas lies over 5100 m above mean sea level and alpine zone exists at a height of 3600 m. From an ecological view point, the zone above the limits of tree growth (timber line) exhibits extreme environmental conditions which greatly influence the biota of this region.

The characteristic features of high altitude environment are following—a low air density, low oxygen and carbon dioxide contents and water vapour, high ozone content, high atmospheric transparency affording greater penetration of light, cold, snow cover, increased rate of desiccation, high wind velocity, insolation of high intensity during the day and rapid radiation during the night, the high glare from the sky and snow, high intensity of ionizing radiation and the exclusion of trees. Alpine zone of Himalayas is characterized by a sparseness of animal groups, the scarcity being relatively important as far as tropical Indian, South Chinese, Indo-Chinese and Malayan derivatives are concerned. Important constituents of the fauna of this zone are cold-adapted palaearctic forms (**Mani**, 1968).

Many invertebrates of alpine zone are predatory and occur in lakes, streams and ponds. Among vertebrates fishes and amphibians are totally lacking and reptilian fauna is greatly impoverished. However, pelobatid frog, *Aleurophryne mammata* is found to occur at 4500 metres of height in Himalayas and viviparous *Mabuya varia* has been reported to live at 4000 metres of height in Mount Kilimanjaro. Some of the representative vertebrates of Himalayan alpine zone are—crow *Corvus corax* ; snow partridge *Lerwa nivicola* ; snow leopard *Felis unica* and *Felis lynx* ; European leech marten *Mustela foina*; the vole *Lagomys*; Tibetan yak *Bos grunnius*; Tibetan sheep *Ovis hodgsoni* ; pamir sheep *Ovis poli* ; ibex *Capra sibricia* ; markhor *Capra falconeri* ; and Persian wild goat *Capra aegagrus.*

A large number of insects and arachnids remain best adapted to Himalayan alpine zone and at 6900 metres height. The torrential streams which are the products of melting snow contain many mayflies, stoneflies and caddisflies. Some of common insects of this zone are—stonefly *Rhabdiopteryx lunata* ; grasshoppers like *Bryodema* and *Gomphomastax* ; apterous species *Conophyma* ; *Dicranophyma* ; *Sphingonotus* ; tettigonid *Hypsinomus fasciata* ; earwigs like *Anechura* ; bugs like *Dolmacoris*, *Tibetocoris*, and *Nysius ericae* ; aquatic beetle *Amphizoa* ; carabids like *Bembidion*, *Carabus*, *Calosoma*, *Harpalus*, *Trechus*, *Calathus*, *Bradytus*, *Broscus*, and *Dyschirius* ; staphylinids like *Atheta*, *Aleochara*, *Geodromicus*, *Oxypoda*, *Philonthus* and *Tachinus*; tenebrionids like *Bioramix* and *Chianalus*; ants like *Formica picea* ; bumble bees like *Lapidariobombus separandus* ; butterflies like *Colias*, *Argynnis* and *Parnassius* ; the fly *Deuterophelebia* ; collembolans like *Entomobrya*, *Isotoma*, *Proistoma*, *Hypogastrura*, *Isotomurus*, *Tomocerus* and *Onychiurus* ; and numerous simulids, blepharocerids, syrphids, anthomyids, tachinids, and sarcophagids.

22-3·3. Forest Biomes

The word **forest** is derived from Latin '*foris*' meaning outside, the reference being to village boundary fence, and must have included all uncultivated and uninhabitated land. Today a forest is any land managed for the diverse purposes of forestry whether covered with trees, shrubs, climbers, etc., or not (see **S.P. Sagreiya,** 1967). The forest biomes include a complex assemblage of different kinds of biotic communities. Optimum conditions of temperature and ground moisture responsible for the growth of trees contribute greatly to the establishment of forest communities. In addition, 50 mm rainfall is a pre-requisite for the growth of trees. The nature of soil, wind and air currents determines the distribution (abundance or sparseness) of forest vegetation. Normally ecologists recognise among forest communities such features as their evergreen nature, whether deciduous or indeciduous, as well as their shape, whether broad-leaved as in temperate forests or more needle-like as in the conifers. On the basis of these features the forest biomes of the world have been classified into following biomes—**coniferous forest, tropical forest**, and **temperate forest.** All these forest **biomes** are generally arranged on a gradient from north to south or from high altitude to lower altitude.

22·3·3A. Coniferous forest. Cold regions with high rainfall and strongly seasonal climates with long winters and fairly short summers are characterized by boreal conifer forest which is transcontinental. For example, adjacent to the tundra region either at latitude or high altitude is the **northern coniferous** or **boreal forest,** which stretches across both North America and Eurasia just south of the tundra (*i.e.*, Canada, Sweden, Finland, Siberia and Mussoorie). The term **taiga** is applied to the northern range of coniferous forests. This is characterized by evergreen plant species such as spruce (*Picea*, etc.), spruce (*Picea glauca*), fir (*Abies balsamca*) and pine trees (*Pinus resinosa, Pinus strobus*) and by animals such as snow shoe hare, the lynx, the wolf, bears, red fox, porcupines, squirrels, amphibians like *Hyla* and *Rana*, etc. Large size is common, both in the trees, which range up to some 40 m in height, and in vertebrates, which include the giants of several groups of animals, such as moose, caribou, elk, grizzly bear, wolverine, beaver, and several large species of birds. Species diversity is low, and pure stands of trees and shrubs are common, as are outbreaks of defoliating insects of several sorts (*e.g.*, bark beetles, saw flies, geometrid moths, etc.). Understory trees are uncommon as a result of the continual low light penetration. Among common understory associates are orchids and ericaceous shrubs like the blue berry. The thalloid mosses and lichens being very rich understory vegetation.

Boreal forest soils are thin podozols, and are rather poor both because the weathering of rock proceeds slowly in the cold environments, and because the litter derived from conifer needle is broken down very slowly and is not particulary rich in nutrients. These soils are acidic and mineral deficient, the result of the movement of a large amount of water through the soil ; in the absence of a significant counter upward movement of evaporation, soluble essential nutrients like calcium, nitrogen, and potassium are leached sometimes beyond the reach of roots thereby leaving no alkaline-oriented cations to encounter the organic acids of the accumulating litter (**Kormondy**, 1976). The productivity and community stability of a boreal coniferous forest are lower than those of any other biome.

22-3·3B. Temperate deciduous forest. The temperate forest biomes are characterized by a moderate climate and broad-leaved deciduous trees, which shed their leaves in fall, are bare over winter and grow new foliage in the spring. These forests are the characteristics of North America, Europe, Eastern Asia, Chile, part of Australia and Japan, with a cold winter and a annual rainfall of 75—150 cm and a temperature of 10—20°C. In these biomes the precipitation may be fairly uniform throughout the year. In India, at elevations of 9009—12,000 feet in Himalayas occur temperate vegetation including pines, fir, yew and juniper trees with an undergrowth of scrubby rhododendrons.

Soils of temperate forests are podozolic and fairly deep. Trees are quite tall—about 40—50m in height and their leaves are thin and broad. The predominant genera of this biome are maple (*Acer*), beech (*Fagus*), oak (*Quercus*), hickory (*Carya*), basswood (*Tilia*), Chestnut (*Castnea*), cottonwood (*Populas*), sycamore (*Platanus*), elm (*Ulmus*) and willow (*Salix*). In some locations, coniferous vegetation may be quite predominant and that includes white pine (*Pinus strubos*), hemlock (*Tsuga canadensis*) and red cedar (*Juniperus virginianus*). There are not many epiphytes or lianas save for some species of mosses, algae and lichens growing on tree trunks, and a few vines, notably *Vitis*, the grape. The understory of shrubs and herbs in the deciduous forest is typically well-developed and richly diversified, with a considerable portion of the photosynthesis and flowering attuned to the short days of the spring season, prior to the leafing out of and consequent shading by the tree canopy. Accordingly, there are often two separate herb assemblages. One consists of spring flowers, which bloom before the trees have expanded their leaves and are gone by summer, and the other is adapted to the low light levels of the forest floor and lasts into the fall.

The animals orginally present in temprate forests are deer, bears, squirrels, gray foxes, bobcats, wild turkey and wood peckers. Other common animals of this region are invertebrates like earthworms, snails, millipedes, Coleoptera and Orthoptera and vertebrates like amphibians such as newts, salamanders, toads, and cricket frogs; reptiles such as turtles, lizards and snakes ; mammals such as racoon opossum, pigs, mountain lion, etc., and birds like horned owl, hawks, etc. All these animals and plants show a profound seasonality; some may even hibernate throughout the winter. The range of animal size and adaptations is wide ; the largest animals include such forms as the deer and black bear. The dominant carnivores are large, including the wolf and mountainlion, although smaller carnivores such as fox and skunk are also common. Diversity of fauna is lower than in any of the rain forests and a few species seems clearly to be dominant.

22-3·3C. Temperate evergreen woodland. Many parts of the world have a Mediterranean-type climate with warm, dry summers and cool, moist winters. These are commonly inhabited by low evergreen trees with small hard needles or slightly broader leaves. The most important area of tropical evergreen woodland in North America is the **'Chaparral'** of the Pacific Coast, the Mediterranean **'maquis'**, Spanish **'encinar'** and **'melle scrab'** on Australia's South Coast are the same type of community. In such a woodland, trees are essentially lacking, although shrubs may range upto 3—4 m in height. Species diversity is roughly intermediate between that of a temperate deciduous forest and a drier grassland. Fire is an important factor in this ecosystem, and the adaptations of the plants enable them to regenerate quickly after being burned. The characteristic animals of temperate evergreen woodland or chaparral are mule deer, brush rabbits, wood rats, chipmunks, lizards wem-tits and brown towhees. Small-hooved cursorial ungulates are the dominant herbivores. Saltatorial (jumping) animals and many fast moving ungulates are also common in this fauna.

22-3·3D. Temperate rainforests. The temperate rainforest is a colder ecosystem than any other rainforest. Such a forest has a definite seasonality, with both temperatures and rainfall varying throughout the year. Rainfall is high, but fog may be very heavy and actually more important as a source of water than rainfall. The diversity is much lower, both in plants and animals, in comparison to warmer rainforests, yet it remains still higher than other temperate forest types. The dominant trees (canopies) are coast redwood (*Sequoia sempervirens*) of the Pacific coast of North America and the alpine ash (*Eucalyptus regnans*) of Australia and Tasmania, both of which reach more than 100m in height. Epiphytes and lianas are common but are not abundant like those of other rainforests. The animals of temperate rainforests are similar to those of deciduous forests, but show a somewhat higher diversity.

22-3·3E. Tropical rainforests. Tropical rainforests occur near the equator in Central and South America, Central and Western Africa (Congo, Zambesi river), Southeast Asia (parts of India and Malaysia), Malaya, Borneo, New Guinea and Northwest Australia. Tropical rainforests are among the most diverse communities on earth. Both temperature and humidity are high and constant. The annual rainfall which exceeds 200 to 225 cm is generally evenly distributed througout the year.

The flora is highly diversified : a square mile may contain 300 different species of trees, a diversity unparalleled in any other biome. The extremely dense vegetation of the tropical rainforests remains vertically stratified with tall trees often covered with vines, creepers, lianas, epiphytic orchids and bromeliads. Under the tall trees is a continuous evergreen carpet, the canopy layer, some 25 to 35 metres tall. The lowest layer is an understory of trees, shrubs, herbs like ferns and palms, all of which become dense where there is a break in the canopy. Nearly all plants are evergreen, and those that do lose their leaves entirely do so at irregular intervals with no apparent regard to the climatic regime. The leaves of most plants are of moderate size, leathery and dark green in colour. Their roots are often shallow and have swollen bases or flying buttrerses.

Soils of tropical rainforests are red latosols, and they may be exceedingly thick. The high rate of leaching makes these soils virtually useless for agriculture purposes, but if they are left undisturbed the extremely rapid cycling of nutrients within the litter layer (due to decomposition) can compensate for the natural poverty of the soil. It is the nature of the soil, both its potential for high leaching as well as its chemical composition that promotes a rock-like quality when exposed to air, that largely has prevented western style agriculture from being applied to the tropical forests (**Kormondy**, 1976).

Invertebrate density and abundance are very high in tropical rainforests, but while vertebrates are diverse, they are not as abundant as in many other communities. The common invertebrates of these forests are worms, snails, millipedes, centipedes, scorpions, isopods, spiders, insects, planarians and leeches. Among insects, heteropterans, orthopterans, blattids, mantids, phasmids, bees, termites and ants are most common. The common vertebrates of tropical rainforests are the arboreal amphibian *Rhacophorus malabaricus* ; aquatic reptiles, chameleons, agamids, geckoes, and many species of snakes ; many species of birds, social birds being predominant ; and a variety of mammals. Nocturnal and arboreal habits are most common in many mammals such as insectivores, leopards, jungle cats, ant-eaters, giant flying squirrels, monkeys and sloths. But in New Guinea and Northern Queensland, where monkeys are absent, there are arboreal kangaroos (*Dendrolagus* sp.) despite the fact that the basic anatomy of the kangaroo is not particularly well-suited to arboreal life. Further, in the foot hills of the forest zone of peninsular India covered with dense tropical vegetation, we have the tiger (*Panthera tigris*), the elephant (*Elephas maximus*), samber deer (*Rusa unicolor*), muntjac (*Muntiacus muntjak*), the gaur (*Bibos gaurus*), the chital or spotted deer (*Axis axis*), and the swamp deer (*Rucervus duraucelli*) as the major ground dwelling mammals.

22-3·3F. Tropical seasonal forests. Tropical seasonal forests occur in region whose total annual rainfall is very high, but segregated into pronounced wet and dry periods. Tropical seasonal forests are found in Southeast Asia, Central and South America, Northern Australia, Western Africa and the tropical islands of the pacific as well as India and Southeast Asia. In exceedingly wet tropical seasonal forests, commonly known as **monsoon forests**, the annual precipitation may be several times that of the tropical rainforests. Trees may reach heights over 40 m, but are more commonly 20—30m high. Stratification is of a relatively simple type with a single understory tree layer, canopy is deciduous and understory is evergreen. Teak is often a major large tree in the best-known tropical seasonal forests, those of India (Central India) and Southeast Asia. Bamboo is also an important climax shrub in these areas although in other areas it is important only in earlier stages of the succession.

22·3·3G. Subtropical rainforests. In regions of fairly high rainfall but where temperature differences between winter and summer are less marked, as in Florida (USA), the broad-leaved evergreen subtropical biome is found. The vegetation includes mahogany, gumbo limbo, bays, palms, oaks, magnolias, tamarinds, all laden with epiphytes (of pineapple and orchid families), ferns, vines and strangler fig (*Ficus aureus*). This stratification is simpler, with only one understory tree horizon. All these plants tend to be evergreen, but may lose their leaves during the dry season. Animal life of subtropical forest is very similar to that of tropical rainforests.

22·3·4. Tropical Savanna Biomes

Savannas are tropical **grasslands** with scattered, drought-resistant trees which generally do not exceed above ten metres in height and do not form a canopy. Thus, a savanna is an intermediate between a forest and a grassland. Savannas constitute extensive areas in Eastern Africa, which support the richest diversity of grazing mammals in the world, and also occur in Australia, South America and Asia. The climate is generally characterized by a rainy (May through October) and dry (November through April) season; in the llanos of Venezuela, for example, nearly 90 per cent of the annual rainfall of 130 cm falls during the rainy season. The latosol soil of savannas are nutrient poor owing a heavy leaching. Quite widely distributed in savannas are soils called laterite which when dried harden to rock-like consistency thereby precluding their use in western style agriculture (**Kormondy**, 1976). Although climate and soil are significant regulating factors in this biome, the controlling factor appears to be fire, which gives grass and certain species of trees a powerful advantage over other tree species. As a result of this latter factor, species diversity is quite low in comparison to adjacent tropical forests ; in some situation a single species of both grass and tree may be dominant over large areas.

The trees of savannas are resistant to desiccation, and may be either deciduous or evergeen. Their leaves are often hard and drought-resistant. Grasses are the most conspicuous plants, and may reach heights of 1½—2m. Gigantism of certain animal groups in these tropical savannas is as pronounced as it is in the boreal forest including such giants as many antelopes, giraffes, elephants, buffalo and lions (**Clapham, Jr.**, 1973). In addition, a rich insectan fauna especially those with strong mandibles capable of mastication are conspicuous in this biome. Grasshoppers and termites are encountered in large numbers.

22-3·5. Grassland Biomes

The grassland biome is found where rainfall is about 25 to 75 cm per year, not enough to support a forest, yet more than of a true desert. The seasonality of grasslands is pronounced, both with respect to rainfall, which is concentrated in the summer and to temperature. Grasslands typically occur in the interiors of continents and include the **tall grass prairies, short grass prairies,** or great plains and **arid grassland** of North America as well as the **steppes** of Eurasia (Southern Russia, Siberia and Asia), the **veldt** of Africa and the **pampas** of South America (Argentina). The grassland communities are open land communities with limited moisture conditions, irregular rainfall, sharp seasonal and diurnal variations and very high radiation. Since tall trees or other thick vegetations are excluded from these communities there is a free movement of air. These winds carry the particles of sand or dust. These open habitats provide natural pasture for grazing animals (herbivore) which are excluded from predation by predators which hide in thick vegetation and prey upon them.

In grasslands, the stratification is reduced essentially to a single story, but within that level, species diversity may be as high as in a deciduous forest-especially for the tall grass prairies. Only along the streams are trees to be found, but the "gallery foresets" within a few metres of stream bank are characteristics of grasslands. The grasses comprising of the grasslands can be divided into two basic groups, the tall grasses more than 1m high, which are found in moist portions of the grassland, and the short grasses less than 1 m high, which are found in the drier regions. For example, prior to its conversion to agriculture and urban development, the tall grass prairie toward east of North America was dominated by species of bluestern (*Andropogon*) forming dense covers four to six feet tall. Westward, Buffalo grass (*Buchloe dactyloides*) and other grasses but a few centimetres high dominated the landscape. Flowering herbs including many kinds of composites are common, but much less important than grasses.

Since the precipitation – evaporation ratio is below 1 in the grassland, leaching is considerably less. The soils of the grasslands are rich, fertile prairie soils (black-earths) and chernozems. Organic matter (humus) accumulates in the upper portion of the soil, rendering it dark ; this upper portion remains neutral to slighily alkaline because of the continued replenishment of cations like calcium and potassium through the upward movement associated with evaporation.

Typical animals of grasslands tend to be quite small, with the exception of a few very large cursorial herbivore mammals such as the bison and pronghorn in North Amarica, the wild horse, ass, and saiga antelope of Eurasia and zebra and antelops of Southern

Africa. The large herbivores are no where near as diverse as they are in savanna areas. Likewise, canivores are relatively small such as are coyotes, weasels, badgers, foxes, ferrets, owls and rattle snakes, Rodents such as prairie dogs, rabbits, and ground squirrels are common. Most herbivores characteristically aggregate into herds or colonies; this aggregation provides some protection against predators. The characteristic birds of grasslands are prairie chickens, meadow larks, and rodent hawks. Further, saltatorial motion is widespread, not only in mammals such as rabbits and kangaroo rats, but also in insects, such as grasshoppers and crickets. This probably has to do with increasing rate at which animal can move through or over the tall grassess, as well as making it harder for a predator to catch the animal. Visibility in the grasslands is exceedingly good and the premium placed on efficiency including predators is very high.

The common insects of steppes are termites, locusts (*Locusta migratoria*, *Schistocerca gregaria* and *Melanoplus* sp.), bees, burrowing wasps, mutilid wasps, bumble bees and blister bees. The steppes also harbour an undisturbed reptilian fauna. Lizards and snakes are met with in large numbers and possess, remarkable diversity. Most of them are fossorial, insectivorous and carnivorous.

22-3·6. Desert Biomes

Deserts are the biome formed in the driest of environments. Temperature may range from very hot as in **hot deserts** to very cold as in **cold deserts**. Major hot deserts of the world are situated near the tropics of cancer and capricon, with a rainfall of less than 10 mm. The most important hot deserts of world is the Sahara-Arabia-Gobi desert complex extending from Africa to Central Asia and contains highly irregular and very insignificant rainfall, and low humidity due to excessive evaporation. Fairly extensive hot deserts also occur in India (Sind-Rajasthan deserts), South America (Chile), North America and Australia. The cold deserts occur at high elevations where the temperatures are low and rainfall scanty as the air losses all its moisture content as it ascends higher and higher. Cold deserts occur in Ladakh regions of Himalayas, Tibet, and Bolivia Arctic. The hot and cold deserts may also be distinguished by differences in plant population which are mostly succulent type (*e.g.*, cactus, palo verde trees, creosote bush, etc.). Most cold deserts have sage bush.

Low erratic precipitation coupled with soil and air temperatures that are extremely high by day and drop abruptly by night, low humidity, and high insolation are the major desiccating environmental factors to which desert vegetation and animals have adapted. Desert plants which tend to be shrubs are adapted to drought conditions through reduced leaf size and the dropping of leaves in dry conditions, both reducing water loss via evapotranspiration. The roots of most desert plants remain well developed and occur in the top metre of the soil to take maximum possible advantage of any rainfall. Further, the root hairs on many desert plants are ephemeral, drying back under drought conditions and thereby reducing potential water loss by osmosis. Yet other species are short lived annuals that complete their life cycles during the short-moist period. In most hot deserts, there occur plants such as cacti, water-storing succulents such as acacias, euphorbias, cacti, prickly pears, etc., which are adapted by their protoplasmic colloids, which enable the accumulation of substantial water reserves, as well as by a reduced leaf surface, which obviates water loss via evapotranspiration.

The animals present in the desert are reptiles, insects and burrowing rodents. All these animals possess special morphological, physiological and ethological adaptations for deserts. In general, large animals are very uncommon except mule, deer and some species of gazelle and all animals have cursorial, fossorial and or saltatorial adaptations. Some desert animals are nicely adapted for high extremes of temperature. For example, the lethal the temperature for different species of insects found to be following—for the canal spider *Galeodes granti* it is 50°C, for the *Gryllus domesticus* it is 40°C and for the forficulid *Labidura riparia* it is 38 C (**Cloudsely-Thompson**, 1962). Diurnal rhythms are perhaps the best method of avoiding the heat. While some desert plants close their petals at night, many blossom only at night. Some insects like tenebrionid bettle *Akis spinosa* remain active during the day, and the centipede *Scolopendra clavipes* restricts its activities to night time.

Further, certain reptiles and certain insects are well adapted for survival in deserts because of their thick, impervious integuments and the fact they excrete dry waste matter. A few species of mammals have become secondarily adapted to the desert by excreting very concentrated urine. They avoid the sun by remaining in their burrows during the day. Kangaroo rat and pocket mouse, both are able to live without drinking water by extracting the moisture from the seeds and succulent cactus they eat. The camel and the desert birds (Ostrich, etc.) must have an occasional drink of water but can go for long periods of time using the water stored in the body. Most insects of deserts are herbivores and as a correlation the number of small insectivorous lizards found in the desert is usually high.

Fauna of Indian deserts. **Prakash** (1974) have recognized four habitat types in the Indian desert based on land forms namely **aquatic**, **sandy**, **rocky** and, **riparian habitat** (river banks). The aquatic habitats of Indian deserts which remain confined to the perennial lakes, are inhabited by various species of fishes like *Labeo nigripinnis*, *Oxygaster clupeoides*, *Tor khurdee*, *Puntius amphibia*, *Neomacheilus denisonii*. Among amphibians of the Indian desert are a species of toad (*Bufo andersonii*) and five species of frogs (Indian bull frog *Rana tigrina*, cricket frog *Rana limnocharis* and burrowing frog *Rana breviceps* being dominant species). Among reptiles, there occur one species of crocodile (*Crocodilus* *palustris*), two species of testudines (Loricata), 18 species of lizards and 18 species of snakes. Of the lizards, some species like *Acanthodactylus cantoris*, *Calotes versicolor*, *Uromastix hardwicki*, *Ophiomorus tridactylus* are predatory on the desert locust inhabiting localized areas in the Thar desert. The avian fauna of the Indian desert includes species like painted partridge *Francolinus pictus pallidus*, grey partidge *Francolinus pondicerianus*, grey quail *Coturnix coturnix*, black breasted or rain quail *Coturnix coromandelicus*, rock bush quail, the little bustard quails *Lurnix sylvatica*, button quail *Turnix tanki*, great Indian bustard *Choriotes nigriceps*, florican *Sypheotides indica*, Indian sandgrouses like *Petrocles exustus erlangeri*, *Petrocles alchata*, *P. senegallus*, and *P. indicus indicus*. During winter numerous aquatic birds become abundant in Indian deserts. Such birds are white fronted goose *Anser albifrons*, wigeon *Anas penelope*, garganey *Anas querguedrila*, red crested pochard *Netta fufina* and tufted duck *Aythya futigola*. Among the predominant predatory birds are two species of the vultures namely *Gyps bengalensis* and the white vulture *Neophron perenopterus*.

The mammalian fauna of Indian deserts includes many species, some of which have been tabulated in the following table :

Table : 22·2. Certain common mammals of **Indian deserts (Prakash,** 1974).

	Order	Species	Popular Name
1.	Chiroptera	*Rousettus arabicus*	Fruit bat
		Pteropus giganteus giganteus	Fruit bat
		Rhinopoma kinneari	Rat-tailed bat
		Rhinopoma hardwicki	The lesser rat-tailed bat
		Taphozour perforatus	Tomb bat
		Megaderma lyra	Indian false vampire
		Rhinolophus lepidus lepidus	Horse shoe-bat
2.	Insectivora	*Hemiechinus auritus collaris*	Long eared hedgehog
		Paraechinus micropus	Long eared hedgehog
		Suncus murinus sindensis	House Shrew
3.	Pholidota	*Manus crassicaudata*	Scaly ant eater
4.	Lagomorpha	*Lepus nigricollis caudatus*	Hare
		Lepus nigricollis dayanus	Hare
5.	Rodentia	*Funambulus pennanti*	Northern palm squirrel
		Hystrix indica indica	Indian crested porcupine
		Rattus rattus	House rat
		Mus musculus rufescens	House mouse
		M. musculus bactrianis	House mouse
		Gerbillus gleadovi	Indian hairy footed gerbil
		Tatera indica	Gerbil
		Merriones hurrianae	Gerbil
6.	Artiodactyla	*Sus scrofa cristata*	Wild boar
		Boselaphus tragocamelus	Blue bull

Order	Species	Popular Name
	Antilope cervicapra rajputani	Black buck
	Gazella gazella bennetti	Gazelle
	Cervus unicolor	Sambhar
	Axis asis	Chital (spotted deer)
	Tetracerus quadricornis	Chousingha
7. **Carnivora**	*Canis lupus*	Wolf
	Canis aureus	Asiatic jackal
	Vulpes vulpes	Red fox
	Vulpes bengalensis	Bengal fox
	Lutra persptcillata	Indian otter
	Viverricula indica	Small India civet
	Herpestes aeropunctatus	Small Indian mongoose
	Felis chaus prateri	Jungle cats
	Felis libyea ornata	Jungle cats
	Panthera pardus sindica	Panther
	Melursuu ursinus	Sloth bear

All these mammals have well specialized adaptations for survival in thermal extremes and low humidity. Many of them possess diurnal rhythmic activities.

22-4. REVISION QUESTIONS

1. Define biome and describe the physico-chemical characteristics and biota of tundra biome.
2. Describe the biota of desert biome.
3. Write short notes on the following :
   (a) Alpine biome ;
   (b) Forest biome ;
   (c) Grassland biome.

23

Biogeography

(DISTRIBUTION OF ANIMALS AND PLANTS)

No organism occurs in uniform numbers throughout the world. Rather, specific organisms are restricted to specific communities or groups of communities. Three aspects of the distribution of an organism are generally recognized : **geographic range**, or the specific extent of land or water area where the organism normally occurs ; the **geologic range**, or the distribution in time, past and present ; and the **ecological distribution**, or the major biotic communities (*i.e.*, marine biome, freshwater biome and terrestrial biome) of which the organism is a member. Certain biologists have also made distinction between geographic distribution (horizontal or supeficial distribution) and **bathymetric distribution** (vertical or altitudinal distribution). Bathymetric distribution includes the following three realms : (i) **Halobiotic**, or vertical distribution of organisms in marine (sea) habitat, (ii) **Limnobiotic**, or vertical distribution of organisms in fresh water habitat ; and (iii) **Geobiotic**, or altitudinal distribution of organisms on land. All the living organisms in a given region are termed the **biota** of that region. The animals of a given region are collectively termed the **fauna**, and the plants of a given region, the **flora** (*i.e.*, fauna+flora=biota). Here, a distinction can be made between the following two terms—flora and vegetation. **Flora** mainly refers to the botanical composition of a place, *i.e.*, the names of different plant species, while **vegetation** means the totality of forms in which the emphasis is not on names of different plants but their life forms, number and coverage. The studies of the distribution of biota are collectively called **biogeography** ; of animals only, **zoogeography** ; and of plants only, **plant geography** or **phytogeography**. There are two major approaches to the study of biogeography or **geographical ecology (Kendeigh**, 1974) : (i) **descriptive** or **static biogeography** which deals with the description of biota of different botanical and zoological areas of earth ; and (ii) **interpretative** or **dynamic biogeography** which describes the forces which have brought about plant and animal distribution.

23-1. PATTERNS OF DISTRIBUTION OF BIOTA

The plants and animals **exhibit** the following specific patterns of distribution over the globe :

23-1·1. Continuous Distribution

A taxon distributed throughout the world in all climatic zones or in all the continents in at least one climatic zone is said to have continuous distribution. It is of the following types :

23-1·1A. Cosmopolitan distribution. Such a taxon or a species occurs in all climatic zones. The common plant species of nearly cosmopolitan distribution are *Chenopodium album*, *Taraxcum officinale*, *Phragmites communis*, *Urtica dioica*, *Poa annua*, etc. Among the animals the following species have cosmopolitan distribution : *Tyto alba*, *Ardea cineria*, *Falco pereginus*, *Vanessa cardui*, *Nonnophila noctuella*, and *Macrobiotus hufelandi*.

23-1·1B. Circumpolar distribution. Certain species such as *Saxiphraga oppositaefolia*, *Carex lapponica*, *Ranunculus nivalis*, etc., are distributed in a belt around the north-pole.

23-1·1C. Circumboreal and circumaustral distribution. The species which are distributed in a near continuous belt in the temperate region of northern or southern hemisphere are said to have circumboreal and circumaustral distribution respectively. The genera *Alnus*, *Draba*, *Acer*, etc., are circumboreal in distribution and *Danthonia* is circumaustral in distribution.

23-1·1D. Pantropical distribution. Certain plant species like *Bauhinia*, *Dalbergia*, *Dioscorea*, *Corchorus*, *Ocimum*, *Cassia*, *Eugenia*, *Phyllanthus*, etc., are distributed throughout the tropical belt.

23-1·2. Discontinuous Distribution

A taxon distributed in two or more widely separated geographical areas, is said to have discontinuous distribution. Such taxa may occur in several small areas in the same continent or in two different continents of the same or different hemispheres. For example, two plant species—*Saxifraga* and *Silene* are distributed in Arctic region and high altitudes. Likewise, three genera of Dipnoi fish, *Epiceratodus*, *Protopterus* and *Lepidosiren* are distributed in Australia, Africa and South America. There are various other examples in plants and animals which show discontinuous distribution.

23-1·3. Endemic

The endemics are taxa of very restricted distribution in small areas. Numerous plant families, genera and species are found to be highly endemic. For example, a great proportion of the taxa of the flora of oceanic islands is endemic to those islands only. Some important endemic plant genera of Indian deserts are *Omania*, *Xerotia*, *Leptadenia*, *Daemia*, etc., and of India are *Amphicome*, *Dittoceras*, *Dodecania*, *Ulteria*, *Cruddasia*, *Heylandia*, *Lagenandra*, *Zeylandium*, *Hitchenia*, *Blepharistemma*, etc. Certain important examples of endemic Indian plant species include *Ficus religiosa*, *Ficus bengalensis*, *Aegle marmelos*, *Artocarpus nobilis*, *Crotalaria juncea*, *Datura metel*, *Indigofera tinctoria*, *Elettaria*, *Eleusine coracana*, *Piper nigrum*, *Piper longum*, *Sesamum indicum*,

Hibiscus abelmoschus, *Butea monosperma*, *Beaumontia grandiflora*, *Memecylon umbellatum*, *Holmskioldia sanguinea*, *Feronia elephantum*, *Saraca indica*, *Shorea robusta* and *Caryota urens*.

The endemics are divided into two types : **1. Palaeoendemic** or **epibiotic** or **relics** : These endemics are supposed to have been the remnants of a once widely-distributed taxon in the past, e.g., *Gingko biloba*, *Sequoia semipervirens*, *Trapa natans*. **2. Neoendemics** or **microendemics** : These endemic taxa are supposed to have evolved only during the recent times and did not have sufficient time to extend their ranges of distribution.

23-2. DESCRIPTIVE PHYTOGEOGRAPHY

With a descriptive approach of phytogeography, will be discussed the major plant communities (biomes) of the world, and different vegetational belts of the earth constituded by these biomes together with characteristic climatic conditions of the area.

23-2·1. Major Plant Communities (Biomes) of the World

As already discussed in sufficient details in chapters 20, 21 and 22, major plant communities of the world are classified chiefly on the basis of the kinds of habitat and environmental conditions into the following chief types :

Plant communities

- Aquatic
  - Fresh
  - Marine
  - Estuarine
- Terrestrial (Biomes)
  - Tundra
  - Forest
  - Grassland
  - Desert

23-2·2. Chief Vegetational Belts of the Earth

On the basis of climatic and geographical conditions, the earth is usually divided into the following four broad vegetational belts :

23-2 2A. Arctic zone. This zone is divided into the following two types :

(1) Arctic proper. This zone occurs around the north pole and remains covered with ice throughout the year. This zone infact is Tundra biome and its chief components of vegetation include some algae, annual flowering plants, mosses and lichens.

(2) Subarctic. This is a less defined zone which extends from southern arctic to the northern limits of temperate zone. This region is very cold, contains abundant bogs and have vegetation including small height trees, shrubs and herbs in the months of June and July. Ground vegetation often includes some pteridophytes, orchids, insectivorous plants, mosses and lichens.

23-2·2B. North temperate zone. This zone extends between 30°N latitude and 55°N latitude and includes following two major zones :

(1) North temperate of the eastern zone. This zone is further subdivided into the following four zones : (i) **Western and central Europe.** This zone is demarcated in north by the subarctics and in south by Alps and British Islands. Its forests are dominated by several gymnospermous tall trees like *Pinus*, *Picea* and *Abies*, and angiospermous trees like oaks, maple and chestnuts. Ground vegetation comprises orchids, wild roses, buttercups, *Viola*, *Salvia Dianthus*, etc. At high altitudes of this zone trees are replaced by grassy vegetation with some herbaceous flowering plants. (ii) **Mediterranean.** This zone extends between 30°N and 40°N latitudes, south of mountain ranges in Europe and in Asia around Mediterranean sea and is characterised by warm temperate type climate. Vegetation is chiefly composed of fruit trees, olives, nut trees, oranges, and also some foreign palms, cacti, acacias, etc. In the Asian region of Mediterranean as in Arab countries, rainfall is low, deserts are common and sparse vegetation include species like *Atriplex*, *Alhagi*, *Polygonum* and *Phoenix dactylifera*. (iii) **Northern Africa.** This zone includes the northern parts of Morocco, Algeria, Libya and Egypt. It is characterized by scanty rainfall and sparse vegetation. In cooler areas (*i.e.*, mountains) some conifers and broad-leaved oaks are common. In deserts (including some portion of Sahara desert) some herbs, shrubs, woody acacias and succulent xerophytes occur. (iv) **Himalayas, eastern Asia and Japan.** Tibet, China and Japan have different type of vegetation. In China and Japan conifers like *Cryptomeria*, *Sciadopitys*, *Cephalotaxus*, *Ginkgo biloba* and *Cycas* and angiosperms like *Rhododendrons*, *Cinnamomum camphora* and *Begonia* are common. The vegetation of Himalays will be described later in this chapter.

(2) North temperate of the western hemisphere. It includes the parts of United States and Canada lying mostly between north latitudes 30° to 55°. The eastern coastal regions of these countries in the temperate belt have some very characteristic plant species like tropical fern (*Schizaea pusilla*). The forest communities are composed of conifers and deciduous trees. On lower altitudes some wild cherries, plums, roses and orchids are abundant. In the New England region trees of *Ulmus americana* and *Castania dentata* are abundant. Forests of conifers are common in southern parts and on western parts of Rocky mountains of USA. In north California grows *Sequoia sempervirens*, the tallest tree of world. Ground vegetation is composed of *Salicornia herbacea*, *Rumex maritima*, *Monotropa uniflora*, *Saxifragea*, *Primula*, etc. In the Colorado desert of Arizona and south-eastern California, there are several types of xerophytic plants.

23-2·2C. Tropical zone. This zone is divided into two subzones :

(A) Palaetropic. It comprises old world or eastern tropics and has the following two botanical regions : (1) **Tropical Africa.** This is a large-sized landmass of varied topography. It includes high altitudes and Sahara deserts which have little or no rainfall. In Africa, most remarkable plant is *Welwitschia mirabilis*, Eastern part of Central Africa has India-like vegetation of *Borassus flabelliformis*, *Tamarindus indica*, *Ficus*, *Aspargus*, *Clematis*, *Phaseolus*, *Cassia fistula*, *Erythrina*, *Acacia*, *Albizzia*, *Zizyphus*, *Bauhinia*, etc. (2) **Tropical Asia.** It includes Arabia, Pakistan, India, Burma, Ceylon, Thailand, Indonesia, Philippines and Islands of Indian sea. In Arabia, the rainfall is low and most of the desert species are found *Coffea arabica* is a native plant of Arabia. Ceylon is rich in species diversity and ferns are the chief components of its sparse natural vegetation. Malaya, Java and Sumatra are characterized by heavy rainfall and their vegetation comprises varied types of palms, some types of ferns, tall trees, lianas and insectivorous plants. Some important plants of Java are *Albizzia*, *Pterocarpus*, *Tamarindus*, *Bombax*, *Cassia*, *Dendrocalamus*, etc. The common trees of Burma and Thailand are jack fruit, orange, mango, banana, betelnut, etc.

(B) **Neotropics.** It comprises Mexico and major part of South America. The low-rainfall areas of Mexico are rich in xerophytes. At higher cooler altitudes there is a forest of conifers like *Pinus*, *Spruce*, *Quercus* and *Populus*. On mountain peaks grasses are most common. In wet areas of Mexico, there are mosses, palms, bamboos, orchids, etc. In South America, there are extensive forests of flood-resistant trees like *Bertholletia excelsa*, *Maximiliana regia* and also mangrove vegetation. There are also found many epiphytes.

23·2·2D. **South temperate zone.** This zone includes extreme southern region of Africa, Australia and New Zealand. In African area, the vegetation chiefly made up of ferns and gymnosperms. On the hills conifers are present. Its lower wet regions contain *Salix* and *Phragmites* and dry regions have grasses like *Andropogon* and trees like *Acacia*. The vegetation of northern part of Australia is similar to that of south east Asia and includes trees of palms, nuts, *Eucalyptus*, *Acacia* and *Casuarina*. Some pteridophytes are also found in the ground vegetation. In the South Australia *Araucarias* are common. New Zealand forests are mostly made up of conifers together with ferns, palm like *Rhopalostylis*, many species for *Metrosideros*. New Zealand has richest bryophytic flora.

23·2·3. Phytogeography of India

Indian subcontinent lying between 8° and 37° N, and 68° and 97° E has its own peculiar physiographic, climatic and biotic features. It is surrounded on its south, east and west by oceans and in the north by mountaineous chains highest in the world. The subcontinent stretches out between tropical and subtropical belts. Its climate is chiefly modified by oceans and mountains. In the south and in the far east the climate is typically tropical, while it is temperate in the north, and highly arid in the north-west. The general climate of India is of monsoon type. In India, the following three distinct seasons occur : 1. Cold weather season during December to February when the mean temperature falls very low. In the north snowfall occurs on several days and the temperature remains below 0°C for many weeks, while in the south the temperature may remain above 20°C. Some rainfall occurs during this season. 2. From March to mid-June, the cold weather season is followed by the hot dry summer as the temperature gradually rises, and relative humidity decreases. The change of season is marked by leaf-fall and blooming of many trees particularly in northern India. The summer in the north often becomes quite severe with average temperatures above 40°C or even upto 50°C in the arid western region. Hot dusty winds commonly blow at high velocity during this season. 3. Towards the end of the summer by late-June, the monsoon winds bring rain and the temperature gradually falls down. The rainfall starts little earlier in the eastern and southern parts of India. The monsoon winds moving from the Bay of Bengal upto the Gangetic plains of north to the west bring more rains in the east (upto 200 cm) and much lesser in the west (100 cm), till in the western most part of the country, less than 50 cm rainfall is left. By the end of September the monsoon starts retreating, the sky becomes clear and the temperature gradually falls down upto the December. This variability in the temperature and rainfall patterns produces a great diversity in the vegetation of the subcontinent.

23-2·3A. Vegetation of India. The most important factors used in the classification of Indian vegetation are rainfall, temperature, edaphic factors, biotic influences and life forms. There are two most common types of plant formations in India—forest and grassland.

(a) **Forest vegetation. Champion** and **Seth** (1968) have recognised the following 16 major types of forests in India.

I. MOIST TROPICAL FORESTS

(1) Tropical wet evergreen forests. These forests occur on the Western Ghats, Assam, Cachar, parts of Bengal, parts of Mysore and Andamans. These forests are climatic climax forests with very dense growth of tall trees (more than 45 metres high). The shrubs, lianas, climbers and epiphytes are abundant, however, grasses and herbs are rare on the forest floor. The rainfall exceeds 250 cm in these regions. The characteristic flora of these forests include *Dipterocarpus indica*, *Hopea*, *Artocarpus*, *Mangifera*, *Emblica*, *Michelia*, *Ervatamia*, *Lagerstroemia*, *Ixora*, etc.

(2) Tropical moist semi-evergreen forests. These forests occur along Western Ghats, in parts of Upper Assam and Orissa. The rainfall of these regions is usually high, above 200 cm per year. These forests consist of dense and tall (25 to 35 metres) trees along with shrubs. In these forests, because deciduous species like *Terminalia*, *Tetrameles* and *Shorea* grow intermixed with the evergreen species like *Artocarpus*, *Michelia* and *Eugenia*, so they are called semi-evergreen forests.

(3) **Tropical moist deciduous forests.** These forests are distributed in a narrow belt along the foot of the Himalayas, on the eastern side of Western Ghats, Chhota Nagpur and Khasi Hills. These forests have high rainfall (150-200 cm) throughout the year except few very short dry periods. The trees are deciduous remaining leafless for one or two months. The most common species is teak (*Tectona grandis*) in the south and sal (*Shorea robusta*) in the north. Other common species of these forests are *Dalbergia, Cedrela, Salmalia, Albizzia, Terminalia, Cordia, Bombusa, Melia, Dillenia, Eugenia, Dendrocalamus*, etc.

(4) **Littoral and swamp forests.** These forests comprise mostly evergreen species and occur along sea coasts in wet and marshy areas and also in deltas of larger rivers on eastern coast. In saline swamps, mangroves are chiefly composed of *Rhizophora, Bruguiera, Ceriops, Nipa*, etc In less saline areas, *Phoenix, Ipomoea, Phragmites, Casuarina*, etc., are found.

II. DRY TROPICAL FORESTS

(5) **Tropical dry deciduous forest.** These forests are distributed almost throughout the country except Kashmir, beyond Bengal, Rajasthan and Western Ghats. These occur in areas with annual rainfall ranging from 75 cm to 125 cm and with a dry season of about 6 months. The trees are small (10–15 metres) but canopy is open. Shrubs are abundant. These forests include following common plant species *Shorea robusta* (in north), *Tectona grandis* (in south), *Anogeissus, Terminalia, Semecarpus, Carissa, Emblica, Acacia, Zizyphus, Dalbergia, Dendrocalamus*, etc.

(6) **Tropical dry deciduous forests of South-East Daccan Region.** In some parts of Tamil Nadu and the Karnatic coasts evergreen forests of short (10–15 metres) but dense trees are found though the annual rainfall is relatively small. There are no bamboos but grasses are abundant. The common flora of this region includes species of *Memecylon, Maba, Pavetta, Feronia, Ixora, Sterculia*, etc.

(7) **Tropical thorn forests.** These forests are distributed in areas with very low rainfall as in Western Rajasthan, parts of Southern Punjab, Bundelkhand, Gujrat, Maharashtra, Madhya Pradesh and Tamil Nadu. The trees are short (8–10 metres), sparsely distributed, mostly thorny ; shrubs are more common than trees. Plants remain leafless throughout the year. During rains, grasses and herbs become abundant. The common plants of these forests are *Acacia nilotica, Prosopis spicigera, Capparis, Albizzia, Zizyphus, Euphorbia, Calotropis, Madhuca, Tephrosia, Crotalaria*, etc.

III. MONTANE (MOUNTAINOUS) SUBTROPICAL FOREST

(8) **Subtropical broad-leaved hill forests.** These forests are dense, have predominantly evergreen species, and occur in relatively moist areas as lower slopes of eastern Himalayas, Bengal, Assam and on hill ranges as Khasi, Nilgiri, Mahabaleshwar, etc. In the north occur *Quercus. Castanopsis, Schima* and some temperate species. In eastern Himalayas due to higher humidity, bamboos,

many epiphytes including orchids and ferns become abundant. In the south common trees are *Eugenia, Actinodaphne, Randia, Ficus, Populus, Canthium, Mangifera* and climbers are *Piper, Gnetum, Smilax*, etc.

(9) Subtropical pine forests. In Western Himalayas and Khasi Jayantia hills of Assam, upto an altitude of 1875 metres, open forests composed mainly of pine species like *Pinus roxburghi* and *P. khasya* are found.

(10) Subtropical dry evergreen forests. Lower elevations of Himalayas have xerophytic forests containing mainly thorny and small-leaved evergreen species. These areas are characterised by low rainfall and low temperature. The common species are *Acacia modesta, Dodonea viscosa, Olea cuspidata*, etc.

IV. MONTANE TEMPERATE FORESTS

These forests are found at 1800 to 3800 metres altitude in the Himalayas.

(11) Montane wet temperate forests. These forests are found in eastern Himalayas with high rainfall and also in Nilgiris in the south. The forests in the south are evergreen and are known as **Sholas**. The trees are 15 to 20 metres high with a dense growth forming closed canopy. Epiphytes are abundant. The common plant species of these forests are *Rhododendron nilagircum, Hopea, Artocarpus hirsuta, Salmalia*, etc.

In the north, the west temperate forests extend from eastern Nepal to Assam at an altitude of 1800 to 2900 metres. These forests are evergreen, dense and high (upto 25 metres). The common plant species include *Quercus, Acer, Prunus, Ulmus, Begonia, Loranthus*, etc. In the western Himalayas *Quercus* species dominate all over. Conifers as *Abies, Cedrus, Picea*, etc., and broad-leaved species as *Betula, Alnus, Ulmus, Acer, Populus*, etc., are also widely distributed.

(12) Himalayan moist temperate forests. Central and Western Himalayas, have the forests of conifers or/and oaks at an altitude of 1500 to 3000 metres. The trees are tall (upto 45 metres), but undergrowth is thin and deciduous. These forests have plants like *Cotoneaster, Berberis, Spiraea, Quercus, Cedrus, Pinus, Picea, Abies, Tsuga, Cupressus*, etc.

(13) Himalayan dry temperate forests. These forests occur in a narrow belt in the western Himalayas extending from parts of U.P. through Himachal Pradesh and Punjab to Kashmir. They are open, evergreen forests with scrub undergrowth. Oaks and conifers are dominant. In western drier areas *Pinus gerardiana* and *Quercus ilex* are common, while in the eastern moist zone *Abies, Picea, Larix, Juniperus* are found. Other associated plant species are *Cedrus, Picea, Daphne, Artemisia, Fraxinus, Alnus*, etc.

V. ALPINE FORESTS

(14) Sub-alpine forests. These evergreen forests are open and occur throughout the Himalayas at high altitudes (above 3000 metres up to tree line). They comprise mostly conifers like *Abies spectabilis*, broad-leaved species like *Betula* and *Rhododendron* and shrubs as *Cotoneaster*, *Rosa*, *Lonicera*, *Smilax*, etc.

(15) **Moist-alpine scrub**. Throughout the Himalayas, above the timber line upto the height of 5500 metres, dwarf evergreen shrubby growth of conifers (*Juniperus*) and broad-leaved species (*Rhododendron*) are found.

(16) **Dry-alpine scrub**. In areas with low rainfall (below 35 cm) open xerophytic scrubs like *Juniperus*, *Caragana*, *Eurotia*, *Salix*, etc., are found upto 5500 metres.

(b) **Grassland vegetation**. In India, natural grasslands (as climax formation) are not present but occur only during succession. The following three types of grasslands occur in India : (1) **xerophilous** that occur in dry regions of north-west India under semi-desert conditions ; (2) **Mesophilous** which are extensive grass flats or **savannas** and occur in moist deciduous forests of Uttar Pradesh; and (3) **Hygrophilous** which are called wet **savannas**. These three basic types of Indian grasslands have been further subdivided into the following eight subtypes, each of which is named on its dominant species: *Sehima—Dichanthium*, *Dichanthium–Cenchrus*, *Phragmites–Saccharum*, *Bothriochloa*, *Cymbopogon*, *Arundinella*, *Deyeuxia–Arundinella*, and *Deschampsia—Deyeuxia*.

23-2·4. Floristic (Botanical) Regions (Provinces) of India

The Indian sub-continent is characterised with a variety of climate types and the flora of the country is also correspondingly of different types in its different parts. The country has been divided into the following nine **floristic regions** for the study of flora :

23-2·4A Western Himalayas. It extends from Kumaon to Kashmir and has annual rainfall upto 200 cm. Altitudinally there are following three zones of vegetation corresponding to three climatic belts :

(i) **Submontane zone**. This extends upto 1500 metres altitude and comprises mostly of Siwalik ranges. The forests are tropical and subtropical having trees like *Shorea robusta*, *Dalbergia sissoo*, *Cedrela toona*, *Ficus glomerata*, *Eugenia jambolano*, *Acacia catechu*, *Butea monosperma* (dhak), *Zizyphus* and thorny succulent enphorbias on slopes.

(ii) **Temperate zone**. Above submontane zone extend montane temperate forests upto 3500 metres altitude. They are dominated by plant species like *Quercus*, *Acer*, *Ulmus*, *Rhododendron*, *Betula* (birch), *Salix* (cane), *Populus*, *Cornus*, *Prunus*, *Fraxinus*, *Pinus*, *Cedrus*, *Picea* and *Taxus*.

(iii) **Alpine zone**. This zone extends from 3500 to 4500 metres altitudes (snow line) **and** is characterised with alpine forest vegetation with scrub and meadows. Most common tree species are *Abies*,

Betula, *Juniperus* and *Rhododendrons*. The herbs which occur near the snow line include species like *Primula*, *Potentilla*, *Polygonum*, *Geranium*, *Saxifraga*, *Aster*, etc.

23·2·4B. **Eastern Himalayas.** It includes regions of Sikkim and NEFA and is characterised by more rainfall, less snow and higher temperature. This is also divided into the following three zones altitudinally :

(a) **Tropical zone.** Upto 1800 metres altitudes, this zone has tropical semi-evergreen or moist deciduous forests. These forests comprise the plants like *Shorea robusta*, *Acacia catechu*, *Delbergia sissoo*, *Terminalia*, *Albizzia*, *Cedrela*, *Dendrocalamus* (bamboo), etc.

(b) **Temperate zone.** This zone extends between 1800 metres to 3800 metres altitudes and has typical montane temperate forests which are dominated by oaks like *Michelia*, *Quercus*, *Pyrus*, *Symplocos*, *Eugenia*, etc., at lower levels and by conifers as *Juniperus*, *Cryptomeria*, *Abies*, *Pinus*, *Larix* and *Tsuga* and also *Salix*, *Rhododendron* and *Arundinaria* (bamboo) at higher cooler levels.

(c) **Alpine zone.** Beyond the temperate zone, extends alpine zone upto 5000 metres altitudes. It has alpine vegetation including *Juniperus* and *Rhododendron* with its other typical flora.

23·2·4C. **Indus plains.** This zone includes the arid and semi-arid regions of Punjab, Rajasthan, Kutch, part of Gujrat and Delhi. The rainfall is less than 70 cm. The vegetation is tropical thorn forest in semi-arid region and is typical desert in the arid region. The plants of this zone are primarily xerophytic. The common plant species of this zone are *Acacia nelotica*, *Prosopis* sp., *Salvadora*, *Tecomella*, *Capparis*, *Tamarix*, *Zizyphus*, *Calotropis*, *Panicum*, *Saccharum*, *Cenchrus*, *Euphorbia*, etc.

23·2·4D. **Gangetic plains.** This region extends over Uttar Pradesh, Bihar, Bengal and part of Orissa and is characterised by moderate amount of rainfall and most fertile (*i.e.*, alluvial) soils. Vegetation of this zone is chiefly of tropical moist and deciduous and dry deciduous forest type. The common plants of this zone are *Dalbergia sissoo*, *Acacia nelotica*, *Saccharum munja*, *Butea monosperma*, *Madhuca indica* (mahua), *Terminalia arjuna* (arjuna), *Buchanania lanzan* (chiraunji), *Diospyros melanoxylon* (tendu), *Cordi myxa* (lisora), *Acacia catechu* (khair), *Azadirachta indica* (neem), *Ficus bengalensis* (bergad), *Ficus religiosa* (pipal), *Mangifera indica*, and weeds and grasses like *Xanthium*, *Cassia*, *Argemone*, *Amaranthus*, etc. In Gangetic delta (South Bengal) mangrove vegetation is common.

23·2·4E. **Central India.** It comprises Madhya Pradesh, parts of Orissa and Gujrat. The rainfall is 150–200 cm and its vegetation is thorny, mixed deciduous and teak type. The chief plants of this region are *Tectona grandis*, *Madhuca*, *Diospyros*, *Butea*, *Dalbergia*, *Terminalia*, *Carissa*, *Zizyphus*, *Acacia*, *Mangifera*, etc.

23·2·4F. **Malabar (west coast).** This region include western coast of India from Gujrat to Cape Comorin and has heavy rainfall.

The forests are tropical evergreen in extreme west, semi-evergreen towards interior subtropical or montane temperate evergreen forests in Nilgiris and mangroves near Bombay and Kerala coast. The species composition of these forests has already been discussed under section 23-2·3A.

23-2·4G. Deccan Plateau. This region extends all over peninsular India (*i.e.*, Andhra Pradesh, Tamil Nadu and Karnataka) and has rainfall upto 100 cm. Its central hilly plateau has tropical dry deciduous forests of *Boswellia serrata*, *Tectona grandis* and *Hardwickia pinnata*, while, the low eastern dry Coromandal Coast has tropical dry evergreen forests of *Santalum album* (chandan), *Cedrela toona* and plants like *Acacia*, *Prosopis*, *Euphorbia*, *Capparis*, *Phyllanthus*, etc.

23-2·4H. Assam. This region is characterised by heavy rainfall (200 to 1000 cm). The vegetation is either dense evergreen forest or sub-tropical. The evergreen forests include trees like *Dipterocarpus macrocarpu*, *Mesua ferrca*, *Shorea robusta*, *Ficus elastica*, etc., bamboos as *Bambusa pallida*, *Dendrocalamus hamiltonii*, etc., grasses like *Imperata cylindrica*, *Saccharum* sp., *Themeda* sp., insectivorous plants as *Nepenthes* sp., and also epiphytes (ferns and orchids). In the northern cooler regions, wet hill forests include plants like *Alnus*, *Betula*, *Rhododendron*, *Magnolia*, etc., are found. The hilly tracts also have pine forests of *Pinus khasiya* and *P. insularis*.

23-2·4I. Andmans. This region possesses a varied type of vegetatian: mangroves and beech forest at its coasts and evergreen forests of tall trees in the interior. Important plant species of this island are *Rhizophora*, *Mimusops*, *Calophyllum*, *Lagerstroemia*, etc.

23-3. DESCRIPTIVE ZOOGEOGRAPHY

As discussed in the section 23·1, a few groups of animals are cosmopolitan in having world-wide distribution and most animals are restricted in distribution beause of same kind of barrier or because of past history of origin and dispersal. An understanding of their present-day distribution takes us into zoogeography. Major units of distribution are the **zoological region** or **zoogeographical regions**, areas defined largely by the past and present relations of the continent to each other. Each region is further subdivided into **faunal** or **ecological units**, depending on the criteria used.

23-3·1. Zoogeographical Regions

There are six zoogeographical regions, each more or less embracing a major continental land mass and all have long been separated from one another by water (oceans and fresh-water bodies), mountain ranges or desert, so that, each region has evolved its distinctive and characteristic orders and families of animals. These major distributional units were first recognised by **Sclater** (1858), modified by **Huxley** (1868), extended by **Wallace** (1876) and best described in a modern way by **Darlington** (1957) and others. The six zoogeographical regions are the **Palaearctic**, the **Nearctic**, the **Neotropical**, the **Ethiopian**, the **Oriental** and the **Australian**. Because some zoogeographers consider the Neotropical and the Australian regions to be so different from the rest of the world, these two are often considered as regions or realms equal to the other four combined. They are classified as **Neogea** (the Neotropical), **Notogea** (the Australian) and **Metagea** (the Palaearctic, Nearctic, Ethiopian and Oriental). Moreover, two zoogeographic regions, namely the Palaearctic and the Nearctic are quite closely related, so the two often considered as one, the **Holarctic**. In fact, these two regions are similar in climate, vegetation, and in their faunal composition having animals like wolf, hare, moose (called elk in North America), caribou, wolverine and bison.

23-3·1A. Palaearctic region. This largest region comprises the whole of Europe, Soviet Russia, Northern China, Japan, Northern Arabia (Persia) and narrow strip of coastal North Africa. It is subdivided into **European**, **Mediterranean**, **Siberian** and **Manchurian** subregions. Its eastern parts have characteristic fauna of the following animal species : *Rhacophorus* (flying frog), *Bombinator* (fire-bellied frog), *Agkistrodon halys* (pit viper), *Strix uralensis*, *Grus japonensis* (crane), *Aix galericulata*, *Erinaceus* (hedgehog), *Podoces panderi*, *Capricornis sumatraensis*, *Camelus ferus* (camel), *Bos mutus*, *Ailuropus* (great panda), *Panthera tigris altaica* (tiger), *Uncia uncia*, *Phoca sibirica* (fresh-water seal), *Pteromys volans*, *Equus hemionus* (donkey) and *Macaca fuscata* (rhesus monkey). The common fauna of Western Palaearctic include animal species like *Clupea harengus* (herring), *Sardina pilchardus* (fish), *Gadus morrhua* (codfish), *Vipera ammodytes* (viper snake), *Chamaeleon chamaeleon* (arboreal lizard), *Pelecanus onocrotalus* (pelecan), *Fratercula arctica*, *Phoenicopterus ruber*, *Rangifer tarandus* (caribou), *Ovis ammon*, *Dama dama*, *Capra aegagrus*, *Odobenus rosmarus* (walrus), *Ursus arctos*, *Lynx lynx* (wild cat), *Castor fiber* (beaver), *Hystrix cristata* (porcupine), and *Panthera pardus* (leopard).

23-3·1B. Nearctic region. This region comprises the North American continent south to the Tropic of Cancer (*i.e.*, Greenland and North America). It is subdivided into **Californian**, **Rocky mountain**, **Allegany** and **Canadian** subregions. The Nearctic is the home of many reptiles and has more endemic families of vertebrates. Characteristic fauna of this region include animal species as *Amblystoma tigrinum*, *Crotalus adamanteus* (rattle snake), *Branta canadensis* (Canadian goose), *Larus schistisagus* (gull), *Bucephala albeola*, *Meleagris gallopavo* (turkey), *Didelphis marsupialis* (opossum), *Ovibos moschatus* (musk-ox), *Rangifer tarandus arcticus* (Caribou), *Bison bison* (North American bison), *Odobenus rosmarus divergens*, *Cystophora cristata*, *Martes pennanti* (fisher), *Lutra canadensis* (otter), *Dasypus novemcinctus* (armadillo), *Canis latrans*, *Procyon lotor* (racoon), *Lagenorhynchus acutus*, *Delphinapterus leucas* and *Enhydra lutric*.

23-3·1C. Neotropical region. It includes Central America, South America, part of Mexico and the West Indies. This region lacks well-developed ungulate fauna of the plains, but have rich, distinctive, varied endemic fauna of vertebrates. Characteristic animals of this region are *Lepidosiren* (lung-fish), *Lepidosteus* (Gar-pike), *Histrio histrio*, tortoises (*Dermatemys, Stourotypus, Peltocephalus*), *Heloderma horridum, Iguana iguana, Amazilia yucatanensis, Phoenicopterus ruber, Pelicanus occidentalis* (brown pelecan), *Phalacrocorax harrisi, Sarcoramphus papa, Amazona amazonica, Spheniscus magellanicus, Chironectes minimus, Desmodus rotundus, Myrmecophaga tridactyla* (giant anteater), *Tapirus terrestris* (tapir), *Lama vicugna, Lama guanicoe, Panthera onca, Arctocephalus australis, Cavia aperea, Bradypus tridactylus* (three-toed sloth), *Vultur gryphus, Pterocnemia pennata* and *Ara macao*.

23·3-1D. Ethiopian region. The old world counterpart of the Neotropical is the Ethiopian, which includes the continent of Africa, south of the Atlas mountain and Sahara Desert, Southern Arabia, Madagascar and Mauritius. It embraces tropical forests in central Africa and in the mountains of East Africa, Savanna, grasslands and deserts. It includes a varied vertebrate fauna and several endemic families. The characteristic animals of this region are *Protopterus* (lung fish), *Crocodilus niloticus, Sagittarius serpentarius, Struthio camelus* (ostrich), *Oryx gazella* (African antelope), *Giraffa camelopardalis* (giraffe), *Loxodonta africana* (Africana elephant), *Equus quagga, Hippopotamus amphibius, Daubentonia madagascariensis* (ayeaye), *Lemur catta, Macaca sylvana, Pan troglodytes* (chimpanzee), *Gorilla gorilla, Acinonyx jubatus, Panthera leo* (lion), *Hyaena hyaena*, and *Delphinus delphis*. It lacks deer and bear among mammals and salamanders and tree-frogs among amphibians and contain several endemic families of birds.

23-3·1E. Oriental region. It includes India, Indochina, South China, Malaya, and the western islands of the Malay Archipelago. It is bounded on the north by the Himalayas and on the other sides by the Indian and Pacific oceans On the southeast corner, where the islands of the Malay Archipelago stretch out toward Australia, there is no definite boundary, although **Wallace's line** is often used to separate the Oriental from the Australian regions. This line runs between the Philippines and the Moluccas in the north, then bends southwest between Borneo and the Celebes, then south between the islands of Bali and Lombok. A second line, **Weber's line**, has been drawn to the east of Wallace's line ; it separates the islands with a majority of Oriental animals from those with a majority of Australian ones. Since the islands between these two lines form a transition between the Oriental and the Australian regions, some zoogeographers call the area **Wallacea (Smith**, 1977).

The Oriental region is divided into the following four sub-regions : **Indian sub-region, Ceylonese sub-region, Indo-Chinese sub-region** and **Indo-Malayan sub-region.** Its characteristic fauna includes the following animal spicies : *Rhacophorus pardalis, Gavialis* *niloticus* (gharial), *Calotes versicolor* (garden lizard), *Draco volans* (flying lizard), *Draco dussumieri* (flying lizard), *Python reticulatus*, *Naja naja* (Indian cobra), *Ophiophagus hannah*, *Gallus gallus* (jungle fowl), *Pavo cristatus* (peacock), *Milvus migrans* (pariah kite), *Eudynamys* (Koel), *Psittacula krameri* (rose-ringed parakeet), *Bubo bubo* (great horned owl), *Coracias benghalensis* (blue jay), *Dinopium benghalense* (golden-backed woodpecker), *Corvus splendens* (crow), *Argusianus argus*, *Elephas maximus* (elephant), *Rhinoceros unicornis*, *Antelope cervicapra* (black buk), *Axis axis* (spotted deer), *Boselaphus tragocamelus* (blue bull), *Selenarctos tibetans* (black Himalayan bear), *Melursus ursinus* (sloth bear) *Sus cristatus* (wild boar), *Hystrix leucura* (Indian porcupine), *Herpestes* (mongoose), *Manis* (pangolin), *Soriculus* (Indian shrew), *Hyaena striata*, *Panthera tigris tigris*, *Panthera pardus*, *Babyrousa babyrussa*, *Macaca mulatta*, *Presbytis entellus* (langoor), *Pongo pygmaeus* (orangutan), *Ailurus fulgens*, and *Hylobates lar* (gibbon).

23·3·1F. Australian region. This region includes Australia, Tasmania, New Guinea, and a few smaller islands of the Malay Archipelago. New Zealand and the Pacific Islands are excluded, for these are regarded as oceanic islands separate from the major faunal regions. Partly tropical and partly south temperate, the Australian region is noted for its lack of a land connection with other regions ; the poverty of fresh water fish, amphibians and reptiles ; the absence of placental mammals and dominance of marsupials. Included are the egg-laying mammals (monotremes) and the spiny ant-eaters. The characteristic fauna of this region includes following animal species : *Neoceratodus* (fish), *Chelmon rostratus*, *Phyllopteryx eques*, *Ornithorhynchus anatinus* (platypus or duck mole), *Zagoglossus bruijni* (proechidna), *Techyglossus aculeatus* (echidna, spiny ant-eater), *Macropus giganteus* (kangaroo), *Notoryctes typhlops*, *Macrotis lagotis*, *Dromaius novaehollandiae*, *Petaurus austalis*, *Paradisaea rubra*, *Pteridophcra alberti* and *Phaseolarctos cinereus* (koala).

New Zealand has *Varanus komodensis*, *Sphenodon punctatus*, *Abteryx owenii haasti* (kiwi), etc. Sometime, south polar region called the **Antarctic** or **Archinotic** is also included in these major distributional units. Antarctic region has the following characteristic animal **species** : *Balaenoptera musculus* (blue whale), *Orcinus orcam*, *Megadyptes antipodes*, *Aptenodytes patagonica* (penguin), *A. foresteri* (emperor penguin), *Pygoscelis adeliae*, *Diomedea exulans* (albatross), *Chionis alba*, and *Lobodon carcinophagus*.

23·3·2. Life Zones, Biotic Provinces and Biomes

Each major distributional unit (*i.e.*, Palearctic, etc.) is further subdivided by secondary barriers such as vegetation types and topography. The **life-zone** concept, restricted to North American, divides the continent into broad transcontinental belts ; the plant and animal differences between these belts are governed chiefly by temperature. The **biotic provinces** approach divides the North American continent into **continuous** geographic units that contain ecological associations different from those of adjacent units, especially at the species and subspecies levels. The **biome system** groups the plants and animals of the world into integral units characterised by distinctive life forms in the climax. Boundaries of biomes or major life zones coincide with the boundaries of the major plant formations of the world. Thus, by including both plants and animals as a total unit that evolved together, the biome permits the recognition of the close relationship that exists among all living things.

23-4. DYNAMIC BIOGEOGRAPHY

The explanation of why a particular species has come to occupy a particular niche in a particular part of the world requires a knowledge of where it originated, how it dispersed, and how it evolved its present adjustments and characteristics. When populations invade new regions and environments and become geographically isolated, they commonly differentiate into new species. Each species thus becomes the product or visible expression of a particular combination of environmental factors, interactions and locality, and as such the species is the practical taxonomic unit with which the ecologist has to deal with.

23-4·1. Dispersal Dynamics

Dispersal is the spread of individuals away from their home-sites. Dispersal movements are usually slow, and cover relatively short distances in the life time of an individual. The cumulative result of short dispersals by successive generations, however, may become conspicuous in the course of years, decades, or centuries, when it constitutes **range expansion** of the species into a new habitat or area. For example, house sparrow was introduced in North America in 1852-53 and now they have spread throughout the continent. Thus, once man had helped them to overcome the ocean barrier that previously held them in check, these species spread unusually rapidly because of the optimum environment on the continent.

23-4·1A. Manner and means of dispersal. Animals find suitable habitat and niches in various ways. In an uniform environment, dispersal movements radiate in all directions from the home area. The greater the density of individuals in the home area, the more quickly distant areas are invaded, and the farther away do individuals move.

Dispersal is usually performed by immature stages of the individuals. A common method of achieving dispersal is the **broadcasting** of enormous numbers of eggs, spores, encysted stages, or young so that they scatter randomly into a wide variety of places. Those that come by chance into suitable environments persist and become established ; those that enter into unfavourable locations are destroyed or never develop. Broadcasting is a wasteful procedure ; it is especially characteristic of aquatic species. The fresh-water clam, for example, annually produces hundreds of eggs. Only two fertilised eggs need maturation that the two parents be numerically replaced when they die, and thus the population of the species be maintained at a constant level. Contrastingly, in those animals which have developed high degree of parental care such as birds and mammals the number of eggs or young ones annually produced is commonly a half-dozen or less, and the offspring exercise considerable discrimination in their choice of suitable habitat.

During broadcasting, the immature stages are either conveyed passively (*i.e.*, **passive conveyance**) by wind, current, or other animals, respond to environmental factors by directed movements or **taxes,** or find their way by trial and error. Dispersal continues at measurable rates until a physiographic, climatic or biotic barrier is reached.

23-4·1 B. Barriers. External factors that limit the distribution of an organism are termed **barriers.** Barriers are always relative and a barrier for one species may well be a main dispersal route for another. For example, water might well be a barrier for a terrestrial species but normally is not for a aquatic organism as fish. Three classes of barriers are generally recognised : (1) **Physiographic** or **physical barriers** such as land, water, elevation, soil types (sand. rock, etc.) ; (2) **Climatic barriers** such as temperature, humidity, rainfall, sunlight, and others ; and (3) **Biotic** or **biological barriers** such as lack of food and the presence of enemies or effective competitors.

23-4·1C. Causes of dispersal. Factors that induce dispersal include population pressure, failure of food supplies or loss of favourable homeostasis, opening up of new areas elsewhere and pre-adaptation for new and different niches.

23·4·2 Dispersal Pathways

An understanding of how animal and plant groups have dispersed over the face of the earth involves the controversy as to whether all the continental land masses were at one time intimately connected and, if so, how long ago they drifted apart to their present locations. Separated continents need at least temporary land bridges if there is to be dispersal and intercharge of fauna between them.

23-4·2A. Continental drift theory. This theory postulates that throughout the Palaeozoic and much of the Mesozoic the presently distributed continents were grouped into two great land masses : a northern **Laurasia** was separated from a southern **Gondwana** by the vast sea of Tethys. In the Jurassic the land masses fragmented, and the fregments subsequently drifted apart (the solid continental crust presumably "floats" upon the deeper, liquid basalt layers of the earth). Laurasia is supposed to have split into North America, Greenland, Europe and most of Asia ; Gondwana, into South America, Africa, Arabia, Madagascar,

India, Australia and Antarctica (**Wegener**, 1924). This theory has been supported by various evidences.

23-4·2B. Land bridges. An alternative theory is that the continents have been situated in their present positions throughout geological time. They have changed their boundaries with rise and fall of ocean levels and with upheaval and subsidence of land masses but there has been no significant lateral drifting. Regardless of their ancient history, the continents have been largely separated during the last 60 to 100 million years, and movements of animal and plant groups during this latter period have been most important in **determining** their present distribution.

The separated continents, however, have been connected by at least temporary land bridges at various times during the period. A land bridge now connects North and South America at the isthmus of Panama, and Eurasia and Africa are connected by the isthmus of Suez. One of the best known and generally accepted land bridges of the past, but not now in existence, connected Asia and North America across the Bering strait and Bering sea. Possibly extensive archipelagoes occurred in the Atlantic ocean area in the past have permitted dispersal of organisms.

Land bridges serve as important dispersal routes for those land animals and plants which are able to cross them. When a land bridge allows free passage of most animals and plants in either or both directions, it is called a **corridor**. If the land bridge is narrow, has an unfavourable climate, a lack of suitable niches, or too many competitors, it is called **filter**, since only a few specially adapted species are able to pass over it. A **sweep-stakes route** of dispersal is one over which only a few species pass, more or less by chance. Because of its **general** unfavourableness, it is generally only one way. In fact, the fauna of continental islands is derived mainly from the adjacent mainland over such land bridges, but the faunas of the more distant oceanic islands are often unique and unbalanced and dependent on accidental **dispersal** of miscellaneous species.

According to **Matthew** (1915), the major vertebrate groups (mammals, etc.) first evolved in temperate Asia (Eurasia) and then spread to other parts of the world. However, **Darlington** (1957) proposed that cold-blooded animals (fish, amphibian and reptiles) and warm-blooded animals evolved in the oriental region and then spread to other parts of the world.

23-4·2C. Centres of origin. The tracing of dispersal routes presupposes a starting point where the taxonomic group first evolved. These starting points are called **centres of origin** and are usually recognised accoring to the following two criteria ; 1. Location of greatest differentiation of the type or the greatest variety of endemic races, species, and genera, including primitive forms or fossils. The older a group is and the longer it has occurred at a particular location, the more chance it has had to radiate into different habitats, become isolated, and evolve into new varieties. 2. Continuity and convergence of lines of dispersal. Dispersal from a centre radiates in all directions in which conditions are favourable and until insurmountable barriers are encountered. Lines of dispersal may be readily distinguished where one or more taxonomic characters can be traced from primitive or generalised types through more and more specialised types the greater the distance involved.

23-4·3. Migration

Migration, like dispersal, involves movements and the invasion of new areas. However, migration differs from dispersal in that it is **periodic** movement back and forth between two areas (dispersal being one-way outward movement). Migratory invasions of areas are temporary and repetitive, but invasions resulting from dispersal may be permanent.

Migrations may be annual or daily, or may be in the form of changes of habitat at different stages in a life cycle (the latter is called **metamorphic migration** and occurs in dragonflies, anadromous Pacific Salmon and catadromous eels of Western Hemisphere). Annual migration may be latitudinal, altitudinal, or local. Annual migration is best developed in birds but also occurs in mammals, fish and insects.

23-4·4. Ecesis

Ecesis is the establishment of organisms in an area into which they have come by dispersal or migration. This involves the establishment of individuals, the growth of populations, the invasion of more and more species, and finally the development of mature (climax) communities. The growth of species populations and of complex communities commonly follow a sigmoid curve. The disappearance of populations, however, follow an inverse or negative growth curve.

23-5. REVISION QUESTIONS

1. Describe different patterns of distribution of biota.
2. What is phytogeography ? Describe chief vegetational belts of Earth.
3. Write an essay on the phytogeography of India.
4. What is zoogeography ? Describe different zoogeographical regions of Earth.
5. Describe the process of dispersal of organisms.
6. Write short notes on the following : Endemic species ; Oriental region; range expansion; broadcasting; barriers; continental drift theory; land bridges; migration; and ecesis.

Part 6

APPLIED ECOLOGY

Man is a vertebrate animal (mammal) and everywhere he lives in interspecific animal and plant communities. He depends on his environment no less than any other organism for food, water, air, space and shelter and in fact, he forms one of the most dominant component of almost all natural ecosystems. In his nearly 2 million years of occupancy of the earth, man has been moulding the planet to his own designs. Domination of the human species is revealed by the layers of dust and smog that blanket much of the earth and the amoeboid growth of human populations stretching across the land. Large areas of natural vegetation have been transformed into crop lands; free-flowing rivers have been dammed to create enormous lakes; mountain and hills have been pared away for the exploitation of minerals ; wild-life has been coldly destroyed and forced to almost extinction for petty gains; and refuse and debris of modern civilisation litter the planet. He has introduced excessive quantities of certain toxic elements such as mercury, lead and cadmium into the biogeochemical cycles. He has added certain synthetic materials such as chlorinated hydrocarbons to which organisms had never been exposed. In his domination of earth, he is causing his environment to deteriorate and is exhausting exploitable resources and he is allowing his population to push upward to the ultimate carrying capacity of the earth and there is no place into which it can expand. Now man finds himself on the verge of an ecological catastrophe. Fortunately, recently man has realised that he is a functional unit of many ecosystems and that his activities must be in harmony with their functioning. If man is to exist, he has to work within the framework of the ecosystems of which he is a part. In other words, he needs to develop an ecosystem approach to the management of the earth (**Smith**, 1977).

The branch of ecology which deals with the application of the ecological principles in the management of nature and its resources. is called **applied ecology**. In this section various aspects of applied ecology such as ecology of human population, pollution, ecology and human welfare (conservation of natural resources) and wildlife management will be described.

24

Ecology of Human Population

There has been a great deal of talk in rec nt years about the "**population explosion**". It is said that the human population is increasing so rapidly that widespread famine is inevitable in the near future. There is much fierce debate about the methods that should be used to slow down or stop the growth of the human population. This is a very controversial topic, as it is linked with the individual's sentiments, religion and economical, educational, social and political awareness. People have strong feelings about the right to have children, or the need to limit population growth. Many feel that any interference with the natural birth rate is immoral. Others see the population-control movement as an attempt of wealthy or super-cast people to keep what they have and hold down the rest of the world in the process. On the other side, some maintain that continued high birth rates are immoral and even criminal, for they will lead to starvation and endless misery for millions. There are no simple answers to the problems of human population.

In this chapter, we will simply look at some of the available information about what has happened to the human population in the past and then we will try to apply some of the principles of ecology to explain the changes in population and talk about the possibilities for the future.

24-1 DEMOGRAPHY

Demography is the description and statistical analysis of the vital processes and movements of human populations. Most of the measures of human populations are the same as those we have already applied to animal populations in general (Chapter 15). By considering the total human population of world one finds that the generalised population formula is applicable : population equals biotic potential minus environmental resistance. According to usage by the demographers, biotic potentials can be expressed in terms of **fecundity**, which is defined as the theoretical or maximum capacity of the individual or population to reproduce. **Fertility** is the actual reproductive performance of a given population. Fecundity in human population has been stable for a long time, but fertility, being dependent upon ecological and social factors, has varied greatly.

Human **birth rate** (natality) is the fertility expressed as the average number of live births per thousand population per year. Since not all age groups in a population have equal fertility, domographers often use more specialised measures of birth rates that relate to age groups.

Human **mortality rate** is the average number of individuals who die per thousand population per year. Again, demographers often use more sensitive measures, for different mortality rates. Both the very young and the very old segments of the population have higher mortality rates than are found in other age groups.

Human population shows uneven or clumping pattern of distribution on earth. The **density** of human population in a village, district, city, province, country or any area can be obtained by dividing the total number of persons living in the given region by the total land area of that region The average number of people per square unit of land area tells us how dense or sparse is the population in a given area. The average population density of the world is calculated about 27 persons per square kilometre.

The information secured from the study of human populations is known as **vital statistics**. These have many applications in our modern society. For example, life insurance is based upon the computation of **actuarial tables** which express the incidence of death by sex, age, and cause. Some general findings resulting from an analysis of vital statistics are as follows :

1. Birth rates are higher in rural areas and among farming people than in cities. Furthermore, the larger the city, the lower the birth rate.
2. Within a given region, such as city or a farming area, the lower economic classes have the higher birth rates.
3. Birth rates decrease during economic depressions and increase during times of prosperity.
4. More males than females are born.
5. Social and economic factors which produce low birth rates also produce low mortality rates. Mortality rates are lower in times of depression, in higher economic groups and in city dwellers.
6. Women have lower mortality rates than do men.
7. Both marriage and divorce rates increase during times of prosperity.
8. Both married men and married women live longer than single persons.

Table 24·1 gives population size, birth rates and death rates for the human populations of different countries of world. Note that these are far from uniform and that, in general, the more highly industrial countries have lower birth rates and lower death rates.

Table 24·1. 1975 population statistics for various countries (from McCombs and Rosa, 1978).

Country	Population (Millions)	Population Density (Persons per Sq. Km.)	Annual Population Growth	Crude Birth Rate	Crude Death Rate
1. United Kingdom	56	229	0·2%	12	11
2. France	53	97	0·8%	16	9
3. Italy	55·8	185	0·8%	16	9
4. United States	213·6	23	0·8%	16	9
5. Poland	34	109	0·9%	17	8
6. U.S.S.R.	242·7	11	0·9%	17	8
7. Spain	35·4	70	1·0%	18	8
8. Japan	111·9	298	1·2%	19	7
9. Argentina	25·4	9	1·3%	22	9
10. Canada	22·8	2	1·4%	15	7
11. China	838·8	87	1·4%	25	10
12. South Korea	34·7	345	1·6%	28	9
13. India	598·1	182	2·1%	36	16
14. Egypt	37·2	37	2·2%	36	13
15. Turkey	39·1	50	2·4%	40	13
16. Burma	31·2	46	2·5%	40	16
17. Ethiopia	27·9	23	2·6%	50	26
18. Iran	33	20	2·9%	41	15
19. Philippines	42·5	142	2·9%	44	11
20. Brazil	107·1	13	3·0%	37	9
21. Pakistan	70·3	87	3·0%	48	17
22. Thailand	42·3	82	3·1%	44	11
23. Columbia	29·7	22	3·2%	42	9
24. Syria	7·4	40	3·3%	46	15
25. Mexico	60·1	30	3·5%	43	8
26. Kenya	13·4	23	3·6%	50	16
27. Libya	2·4	1	4·2%	46	15

24-2. HISTORY OF HUMAN POPULATION GROWTH

Human population increased gradually in size from the time of primitive man until about the mid 1800's. Since then human populations have increased rapidly. Population increase has been associated with major shifts in man's relation to his environment. The ecology of man and trend of growth of human population can be studied at the following three levels : primitive man, agricultural man and industrial man.

24-2·1. Primitive Man

It is believed that the population of our ancestors, the **hominoid apes** or *Australopithecus* and relatives, a few million years ago, were confined to Africa and total number of individuals was 125,000 only. During man's evolutionary history, the size and functioning of human brain has increased considerably. (The *Australopithecus* had small brain with an average volume of only about 450 to 600 cubic centimetres). Larger brains in turn increased the potential store of cultural information and resulted in self-reinforcing and coupling of the growth of culture and brain size. This trend continued until perhaps 100,000 to 35,000 years ago, when growth of brain size stabilised at an average of some 1500 cubic centimetres and man constituted the modern species, the *Homo sapiens*.

For a hundred thousand generations, our ancestors lived in small, isolated, nomadiac groups as hunters and food-gatherers, spacing themselves according to the availability of wild food. The numbers of people in these scattered foraging bonds probably grew slowly, if at all. Around 30,000 B.C. mankind has spread (emigrated) from Africa to every other part of this earth. Around 9000 years ago, the total human population was low, usually estimated as less than 20 million.

Ecologically, man's primitive position in natural communities was that of a particularly large, vicious land animal with an unusually wide range of food. He was and still is omnivorous structurally, physiologically and psychologically. In other words, he was a consumer, a primary consumer as well as a secondary or tertiary consumer. It does not mean that man could eat literally everything, but that he could eat almost any kind of animal food and any kind of concentrated plant food, especially fruits, seeds and tubers or starchy roots. Biological consequences of this breadth of adaptation were that man could find food in almost any of the endlessly

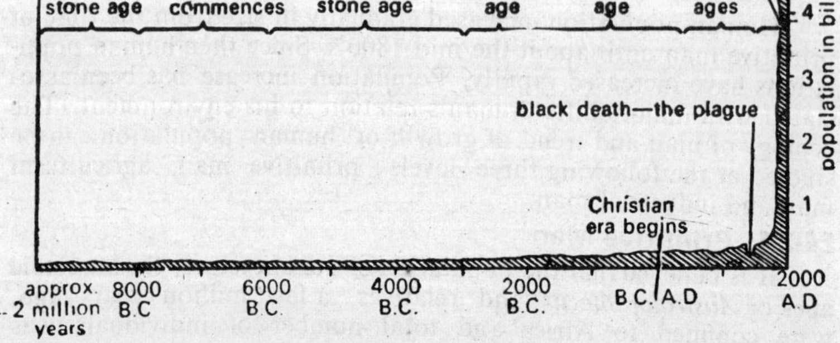

Fig. 24·1. Growth of the human population through history (after Herreid II, 1977).

diverse natural communities of the earth and that in each community his status was complex. He became part of almost all food chains that could include a large animal. In fact, as long as man remained in this stage of cultural development, he was just another member of the complex food-web of the community and populations soon reached some maximal population levels in balance with the rest of the community. All his energy requirements were furnished by the **renewable resources** in other words, by plants or animals that are "renewed" each year by biological growth. Even when primitive man began to use fire, his fuel was wood from trees that were renewed.

24·2·2. Agricultural Man

Some time between 7000 B.C. and the old kingdom of Egypt (which began prehaps about 5500 B.C.) certain groups of primitive man began modifying their environments by cultivating specific kinds of plants (especially **grains**) and domesticating food animals. Ecologically speaking, this was accomplished by shifting the relative numbers of individuals of various species in the food web. Agricultural man further shifted the balance of nature in the original biotic communities by killing any wild herbivores that fed on his cultivated plants and killing carnivores that preyed on him and his domestic animals.

During this **agricultural revolution**, a sedentary way of life evolved, which allowed the construction of permanent dwellings in which man could seek shelter from the weather. As life became safer and the food supply more stable, a large number of babies survived. Around 8000 B.C., the world population is estimated to be about 5 million. By the time Egypt became well established as an ancient civilisation (perhaps about 4000, B.C.) the world population probably reached the 100 million level. By 1650 A.D., human population had climbed to 453 million. The average rate of increase during this period (7000 B.C. to 1650 A.D.) was about 50,000 per year.

24·2·3. Industrial Man

As human population continued to increase, man continued to modify his environment in such a way as to permit even greater population-size. With the invention of the wheel, transportation devices were developed and utilised for moving food from place to place, thus allowing for larger concentration of people in a given place. The discovery and utilisation of non-renewable resources, such as the various metals and the fossil fuel sources (coal and oil), premitted human population to increase beyond the limits originally imposed by the balance of nature. The so-called industrial revolution saw the development of tools and machines that resulted in the construction of more and better housing, the production of more and better food items and textiles, and the development of food storage, preservation and transportation techniques. Simultaneous acquisition of knowledge and development of drugs and techniques permitted medical science to greatly reduce mortality rates and the agricultural sciences to produce more and better food from cultivated plants and domestic animals.

During this period human population has increased at a rate that has become alarming to many people. By 1830, the population had reached one billion; by 1930, two billion; and by 1961, three billion. Between 1650 and 1830, the rate of increase was just over three million per year; between 1830 and 1930, abont ten million per year; and between 1930 and 1961, about 30 million per year.

In recent years, total population has been increasing at an ever-increasing rate (Fig. 24·1) and is apparently in a logarithmic phase of population increase. The current rate of increase is about 55 million per year ; about 150,000 per day ; 6300 per hour ; or 100 per minute. Remember, this is net increase—total births less total deaths.

If current birth rates and death rates are maintained, human population will reach six billion by 2025 A.D. It is estimated that a human population of only 1 billion people is the maximum supportable at U.S. levels of affluence by present agricultural and industrial resources (**Kendeigh**, 1974). The much larger populations now occurring are possible only because over 80 per cent of the world population is living well below these levels. These underprivilege people seek a larger share of the world's resources and a rise in their standard of living. The option is either ruthless suppression which is unthinkable or accommodation which is applied by present Governments in the form of family planning measures.

24·3. FACTORS REGULATING HUMAN POPULATION

Man surpass animals in being able to control his environment in some degree, by (1) **food production** (agriculture) and **food storage** (warehouses, canning, refrigeration) against seasonal shortage and deterioration ; (2) shift in energy **utilisation from solar energy→wood fire→fossil fuels→electrical** and atomic energy to increase human **efficiency** ; (3) improved **shelter** for adults and offsprings against unfavourable weather and against enemies (housing, flood protection, police protection, national defense); (4) **destruction of animal enemies** (lions, tigers, wolves, snakes, rats and mice, household insects) ; (5) **reducing competition** from other animals (fencing crops and livestock) ; and (6) **control of diseases** that attack human beings (sanitation, immunisation).

However, human beings are reduced in numbers by following agencies :

1. **Food shortage or famines** due to crop failure as from unfavourable weather, especially in densely populated lands such as China and India where food reserves are scant or unequally distributed. Malnutrition may lead to physical impairment with greater

susceptibility to disease and death. **2. Inadequate shelter**, especially during cold of winter outside the tropics is a important mortality factor. **3. Calamities in nature**, like floods, droughts, hurricanes, volcanic erruptions, earthquakes, etc., claim great toll of human population (Table 24·2). **4. Enemies**. Large predatory animals like lion, tiger and poisonous snakes, cause many deaths in less civilised regions. **5. Wars**. Warfare is a form of competition between human races or nations for land, natural resources, trade routes, or other desired features of human environment. Wars have remained a major mortality factor throughout human history, not only by causing direct battle casualities, but also by stimulating disease and famine. **6. Disease** is the largest single factor in reducing populations. Diseases like bubonic plague. cholera, typhus, small pox, malaria, yellow fever, sleeping sickness, etc., have been among the great killers of man throughout history. 7. Explosions, fire, pollution, shipwrecks and transportation accidents cause many death in modern civilisations.

Table 24·2. Human deaths due to noteworthy tropical cyclones. (from Dube and Singh, 1981).

Year	Location	Deaths
1970	Bangladesh	500,000
1737	India	300,000
1881	China	300,000
1923	Japan	250,000
1897	Bangladesh	175,000
1876	Bangladesh	100,000
1864	India	50,000
1833	India	50,000
1822	Bangladesh	40,000
1780	Antilles	22,000
1839	India	20,000
1789	India	20,000
1965	Bangladesh	19,279
1963	Bangladesh	11,520
1961	Bangladesh	11,468
1977	India	10,000
1963	Cuba-Haiti	7,196
1900	Texas	6,000
1960	Bangladesh	5,149
1960	Japan	5,000

By performing experiments on the populations of *Drosophila*, deer, house mice (*Mus musculus*), etc., biologists have predicted that increased size of human population will cause a lowered birth rate. This lowered birth rate might result from a lowered fertility occurring as a part of the **stress syndrome**, or it may result from social and economic changes. One major social factor still to be evaluated is the shift in thought concerning birth control that is now occurring in many religious groups. Others predict the human population size will be limited by intraspecific strife such as wars, revolutions and murders since the stress syndrome (adrenal gland deterioration, high blood pressure, hardening of the arteries, liver disease, and other breakdowns) is known in man, stress will cause increase in the mortality rate that will control population size. As biologists, we can only hope that modern man will solve this new problem by again applying his unique power of thought and reasoning and develop techniques of birth control instead of increasing deaih rates !

24-4. REVISION QUESTIONS

1. Describe the trend of human population growth through history ?
2. Describe the factors which regulate the size of human population ?
3. Why there is a great hue and cry on overpopulation in world?Explain it with sound reasons.

25
Pollution

Recently, first time in his entire cultural history, man has faced one of the most horrible ecological crisis—the problem of pollution of his environment which sometime in past, was pure, virgin, undisturbed, uncontaminated and basically quite hospitable for him. **Pollution** is an undesirable change in the physical, chemical or biological characteristics of our air, land, and water that may or will harmfully affect human life or that of desirable species, our industrial processes, living conditions, and cultural assets (**Odum**, 1971). In other words, pollution is the unfavourable alteration of our environment, largely as a result of human activities (**Southwick**, 1976).

25-1. ORIGIN OF POLLUTION

There exist different views regarding the origin of pollution crisis on the planet earth. Certain authors such as **Lynn White** (1967) and **Ian McHarg** (1969) blamed Judeo-Christian ethic for pollution. According to them this ethic taught man to believe that the earth was made for man to do with as he wished, and thereby encouraged exploitation. However, this view has been contradicted by **Wright** (1970). **Wright** pointed out that the Judeo-Christian religion teaches stewardship and he proposed that it is not religious belief but human greed and ignorance which have allowed our culture to develop an ecological crisis like pollution. **Tuan** (1970) described an ancient oriental culture untouched by the Judeo-Christian tradition which despoiled their forests and faced ecological crisis.

Some authors (such as **Southwick**, 1976), associate the human population explosion with the pollution problem. They point out that with more people there is more sewage, more solid wastes, more fuel being burned, more fertilizers and insecticides being used to produce more food for hungry mouths. But, there are certain writers who have pointed out that in underdeveloped countries, pollution is not the severe problem as it is in technologically developed countries and yet the populations may be very dense. They feel that it is the wasteful aspects of our technology which strive always to produce more convenient products ("disposable" items) which pollute our environment.

Some authors blamed the profit motives and capitalism of the modern economic system of the capitalist countries like USA,

Britain, etc., for pollution. But Russia, according to *Izvestia*, has problems very similar to West countries—severe oil pollution in the Caspian Sea from offshore oil drilling and flushing the holds of their ships, industrial pollution from milk, fish, and linen factories, regional food combines, and industrial complexes (**Benton** and **Werner**, 1974). Thus, it seems that the economic system is not the real culprit of pollution.

Finally, there are certain modern ecologists such as **Odum** (1971), **Southwick** (1976), **Smith** (1977), etc., who sought many factors such as human population explosion, unplanned urbanization and deforestation, profit oriented capitalism and technological advancement, which may have originated pollution crisis on earth. In fact, in countries where there have been the greatest technological advances, the worst pollution occurs. In these countries, whether capitalist. socialist, or communist, there has been an emphasis on growth, and what **W.G. Rosen** (1970) has called **frontier or flat-earth concept** of the world, instead of a spaceship or round-world view. When our population was small, and our society largely pastoral, this attitude did not usually result in obvious pollution, but more people combined with modern technology and a frontier concept has produced our environmental crisis. Thus in well-developed nations on a per capita basis, citizens consume more food, use more pesticides, fertilizers, fuel, minerals, cars, and other manufactured products of all kinds. Most of these products are manufactured in one or other kinds of industries, all of which in their turn add some pollutants in our environment and cause pollution.

25-2. POLLUTANTS : THE CREATORS OF POLLUTION

Every human society, be it rural, urban, industrial and most technologically advanced society, dispose of certain kinds of by-products and waste products which when are injected into the biosphere in quantities so great that they affect the normal functioning of ecosystems and have an adverse effect on plants. animals, and man are collectively called **pollutants** (**Smith**, 1977). Certain common pollutants of well-developed and developing countries are following : 1. **Deposited matter** such as soot, smoke, tar, dust and grit. 2. **Gases** like sulphur dioxide, carbon monoxide. carbon dioxide, nitrogen oxide, hydrogen sulphide, ammonia, fluorine, chlorine, etc. 3. **Chemical compounds** such as aldehydes, arsines, hydrogen fluorides, phosgenes, detergents, etc. 4. **Metals** like lead, iron, zinc, mercury, etc. 5 **Economic poisons** like herbicides, fungicides, pesticides, nematocides, insecticides, rodenticides and other biocides 6. **Fertilizers**. 7. **Sewage**. 8. **Radioactive substances**. 9. **Noise** and **heat**.

However, from the ecosystem viewpoint, these pollutants can be classified into two basic types : **nondegradable pollutants** and **biodegradable pollutants** (**Odum**, 1971). The materials and poisons, such as aluminium cans, mercurial salts, long-chain phenolic chemicals and DDT that either do not degrade or degrade only very slowly in the natural environment, are called **nondegradable pollutants**. Such nondegradable pollutants not only accumulate but are often "biologically magnified" as they move in biogeochemical cycles and along food chains. Also they frequently combine with other compounds in the environment to produce additional toxins. The **biodegradable pollutants** include domestic sewage, heat, etc. The domestic sewage can be rapidly decomposed by natural processes or in engineered systems (such as a municipal sewage treatment plant) that enhance nature's great capacity to decompose and recycle. Problems arise with the biodegradable pollutants when their input into the environment exceeds the decomposition or dispersal capacity.

25-3. TYPES OF POLLUTION

Generally pollution is classified either according to the environment (air, water, soil) in which it occurs, or according to the pollutant (lead, mercury, carbon dioxide, solid waste, noise, biocide, heat, etc.) by which pollution is caused. Sometimes, pollution is also classified into two broad categories: (a) **Natural pollution** which originates from natural processes ; and (b) **Artificial pollution** which originates due to the activities of man.

25-4. AIR POLLUTION

When due to some natural processes or human activities the amount of solid waste or concentration of gases other than O_2 increase in the air which normally has constant percentage of different gases in it the air is said to be polluted and this phenomenon is referred to as **air** or **atmospheric pollution**. Air pollution is one of the most dangerous and common kind of environmental pollution that is reported in most industrial towns and metropolitans of India and abroad such as Delhi, Bombay, Calcutta, Kanpur, Madras, Hyderabad, Jaipur, Ahmedabad, Nagpur, Firozabad and also in London , New York, Tokyo, Pittsburg, etc.

The first case of severe air pollution in modern times were in the Meuse valley of Belgium in 1930 ; a killing smog in Donora, Pennsylvania along the Monongahela River in 1948, in which hundreds of people died ; deadlier smog in London in 1952 in which 4,000 to 5,000 people died from respiratory failure ; "Episode 104" which blanketed all or part of 22 states east of the Mississipi, with air pollution haze in August, 1969 ; a massive hood of stagnant air stretched from Chicago and Milwaukee South to New Orleans and east to Philadelphia, creating dangerous air pollution level in rural as well urban areas (**Brodine**, 1973; **Benton** and **Werner**, 1974 ; **Southwick**, 1976). **Newell** (1971) has estimated that 164 million metric tons of artificial pollutants enter the United States air every year.

25-4·1. Kinds and Sources of Air Pollutants

A major source of air pollution is the particulate and gaseous matter released by the burning of fossil fuels such as coal, petroleum, etc. Out of this comes a variety of emissions : (1) **Fine particles** (less than 100μ in diameter), which include carbon particles, metallic dusts, tars, resins, aerosols, solid oxides, nitrates, and sulphates; (2) **Coarser particles** (over 200μm), largely carbon particles and heavy dust that is quickly removed by gravity from the air ; (3) **Sulphur compounds** ; (4) **Nitrogen compounds** ; (5) **Oxygen compounds** ; (6) **Halogens** ; and (7) **Radioactive substances.**

These pollutants are artificial pollutants and they are poured in air mainly by at least five major fuel-burning sources. **Automobiles** (cars, scooters, motorcycles) are the greatest sources of air pollution. They produce nearly two-thirds of the carbon monoxide and one-half of the hydrocarbons and nitrous oxides. The automobile exhaust also contains leaded gas and particulate lead. **Electrical power plants** burning fossil fuels, particularly coal and sometimes petrol or diesel, produce two-thirds of the sulphur dioxides. **Industrial processors** such as metallurgical plants and smelters, chemical plants, petroleum refineries pulp and paper mills, sugar mills, cotton mills, and synthetic rubber manufacturing plants are responsible for about one-fifth of the air pollution. **Heating plants** for homes, apartments, schools, and industrial buildings are the fourth largest source of air pollution. The **transportation industry**, exclusive of automobiles and including railroads, ships, aircrafts, trucks, buses, tractors, etc., contribute the same type of pollutants as cars.

Other sources of air pollution are minor in quantities but bear significane due to the harmful substances they release, are **agriculture**, which is responsible for pesticides, dust from agriculture practices and field burning, and the construction industry. Nature too adds few natural pollutants such as pollen, hydrocarbons released by vegetation, dusts from deserts, storms, and volcanic activity.

Some of the most common air pollutants, their sources and their effects on human health have been tabulated in Table 25·1.

To measure and control the magnitude of air pollution in various industrial centres of India, National Environmental Engineering Research Institute (NEERI) has set air monitoring stations in Bombay, Calcutta, Delhi, Madras, Hyderabad, Kanpur, Jaipur, Ahmedabad and Nagpur. In one of the Survey conducted by NEERI in 1970 to measure the air pollution by sulphur dioxide (SO_2) and suspended particles in some major cities of India (Table 25·2), it is founnd that Chembur-Trombay area of Bombay has highest SO_2 pollution, while, New Delhi has highest air pollution of suspended particulate matter. In another survey, Calcutta is reported to have highest carbon monoxide pollution during peak traffic hours.

25-4·2. Methods of Detection and Measurement of Air Pollution

Air pollution is usually measured by sampling of air by *thermal* and by *electrostatic precipitation*, *Sonkin impactor* and *electrostatic*

Table 25·1. Common air pollutants, their sources and pathological effects on man (after **Southwick**, 1976).

Pollutants	Where they come from (source)	Pathological effect on man
1. Aldehydes	Thermal decomposition of fats, oil, or glycerol.	Irritate nasal and respiratory tracts.
2. Ammonias	Chemical processes—dye-making ; explosives; lacquer; fertilizer.	Inflame upper respiratory passages.
3. Arsines	Processes involving metals or acids containing arsenic soldering.	Break down red cells in blood, damage kidneys; cause jaundice.
4. Carbon monoxides	Gasoline motor exhausts; burning of coal.	Reduce oxygen-carrying capacity of blood.
5. Chlorines	Bleaching cotton and flour; many other chemical processes.	Attack entire respiratory tract and mucous membranes of eyes; cause pulmonary edema.
6. Hydrogen cyanides	**Fumigation**; blast furnaces; chemical manufacturing; metal plating.	Interfere with nerve cells; produce dry throat, indistinct vision, headache.
7. Hydrogen fluorides	Petroleum refining; glass etching; aluminium and fertilizer production.	Irritate and corrode all body passages.
8. Hydrogen sulphides	Refineries and chemical industries; bituminous fuels.	Smell like rotten eggs; cause nausea; irritate eyes and throat.
9. Nitrogen oxides	Motor vehicle exhausts; soft coal.	Inhibit cilia action so that soot and dust penetrate far into the lungs.
10. Phosgenes (carbonyl chloride)	Chemical and dye manufacturing.	Induce coughing, irritation, and sometimes fatal pulmonary edema.
11. Sulphur dioxides	Coal and oil combustion.	Cause chest constriction, headache, vomiting, and death from respitratory ailments.
12. Suspended particles (ash, soot, smoke)	Incinerators; almost any manufacturing.	Cause emphysema, eye irritations and possibly cancer.

dust collectors. The particulate pollution is measured by the instrument called *deposit gauge* or by *Owen's dust counter*. The thickness of the smoke is measured by *Liegean sphere* and by *Ringelmann chart*. The rough estimation of SO_2 in air can be made by chemical analysis of the dust collected in a deposit gauge or by a *bubbler method*. Fluorides are estimated by colour reactions.

Table 25·2. Levels of sulphur dioxide (SO_2) and suspended particulate matter in air during 1970 in some Indian cities (from Seth, 1976).

City	Mean Value of SO_2 microgram/cubic metre	Suspended particulate matter microgram/cubic metre
1. Bombay	47·1	240 8
2. New Delhi	41·4	601·1
3. Calcutta	32·9	340·7
4. Kanpur	15·9	543 5
5. Ahmedabad	10 7	306·6
6. Madras	8·3	100·9
7. Nagpur	7·7	261·6
8. Hyderabad	5·1	146·2
9. Jaipur	4·2	146·1

25-4·3. Ecology of Air Pollution

Once injectd into the atmosphere, pollutants enter the biogeochemical cycles by different routes. The air above many cities can assimilate and disperse great quantities of fine particulate and gaseous pollutants as long as air can move and disperse. But if air masses over cities become stagnant, pollutants accumulate quickly and deteriorate air quality which cause many respiratory diseases in man and other animals. Air pollutants also accumulate during **temperature inversions**, when cooler surface layers of air become trapped under warmar upper layers. In these situations, the upper layers of warm air prevent the vertical rise and dispersal of pollutants which are held near the ground. Temperature inversions commonly occur in cities surrounded by mountains or bordered by mountains on the leeward side.

Further a portion of air pollutants reaches land as dry fallouts; it may then enter various nutrient cycles and food chains through water and soil. Other contaminants of air react chemically or photo-chemically with each other and produce such secondary pollutants as sulphuric acid, ozone, and peroxyacetyl nitrate or PAN. Aerosols and other forms of fine particulate matter act as **condensation nuclei**, to which water vapours present in the air are quickly surround to form droplets of fog or rain.

Moreover, different air pollutants adversely affect flora, fauna and climate of a given area variously and some of the common air pollutants and their specific effects on man, vegetation, climate, etc., have been discussed as follows :

A. **General pathological effects of air pullution.** Severe air pollution affects human health and causes many fatal diseases in them. For example, there occur lung diseases in workers exposed to occupational hazards, such as **black lung disease** among coal miners who inhaled mine dust for many years; or **asbestosis** among pipe fitters and insulation workers, exposed to airborne asbestos fibers. As listed in table 25·1, a variety of air pollutants have been found to cause many human diseases such as **emphysema, chronic bronchitis, pollen allergies, lung cancer**, especially in city dwellers. **Mountain** *et al.*, (1968) reported that air pollution due to particulate matter and carbon monoxide in New York city causes respiratory problems in children under 8 years of age. **Becker** *et al.*, (1968) investigated that in many American cities along the Eastern Seaboard increasing frequencies of **bronchitis, cough, sore throat, wheezes, eye irritations**, and general ill health in people occurred as air pollution level increased.

Some of the significant air pollutants which directly or indirectly cause economic loss to man are following :

1. **Sulphur dioxide**. One of the common gaseous air pollutant which is known to be injurious to human health is **sulphur dioxide** or SO_2. It is originated primarily from the combustion of coal and petroleum and generally, it irritates the respiratory epithelium **and** impairs normal breathing. SO_2 also causes in an increase in coughs, pharyngitis, eye irritation, and headache in human beings. When there occurs severe pollution of SO_2, the death rate and bronchial asthema are found to increase and in past it caused such disasters as Meuse Valley in Belgium in 1930; Donora, in 1938; London, in 1952; and New York and Tokyo in 1960s (see **Southwick**, 1976).

Further, in the atmosphere, SO_2 does not remain in the gaseous state for long time, but very soon it reacts with moisture to form sulphuric acid or H_2SO_4. Sulphuric acid causes many respiratory diseases in man and also produces **acid rainfalls** over parts of the earth. In Scandinavia, downwind from the industrial centres of Britain and the Ruhr Valley, the acidity of the rainfall has increased 200 fold since 1966, with pH values as low as 2·8 being recorded (**Oden** and **Ahl**, 1970). This acid rainwater has increased the acidity of Scandinavian streams, interfering with salmon reproduction and destroying salmons runs. It reduced forest growth and increased the amount of calcium and other nutrients leached from agricultural soil.

Plants exposed to atmospheric sulphur are injured or killed outright. Exposure of plants to low pollution of SO_2 can produce both acute and chronic injury. Injury to plants is caused largely by acidic aerosols during periods of foggy weather, light rains or periods of high relative humidity and moderate temperatures. Pines are more susceptible than broadleaf trees and react by partial defoliation and reduced growth.

2. Carbon monoxide. Another significant gaseous air pollutant which is injurious to human health, is **carbon monoxide** or CO. It is mainly released from gasoline engine and burning of coal. Carbon monoxide combines with haemoglobin in human blood to form carboxyhaemoglobin, which impairs oxygen transport. Nervous system function can be affected at levels of 2 per cent to 5 per cent carboxyhaemoglobin which occur after breathing air with only 30ppm carbon monoxide (**Bodkin**, 1974). The problem is considerably increased by cigarette smoking. The symptoms of low-level CO poisoning are reduced reaction times, psychomotor impairment, headaches, dizziness, and lassitude. At more advanced stages nausea, ringing in the ears, heart palpitations, pressure in chest, and difficulty in breathing occurs.

3. Nitrogen oxides and photochemical smog. The nitrogen oxides are most important gaseous air pollutants which arise due to burning of fossil fuels in automobiles and power plants. The most common type of nitrogenous air pollutant is **nitrogen oxide** or NO_2. In the atomosphere nitrogen dioxide is reduced by ultraviolet light to nitrogen monoxide and atomic oxygen :

$$NO_2 \rightarrow NO+O$$

Atomic oxygen reacts with oxygen to form ozone :

$$O_2+O \rightarrow O_3$$

Ozone reacts with nitrogen monoxide to form nitrogen dioxide and oxygen, thus closing the cycle :

$$NO+O_3 \rightarrow NO_2+O_2$$

Sometimes, in the presence of sun light, atomic oxygen from the photochemical reduction of NO_2 also reacts with a number of reactive hydrocarbons (such as methane, ethane, toluene etc., all of which originate from burning of fossil fuels or directly from plants) to form reactive intermediates called **radicals**. These radicals then take part in a series of reactions to form still more radicals that combine with oxygen, hydrocarbons, and NO_2. As a result nitrogen dioxide is regenerated, nitric oxide disappears, ozone accumulates and a number of secondary pollutants are formed such as formaldehyde, aldehydes and peroxyacetyl nitrate or PAN ($C_2H_3O_5N$). All of these collectively form **photochemical smog**.

Nitrogen oxide and secondary pollutants are harmful to both man and plants. NO_2, a pungent gas that produces a brownish haze, causes nose and eye irritations and pulmonary discomfort. Lower concentrations of ozone irritates the nose and throat, while, higher concentrations of it cause dryness of the throat, headaches and difficulty in breathing.

Ozone, PAN and nitrogen dioxide severely injure many forms of plant life, destroying the cells of leaves, damaging the chloroplasts, and interfering with the plant's metabolic processes.

4. Lead. Lead is injected in the atmosphere mainly from automobile exhaust and it is found to cause long-term environmental pollution. Automobile gasoline contains tetra-ethyl lead

[$(CH_3CH_2)_4Pb$] which when is burned enters the atmosphere. In the rainfall and soil samples of urban areas large concentrations of lead have been reported. Roadside plants and meadow mice living along major highways are found to contain high concentrations of lead in their tissues, and this has a sublethal effect on the health and longevity of the animal. Traffic policemen and others who are exposed for long periods to heavy traffic have higher than average levels of lead in their blood. It is estimated that 30 to 50% of the lead inhaled is absorbed into the body and these airborne lead compounds are found to cause lead poisoning.

Further, the use of lead-lined vessels for cooking and the storage of wine resulted in heavy lead burdens in the bodies of Roman citizens. Some have attributed the decline of the Roman empire to chronic lead poisoning. In fact, analysis of bones of Roman citizens have revealed high lead concentrations (see **Kimball**, 1975).

Also, certain paints and putty have major components of lead and there are cases in which lead poisoning occur due to occupational exposure of painters and also to children who habitually or accidentally nibbled cracks and peels of old paints. Although 90–95% of the lead that is ingested is insoluble and is quickly eliminated, the remainder does enter the blood and tissues, including bone. Lead levels of 20–40 μg per 100 g of blood (0·2–0·4 ppm) are considered normal and harmless for city dwellers. Both 0·8 ppm lead levels in adult human blood causes overt symptoms such as anaemia, kidney disease and convulsions. However, in children 0·6 ppm level of lead in blood may cause lead poisoning and ultimate death. For a instance, from 1954 to 1967, 2018 children in New York city were treated for lead poisoning. Of this group, 128 died and many others sufferred some degree of irreversible damage to the central nervous system.

Lastly, the accumulation of lead in the snow layers of the Greenland ice sheet over a 200 year period has provided a dramatic example of increasing pollution with the development of industry and transportation (**Southwick**, 1976).

5. **Dust pollution.** Dust is found to travel several thousands of kilometres, across deserts and seas. Air-borne particles of the Saharan sand cross the Arabian sea and reach India. Though dust particles provide nuclei for cloud formation, they can be nuisane to certain industries requiring aseptic and clean environment like drug-industries and food-processing plants. They sometime become health hazards as they may lead to diseases like allergic asthma, bronchitis, emphysema and even fibrosis of the lungs (**Das** *et al.*, 1981).

However, dust pollution of air has been found to be controlled by certain evergreen plants, grasses and epiphytes like orchids. Pollution Research Laboratory, College of Agriculture, University of Calcutta has reported that certain plants have remarkable dust-filtering, air-cleaning and air-purifying capacities. In one of the study, it has been investigated that certain plants with simple leaves such as peepal (*Ficus riligiosa*), pakur (*Ficus infectoria*), banyan (*Ficus benghalensis*), teak (*Tectona grandis*), sal (*Shorea robusta*), arjuna (*Terminalia arjuna*), mast (*Polyalthia longifolia*), mango (*Mangifera indica*); etc., are better dust collectors than the plants with compound leaves like gul mohar (*Poinclana regia*), tamarind (*Tamarindus indica*), *Cassia fistula*, neem (*Azadirachta indica*) (**Das** *et al.*, 1981).

B. Certain miscellaneous effects of air pollution : Besides its pathological effects on man, air pollution causes following kinds of damages to man :

1. The increased amount of ground haze and air pollution domes over cities have caused obstruction to vision of pilots and have remained the major contributors of air craft accidents.

2. In many areas, air pollution has dramatic injury to both agricultural crops and natural plants communities (**Waggoner**, 1971). Smog and air pollution are found to specifically damage pine forest, truck garden crops, citrus groves, onion and celery fields and field crops of alfa-alfa and sweet corn. **Turk** *et al.*, **1974**, have estimated an annual loss of one billion dollors due to destruction of vegetation by air pollution in U.S.A.

3. There has been injury to agricultural livestocks, particularly from air borne fluoride and arsenic pollution. Excessive fluorine compounds which have contaminated forage by fall-out have sometimes caused cattle to develop **fluorosis**, an abnormal calcification of bones which leads to lameness (**Turk** *et al.*, 1974).

4. Air pollution also takes its toll of buildings and other man made objects. When moisture accumulates in polluted air, the oxides of sulphur, carbon and nitrogen form weak sulphuric, carbonic and nitric acid, which are corrosive to metals, stone, paint, rubber, textiles and even some plastics. Throughout Europe and in major Indian metropolitans and cities such as Agra, Delhi, Lucknow, Calcutta and Bombay, many famous buildings, monuments and art treasures are deteriorating at an alarming rate because of the erosional effects of air pollution.

C. Effects of air pollution on weather, climate and atmospheric processes. At gross level, air pollution causes two worldwide problems—contamination of the upper atmosphere and the alteration of weather and climate. In fact, pollution and population concentrations influence local weather patterns, as in the well-known phenomena of "**heat islands**" around cities. Since local rainfall patterns are altered by the distribution and abundance of particulate nuclei in the lower atmosphere (**Brodine**, 1973), there is a significant increase in precipitation in and around cities that is due to air pollution (**Thompson**, 1975).

Air pollution also affects weather on a continental or global basis. Many gaseous pollutants and fine aerosols reach the upper atmosphere, where they have basic effects on the penetration and absorption of sunlight. **Brodine** (1973) and certain other modern environmental biologists feel that increasing particulate pollution may be reducing the amount of sunlight energy reaching the earth's surface, thereby lowering solar radiation at the earth's surface and producing a cooling effect on world climates which could ultimately trigger another ice age. In fact, **Thompson** (1975) has reported a decrease in mean annual temperatures in the northern hemispheres and an increase in the north polar ice cap.

Green house effect. Carbon dioxide is a natural constituent of the atmosphere, but, its concentration is increasing in the air with an alarming rate. A by-product of the burning of fossil fuel, it is not necessarily a pollutant. It produces adverse physiological effects only at very high levels. It is estimated that approximately onehalf of the input stays in the atmosphere and other half of it is removed by the oceans and by plants. The increased amount of CO_2 in atmosphere is found to increase the temperature of earth.

The spectral properties of CO_2 in the atmosphere are such that it tends to prevent the long wave radiations (*i.e.*, infra-red heat radiation) from earth from escaping into outer space and deflect it back to earth. The latter has an increased temperature at surface (**Turk** *et al.*, 1974). This phenomenon is called **atmospheric effect** (**Lee**, 1974) or **greenhouse effect** (see **Southwick**, 1976, **Smith**, 1977).

The simultaneous cooling and heating effects of air pollution on earth have increased variability in the world-wide weather patterns which may be a serious threat to global food production (**Thompson**, 1975). Recently, certain ecologists have tried to correlate air pollution with serious and prolonged droughts, heavier rains and floods, and more serious hurricanes and tornadoes (see **Southwick**, 1976).

Peeling of ozone umbrella by CFMs. Certain fluorocarbon compounds which are called chlorofluoromethanes or CFMs or "freon" are used as propellants in pressurized aerosol cans. They are inert in normal chemical and physical reactions, but they get accumulated in greater amounts at high altitudes and there in the stratosphere these inert gaseous compounds (*i.e.*, CFMs) release chlorine atoms under the influence of intense short-wave ultraviolet radiation. Each atom of chlorine chain then reacts with more than 1,00,000 molecules of ozone, converting ozone to oxygen. The reduction in stratospheric ozone permits greater penetration of ultraviolet light, which intensifies UV radiation at the earth's surface. Some scientists such as **Ahmed** (1975), **Brodeur** (1975), and **Russell** (1975) feel that this intensified radiation will cause a significant increase in skin cancer and eventually have a lethal effects on many organisms, including man.

The protective ozone layer of the stratosphere is also considered by many ecologists to be endangered by supersonic jets, the SSTs. The jet engines of supersonic aircraft flying at high altitudes release nitrogen oxide (NO_x) which catalytically destroy ozone molecules (see **Southwick**, 1976).

25-4·4. Control of Air Pollution

Most kinds of air pollutions can be controlled by modern technology, but the costs ultimately be borne by the public in the form of higher prices for manufactured goods, higher taxes, reduced profit margins in industry, and more restrictions on individual activities such as burning leaves and trash and use of automobiles. The benefits involve not only improved environmental quality, but improved health, improved agriculture and plant growth, and reduced deterioration of material goods.

In USA, various devices such as the positive crankcase ventilation valve and catalytic converter, have been developed to reduce exhaust emissions by automobiles, but these devices are not always fully maintained by the public. Likewise, particulate pollution from industry and power generation can be controlled by electrostatic precipitators which are capable of dramatically reducing smoke and dust. Gaseous pollutants of industry and power station can be removed by chemical means, *i.e.*, differential solubility of gases in water. A fine spray of water in a device known as a "scrubber" can effectively separate many gases such as ammonia and sulphur dioxide. Other gases may be removed by filtration or absorption through activated carbon, and still others by chemical conversion to inert or innocuous materials.

Legal control of air pollution. In India, air pollution control legislation envisages the formation of **air pollution boards** at the Central and State levels "with powers to issue and revoke licences to polluting industries, enforce emission standards and frame rules and regulations for the control of air pollution". The legislation is primarily directed the highly polluting industries such as iron and steel, textiles and power plants. The Board will have power to prohibit certain trades and manufacturing processes in notified areas and prescribes emission standards in scheduled premises. The legislation is also understood to ban the burning of garbage and other waste products in urban areas as well as the fouling up of air by burning smoking fuels for domestic purposes. Recently, **Bombay Smoke Nuisanse Act** has been enforced only in case of smoke emanating from the chimneys of industrial units. However, there is a need of legislation to deal with fumes of petro-chemical units, ash, carbon particles, unpalatable smell and even noise from industrial units (see **Gopal Bhargava**, in the Hindustan Times Magzine, pp. 1, 9th April, 1978).

25-5. WATER POLLUTION

The great solvent power of water makes the creation of absolutely pure water a theoretical rather than a practical goal. Even the highest-quality distilled water contains dissolved gases and to a slight degree, solids. The problem, therefore, is one of determining what quality of water is needed to meet a given purpose and then finding practical means of achieving that quality. The problem is further compounded because every use to which water is put—washing, irrigation, flushing away wastes, cooling, making paper, etc., adds something to the water. In fact for centuries rivers and lakes have been used as dumping grounds for human sewage and industrial wastes of every conceivable kind, many of them are highly toxic. Added to this are the materials leached and transported from land by water percolating through the soil and running off its surface to aquatic ecosystems.

The term **water pollution** is referred to any type of aquatic contamination between following two extremes : (1) a highly enriched, over productive biotic community, such as a river or lake with nutrients from sewage or fertilizer (**cultural eutrophication**), or (2) a body of water poisoned by toxic chemicals which eliminate living organisms or even exclude all forms of life (see **Southwick**, 1976).

25·5·1. Types of Water Pollution

Types of water pollution may be classified by the *medium* in which they occur, such as **surface water pollution, ground water pollution, soil water pollution,** etc. ; the *habitat* in which they occur, such as **river pollution, lake pollution, estuarine pollution, coastal water pollution, open ocean pollution,** etc. ; and *source* or *type* of contamination, such as **nutrient pollution, bacterial pollution, viral pollution, metallic pollution, petrochemical pollution, pesticide pollution, thermal pollution, radioactive pollution,** etc.

25·5·2. Kinds and Sources of Water Pollutants

Pollutants entering water sources are classified broadly into following categories : domestic sewage and oxygen-demanding wastes ; infectious agents ; plant nutrients ; chemicals such as insecticides, herbicides, and detergents; other minerals and chemicals; sediment from land erosions ; radioactive substances ; and heat from power and industrial plants.

These aquatic pollutants come from many sources. Excessive nutrients, such as nitrates and phosphates, commonly originate in domestic sewage, run-off from agricultural fertilizer, wasts materials from animal feed lots, packing plants, etc. These nutrients cause pollution primarly because they stimulate the growth of micro-organisms which often increase the **biological oxygen demand (BOD)** of the water and reduce the amount of dissolved oxygen available for fish, higher animals and other aquatic animals. Toxic chemicals as agents of water pollution originate in industrial operations, acid mine drainage, surface erosion from strip mines, washing of herbicides and insecticides, radioactive fall out from atomic explosion, and commercial accidents such as oil spills or the rupture of chemical tanks. Besides the pollutants which come from point sources such as sewage, factory or industry, there are many pollutants which come from watershed run-off. Urban and sub-urban run-off, for example, contains many pollutants such as oil, pesticides, radioactive dust, salt, fertilizers, miscellaneous chemicals. and nematodes, pathogenic protozoans, bacteria and viruses. However, **Kimball** (1975) recognized only three major sources of pollutions : *domestic*, *industrial* and *agricultural*.

25-5·3. Ecology of Water Pollution

Each type of water pollution affects the abiotic and biotic factors of different aquatic systems in different degrees and its ultimate effect on man remains quite drastic in medical, asthetic, and economical sense. Some of the well known ecological effects of water pollution are following :

25-5·3A. Sewage pollution. Contamination of fresh-waters and shallow offshore seas by sewage is a common occurrence. Domestic sewage and waste-water is about 99·9 per cent water and 0·02—0·04 per cent solids of which proteins and carbohydrates each compirse 40–50 per cent and fats 5–10 per cent (**Simmons**, 1974). In other words, sewage includes mostly biodegradeable pollutants such as human faecal matter, animal wastes, and certain dissolved organic compounds (*e.g.*, carbohydrates, urea, etc.) and inorganic salts such as nitrates and phosphates of detergents and sodium, potassium, calcium and chloride ions. Under natural processes most of the biodegradeable pollutants of sewage are rapidly decomposed, but, when they accumulate in large quantities, they create problem, *i.e.*, when their input into environment exceeds the decomposition or dispersal capacity of the latter. Most cities of well developed countries like USA, Britain, etc., and some cities of developing countries like India have evolved certain engineering systems, such as, **septic tanks, oxidation ponds, filter beds, waste water treatment plants** and **municipal sewage treatments plants** for the removal of many harmful bacteria and other microbes, organic wastes and other pollutants from the sewage, before it is tipped into river or sea.

Sewage treatment is usually performed in following three stages : (1) **Primary treatment**, which removes large objects and suspended undissolved solids of raw sewage and converts them into a biologically inactive and aesthetically inoffensive state, the **slugde**, a valuable fertilizer. (2) **Secondary treatment**, which supplies aeration and bacterial action to decompose organic compounds into harmless substances such as CO_2, sulphate and water. During later stage of secondary treatment, whole waste water is chlorinated (*i.e.*, treated with chlorine) to reduce its content of bacteria. (3) **Tertiary treatment**, which removes nitrates and phosphates and releases pure water. These three stages of sewage treatment have become increasingly expensive and only in most advanced countries all the three treatments of sewage are done. However, most Indian cities either lack any sewage or waste-water treatment plant or have inadequate sewage treatment facilities. Consequently, normally and especially during heavy downpour and floods, raw sewage or incompletely treated sewage is dumped into rivers which cause severe water pollution problems in following ways :

(i) **Bacterial and viral contamination**. Sewage wastes may contain pathogenic bacteria and viruses which are a threat to human health. Waterborne disease such as typhoid, bacillary dysentery, amoebic dysentery, botulism, poliomyelitis, and hepatitis all represent potential health hazards in sewage-contaminated waters. Due to such kinds of sewage pollution waters of many ponds, lakes, rivers, sea-beaches in India and abroad have been prohibited for human use, whether for drinking, bathing, swimming or other sort of water recreation.

(ii) **Eutrophication**. According to **Hutchinson** (1969), the eutrophication is a natural process which literally means "well nourished or enriched." It is a natural state in many lakes and ponds which have a rich supply of nutrients, and it also occurs as part of the aging process in lakes, as nutrients accumulate through natural succession. Eutrophication becomes excessive, however, when abnormally high amounts of nutrients from sewage, fertilizer, animal wastes and detergents, enter streams and lakes, causing excessive growth or 'bloom' of micro-organisms and aquatic vegetation.

Most secondary sewage treatment plants, though, precipitate solids and inactivate most bacteria in domestic sewage, yet they do not remove the basic nutrients such as ammonia, nitrogen, nitrates, nitrites and phosphates. These **nutrients** stimulate algal growth and lead to **plankton blooms.** Some plankton blooms, particularly those of blue-green algae produce obnoxious odours and tastes in waters. Others, such as the dinoflagellate blooms or "the red tide" of southern coastal regions, produce toxic metabolic products which can result in major fish kills.

Plankton blooms of green algae do not always produce undesirable odours or toxic products, but still create problems of oxygen supply in the water while these blooms exist under abundant sunlight, they contribute oxygen to the water through photosynthesis, but under conditions of prolonged cloudiness, they begin to decay and consume more oxygen than they produce, leading to oxygen depletion in the water. Further, bacterial decomposition of organic wastes too requires oxygen and with heavy loads the oxygen contents of the water may diminish below the point where most fish can not survive. As the conditions in the water become anaerobic due to increased oxygen depletion by bacterial decomposition or planktonic blooms, the breakdown products become reduced rather than oxidized molecules, many of which (*e.g.*, hydrogen sulphide) produce offensive odours and tastes.

Because of this relationship between organic pollutants and dissolved oxygen, the most widely measure of water pollution is the **biochemical oxygen demand** or **BOD**. This is a test of the amount of oxygen needed for bacteria to break down the organic matter in a water sample over a period of five days. Typical BOD value for raw sewage run from 200 to 400 mg of oxygen per litter of water (therefore, 200–400 ppm). Water for drinking should have a BOD less than 1.

Excessive nutrient levels in aquatic system can also cause two other kinds of ecological problems. Primarily, they may lead to extensive growth of aquatic weeds such as Eurasion milfoil, water hyacinth, water chestnut, etc. Excessive growth of these weeds can impair fishing, bathing, fish spawning, shell fish production, and even navigation (**Sculthorpe**, 1967). Secondarily nitrates can be converted in the human digestive tract by certain bacteria to nitrites, and the same transition may occur in opened cans of food even if they are subsequently refrigerated. Nitrites react with haemoglobin, forming methaemoglobin which will not take up oxygen. Laboured breathing and occasional suffocation result most severely in human infants. Nitrites may also react with creatinine (present in the vertebrate muscles) to form nitrosarcosine which can be carcinogenic (see **Simmons**, 1974 ; **Southwick**, 1976).

25-5·3B. Industrial pollution. Most of the Indian rivers and fresh-water streams are seriously polluted by industrial wastes or effluents (see Table 25·3) which come along waste waters of different industries such as petro-chemical complexes ; fertilizer factories ; oil refineries ; pulp, paper, textile, sugar and steel mills, tanneries. distilleries, coal washeries, synthetic material plants for drugs, fibres, rubber, plastics, etc. The industrial wastes of these industries and mills include metals (copper, zinc, lead, mercury, etc.), detergents, petroleum, acids, alkalies, phenols, carbamates, alcohols, cyanide, arsenic, chlorine and many other inorganic and organic toxicants. All of these chemicals of industrial wastes are toxic to animals and may cause death or sublethal pathology of the liver, kidneys, reproductive systems, respiratory systems, or nervous systems in both invertebrate and vertebrate aquatic animals (**Wilbur**, 1969). Chlorine which is added to water to control growth of algae and bacteria in the cooling system of power station, may persist in streams to cause mortality of plankton and fish. Heavy fish mortality in river **Sone** near Dehri-on-sone in Bihar is reported to cause by free chlorine content of the chemical wastes discharged by factories near Mirzapur in U.P.

Mercury like other heavy metals such as lead and cadmium has cropped up as a toxic agent of serious nature. Mercury, a by-product of the production of vinyl-chloride, is used in many chemical industries and it is also a by-product of some incinerators, power plants, laboratories and even hospitals, (**Aaronson**, 1971). In Japan, illness and even death occurred in the 1950s among fishermen who ingested fish, crabs, and shell-fish contaminated with methyl mercury from Japanese coastal industries. This mercury poisoning produced a crippling and often fatal disease called **Minamata disease.** Initial symptoms of minamata disease included numbness of the limbs, lips,

Table 25·3. Some Indian rivers and their major sources of pollution

Name of the river	Sources of pollution
1. Kali at Meerut (U.P.)	Sugar mills; distilleries; paint, soap, rayon, silk, yarn, tin and glycerine industries.
2. Jamuna near Delhi	D.D.T factory, sewage, Indraprastha Power Station, Delhi.
3. Ganga at Kanpur	Jute, chemical, metal and surgical industries; tanneries, textile mills and great bulk of domestic sewage of highly organic nature.
4. Gomti near Lucknow (U.P.)	Paper and pulp mills; sewage.
5. Dajora in Bareilly (U.P.)	Synthetic rubber factories.
6 Damodar between Bokaro and Panchet	Fertilizers, fly ash from steel mills, suspended coal particles from washeries, and thermal power station.
7. Hoogly near Calcutta	Power stations; paper pulp, jute, textiles, chemical mills, paint, varnishes, metal, steel, hydrogenated vegetable oils, rayon, and soap, match, shellac, and polythene industries and sewage.
8. Sone at Dalmianagar (Bihar)	Cement, pulp and paper mills.
9. Bhadra (Karnataka)	Pulp, paper and steel industries.
10. Cooum, Adyar and Buckinghum canal (Madras)	Domestic sewage, automobile workshops.
11. Cauvery (Tamil Nadu)	Sewage, tanneries, distilleries, paper and rayon mills.
12. Godavari	Paper mills.
13. Siwan (Bihar)	Paper, sulphur, cement, sugar mills.
14. Kulu (between Bombay and Kalyan)	Chemical factories, rayon mills and tanneries.
15. Suwao (in Balrampur)	Sugar industries.

and tongue, impairment of motor control, deafness, and blurring of vision. Cellular degeneration occurred in the cerebellum, mid-brain, and cerebral cortex and this led to spasticity, rigidity, stupor and coma. In Japan in 1953, due to Minamata disease 17 persons died and 23 were become permanently disabled (see **Southwick**, 1976).

The toxic and pathological effects of some heavy metal water pollutants have been tabulated in Table 25·4.

Table 25·4. Pathological effects of heavy metal water pollutants on man.

Metal	Pathological effects on man
1. Mercury	Abdominal pain, headache, diarrhoea, hemolysis, chest pain.
2. Lead	Anaemia, vomiting, loss of appetite, convulsions, damage of brain, liver and kidney.
3. Arsenic	Disturbed peripheral circulation, mental disturbance, liver cirrhosis, hyperkeratosis, lung cancer, ulcers in gastro-intestinal tract, kidney damage.
4. Cadmium	Diarrhoea, growth retradition, bone deformation, kidney damage, testicular atrophy, anaemia, injury of central nervous system and liver, hypertension
5. Copper	Hypertension, uremia, coma, sporadic fever.
6. Barium	Excessive salivation, vomiting, diarrhoea, **paralysis,** colic pain.
7. Zinc	Vomiting, renal damage, cramps.
8 Selenium	Damage of liver, kidney and spleen, fever, nervousness, vomiting, low blood pressure, **blindness,** and even death.
9. Hexavalent chromium	**Nephritis,** gastro-intestinal ulceration, diseases in central nervous system, cancer.
10. Cobalt	Diarrhoea, low blood pressure, lung irritation, bone deformities, paralysis.

25-5·3C. Thermal pollution. Various industrial processes may utilize water for cooling, and resultant warmed water has often been discharged into streams or lakes. Coal-oil-fired generators and atomic energy plants cause into large amounts of waste heat which is carried away as hot water and cause **thermal pollution** or **calefaction** (warming). Thermal pollution produces distinct changes in aquatic biota. A body of water at 30-35°C is essentially a biological desert and many game fish require temperatures of $\angle 10°C$ for successful reproduction, although they will survive above that temperature. A temperature rise of 10°C will double the rate of many chemical reactions and so the decay of the organic matter, the rusting of iron and the solution rate of salts are also accelerated by calefaction. Since the rate of exchange of salts in organisms increases, any toxins are liable to exert greater effects and temperature fluctuations are likely to affect organisms. Thermal pollution is thus can exert a disruptive effect on aquatic ecosystems.

25-5·3D. Silt pollution. A result of intensive agriculture, earth moving for construction projects, poor conservation practices and downpour with resultant floods, is the increased production of silt in stream and lakes. This load of particulate matter cuts down primary productivity by decreasing the depth of light penetration. Silt may also interrupt or prevent the reproduction of fish, by smothethering eggs laid on the bottom.

25-5·3E. Estuarine and oceanic pollution Often oceans are considered as so vast that they are virtually unlimited in their ability to accomodate the waste products of human civilization. But there are substantial evidence to indicate global pollution of coastal waters and open oceans due to dumping of domestic and industrial wastes, sewage, oil drilling in coastal waters, spilling of oil tankers, etc. The oceans have in fact, become the final settling basin for millions of tons of waste products from human activities. For example, in the late 1960s West Germany was dumping 375 tons of sulphuric acid, 750 tons of iron sulphate, 20 tons of chlorinated hydrocarbons and 16,000 tons of gypsum wastes into the North sea and North Atlantic every day. Table 25·5 shows a few of the industrial and agricultural pollutants reaching the ocean annually in the mid 1970s.

Table 25·5. Examples of industrial and agricultural pollutants discharged annually into the world's oceans (after Southwick, 1976).

Pollutants	Estimated annual discharge, 1970-75 (in metric tons)	Source
1. Petroleum and industrial hydrocarbons	3,405,000	Offshore wells, oil tankers, industrial wastes.
2. Hydrocarbons (airborne)	15,000,000	Vehicles, industries, power plants.
3. Airborne lead	350,000	Vehicles.
4. Mercury	100,000	Industrial operations.
5. Aldrin-Toxaphene (converted to dieldrin)	25,000	Agricultural and public health operations.
6. Benzene hexachloride	50,000	,,
7. DDT	25,000	,,
8. Polychlorinated biphenols (PCBs)	25,000	Plastic industries.

This tremendous burden of pollutants is evidently affecting the health and integrity of the world's oceans. Due to oceanic pollution the marine biota has been seriously affected. **Cousteau** (1974) reported a 30 to 40 per cent decline in the overall productivity of pelagic organisms from shrimps to whales in the past 25 years. He also observed a serious shrinkage of coral reefs in many tropical areas of world, and a displacement of these rich and varied communities with turbid and relatively barren waters. There has occurred a decline in populations of many fishes due to oceanic pollution. Oil spills have killed water birds, mammals, fish and vegetation (**Hunt**, 1965).

25-5·4. Control of Water Pollution

The intensity of water pollution can be minimised by following methods :

1. **Adequate sewage treatment.** Raw sewage should not be dumped in rivers or oceans. Before its disposal into them, sewage should be properly treated in sewage treatment plants.

2. **Treatment of industrial effluents.** The industrial effluents should be cleaned before they are discharged into rivers.

3. **Recycling.** The best method of prevention and control of water pollution is the recycling of various kinds of wastes. Dung of cow and buffaloes can be used for the **production of gobar gas,** a cheap source of fuel and also of manure.

25-6. LAND POLLUTION

The land pollution is caused by solid wastes and chemicals. One of the major pollution problems of large cities is disposal of solid waste material including farm and animal manure, crop residues (agricultural wastes), industrial wastes such as chemicals, fly ash and cinders which are residues of combustion of solid fuels, garbagae, paper, cardboard, plastics, rubber, cloth, leather, construction rubbish, brick, sand, metal and glass resulting from demolition of buildings, dead animals like cattles, dogs, cats, birds, containers, discarded manufactured products such as old refrigerators, washing machines and autos. Huge quantities of unwanted material cause serious disposal problems. The simplest method is crude **tipping** or **open dumping**, a common method used in most Indian cities. More satisfactory is **controlled tipping** or the **sanitary landfill**, which is recenty adopted in Delhi for solid waste disposal. In sanitary land fill, a layer of about 2 metres of refuse is covered by at least 23 cm of earth, ash or other inert material, upto the level of the hole chosen. The surface can then be used for housing or sports fields. Before such filling, the wastes can be pulverized by machines to a uniform particle size : by this means the volume is reduced and thus the life of the tip is extended, and some of the refuge is more quickly biodegraded.

Further, many of the chemicals emitted into the air such as radioactive minerals, sulphur, and lead, eventually come to earth to pollute the soil. Many pesticides and herbicides are applied by aerosol spray and find their way into the soil. These chemicals may have significant effects on plants and animals, causing a disruption in species composition of communities. In forest ecosystems, these chemical pollutants may inhibit processes of soil formation, and reduce the capacity of the forest to maintain fertility of the soil.

25-7. PESTICIDE AND HERBICIDE CONTAMINATION

Pesticides are chemical compounds which are specifically designed and manufactured to destroy animal pests. They have proved tremendously beneficial to human populations, in reducing or eliminating insects, snails, rats, and other animals which transmit disease, destroy agricultural crops, damage homes and stored products, and directly or indirectly affect human health and welfare. Hence, the significance of chemical pesticides in controlling mosquitoes, termites, houseflies, cockroaches, house crickets, weevils, locusts and grasshoppers, borers, snails, rats, rabbits, and a multitudes of other animals has been great, and it is difficult to imagine modern disease control and agricultural programmes without some forms of chemical control. Like pesticides, **herbicides** are specifically designed chemicals for the control of weed pests and unwanted plant growth.

Our dependence on pesticides and herbicides have increased upto the extent that if modern agriculture will attempt to operate without chemical control of any sort, crop production will probably decline in many areas, food prices will soar far higher and food shortage will become even more severe. Ecologically, however, pesticides and herbicides have created two major serious problems which were not previously anticipated. In the first place many of them have persisted and accumulated in the environment and have harmed or contaminated numerous animals or plants not intended to be targets. Secondly, many of them have directly or indirectly affected human health.

Most pesticides such as DDT, DDE, DDD, dieldrin and heptachlor epoxide and most herbicides such as 2, 4-D (2, 4 dichlorophenoxy acetic acid), 2,4, 5-T (2,4,5-trichlorophenoxy acetic acid) and dioxan have been extensively used for the control of disease-carrying and crop-destroying insects, rodents and weeds. These chemicals are non-degradeable pollutants that either do not degrade or degrade very slowly, so, they have seriously contaminated the global ecosystem and entered food chains. Since they become more concentrated at higher trophic levels (biomagnification), these toxicants affect the predacious animals most adversely. Fish eating birds such as grebes, ring billed gulls, herring gulls, robins, etc., are endangered because chlorinated hydrocarbons interfere with their reproductive ability (Fig. 25·1). Certain pesticides are found to have mutational effects on human DNA molecules (**Wurster**, 1974). The small doses of DDT in human tissues are reported to have lethal effects. DDT is suspected to be a carcinogenic in human tissue. Even certain herbicides, such as dioxan is found to

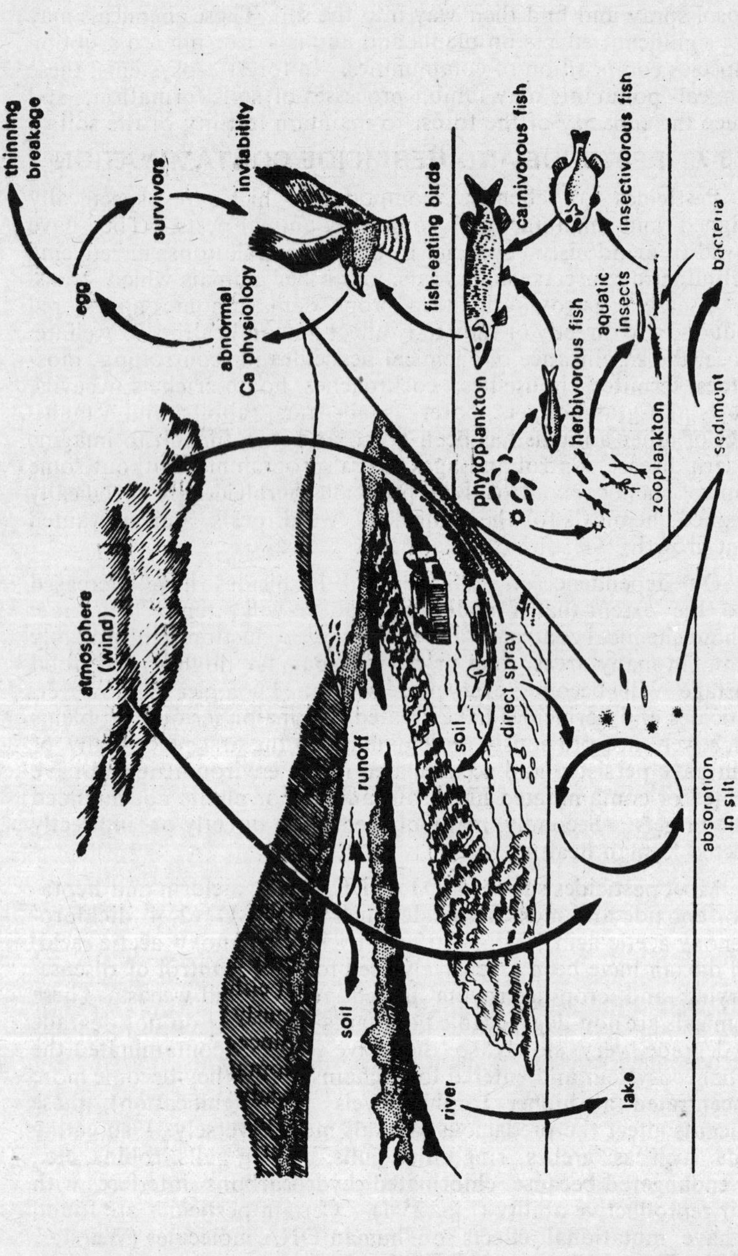

Fig. 25·1. Cycle of DDT and other chlorinated hydrocarbons pesticides and their toxic effects on fish-eating birds due to biomagnification through aquatic food-chain (after Smith, 1974).

cause congenital birth defects in offspring of experimental female who have ingested or inhaled certain concentrations of dioxan at critical periods. Herbicides are also reported to have certain most serious ecological effects during defoliation programmes of USA during Vietnam war : (1) Destruction of 50 per cent upland forest; (2) 90 per cent destruction of mangrove forest; (3) accidental damage to rubber plantations rice crops; (4) loss of insectivorous birds; (5) increased mosquito populations and increased malaria; (6) increased soil erosion; (7) invasion of forests by grasses and bamboo; (8) increase birth defects in both the people and livestock of North Vietnam (**Galston**, 1974).

Visualizing the damaging ecological impacts of pesticide and herbicide pollution, recently biologists became interested in **integrated pest control** (see **Odum**, 1971) The concept of integrated pest control involves coordinated use of a mixed bag of weapons, including old fashioned but common sense cultural practices, judicious of degradable or "short-lived" chemical pesticides, and greater use and stimulation of nature's own control methods, *i.e.*, biological control.

25-8. RADIOACTIVE POLLUTION

Radioactive isotopes, or radionuclides, are forms of elements with unstable atomic nuclei; that is, they decompose with ionizing radiation in the form of alpha or beta particles, or gamma rays. Many radioisotopes, such as radium-226, uranium-235 or 238, thorium-232, potassium-40, or carbon-14, occur naturally in rocks and soil. Other radioisotopes such as those of cesium, cobalt, iodine, krypton, plutonium, and strontium, result primarily as fission products from atomic bomb fallout, nuclear reactors or other radiation sources. Of more than 450 radioactive isotopes which can occur as fission products, only a few are of major environmental concern. These are primarily argon-41, cobalt-60, cesium-137, iodine-131, krypton-85, strontium-90, tritium and plutonium-239 (**Bebbington**, 1973)

Within biotic communities and ecosystems, these radioactive elements may become dispersed or accumulated, depending upon the biological activity of the element and period of radioactivity of the isotope. **Strontium-90** for example, normally occurs in radioactive fallout, has a half-life of 28 years and behaves like calcium in biogeochemical cycles. Thus it is absorbed by plants, ingested by animals and deposited in bone tissues close to blood forming tissue. Strontium-90 can also concentrate in natural biological systems in following method : water→bottom sediments→aquatic plants→fresh-wate clams→minnows and small fish→musk rats. It is demonstrated that due to this food chain musk rats concentrate strontium-90, 3500 times above the levels of the water in which they live. Grazing animals concentrate strontium-90 by ingesting it through grass and forage, and it can then be passed on to humans through milk.

Radioactive phosphate cesium and iodine-132 can also readily accumulate in plants and animals through natural food chains. However, in food chains involving arthropods radioactive isotopes of potassium, sodium, and phosphorus accumulate, but isotopes of strontium and cobalt do not (**Reichle** *et al.*, 1970).

Although isotopes may accumulate in human tissues as well as those of plants and animals, it is not established at the present time whether current levels of isotopes in human tissues represent serious health hazards to man (**Southwick**, 1976 ; **Smith**, 1977). Some medical scientists such as **Gofman** and **Tamplin** (1970) and **Sternglass** (1972), however, feel that man's radiation exposure from artificial sources is already sufficient to produce serious disease problems (leukemia and bone tumors), genetic damage, and infant mortality

25-9. NOISE POLLUTION

Noise is primarily a feature of cities, as exemplified by **J. Caesar's** action in banning chariots from the streets of Rome by day, thus producing insomnia by night. Noise is defined as 'sound without value' or 'any noise that is undesired by the recipient'. Noise levels in many urban-industrialized situations are known to be deleterious to human health and efficiency, with effects on the sense organs, cardiovascular system, glandular and nervous system.

High intensity sound or noise pollution is caused by many machines man has invented during his technological advancement. Thus, there exists a long list of sources of noise pollution including different machines of numerous factories, industries and mills, different kind of auto and motor vehicles such as scooters, motorbikes, cars, tempos, buses, trucks, tractors, aircrafts, motorboats, ships, loudspeakers, social gatherings, loud pop-music, supersonic aircrafts, and others.

25-9·1. Effects of Noise Pollution

Noise pollution has certain well evident ecological and pathological effects on biota and human beings. For instance, the sonic boom path associated with SST projects such as Concorde produces noise of a very different order, in the form of sudden but repeated shock waves, which is suspected to cause disturbance to wild birds as well as domestic stock and buildings (see **Simmons**, 1974).

Robert Alex Baron (1964) for the first time pointed out that noise has damaging physiological and psychological effects on human beings. Noise is measured in the unit of **decibel** (dB), which is a tenth of the largest unit, the **bel**. One decibel is equivalent to the faintest sound that can be heared by the human ear. Some people feel discomfort with sound of 85 dB, whereas most do not feel discomfort with sound of 115 dB. Pain is usually felt at 145 dB.

There is a clear evidence now that the hair cells of organ of Corti of inner ear can be permanently damaged if they are subjected to repeated sounds of high intensity before they have an opportunity to recover. Workers of different industries develop noise

deafness. Rock music (and **Disco music**) too is found to produce temporary deafness in the listeners and also in the members of rock band. Further, noise is found to increase the level of cholesterol in the blood, to increase blood pressure, and to cause headache. Noise pollution also causes the pupils of the eyes to dilate, and mental abnormalities in foetus, hypertension, peptic ulcers and ultimately, may lead to emotional and behavioural problems in man.

Recently many nations has enacted laws to penalise noise production by the vehicles or any type of industry.

25-10. REVISION QUESTIONS

1. Define the term 'pollution'. Write an essay on the 'problem' of ecological pollution and its control methods.
2. What is 'air pollution' ? How it affects the biota including man ? Describe certain recent methods to control the air pollution.
3. What is 'water pollution' ? Describe different kinds of water pollutions and their ecological effects on the aquatic life and also on man.
4. Write an essay on 'the eutrophication of aquatic habitats'.
5. Write short notes on the following :
   (i) Minamata disease ;
   (ii) Greenhouse effect ;
   (iii) Photochemical smog ;
   (iv) Thermal pollution ;
   (v) Pesticide and herbicide contamin ation ;
   (vi) Noise pollution ;
   (vii) Radioactive contamination ;
   (viii) Sewage treatment ;
   (ix) Sanitary landfill ;
   (x) Biomagnification.

26
Ecology and Human Welfare
(CONSERVATION OF NATURAL RESOURCES)

Everywhere, man lives in interspecific animal and plant communities and like any other organism, he depends on his environment for food, water, air, space and shelter. Being a material using animal, everything of his use, from the food that is needed to keep him alive to objects he fabricates, whether tools or sculptures, comes from the substances of the planet earth, on which he lives. His wastes are then returned to the biological and abiotic systems of the earth. However, it is part of his cultural development that he reacts on the environment more widely and more strongly than any other species. Man being a dominant organism of most ecosystems, controls and modifies environments more extensively than any other organism. In fact, there have been significant changes in the natural environments due to man's intervention and his rapid progress in colonization, urbanization, industrialization, agriculture, mining, transportation and technology. Thus, deforestation has provided land for agriculture and rural inhabitation, rural land has been converted into urban settlements and open spaces are fast vanishing. The mass-scale destruction of flora and fauna has become detrimental to ecological balance (**Bhargava**, 1978). Many environments throughout the world have been rendered barren and inhospitable by excessive pressures from the axes, plows, hoofed animals and military machines of man. The pioneer civilizations altered their own biotic and physical environments and displayed man's ability to trigger ecological changes leading to his downfall. Despite man's awareness to his ecological malpractices, still in many tropical lands deforestation is occurring at increasing rates, and the people believe that their forests are unlimited. In India, deforestation, overgrazing and hydroelectric power development schemes are invading the new national parks and forest reserves which still exist. In the Himalaya Mountains, the landscape is being carved within an inch of its life to support a burgeoning human population. In Africa and Latin America, slash and burn agriculture is encroaching upon natural areas which have existed in ecological balance for thousands of years (**Southwick**, 1976).

Quite ridiculously, much of this environmental destruction is done in the name of economic development. The pressures of increasing populations place a great burden than ever on the finite blanket of life that covers this earth. That blanket is getting torn and shredded in thousands of places. Like a wound or burn on the skin, its ability to heal depends upon the extent of the lesion and the health of the surrounding tissue. If man remains unaware, a point can be reached in future, where the healing capacity may be lost.

Apparently, due to his numerous ecological malpractices and short sighted and greedy exploitation of different ecological resources such as air, water, space, minerals, vegetation, etc., modern human civilization has to face many serious ecological problems such as land use, pollution, energy crisis, flooding, erosion, pesticide contamination, radioisotope accumulation, rural and urban blight, population growth, urbanization, crowding, group conflict, and infectious diseases. If some ecologically sound measures are not applied to human utilization of his environment, there may arise several uncontrollable ecological hazards which may become responsible for the extermination, extinction and devastation of human race from the planet earth. Many applied ecologists have shown that ecology may play a significant role in the human welfare. They have suggested following measures by which man may use his environment and its resources fully and without disturbing naturally occurring ecological processes.

26-1. CONSERVATION OF NATURAL RESOURCES

Conservation is one of the most significant application of ecology. It avoids unplanned development which breaks ecological as well as human laws. A conservationist has two folds basic aims (i) to insure the preservation of a quality environment that considers aesthetic and recreational as well as product needs and (ii) to insure a continuous yield of useful plants, animals, and materials by establishing a balanced cycle of harvest and renewal (Odum, 1971). Thus, conservation process remains chiefly concerned with the use, preservation and proper management of the **natural resources** of the earth and their protection from the destructive influences, misuse, decay, fire, or waste. The natural resources include all the land, minerals, water, vegetation, wild-life and sceneries, all of which remain useful to human society in one way or other. Customarily, natural resources are classified into following three categories—(i) **Non-renewable resources** like metals (iron, copper, zinc, etc.), fossil fuels (coal and oil deposits), other minerals and their salts (phosphates, nitrates, carbonates. etc.) and stone. (ii) **Renewable resources** such as living resources like fish, forests, crops, wood, etc. (iii) **Unalterable resources**—the resources that are used outside the human body the gathering of which leaves them unaltered. Such features of our surroundings as scenery, wildlife (if observed rather than hunted) and water, for swimming or sailing, remain unchanged by our use of them for recreational and aesthetic satisfaction. The total flow of a resource from its state in nature

through its period of contact with man to its disposal has been **termed** as a **resource process** by **Firey** (1960).

Different kinds of these natural resources and their conservation and management practices can be studied under following heads :

26-2. MINERALS AND THEIR CONSERVATION

Minerals are often called "**stock**" or "**non-renewable**" resources, because, their 'new' materials can only be extracted from the earth's crust once. But even in the transformed state in which they are used, they are not lost to the planet and so are ideally available for reuse.

26-2·1. Terrestrial Mineral Resources

For his industrial, technological and cultural growth man has required and still needs a great variety of inorganic materials, all of which come from the earth's crust. Chief among these are the ores which are used on a large scale to yield metals like **iron, aluminium, copper, silver, gold, platinum**, etc. To them must be added elements which may not be needed in large quantities but which are indispensable in many modern industrial processes, as for example catalysts and hardners such as **vanadium, tungsten,** and **molybdenum.** Finally there are non-metallic materials which are vital to industrialized nations such as sand and gravel, cement, fluxes, clay, salt, sulphur, phosphorus, diamonds, and the chemical by-products of petroleum refining.

The distribution of minerals in the earth's crust is characterized by discontinuity. There is spatial discontinuity in which deposits rarely coincide with the boundaries of nation states that wish to use them. North America is well supplied with the ore of molybdenum, for example, where Asia is not ; by way of compensation Asia is rich with **tin, tungsten, and manganese.** Between them, Cuba and New Caledonia have half the world's reserve of **nickel**, and industrial diamonds are dominated by Zaire. Such discontinuity are also emphasized by temporal patterns of use—the older industrial countries such as UK are running out of their ore reserves and coming to depend upon imports, and heavy users like the USA face similar problems; in both cases iron ore stands as a good example. Another type of discontinuity is exemplified by the richness of an ore. A few metals show a more or less continuous grading from the richest ores to the poorest, *e.g.*, iron and porphyry copper. On the other hand some ores are either very rich or very poor or both.

26·2·2. Marine Mineral Resources

The sea's mineral resources can be divided into three categories: those which are dissolved in water itself ; sediments present on the sea-bed at various depths ; and those present at some depths below the sea-floor, beyond the sediments of relatively recent origin.

At present the utility of the dissolved elements is in direct proportion to their abundance and to the relative cost from terrestrial sources. **Cloud** (1969) has enlisted certain commonest elements in sea water which could be extracted (Table 26·1). Common salt is one of the resources that has been utilised since prehistoric times for its value in flavouring and meat pieservation. At present only **salt, magnesium** and **bromine** are being extracted in commercial quantities and sea seems to remain inexhaustible for these elements. Except a few other elements, economically feasible recovery processes of most other elements from sea is very low. However, it may be possible to lower feasibility thresholds by investigating the capacity of marine organisms to concentrate desired elements (this is done *in facto* for **nitrogen** used in the form of fish-meal fertilizer and sea-bird guano), and in the possible exploitation of zones along the sea-bed where fractures allow the escape of unusually high concentrations of mineral ions (**Simmons,** 1974).

Table 26·1. Concentration and value of the elements in sea water (source Simmons, 1974). The elements are listed in order of abundance those in the italic type have concentrations valued at \$ 1·00 or more per million gallons of sea water. All others are <0·02 lb/10⁶ gal.

Element	Concentration lb/10^6 gal.	As	Value \$/10^6 gal.
1. *Chlorine*	166,000	NaCl	924
2. Sodium	92,000	Na_2CO_3	378
3. *Magnesium*	11,800	Mg	4,130
4. *Sulphur*	7,750	S	101
5. *Calcium*	3,500	$CaCl_2$	150
6. *Potassium*	3,300	K_2O	91
7. *Bromine*	570	Br_2	190
8. Carbon	250	Graphite	8×10^{-8}
9. *Strontium*	70	$SrCO_3$	2
10. *Boron*	40	H_3BO_3	3
11. Silicon	26	—	—
12. Fluorine	11	CaF_2	0·35
13. Argon	5	—	—
14. *Nitrogen*	4	NH_4NO_3	1
15. *Lithium*	1·5	Li_2CO_3	36
16. *Rubidium*	1·0	Rb	12·5

Element	Concentration lb/10^6 gal	As	Value \$/10^6 gal.
17. Phosphorus	0·6	$CaHPO_4$	0·08
18. *Iodine*	0·5	I_2	1
19. Barium	0·3	$BaSO_4$	0·01
20. *Indium*	0·2	In	4
21. Zinc	0·09	Zn	0·013
22. Iron	0·09	Fe_2O_3	0·001
23. *Aluminium*	0·09	Al	0·04
24. Molybdenum	0·09	Mo	0·004
25. Selenium	0·04	Se	0·2
26. Tin	0·03	Sn	0·2
27. Copper	0·03	Cu	0·01
28. Arsenic	0·03	As_2O_3	0·002
29. Uranium	0·03	U_3O_8	0·3
30. Nickel	0·02	Ni	0·02
31. Vanadium	0·02	V_2O_5	0·04
32. Manganese	0·02	Mn	0·006

Sediments and sedimentary rocks on the continental shelves are sources of certain materials. Placer deposits contain workable quantities of **gold**, **tin**, and **diamonds**, and the other sediments which may be amenable to exploitation include **sand**, **gravel** and **shells**. However, land-use problems by their extraction from the land would largely be obviated by the use of the sea as a source, provided that the ecosystems of the oceans were not too greatly damaged by the recovery processes, which create great qnantities of silt and also eventuate imbalance in the sedimentary systems of the sea-floor. **Phosphates** are found as nodules and in crust. A further resource of the continental shelves is **fresh-water**: large quantities of artesian water may be found in certain aquifers. Finally there are petroleum and natural gas. The deep sea basins are found to contain enormous quantities of minerals such as **manganese, nickel. cobalt and copper**, but their extraction is not economically feasible (**Cloud,** 1969).

26-2·2A. Marine mineral resources of India. The deep sea basins and continental shelf of Indian sea (Arabian sea, Bay of Bengal and Indian ocean, are found to have rich mineral resources of the following three types : 1. **Terrigenous minerals.** Recently

Geological Survey of India (GSI) and National Institute of Oceno-graphy (NIO) have discovered huge deposits of **ilmnite, monazite** and **calcareous** sands along the cost of Maharashtra, Kerala, Tamil Nadu, Andhra Pradesh and Orissa. Ilmnite is used in the manu-facture of titanium (used in supersonic aircrafts) and titanium dioxide (TiO_2) that is used in paint industry. Monazite is used for obtaining thorium. 2. **Biogenous deposits.** Biogenous deposits along the ocean floor consist of shell pieces and skeletal debris of marine organisms and they contain mainly calcium carbonate. Such deposits are common in Kerala's Vembanad Lake, in Gulf of Kutch, Gulf of manar, Palk straits, and along the coasts of Andman and Nicobar and Lakshadweep islands. 3. **Chemo-genous diposits.** The most important chemogenous deposits are manganese nodules. Manganese nodules in the shape of potatoes contain manganese (19%) and iron (12%) They also contain nickle, cobalt and copper in recoverable quantities. These deposits are very important resource for manganese, as terrestrial source of manganese (*i.e.*, Indian reserve of manganese ore is 68 million tonnes) is fastly depleting (**Joshi**, 1978).

26·2·3. Conservation of Terrestrial Mineral Resources

Until recently little attention was paid to conservation of terrestrial mineral resources because it was assumed that there were plenty for centuries to come and that nothing could be done to save them any way. But quite recently **Cloud** (1969) made it evident that both assumption are dead wrong. During his assessment of the situation, he has formulated following two situations about the exploitation of terrestrial mineral resources—(1) The first is the **demographic quotient** or **Q** :

$$Q = \frac{\text{total resources available}}{\text{population density} \times \text{per capita consumption}}$$

As this quotient goes down, so does the quality of modern life ; it is going down at a frightening rate because available supplies can only go down as consumption goes up.

(2) The other concept introduced by **Cloud** is the graphic model of **depletion curve**, as shown in figure 26·1. With the present procedure of "*mine, use and throw away*", a huge boom and bust is projected, as shown in curve A. The time scale is uncertain because lack of data, but the "bust" could begin within this century since certain key metals and fuels such as zinc, tin, lead, copper, ura-nium-235, natural gas could be mined or extracted out within 20 years in so far as the readily exploitable reserves are concerned. If a programme of mineral conservation, substitutions (using less scarce minerals wherever possible), and partial recycle were to be started now, the depletion curve could be flattened as shown in curve B. Efficient recycling combined with stringent conservation and a reduction in per capita use could prolong depletion for a long time, as shown in curve C. It should be noted that even with perfect recycle depeletion would still occur, because small amounts of most metals lose by friction, rust, etc.

Consequently, terrestrial mineral conservationists have suggested following measures for terrestrial mineral conservation : 1. Technological advancement for extraction of minerals from sea which is unlimited supply depot. 2. Extraction of metals from lean ores with the use of vast quantities of cheap atomic energy. **Granite** is suggested as a source for the extraction of many metals. 3. Recovery of mineral elements from scrap and waste (recycling).

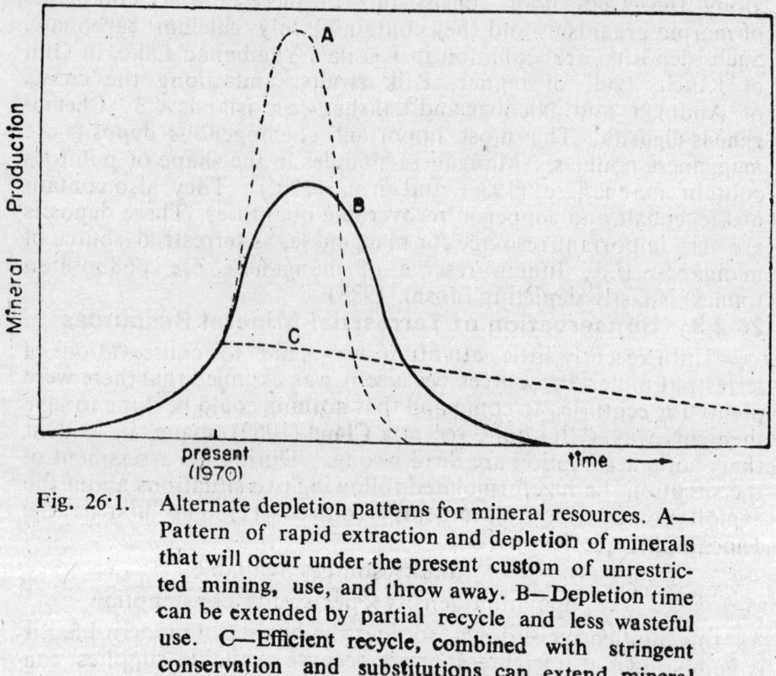

Fig. 26·1. Alternate depletion patterns for mineral resources. A—Pattern of rapid extraction and depletion of minerals that will occur under the present custom of unrestricted mining, use, and throw away. B—Depletion time can be extended by partial recycle and less wasteful use. C—Efficient recycle, combined with stringent conservation and substitutions can extend mineral depletion curves indefinitely (Cloud, 1969).

26·2·4. Ecological Aspects of Mining

Most kinds of mining processes of man have side-effects of an ecological nature. Underground mining may include whole new towns among its surface installations (*e.g.*, the Khateri Copper Project, JhunJhnu, Rajasthan, India). Timber is cut in forested areas, often leading to soil erosion and the tailings and mine-waste have to be discarded. Large-size solid wastes can be used as back-fill or sold for aggregates, but tailings usually yield silt particles to wind and water and are often chemically unstable; the only suitable treatment appeares to be "fix" them with vegetation. Mine waters are often heavily contaminated and have to be treated chemically and physically, or injected into "safe" rock strata.

Further, in most countries open-pit mining is more widespread than extraction by shaft. In such mining waste disposal becomes a major problem to which back fill is the obvious solution ; if the topsoil is saved then restoration of agriculture or recreational use is often possible. The processes of concentration, beneficiation and refining may all create biological changes if various products are released into nearby ecosystems, *i.e.*, they may create environmental pollution which affect adversely the ecosystems.

26-3. ENERGY AND ITS CONSERVATION

Ecologists and economists have looked at energy from different point of view. For ecologists, the study of energy has begun with sunlight, considering first its conversion by green plants into the chemical energy of organic compounds, and then the transfer of this energy through food chains to animals and human populations. For economists, the study of energy has involved the immediate sources of energy of human needs : industry, transportation, agriculture, and domestic activities. The ecologist has started with sunlight and photosynthesis ; the economists with coal mine, the oil well, the power plant, or the nuclear reactor. The coal, oil and lignite are the fossil fuels which contain stored solar energy and are non-renewable energy resources.

The energy utilization of human **populations** has increased steadily in the progression from hunter-gatherer to agricultural and industrial societies. For example, early agricultural societies depended entirely on the energy of sunlight to grow food, and the energy of physical labour for its cultivation and harvest. Domestic animals, introduced later, converted food energy into fats, fibres and proteins and provided greater physical strength to draw ploughs, pull carts and turn water wheels. The invention of electric motor and the internal combustion engine in the nineteenth century changed this orientation completely. These machines boosted the production by enabling a man to do far more work and to cultivate and harvest greater acreage, and they also replaced draft animals with the result that modern agricultural operations became dependent on pertroleum and electricity.

Energy utilization, thus, has become the modern key to human progress. At the present time, modern nations differ markedly in their energy consumption, with industrialized nations such as Europe, United States, Japan, Hong Kong and Singapore, typically using 10 to 50 times as much energy per capita as agricultural and developing nations such as India, Bangladesh, Indonesia and Ethiopia (**Southwick**, 1976). **Higher energy consumption is often associated with higher gross national production and personal income per capita, but it is also associated with higher levels of industrial pollution.**

Human populations have become predominantly dependent upon fossil fuels, but this will have to change in the next 100 years. Reserves of petroleum, the major source of energy utilized at present, are dwindling. Vast coal reserves still exist, but both the mining and the combustion of coal involve serious environmental hazards. For example, underground coal mining leads to the occupational hazards of black lung disease in miners and the accidental dangers of mine fires or collapse. The strip-mining of coal destroys vegetation, alters topography, upsets watersheds, increase erosion, contributes to water pollution and devastates the landscape. Coal combustion is a major source of sulphur dioxides, soot, and fly ash ; all of which contribute air and water pollution. However, recently a technique has been evolved which would permit expanded use of coal in industries without any pollution. It is the development of clean fuels by means of coal gasification. Humidified air is introduced under pressure at the base of a column of pulverized coal with a grannular size of 5 to 15 millimetres. Under controlled temperature and oxygen conditions, various industrial gases are released and can be collected for later combustion. These gases are primarily mixtures of methane, carbon monoxide, hydrogen, and nitrogen. Liquid fuels and gasoline can also be obtained from coal gasification and will provide cleaner fuels than the direct combustion of coal A number of countries such as Germany, Great Britain, Turkey, India, South Africa, Morocco, Korea and Yugoslavia already use coal gasification to produce industrial fuel. Coal has also been used more efficiently in magnetohydrodynamic (MHD) generators which convert heat from combustion gases directly into electricity. MHD generators require less fuel, produce less thermal pollution and generate fewer oxides of sulphur and nitrogen than conventional coal combustion.

Further, many nations of the world made their primary commitment to the development of nuclear energy to replace fossil fuels. Due to inherent danger of nuclear energy in terms of isotope production and transportation, waste disposal, accidental potential, and the threat of nuclear proliferation, black mail, and terrorism, the heavy investments in nuclear energy **may** prove to have been shortsighted and even tragic in the long run (**Southwick**, 1976).

Alternative energy sources, including solar, tidal, wind, waste incineration and hydroelectric have assumed importance in local areas, but they have not been developed adequately on a national or global basis. All of them have high potentials if properly developed. However, greatest potential of all lies in solar energy, for it is an unlimited and essentially non-polluting source of energy. Currently Indian scientists are trying hard to tap certain new non-polluting and cheap sources of energy like biogas, solar energy, hydro-electricity, tidal power and wind power and are getting substantial success in their efforts.

26-4. FOOD, AGRICULTURE AND AQUACULTURE

Humans need food as a source of energy and for tissue replacement, like any other animal. The source of much of the food consumed by man is **terrestrial agriculture**. This represents the most manipulated of all the non-urban ecosystems, in which the energy and matter pathways are directed almost entirely to man and where he maintains a high level of input of matter and energy to keep the system stable in order to yield his preferred crop. Not only is the ecosystem man-made, but the plant and animal components of it have usually been genetically altered by man in the course of their domestication. There are two main types of agriculture : **crop agriculture**, in which plant production is harvested for use by man either directly or after processing ; and **animal agriculture**, where a crop from highly mainpulated ecosystem is fed to domesticated animals. Ecologically, terrestrial agriculture presents man either as a herbivore or as a third trophic level carnivore.

Terrestrial agriculture systems today exhibit a major division into shifting and sedentary types :

26-4·1. Shifting Cultivation

In shifting cultivation, total manipulation of the natural system is practised over a limited area but for only a short (1-5 years) period of time. Thus the agricultural path is spatially and temporally enclosed by wild vegetation. Today, shifting cultivation is largely confined to tropical forests, savannas, and grasslands. In this type of cultivation crops are planted in a mosaic of different heights and times of fruition so that the plant cover of the soil remains as complete as possible throughout the year in order to reduce the leaching effect of heavy rain fall. 'Slash and burn' techniques provide mineral nutrients for their uptake by the crops. The natural diversity of the forests is imitated by the variety of crops which are grown by some shifting cultivators. For example, the tropic crops of Hanunoo of the Philippines include rice, beans, root crops, shrub legumes, tree crops, yam, taro, sweet potato, vines, bananas and sugarcane, together with European-contact crops such as maize, ground nuts, tomatoes, melons and pumpkins.

The major disadvantage of shifting cultivation is the inability to store surpluses, even when partly processed to paste or flour form. The grains are more stable but highly vulnerable to insects. bacteria and scavenging rodents. Ecologists have found shifting cultivation as a well-adapted system in forested lands where the trees regenerate easily when the plots are deserted, and where an equilibrium population has established.

26-4·2. Sedentary Cultivation

Sedentary cultivation represents the permanent manipulation of an ecosystem : the natural biota are removed and replaced with domesticated plants and animals. Competition by the remnants of the original biota or by man-introduced organisms may still remain and considerable effort may be needed so keep these weeds and pests at an acceptable level. **In dry land agriculture** the soil assumes an importance for it now becomes the long-term reservoir of all nutrients and is constantly depleted as crops are harvested and removed. The nutrients must be replenished either by the addition of organic excreta or chemical fertilizers. There is also the **paddy-culture** of rice, where the soil is very largely a mechanical rooting medium for the plants and the water supplies the essential mineral nutrients ; it often contains blue-green algae which fix nitrogen, for example. The variety of crops grown under the various forms of sedentary cultivation is very high and changing patterns of agriculture together with shifting trade flows and altered rates of consumption make a world kaleidoscope of infinite variety.

Ecological training has been widely used in agriculture. To select particular climate and edaphic conditions for a particular crop is primarily an ecologist's job. Ecology plays a vital role in weed control. The nitrogen and phosphorus requirements of soil, crop rotation practice, etc., have been learned from ecology. Due to application of ecologically sound principles agriculture has resulted into **Green Revolution**, *i.e.*, increased agriculture production. In fact, the Green Revolution of the late 1960s, was based on new genetic strains of rice and wheat, improved irrigation, and better application of fertilizers and it has achieved a doubling and tripling of crop yields in many tropical countries But recently, Green Revolution has exhibited certain ecological problems–the breakdown of soil structure in European countries under a regime of continuous cereal cropping ,with the use of physically heavy machinery and its increase dependence upon petroleum and fertilizer. Further, new strains of agricultural crops which are high yielding, protein-rich, and weatner and pest resistant, require more careful management, more water, more fertilizer, more insecticides, and more intensive care.

Lastly, agricultural man set into motion several major forces which had significant geographic and mateorological consequences: deforestation, overgrazing, intensive burning, and land scarring. Geographic consequences came about through increased erosion, soil loss, and declining water tables ; mateorologic consequences resulted from reductions in atmospheric humidity and cloud cover, increased heat reflectivity and a lowering of rainfall.

Though terrestrial agriculture has immense significance in food production for man but surprisingly very little of the surface of the continents is cultivated and problems of starvation and malnutrition have remained unsolved ecological problems still today. Consequently, ecologists have provided certain solutions for agriculture extension and intensification and have suggested certain new food resources for man.

(a) Extension of agriculture. The extension of the world's agricultural area seems an obvious way to resolve some of the problems. Improved machinery, irrigation, better roads, reclamation from the sea, tranformation of other ecosystems, all ought to provide greater agriculture area.

(b) Intensification of agriculture. By this process an improvement of food yields from existing cultivated lands is sought. It is a complex sequence of events which brings a low-yieiding traditional agriculture into connection with the technology and economics of industry. Thus, a great deal of capital and management skill are needed for success in applying developments such as irrigation, flood control, drainage, erosion control, mechanization, fertilizer and biocide use and raising of improved varieties of plant and animal. The keys to intensification are energy availability, both on site and in the places where tractors, pesitides, pumps and the milk are made ; a steady effective demands for the products of the farming system and good communications between the source of supply of the input, the farm, and the consumers.

26·4·3. New Sources of Food

Most modern ecologists hold the view that even if the rapid development of conventional agriculture is sustained, protein deficiencies will continue to exist (**Simmons**, 1974). A search for supplementary sources of plant and animal proteins is, therefore, in progress Animal proteins are more preferred ones because their amino acid make-up is closest to man's requirements. Animal flesh has the further advantage that it is usually the more easily assimilable, since the plant proteins are locked away behind a cell wall of cellulose not easily broken down by the action of the human stomach. Animals thus have been a means of harvesting the plant protein in a digestible form, and so have considerable dietry advantages. If animals are to be avoided, a food source which will yield plant protein in a digestible form or from which the majority of cellulose has been removed is clearly attractive. Fungi appear to be easily assimilated and contain a good deal of protein by comparison with some other sources such as beef, fish, etc. Other advantages are that they do not absorb much human or fossil energy in production, can be grown independently of environmental factors in places such as caves and abandoned railway tunnels, can readily be stored in dried form and require little sophisticated knowledge or technology (**Pyke**, 1970).

Requiring rather more technology readily available are leaves of unused grasses, shrubs or marginal aquatics which contain high protein contents. If this protein can be separated from the fibrous material of the leaf and made palatable, leaves may prove cheap and best source for human dietry protein. Recently algae have received a good deal of attention as possible sources of food, especially the noncellular varieties which under optimal conditions have exceptionally high rates of primary productivity. Since algae would be cultured in tanks, non-agricultural surfaces such as roof-tops might become food producing. However, production of algae would be a very technical process, requiring stirring, bubbling of carbon dioxide, sophisticated machinery and skilled man-power (**Pirie**, 1969).

Animals retain many desirable characteristics as cellulose-converters and as saliva-inducers. As very small number of wild animals have been domesticated by man for milk, wool, skin, food, transportation, etc., recently, certain other wild-animals are increasingly domesticated. Birds such as young colonial seabirds, reptiles, amphibians, insects, crustaceans, molluscs, rodents, and many other animals can be utilized as food. Fresh-water and brackish-water fish are other source of animal proteins. Marine and fresh-water fisheries provide a best sources of animal proteins, the fishes and many nations such as Peru, Japan, USSR, China, Norway, USA, UK, Burma, Thailand and India depend on it for food requirements. Due to overfishing of certain favourite fish species, recently following trends are adopted for continuous abundant sppply of fishes:

26·4·3A. Extension of fisheries. The fisheries have been extended from three sources : the utilization of untapped species, the cropping of hitherto unattractive areas, and the development of more novel methods of culture and harvesting. During recent years many new fisheries have started to flourish, such as Peruvian anchoveta, Alaska pollock, Bering sea flatfishes and herring, and several more. In future fishing method may be extended in the cool temperate parts of southern hemisphere. In heavily fished northern areas these are abundant but little-used species—grenadiers in the north-west Atlantic, sandlance, anchovies and sauries in the Pacific and small sharks like the dogfish in both areas.

26·4·3B. Krill. Uneaten food of whales is also found suitable for human consumption as food. Approximately 80 per cent of the prey of blue and fin whales is **krill**, the shrimp *Euphausia superba*, each of which contains 7% fat and 16% protein.

26·4·3 C. Aquaculture. To increase the yield of certain edible species of fresh-water and marine fishes, molluscs and crustaceans and other aquatic organisms, **fish farming** or **aquaculture** that includes their culturing and herding, has been employed and at an commercial scale in Hongkong, Philippines and Japan. Frameworks are lowered into shallow offshore waters and allowed to colonize with sedentary molluscs like oysters and mussels with sense species, the individuals grow on ropes that hang clear of the action so that they are out of the reach of predators such as starfish. Although productive, such systems are very vulnerable to contamination, and since the organisms filter large quantities of water their ability to concentrate substances toxic either to themselves or to consumers is very high.

True aquaculture involves genetic manipulation of the chosen species by keeping them captive throughout their breeding cycle. The requirements are unpolluted sea water and a suitable coastal site with adjoining land The most efficient plant would be large and would require buildings, stores, hatcheries and covered tank complexes on the land, together with enclosed tidal areas and tanks, and ponds or lagoons in deeper water. For preferable use of all the water areas would be possible by housing together algae browsers such as abalone, pelagic herbivores like grey mullet and bottom dwelling carnivores (C.E. Nash, 1970) Fresh-water aquaculture is most common in Hongkong, where small ponds have become the popular with farmers, who also keep ducks which enrich water and are marketable as well. **Mariculture** (aquaculture in marine or estuarine waters) is recently more advanced in Japan, Indonesia and Australia

(i) **Aquaculture in India**. Aquaculture, the farming of aquatic plants and animals to increase additional food resources for man, is presently in the state of infancy in India. However, certain Indian ecologists like **S.Z. Qasim**, who is an eminent and dynamic oceanographer of India and Director of National Institute of Oceanography, Dona Paula, Goa, have done great efforts for the development of aquaculture in India. In one of his article—'*Role of Aquaculture in Rural Development*', **Qasim** (1977) has considered various aspects of aquaculture including its advantages, disadvantages, economic viability, techniques, gains and scope. According to him, in developing countries like India, herbivores or plankton-feeders are most suitable for aquaculture, because their harvest in terms of surface area or volume of water, utilization of raw materials like solar energy, and natural or artificial fertility is greater than carnivores. Thus, in India, the culture of fish like milk-fish, mullet, and eel has been found suitable and economical. Among crustaceans, marine animals like shrimps, lobsters and crabs have been found most suitable and rewarding for cultivation. So far, in India, maximum success in aquaculture has been achieved in the culture of molluscs like mussels, however, clams, edible oysters and window-pane oysters have shown great possibilities for their economical cultivation. In recent years, the culture of pearl-oysters on wooden rafts using sandwich-type frame-nets has given excellent results. **Qasim** has suggested that, culture of other organisms of economic value like sponges, holothurians, turtle and seaweeds can be done by adopting similar aquaculture techniques.

(ii) **Waste-management parks**. One of the most promising opportunities for aquaculture in an affluent country is to tap the effluents from waste-treatment. Heated water from power plants and certain types of diluted or treated domestic or industrial organic wastes flowing through a series of ponds can provide energy subsidies for adapted species of fish, molluscs, crustaceans that could provide human and animal food or other useful products. Waste heat may maintain constant water temperatures in coastal temperate zones and there certain tropical fishes of high productivity can be raised besides native fishes which are also benefitted from heated water. Further, sewage and eutrophicatory waste products might be used as the basis for algal production which forms the food of oysters which then filter the water as well. The oyster droppings are eaten by worms which are the prey of bottom living fish, whose nitrogenous excretions nourish water weeds which oxygenate water. In another scheme, CO_2 from a power-station chimney is used to enhance production of algae which are fed to clams that then grow rapidly in heated water which is also the output of the power station.

26·5. FORESTRY

About one-third of the world's land surface (4,028 million hectares) is covered with forests, which are ecosystems that are dominated by trees. Forests are the vast natural resources for man for fuel, timber, paper, fodder, water yield and animal products. Ecological studies have shown that forests help to maintain the level of rainfall necessary for their own existence and the indiscriminate deforestation by agricultural man for land. for cultivation or by modern industrial man for dam, mining, road, etc., results not only in an immediate lowering of ground water levels, but also in long-term lowering of rainfall. Forests recycle moisture back into their immediate atmosphere by transpiration where it again falls as rain. Due to deforestation, this natural reuse cycle is broken and water is lost through rapid runoff. Forests are also necessary to check the floods and soil erosion, and are important for wild-life, human recreation, air and water sheds, etc.

To maintain them certain forest conservation and management processes have to be employed to the forests. In India, at Forest Research Institute, Dehradun, a team of scientists is engaged with such a viewpoint. To get desired quality of timber or pulp of paper industry, monoculture forests of fast-growing trees such as poplars, certain conifers, teak and eucalyptus have been cultivated by man. Existing forests are strongly manipulated in order to increase their yield of desired benefits It includes **weeding** (the elimination of species which might compete with the seedlings of desird species), **thinning** (eradication of individuals of the same species) and **brashing** (removal of leafless lower branches especially in conifers). Chemical technology may be applied to forestry to control insect parasites and damaging fungi. Forest management also includes the controlling of forest fires.

Recently, the widely adopted practice of monoculture of forestry has posed certain new and serious problems. The monoculture drain the soil of specific nutrients required by the species in greater quantities. For example, teak plantation requires more calcium thereby leaving the soil impoverished. Once the plantation has been cut, the land becomes generally unfit for any other vegetation and fertilizers have to be applied. Likewise, monoculture of *Eucalyptus*—a fast growing species, withdraws large quantities of underground water and also removes nutrients in large quantities. Its litter is negligible in quantity, does not decompose rapidly and trees do not support any animal life which may bring even small quantities of organic matter to the soil.

Silviculture, horticulture and arboriculture. Silviculture is that branch of forestry which remains concerned with the establishment, development, care and reproduction of monocultures of valuable timber trees such as teak. *Sal*, *Sheesham Kel* etc Arboriculture and horticulture remain concerned with the cultivation of fruit trees and flower gardens and consequent beautification of a landscape. All these require ecologically oriented managements and conservations which may include plantation of mixed forests and recycling of forest products.

26-6. RANGE MANAGEMENT (GRASSLAND MANAGEMENT)

Grassland biomes are important to maintain the crops of many domesticated and wild herbivorous mammals such as horses, mules, asses, cattle, pigs, sheeps, goats, buffaloes, camels, deers, zebras, etc., all of which provide food, milk, wool, hide, or transportation, etc., to man. The range management involves an important application of ecological principles in maintaining the grassland biomes. The objective here is to preserve grasslands for maximum forage, *i.e.*, food for cattle In India, most of the grasslands represent the seral stages in succession and if they are not maintained by grazing and fire they would develop into forest communities. A team of scientists at Indian Grasslands and Fodder Research Institute, Jhansi and Central Arid Zone Research Institute, Jodhpur, is mainly entrusted with such responsibility.

During range management a suitable degree of grazing is to be maintained. Some species are more palatable to the stock and are very sensitive for grazing and thus overgrazing usually results into their disappearance from the area where now some unpalatable, annuals, weeds, shrubs start to grow and turn the area into man-made desert. The palatable plant species, which are sensitive to grazing, have been called the **decreasers** whose disappearance from the area is an indicator of grazing stress and warning signal for range managers.

Overgrazing has certain other ecological effects—reduction of the mulch cover of the soil occur, microclimate becomes more dry and severe and is readily invaded by xerophytic plants. Due to absence of humus cover, mineral soil surface is heavily trampled when wet and produces puddling of the surface layers, which in turn reduces the infiltration of water into the soil and accelerates its runoff, producing drought. These changes all contribute to the reduction of the rate of energy flow, and the disruption of the stratification and periodicity of the primary producers results in a breakdown of the biogeochemical cycles of water, carbon and nitrogen. Water and wind erosin completely breakdown a very dry grassland-microclimate. Further, intensive grazing which results in increased areas of bare soil, creates a new habitat for burrowing animals such as mice, jackrabbits, gophers, prairie dogs, locusts, etc., which render sterile much areas of forage lands.

In range management, fire plays an important role. Under moist conditions fire favours grass over trees and under dry conditions fire is often necessary to maintain grasslands against the invasion of desert shrubs. Burning of *Cynodon dactylon* increases forage yields.

26-7. WILD-LIFE MANAGEMENT

Wild-life management remains chiefly concerned with the protection, preservation, perpetuation and judicious control of populations of rare species of plants or animals in their natural habitats. In fact population of many animal species has been changing since prehistoric time. Over the past two thousand years about 106 species of animals and 139 species of birds have become extinct due to climatic and geographical changes and also by overhunting by man for food, fur, or other reasons. According to **Red Data Book** of ecologists about 600 more species of animals and birds are going to be extinct, if not protected by wild-life management. Wild-life management includes following fundamental approaches :

(i) **Enaction of wild-life acts** to prevent the hunting of certain rare species of wild-life. For example, certain Indian animal species such as python, leathery turtle, marsh crocodile, *Gharial*, Bustard, Peafowl, Black buck, elephants, Eastern pangolin, Snow leopard, etc., are protected by Law.

(ii) **Species preservation**. Nature reserves are usually designated in order to give protection to a species of plant or animal which is rare. In Britain certain sea birds, wild birds and plant species are protected. The Farne Islands off the coast of Northumberland (UK) are a sanctuary for the grey seals. In the northern Alberta and the North-West Territories of Canada, Wood Buffalo National Park preserves in a wild and almost roadless terrain one of the few remaining herds of North American bison. To small islands in Sabah, Malaysia, protect frigate birds, and Udjung Kulon —Panailan in Western Java preserves 25 to 40 Javan rhinos. Kaziranga Sanctuary (Assam) shelters one horned rhinoceros and Gir forests (Gujrat) preserve Gir lions.

(iii) **Assemblage protection.** Most commonly an assemblage of species is protected. The assemblages may have some linked affinity —wildfowl refuges of USA cater for the nesting or migration of many species of ducks, geese and waders. High mountain reserves often protect a very diverse suite of alpine plants sometimes with the attendant fauna. In India Delhi Zoo and Bharatpur Bird Sanctuary are specifically managed for the assemblage of migratory birds. The management of assemblage protection is designed to provide maximum cover and food supply for the migratory birds and to provide refuge from hunting.

(iv) **Habitat preservation.** Reserves which are large and diverse enough to protect whole sets of ecosystems which are eithr rare on a national or world basis, are often designated as **National Parks**. Indian subcontinent has more than 50 sanctuaries, few national parks. protected areas and reserves (**Natrajan**, 1977). Indian sanctuaries possess unique landscapes, broad-leaved forests, mountain forests and virgin bushlands in deltas of big rivers (See Section 27-4·3).

26-8. WATER CONSERVATION

Water occurs naturally in gaseous, solid and liquid phases ; man's use of it is nearly all concerned with the last state and is also dominated by his demand for water relatively low in dissolved salts, *i.e.*, **freshwater**. Two Russian geographers and hydrologists, **Kalinin** and **Bykov** (1969) have estimated world's total water resources. Of $1{\cdot}46 \times 10^9$ cubic kilometers of water in the world, 93 per cent is in the oceans, 4·1 per cent is in the earth's crust, 2·0 per cent is in glaciers and polar ice caps, and only 0·052 per cent is in fresh-water lakes, rivers, and atmospheric moisture. The annual production of freshwater by evaporation and precipitation is estimated to be 37,000 cubic kilometers. At the present time, the world's total water use is less than 10,000 cubic kilometers, but this is estimated to increase to approximately 18,700 cubic kilometers by the year 2000. Though there is no alarming shortage of water resource, but, there are two practical difficulties in its proper utilization–its distribution and quality. A large proportion of the world's fresh-water falls and resides in relatively unpopulated portions of the globe, particularly Siberia and Northern Canada. Here, man's activities are limited by other factors, and he cannot easily obtain these vast stores of water. Secondly, the matter of water quality is of immediate concern. The principle of water reuse is now becoming accepted, so that people recognize that much of their water has been used previously by other municipal systems. Each reuse cycle, however, is increasingly become costly in terms of treatment, and each reuse cycle requires disinfectants and other industrial chemicals. Thus, water problem of the future is not going to be one of a global water shortage on a planetary basis, but one of local crisis due to concentrations of people, industry and water use, and the vagaries of weather.

Certainly, the immediate concern is finding local water resources of adequate quality to meet the burgeoning needs of urban populations and intensive agriculture. The fresh-water resource can also be put under conservation and management by following methods : weather modification by cloud seeding; watershed management by manipulation and alteration of vegetation of a catchment area or watershed, in order to produce more runoff and by increasing of the storage capacity in the runoff phase so that water may be held for use in dry seasons or to prevent floods ; managing water flow over land surface to get rid of any excess by drainage ; controlling water supplies by storage of it by dams ; controlling of floods by constructing floodcontrol dams; channel widening, channel straightening and deepening and by construction of by pass channels; pumping out of ground water; demineralizing salt and freshwaters and lastly, melting down of glaciers and polar ice caps.

26-9 SPACE UTILIZATION ON EARTH

The earth's total **surface** is 196,940,400 square miles, of which about 30 per cent or 57,230,000 square miles is land surface. Of this total land surface, approximately 5,000,000 square miles are in Antarctica and are too cold to be inhabited except by scientists and explorers ; 9,000,000 square miles are in deserts, and 11,000,000 square miles are in mountanious regions. At least one-fourth of the land area of deserts and mountains cannot be inhabited, so this eliminates another 5,000,000 square miles of land surface for human habitation. We are left with approximately 47,000,000 square miles of land surface, at the most, for the world's human population (**Southwick**, 1976).

The world now has about 4 billion people, and will probably reach 8 billion within the next 32 years. Theoretically this means, we now have about 85 people for every square miles of inhabitable land surface on the face of the earth, and this will approach 150 people per square miles by the year 2000 A.D. Practically, however, the human population is not evenly distributed. In the great urban concentrations of the world, population densities often exceed 50,000 people per square miles. The amount of living space per person in some cities like Calcutta is less than 20 square feet per person. Further, economic and behavioural needs of man are such that he must cluster together, but in doing so, he creates super densities such as Calcutta, Cairo, London, New York, Paris, Tokyo, Shanghai and Buenos Aires—the slums of which are definitely deleterious to human health and welfare.

Space beyond earth. Visualizing overpopulation on earth's inhabitable space, overcrowding and consequent phycological and behavioural diseases, possible exhaustation of four life supporting requirments of man—food, water, air and space in future, other terrestrial ecological hazards such as pollution, colonization of extra-terrestrial space (*e.g.*, sea) and other planets has proposed as a long-range solution to the world's population problem. But it will not scientifically and economically feasible for many years. The efforts of mere putting astronauts into orbiting earth stations or on the moon for short periods have proved very expensive and tiresome affairs for the countries like USA and USSR. No doubt interplanetary travel will become a routine matter in further decades, but it certainly will not offer significant emigration opportunities in the twentieth century. However, space science does help ecologists in surveying different ecological aspects of earth and in future, it is hoped, it may provide some answers to certain global ecological problems such as climate change, alterations in the stratopheric ozone layer, agricultural yields, soil conditions and sea pollution (**Southwick**, 1976).

26-10. LAND USE

Land use is everybody's concern and the application of ecological principles to land-use planning is now the most important application of ecology. Land-use planning became necessary due to man's severe damage to landscape. Proper ecological knowledge make full use of the land. In USA the land has been classified into eight land types according to natural ecological features such as soil, slope and natural biotic communities—land types I and II comprise level areas with good agricultural soils, that can be regularly cultivated with crop rotation and strip cropping ; land types III and IV (steeper slopes) requiring greater restrictions if cultivated, type V through VII, not suitable for cultivation, used for permanent pasture, tree crops or retained in natural state for forestry and wild-life, and type VIII (steep slops, with thin soil) which is productive only in its natural state, as habitat for game and recreation, fur bearers, forest products, etc.

26-11. SOIL CONSERVATION

Soil is most important natural resource as it provides habitat for plant growth. It is lost easily by soil erosion as has been discussed in previous sections of this chapter. Unless this loss of soil is checked, it would amount to a great loss for mankind. To check soil erosion and for the preservation of soil (soil conservation) following methods are adopted by different ecologists—1. Protection of soil from impact of rain drops. 2. Prevention of the water from concentrating and moving down the slope in a narrow path. 3. Slowing down of the water movement when it flows along the slope. 4. Encouraging more water to enter the soil. 5. Increasing the size of soil particles. 6. Reduction in the wind velocity near the ground by growing vegetation cover, ridging the land, etc., and, 7. Growing of stubble or other vegetative cover, which might catch and hold the moving particles of soil.

26-12. REVISION QUESTIONS

1. Write an essay on ecology and human welfare.
2. What are the natural resources ? Give an account of different natural resources and the means of their conservation.
3. Write short notes on the following :
   (i) Aquaculture;
   (ii) Wildlife management;
   (iii) Soil conservation;
   (iv) Silviculture;
   (v) **Waste-management parks.**

27
Wild Life Management in India

It has already been discussed in the previous chapter that wild life is a renewable resource like soils, water and forests as it can be used time and again without being destroyed and there is a great need of conservation of these renewable resources. By conservation, we do not mean merely preservation. Conservation, in contrast to preservation, means the utilisation of renewable natural resources in such a way that they are not destroyed but are to be used later. **Wild life management** which is a important branch of conservation, is concerned with assuring the maximum possible populations of game animals cosistent with other land uses in the same area and with the number that the given habitat will support. This is attained by manipulating the balance of nature in such a way that the desired game species are favoured. Many sportsmen have the idea that the way to increase game animals is to kill off all of their predators. Wild life managers, of course, are well aware that predator elimination may well be the very thing that should not be done. Successful game management depends upon a number of factors including an understanding of ecological principles, an appreciation of the conflicts between game species and agricultural uses of land, and educating both the landowner and the hunter concerning the objectives, techniques and limitations of game management.

The idea of wild life management is not new to Indians. The *Vedas* contain hymns in praise of animals. *Sanatan Dharma* have sought the best way of conservation of wild life by linking some animal with the specific god or godess and thereby rendering it pious and protected. For example, python has been associated with god Vishnu, snake with god Shiva, peacock with godess Sāraswati and Lord Krishna, owl with godess Laxmi, lion with godess Kali and Narsing and so on. *Mahabharata* contain many references indicating how rishis and munis conserved wild life fauna (mostly deer and birds) around their ashrams. In the classical book *Arth Shastra* Chanakya had imposed severe penalties for killing, entrapping or molesting deer, bison, birds and fish in protected areas. Certain other indigenous religions of India like Budhism and Jainism strongly advocated the idea of non-violence and thus, partially check the destruction of wild life by hunting Ashoka's Fifth Pillar Edict declared game and fisheries laws as early as third century B.C. During the medieval period, even Mughal Emperors exhibited their deep interest in wild life of India. For the protection and preservation of game animals, they created hunting reserves called "Shikar Gaha." However, British rulers and certain Indian rulers caused ruthless destruction of Indian wild life during 18th and 19th centuries. Thus, unscrupulous hunting of game animals throughout the man's cultural evolution and growth of civilisation for recreation, food, hide, horn, musk and tusk ; and deforestation (*i.e.*, the act of destruction of natural abode of game animals) for crop culture, building of roads, dams or human dwellings, etc., have caused gross destruction of Indian wild life. About 200 species of wild animals of India have already been extincted and 250 species are at the verge of extinction, if not saved.

27·1. WILD LIFE OF INDIA

Indian subcontinent is unique in having immense natural beauty in its different biomes and also in possessing a rich and diverse wild life fauna (see chapters 22 and 23). It includes about 123 families of terrestrial vertebrates. According to an estimate, there are 400 species of mammals, 1200 species of birds, 350 species of reptiles and more than 29,70,000 species of insects (**Khajuria**, 1957). The animals like black buck, Nilgiri tahr, pigmy hog, golden langur, lion-tailed macaque, etc., are unique to India The other typical wild animals of India are elephant ; rhinoceros ; deer such as musk deer (kastura), barking deer (kotra), spotted deer (cheetal), hog deer (hoghiran), mouse deer (Indian chevrotain), swamp deer (Bara singha), dancing deer (sambhar), Thamin, and kashmir stag; antelopes like black buck, four-horned antelope (chausingha), Indian gazelle (chinkara), and blue bull (nilgai) ; bison ; wild buffalo; Himalayan ibex or wild goat (*Capra siberica*) ; wild boar ; wild ass ; Nilgiri tahr ; carnivores such as big cats like lion, tiger and leopard ; striped hyaena ; wolf ; bears like black Himalayan bear and sloth bear; monkeys and apes like Nilgiri langur (*Ceropithecus johni*), lion tailed monkey (*Macaca silensus*), rhesus monkey (*Macaca mulatta*), and Hanuman monkey (*Semnopithecus entellus*) ; Indian giant squirrel (*Ratufa indica*) ; porcupine ; pangolin ; birds like peafowl (*Pavo cristatus*), jungle fowl, partridge, quail, great Indian bustard (*Choriotes nigriceps*), duck, pigeon, sandgrouse, storks and egrets, pelican, eagle, crane owl, hornbill, etc., and reptiles such as crocodiles, gharials, lizards (*Uromastix*) and enormous varieties (about 216 species) of snakes. Wild life management requires autecological informations about game animals and some basic facts of game species are the following :

27·1·1. Deer

The deer belong to the family 'Cervidae' of the mammalian order 'Artiodactyla' and they are herbivores and form significant component of grazing food-chains of the Indian forests. They control the overgrowth of forest vegetation and themselves become food of carni ores. Due to limited shelter and food supply, deforestation, urbanisation, mass hunting, etc., there has been a great reduction in the number of deer population in India. Sangai of Manipur or brow-antlered deer (*Cervus eldi eldi*) is at the verge of extinction.

The musk deer (*Moschus moschiferus*) is found at 2500 to 4000 metres altitudes in Himalayas, Himachal Pradesh, Northern Uttar Pradesh, Nepal and Sikkim. It is about 50 cm high, has long ears, short tail and gall-bladder, but no antlers and no pit-glands (suborbital glands) below the eyes. Male or buck has downwardly projecting canines of the upper jaw. It is a shy animal and lives a solitary life. This deer has been almost hunted out mainly for musk which is a **valuable** product used as perfume fixative and as important ingredient of several Aurvedic and Homeopathic drugs and is obtained from the musk pouch existing below the skin of abdomen of buck. For the conservation of musk deer a National Park of 1,000 sq. km has been established at Manali in Kulu Valley by Himachal Pradesh Government.

The barking deer or rib-faced deer (*Cervulus muntjac*) is found at the altitudes of 2000 to 3000 metres and has four subspecies in India. It is 61 cm in height and has suborbital glands below eyes. Buck has **antlers** and non-protruding canines. It lives in pairs or small families in thickets and being diurnal, grazes in the morning and evening hours. The dancing deer or sambhar (*Cervus unicolour*) is large- sized (150-160 cm high), browsing and gregarious deer, occurring in the herds of 4 to 12 animals and is found in forests all over the country. It has very long (upto 100 cm) antlers. The spotted deer or cheetal (*Axis axis*) is 76 to 91 cm high, grazing and gregarious deer, living in the herds of 20 to 30. It is spotted with white on rich brown coat. The buck has long slender antlers, each of which has three branches or tines and is shed annually. It is mostly found in the forests of Indian plains and in hills upto 1333 metres but are not found in Assam, Punjab and Rajasthan. It prefers open type forests with good grazing sites and running streams. The hog deer (*Axis porcinus*) is smaller and stouter in built and has hog-like appearance. It lives either solitary or in pair and prefers to live in grass patches bordering on forests. It is diurnal and comes out to feed only during morning and evening.

27-1·2. Antelopes and Other Herbivores

27-1·2A. Antelopes. In India there are four species of antelopes : the Indian gazelle, nilgai, black buck and chowsingha.

The Indian gazelle or chinkara (*Gazella dorcas bennetii*) is mostly found throughout desert region of southern part of Uttar Pradesh and Rajasthan. It lives in herds of 3 to 5 animals and has adaptations to live in extremely arid conditions, kilometres away from drinking water sourees (**Prakash**, 1977). It prefers to live in open grasslands with thick bush cover. They feed on a variety of grasses and browse on shrubs.

The nilgai or blue bull (*Boselaphus tragocamelus*) occurs in plains and Rajasthan deserts but not in Bengal and Assam, living in small herds of 4 to 10. It prefers to stay near cultivated lands but is also found in desert grasslands. It is 131—137 cm high, females being smaller. Young bulls and cows are tawny, but old males have iron gray colour. Both sexes have a short dark mane on the neck. Male has small cone-shaped curved horns. In Thar desert it is regarded sacred like 'cow', so is spared from hunting. However, it is killed under crop-protection act, because herds of nilgai inflict severe damage to crops.

The black buck has its four subspecies in India : *Antelope cervicapra rajputanae*, *Antelope cervicapra cervicapra*, *Antelope cervicapra rupicapra* and *Antelope cervicapra centralis*. Black buck is a graceful and fast-running animal which occurs in grassy plains of India and avoids forests. It inhabits regions having scrub vegetation in the vicinity of tanks and lakes where they pass the day hiding under bushes and become active during nights. The male antelope is brown or black in colour while doe and fawns are yellowish brown in colour. Male has spirally ringed, unbranched, long (46 cm) and permanent horns. There occur large preorbital glands below the eyes of both sexes. Black buck is a polygamous and gregarious animal and its herd often include single male, several (50 or more) females and a few fawns. Almost all subspecies of black bucks have been hunted out from most part of India and now they occur in appreciable numbers only in certain pockets of Rajasthan, especially around the villages of 'Visnoi caste' of Hindus, who protect them from hunting.

The four-horned antelope or chausingha (*Tetracerus quadricornis*) has two pairs of horns, the front pair being shorter. It occurs in the foot hills of the Himalayas and in Bundel Khand region. It mostly leads a solitary life.

27·1·2B. Wild buffalo. The wild buffalo (*Bubalus bubalis*) is a large-sized, robust animal having streamlined body. It prefers tall grassy forests close to marshy areas in the vicinity of rivers and lakes It was once distributed over the grass jungles and river in forests of the Gangetic plain, the terai, Assam and eastern peninsular India down to the river Godavari. Great hunting of wild buffalo has restricted them now only to Assam (Manas and Kaziranga sanctuaries), the Nepal Terai and Bastar district in Madhya Pradesh. According to **Sheshadari** 1969), there are only 1425 wild buffaloes in India.

27·1·2C. Bison. Mithun, bull gaur or bison (*Bos gaurus*) is a relative of ox and is found in Andhra Pradesh, Orissa, Madhya Pradesh and Bihar. Due to certain epidemic diseases such as rinderpest and enteric viral disease numerous individuals were wiped out some years ago and now a few bisons are mainly confined to Bandipur in Karnataka.

27·1·2D. Wild goat and sheep. In India, wild goat and sheep are restricted to north. Ibex is found in the Himalayas. The Nilgiri tahr (*Hemitragus hylocrius*) is the only one to be in south, in Nilgiris, Tamil Nadu. It moves about in herds and grazes on grasses. This species is greatly endangered due to habitat disturbance, predation by black panther and wild dogs (dhole) and hunting by poachers and only 450 tahr are present today (**Davidar**, 1975).

27·1·3. Elephant

The Indian elephant, *Elephas maximus* is a largest terrestrial mammal and is confined to the Terai and the foot-hills because of its dependence on succulent grass, bamboo and plenty of water. It is gregarious and its herds move constantly in the search of new feeding grounds. Its population is hunted out mainly for tusk and now Periyar Sanctury (Madras) has some elephants.

27·1·4. Rhinoceros

The Indian rhinoceros (*Rhinoceros unicornis*) is found in the grasslands and jungle area of the foot hills of the Himalayas (central Nepal) and certain pockets of plains of West Bengal and Assam. It prefers swamps and open savannah, covered with the tall elephant grass. A full grown rhinoceros is 1·8 metres in height and 3·9 metres in length. It has a single horn on the nose. The horn is made of keratinized skin and matted hair, it is said to have aphrodisiac and other medicinal values, so that large sums are paid for it, this result in indiscriminate shooting and its near extinction in India. It is a slow, solitary animal and is strictly territorial.

27·1·5. Wild Ass

The wild ass (*Equs hemionus khur*) had a fairly wide distribution in the dry regions of North-West India (Jaisalmer and Bikaner), but their distribution now become restricted to the Little Rann in the southern part of the Thar desert, where **Gee** (1963) estimated their number to be 870 In past, the population of wild ass severly reduced due to hunting, catching for breeding mules, surra disease and south African Horse sicknees disease. However, people around the Little Rann are orthodox and vegetarian and do not molest the wild ass inspite of the fact they inflict serious damages to crops. Because it is the only true 'wild ass' of the world, so it needs protection and conservation (**Prakash**, 1977).

27·1·6. Wild Boar

The wild boar, *Sus scrofa cristatus*, has a long mobile snout with terminal nostrils. The canines of both jaws grow continuously and form long triangular and up curved tusks for defence and digging roots, they are better developed in males. The wild boar or pig is omnivorous, it feeds on carrion, snakes, insects, roots, tubers and cultivated crops. It is gregarious living in families in grass or bush of marshy places and is now rare and is mostly found in small numbers along the Aravalli ranges in the eastern most part of Thar desert where they inhabit rocky slopes

27·1·7. Big Cats and Other Carnivores

India has a variety of world-known large-sized cats and other carnivores which are as follows :

The Asiatic lion, *Panthera leo persica* was found in the Thar desert, in the arid plains of Sind, and Rajasthan and Punjab. It is on record that the last lions were shot at Anadra and at Jaswantpura of Rajasthan desert in about 1876. In India lions are now preserved only in the Gir sanctuary, in Gujrat, on the South-West of the Thar desert. Its extermination is well known.

The cheetah, *Acinonyx jubatus venaticus* was found in Central India, Deccan region, and in Thar desert upto Jaipur in Rajasthan. **Sheshadri** (1969) stated that it disappeared from Rajasthan about the turn of this century, from Central India in 1920's and in Deccan it was last reported in 1952. Now the cheetah, one of the most beautiful, agile and fastest animal, is regarded to be extinct from India.

The tiger *Panthera tigris tigris* is a gracious carnivore which is distributed in Uttar Pradesh from the Himalayas to the Vindhya forests in the South. In Himalayas it is found upto an elevation of about 2,000 metres. It lives in a variety of habitats, from thorn forests to the dense terai forests. According to an estimate, in 1948, the tiger population was 20,000 to 25000, but in 1958, about 4000 and in 1970 less than 3000 (merely 1540 ; **Daniel**, 1970). To save the Tiger from extinction in India, "**Project tiger**" has been launched in 1972 ; this project planned to create Tiger reserves in selected areas of India. Due to this effort, considerable improvement was observed in tiger population in 1973 (Table 27·1).

Table 27·1. Increase in Tiger population due to 'Project Tiger' (News Letter 1977, Vol I, No 1).

Tiger Reserve	Population in 1972	Population in 1973
1. Manas	31	41
2. Palamaus	22	30
3. Simplipal	17	50
4. Corbett National park	44	55
5. Ranthambore	14	20
6. Kanha National park	43	48
7. Melghat	27	32
8. Bandipur	10	19

The leopard, ***Panthera** pardus* is the smaller spotted cousin of tiger which can live in all types of forests. Once it was found all over India but now its number has been drastically reduced. Leopard is given full protection in the Mount Abu sanctuary, however, can be hunted under licence from the Indian wild Life Act. The snow leopard (*Panthera uncia*) is found in the high Himalayas from

Kashmir to Sikkim near snow line. It has a creamy grey coat with large black rings. It is greatly hunted for its fur and now is endangered species. Other important cats are clouded leopard (*Neofelis nebulosa*), leopard cat (*Felis bengalensis*), pallas's cat (*Felis manul*), caracal (*Felis caracal*), etc. Caracal once was common in Rajasthan deserts, now it has become extremely rare and is suspected to occur only in the Sirohi-Jalore and Bikaner regions (**Prakash**, 1977).

The sloth bear, *Melursuu ursinus*, is a stout carnivore which is found in the hilly forests like Mount Abu and Erinpura ranges. It lives mostly on fruits, honey and insects, it sucks up white ants from their hills which are broken by its claws with great speed. It is rapidly vanishing due to ruthless hunting and it needs conservation. The black Himalayan bear, *Selenarctos tibetanus*, is larger than a sloth bear, has V-shaped white mark on the chest and is carnivorous.

Dolphin. In the Indus river, in the North-West and in the rivers Chenab and Sutlaj in the north of the Thar desert, the fresh-water gangetic dolphin, *Platanista gangetica* (Cetacean mammal) was once found in good numbers (**Murray**, 1884), but is now totally extinct (**Prakash**, 1977).

27·1·8. Birds

The important wild bird-fauna of India include the following species : pea fowl (*Pavo cristatus*), grey jungle fowl (*Gallus sonnerati*), red spur fowl (*Galloperdix spadicea*), ducks (*Anser indicus, Sarkidiornis melanotos, Anaspoecilorhyncha*), pigeons (*Columba livia*), sand grouses (*Pterocles exustus, P. orientalis, Streptopelia decaocta*), storks (*Leptotilos dubuis*), egrets (*Egretta garetta*), grey partridge (*Francolinus pondieirianus*), black partridge (*Francolinus francolinus*), golden eagle (*Aquilla chrysaetos*), vultures (*Gyps bengalensis, Neophron percnopterus*), spotted billed pelican (*Pelacanus philippensis*), sarus crane (*Grus antigone*), quails (*Coturnix coturnix, C.coromandelica*), and great Indian bustard (*Choriotes nigriceps*). Certain birds like osprey, great Indian bustard, cranes, mountain quail, horned owl, red start avocet, pinkheaded duck, etc., are on the verge of extinction, if not saved.

The great Indian bustard is a large Indian game bird having a heavy body, long neck and long bare running legs. It is essentially an inhabitant of wide, open, dry scrubby plains and waste, broken undulating lands of the Rajasthan, West Punjab and Gujrat. It is ruthlessly hunted out for its delicious flesh and is almost at the verge of extinction. In 1980, an International Conference was held at Jaipur to assess its present position and to suggest conservation steps to save it from extinction.

27·1·9. Crocodiles and Other Aquatic Reptiles

The following three Indian aquatic reptiles have been found endangered : the estuarine crocodile (*Crocodilus porosus*) ;

subnosed crocodile, freshwater or marsh crocodile or muggar (*Crocodilus palustris*) and gharial (*Gavialis gangeticus*). The estuarine crocodile is found in Bhitarkanika Island and adjacent mangrove areas of Orissa and in the Sunderbans of West Bengal. The marsh crocodile is widely distributed in India and has been greatly reduced in number in the past due to indiscriminate shooting for its valuable hide. *Gavialis* was once common in most large Indian rivers like Indus, Ganga, Brahmputra and Mahanadi, but now has very sparse distribution in Ganga and larger feeders in the Nepal-Terai, Bihar and Bengal. According to **Bustard** (1974) Indian gharials are on the verge of extinction and there are only 143 living gharials (**Subba Rao**, 1977). Mugger is also fast depleting and may be endangered, if not saved. In 1974, under the guidance of **Dr. H.R. Bustard** the project called 'save the crocodiles' was started for the conservation and management of crocodiles, through the incubation of eggs, rearing and release in preserves. Projects for preservation of muggar, estuarine crocodile and gharial are now being undertaken in Andhra Pradesh, Bihar, Gujrat, Kerala, Orissa, Rajasthan, Tamil Nadu and Uttar Pradesh.

27-2. REASONS FOR DEPLETION OF WILD LIFE

Many wild animals became extinct due to various human and natural activities : 1. Absence of cover or shelter to wild animals. 2. Due to deforestation for cultivation, road-building, railway routes, dam construction or for urbanisation, reduction in the area for free movement of wild animals which retard reproductive capacity of certain wild animals like deer, bison, rhino, tiger, etc. 3. Destruction of wild plants of forests for timber, charcoal and fire wood often deprive wild animals their most palatable food and affects their survival. 4. Noise pollution by different transporting media (trucks, buses, rails, aeroplanes, etc.) and polluting river water have adversely affected wild animals. 5. Various natural calamities such as floods, droughts, fires, epidemics, etc., have also caused great destruction of wild life. 6. Hunting methods of all kinds and for any purpose (*i e.*, for food, recreation, hide, fur, plumage, musk, tusk, horn, etc.) have caused destruction of wild life.

27-3. NECESSITY FOR WILD LIFF CONSERVATION

The conservation of wild life is required for the following benefits: 1. The wild life helps us in maintaining the 'balance of nature'. Once this equilibrium is disturbed it leads to many problems. The destruction of carnivores or insectivores often leads to the increase in herbivores which in turn affects the forest vegetation or crops. 2. The wild life can be used commercially to earn more and more money. It can increase our earning of foreign exchange, if linked with tourism. 3. The preservation of wild life helps many naturalists and behaviour biologists to study morphology, anatomy, physiology, ecology, behaviour biology of the wild animals under their natural surroundings. 4. The wild life provides best means of sports and recreation. 5. The wild life of India is our cultural asset and has deep-rooted effect on Indian art, sculpture, literature and religion.

27·4. MODES OF WILD LIFE CONSERVATION

In India, the following measures have been under taken for the wild life management :

27·4·1. Protection by Law

India was probably the first country to enact a Wild Life Protection Act. The Wild Birds and Animals Protection Act was passed in 1887 and repealed in 1912. For game protection in the states, in 1927, the Forest Act XVI was enacted. Indian Board for Wild Life was established in 1952 and this was followed by setting up of Wild Life Boards in different states. In 1972, new wild life Protectoin Act was passed. Under this Act, possession, trapping, shooting of wild animals alive or dead; serving their meat in eating houses; their transport and export are all controlled and watched by special staff (Chief Wild Life Warden and authorized officers). This Act prohibits hunting of females and young ones. Under this Act threatened species are absolutely protected and the rest afforded graded protection according to their state of population size.

27·4·2. Protected Species of Indian Wild Life

The following wild animals have been enlisted as threatened and protected species : White eyed buck, Indian antelope or black buck, Four-horned antelope, Bharal, Swamp deer, Eastern pangolin, Elephant, Indian gazelle or Chinkara, Musk deer, Serow, Tahr, Golden cat, Snow leopard, Great Indian Bustard, Pink headed duck, Peafowl, Monal pheasant, Koklas pheasant, Chir pheasant, Horned pheasant, Monitor lizard, Water lizard, Gharial, Marsh crocodile, Python, and Leathery turtle. Albino and melanic individuals of all species of wild animals are also protected.

27·4·3. Establishment of Sanctuaries and National Parks

In India, nearly 200 sanctuaries and national parks have been established for wild life management. **Sanctuaries** are places where the killing or capturing of any animal is prohibited except under orders of the authorities concerned. They provide protection and optimum living conditions to wild animals. **National Parks** are set up for preserving flora, fauna, landscapes and historic objects of an area. Some well known wild life sanctuaries and national parks of India are as follows :

1. **Kaziranga wild life sanctuary**. It is established in 1926 in district Sibsagar, subdivision Jorhat (Assam) on the south bank of the Brahmputra river. It consists of 430 sq km of forest grasslands and swamps and supports a fauna of 700 rhinoceros, besides elephant, wild buffalo, bison, tiger, leopard, sloth bear, sambhar, swamp deer, hog deer, barking deer, wild boar, gibbon and birds like pelican, stork and ring-tailed fishing eagles.

2. **Manas wild life sanctuary and Tiger reserve**. It is located in district Kamrup in Assam, has an area of 540 sq km and is

situated at an altitude of 80 metres. River Manas passes through it. It contains the following wild animals : tiger, panther, wild dog, wild boar, rhinoceros, gaur, wild buffalo, sambhar, swamp deer and golden langur.

3. **Jaldapara wild life sanctuary.** It is situated in Jalpaiguri district of West Bengal and is a 65 sq km stretch of grassland. Its wild life fauna includes animals like rhinoceros, gaur, elephant, tiger, leopard, deer, and a variety of birds and reptiles.

4. **Palamau national park.** It is situated in Daltongunj district in Bihar and has an area of 345 sq km. Its flora is thick tropical forests. The fauna of this park includes tiger, panther, sloth bear, elephant, chital, gaur, nilgai, chinkara, chowsingha and mouse-deer.

5. **Hazaribagh national park.** This national park was established in 1954 in Bihar. It has an area of 184 sq km of thick tropical forests. The typical fauna of this park includes wild boar, sambhar, nilgai, tiger, leopard, sloth bear, hyena and gaur.

6. **Simlipal national park.** It is situated in district Mayuri Bhanj in Orissa and has an area of 2750 sq. km. It is covered over by dense sal forests and is chosen for project tiger. Its typical fauna includes tiger, elephant, deer, pea fowl, talking mynas, chital, sambhar, panther, gaur, hyena and sloth bear.

7. **Chilka lake.** This largest inland lake has an area of 1000 sq km and is about 100 km from Bhubaneshwar (Orissa). Its typical fauna includes water fowls, ducks, cranes, ospreys, golden plovers, sandpipers, stone curlews, flamingoes, etc.

8. **Kolameru bird sanctuary.** This is a small bird sanctuary near Tadepallegudam in Andhra State. It is a breeding place for pelicans and many marine birds visit this place.

9. **Vedanthangal bird sanctuary.** This small-sized but very old sanctuary extend over a lake of about 0·30 sq km, 85 km south of Madras. Many migratory birds regularly visit this temporary lake. The regular visitors of this lake are birds like spoon bill, open billed stork, egrets, ibis, cormorant, darter, grey heron, pelican, snipes and dab chick.

10. **Guindy deer park.** It is situated near Madras city and has mainly chitals and black bucks. A few albinos of black buck are also found here.

11. **Point calimera wild life sanctuary.** This is situated at the southern tip of the Thanjavur district of Tamil Nadu abutting the Palk strait. Its back water and lagoon is visited by flamingos and pelicans. Its nearby vedaranyam forests has a fauna of numerous black bucks, chitals and wild boars.

12. **Mundanthurai sanctuary.** It is established in 1962 in Tirunelveli district of Tamil Nadu and has an area of 520 sq km It has evergreen forests and Tamaraparani river flows through this sanctuary. Its typical fauna includes panther, tiger, sambhar and chital.

13. Periyar wild life sanctuary. This is situated in Kerala State and has an area of 777 sq km. It was established in 1940 around the artificial lake which arose behind the dam built across the Periyar river in 1900. This sanctuary of great scenic beauty supports a fauna of wild elephants, gaurs, leopards, sloth bear, sambhars, barking deer, wild dogs, wild boars, black Nilgiri langur and water birds like grey hornbills, egrets.

14. Mudumalai wild life sanctuary. This wild life sanctuary was established in 1940 in North Western part of Nilgiris in Tamil Nadu. It is known for its rich forests and diversity of fauna that includes wild elephant, gaur, sambhar, chital, barking deer, mouse deer, four-horned antelope, tiger, panther, bonnet monkey, langur, giant squirrel, flying squirrel, wild dog, jackal, wild cat, sloth bear, porcupine, pangolin, flying-lizard, monitor-lizard, rat snake, python and various birds.

15. Ranganthittoo bird sanctuary. It covers an area of 166 sq km and includes a series of islands in the Cauvery river 15 km off the Banglore-Mysore Road, near Shrirangapattnam. Its avial fauna includes open bill strock, white ibis, egret, spoon bill, wild duck, pea fowl, night heron, etc.

16. Bandipur wild life sanctuary. This sanctuary was established in 1941 by the then ruler of Mysore (Karnataka State). It is situated 80 km south of Mysore city enroute to Ootacamund. It has an area of 874 sq km and is at an altitude of 1454·4 metres. Its forest is very thick and has plenty of rain fall. Its wild life fauna includes plenty of gaurs and animals like elephant, leopard, sloth bear, wild dog, chital, panther, barking deer, porcupine and langur.

17. Cotigao wild life sanctuary. It is located in South Goa and has an area of 105 sq km. It has wet evergreen forest of Elve and Berla trees and supports rich fauna of the following animals : gaur, sambhar, chital, hog-deer, barking deer, wild boar, hyaena, panther, leopard, jackal, otter, porcupine and birds like parakeet, lorikeet, woodpecker, kingfisher, bulbul, jungle fowl, egret, etc.

18. Bhagwan Mahadev wild life sanctuary. It is located in North Goa and has an area of 240 sq km and supports fauna almost similar to cotigao wild life sanctuary.

19. Sesan Gir. This famous wild life sanctuary for the Asiatic lion is situated in Gujrat state, 468 km from Ahmedabad and 43 km from Veraval. It has an area of 1295 sq km of semiarid country with patches of thorn scrub and deciduous trees. Its fauna includes Asiatic lion, spotted deer, blue bull (nilgai), four-horned antelope, chinkara, striped hyena, wild boar, porcupine, langur, python, crocodiles and birds like green pigeon, partridge, rock-grouse, etc.

20. Kanha national park. This national park was established in 1955 in former Banjar Valley Reserve (Madhya Pradesh). This park has an area of 939·94 sq km and includes hilly terrains and streams. It is 175 km away from Jabalpur and has forests of sal trees. Its typical fauna includes animals like tiger, chital, panther, sambhar, black buck and bara singha.

21. **Tandoba national park.** It is located in Chandrapur district (Maharastra) and has an area of 116 sq km. Its typical fauna includes tiger, sambhar, sloth bear, bison, chital, chinkara, barking deer, blue bull, four horned deer, langur, peafowl and few crocodiles.

22. **Sariska.** This is one of the most beautiful wild life sanctuary of Rajasthan near Alwar. It has an area of 800 sq km and has dense Dohokra and Solar forests. Its typical fauna includes tiger, leopard, spotted deer, jungle cat, four-horned antelope, langur, porcupine, hedgehog, pea fowl, etc.

23. **Bharatpur bird sanctuary (Koeldeo Ghana).** It is located at Bharatpur (Rajasthan), has an area of 29 sq km and harbour all kinds of indigenous nesting water birds, water side birds and migratory birds. More than 328 varieties of birds including cormorants, spoonbills, white ibis, Indian darters, egrets, painted storks, open billed storks, great black necked storks, etc. Many migratory birds like ducks, geese, siberian cranes, etc., regularly visit this sanctuary. Drier parts of this marshy sanctuary have animal like spotted deer, black buck, sambhar, blue bull, wild boar and python.

24. **Sultanpur lake bird sanctuary.** This small sized (2 sq km area) bird sanctuary is located in Gurgaon district (Haryana) about 30 km from Delhi. Its typical avian fauna includes crane, sarus, spot-bill, rudyshel and drake.

25. **Shikari Devi wild life sanctuary.** It is located in Mandi district in Himachal Pradesh. It has an area of 213 sq km and has the following animals : snow leopard, flying fox, black bear, barking deer, musk deer, chakor, partridge, etc.

26. **Bir Motibagh sanctuary.** It is located near Patiala in Punjab. The typical fauna of this wild life sanctuary includes black buck, blue bull, hog deer, hare, jackal and birds like peafowl, partridge, sparrow, babbler, myna, parakeet, pigeon, dove, etc.

27. **Dachigam wild life sanctuary.** It is established in 1951 in Kashmir, 26 km away from Srinagar. It has an area of 89 sq km and has two levels: upper Dachigam at 3692·3 metres altitudes and Lower Dachigam at 1846·2 metres altitudes. It mainly preserves hangul or Kashmir stag, but also have animals like musk deer, leopard, black bear, brown bear and baboon.

28. **Corbett national park.** It is one of India's most famous wild life sanctuary and was constituted in 1935 as the first national park of India. It is situated between Nainital and Garhwal districts in Uttar Pradesh. It has an area of 525 sq km and is located within west-to-south bend of river Ramganga. It supports a rich and diverse fauna of the following animals : tiger, panther, sloth

bear, hyaena, elephant, blue bull, swamp deer, barking deer, Indian antelope, porcupine, birds like bulbul, woodpecker, barbet, babbler and bee eater, and reptiles like crocodile, python, etc.

29. Shivpuri sanctuary. It is located in Madhya Pradesh and is an asylum for tigers.

30. Anamalai sanctuary. This sanctuary was established in 1972 in the southern part of Coimbatore District in Tamil Nadu. It has a vast area of 958 sq km and supports rich fauna of animals like elephant, gaur, sambhar, spotted deer, barking deer, Nilgiri tahr, lion-tailed macaque, tiger, panther, sloth bear, langur, porcupine and pangolin.

27·4·4. Other Conservation Measures

Other important steps needed to preserve wild life include the following measures : (i) For the preservation of species wild life management staff should have a correct idea about the exact habitat which the species under consideration needs. (ii) Natural habitat of wild animals should be carefully protected. (iii) Habitats of wild life should be improved by constructing water holes, salt-licks and by raising plantation of better and nourishing fodder grasses and trees. (iv) Effective means of census operation should be adopted to measure population sizes of various wild animals. (v) The enforcement of Wild Life Protection Acts should be observed strictly. (vi) Shooting and hunting of endangered species should be totally banned. (vii) To take care of wild life in case of epidemics, veterinary efforts should be made. (viii) To understand the biology and behaviour of wild animals research on wild life should be encouraged. (ix) Public should be educated about the advantages and disadvantages of wild life.

27·5. REVISION QUESTIONS

1. **Write an essay on the wild life of India.**
2. **Describe, why should wild life of India be protected ?**
3. **Describe various measures of wild life management.**
4. **Define the terms wild life sanctuary and national park and describe some important wild life sanctuaries and national parks of Indian Subcontinent.**

Glossary

Abiotic factors—Non-living ecological components of an ecosystem, such as water, oxygen, sodium chloride, nitrogen, carbon dioxide, etc., and other physical and chemical influences such as temperature, light, humidity, pH, etc.

Abyssal—Refers to ocean depths.

Accidental species—The species that occur with a low degree of fidelity in a community type.

Accumulators—Term used for pioneer plants whose activities enrich habitat with nutrients.

Aeolian—Related to wind ; produced or blown by the wind.

Aestivation—A condition in which an organism may pass an unfavorable season and in which its normal activities are greatly curtailed or temporarily suspended.

Aggregation—It is coming together of organisms into a group, as in locusts.

Allelopathy—The influence of plants, exclusive of micro-organisms, upon each other, caused by products of metabolism ; "antibiotic" interaction between plants.

Allergen—Substance that producing a pathogenic sensitivity within an organism.

Ambient—Existing conditions ; encompassing on all sides.

Anadromous—Refers to animals that live in the sea and breed in fresh-water, such as salmon.

Anaerobic—Capable of living without oxygen.

Antibiosis—A process fatal or injurious to any form of life.

Association—A major unit in community ecology, characterized by essential uniformity of species composition.

Aufwuchs—Organisms that live on stem of submerged plants. Synonym "periphyton".

Autecology—Ecological study of the individual organism in relation to environmental condition.

Autotrophic—Organisms capable of producing organic materials from inorganic chemicals by means of energy conversion *e.g.*, green plants.

Benthic—Refers to the bottom layer of any body of water, and the organisms therein.

Biochemical oxygen demand (BOD)—The amount of dissolved oxygen required by micro-organisms in water.

Biocide—Any chemical that kills living organisms.

Biodegradeable—Capable of being decomposed by bacterial action.

Biogeography—A branch of biology that deals with the geographic distribution of plants and animals.

Biological clock—The rhythmic occurrence of processes within organisms therein.

Biological control—Use of organisms or viruses to control parasites, weeds, or other pests.

Biomass—The total weight of living organisms per unit area.

Biome—A major terrestrial biotic community characterized by distinctive life forms ; *e.g.*, temperate deciduous forest, grassland, coniferous forest.

Biometry—The application of statistics to biology.

Biosphere—The part of the earth in which life exists.

Biota—Species of all plants and animals occurring within a certain area or region.

Biotic factors—Living ecological components of an ecosystem such as *producers* (chemosynthetic bacteria and photosynthetic green plants), *consumers* (heterotrophic animals) and *decomposer organisms* (bacteria, fungi, and other saprophytes).

Calorie—The amount of heat energy capable of raising the temperature of one gram of water 1°C.

Canonical distribution—A particular configuration of the log-normal distribution of species abundance.

Carnivore—The animals which feed on the flesh of other animals ; the consumers of secondary, tertiary and quarternary grade, such as, a hawk feeding on a weasel which in turn consumed a mouse, a primary consumer or herbivore.

Carrying capacity—The ability of landscape to support any given animal or plant species or groups of species.

Catastrophic agents—A term used by Howard and Fiske to describe agents of destruction in which the percentage of destruction is not related to population density ; synonymous with density-independent factors.

Characteristic species—The species that are rigidly limited to certain communities and thus can be used to identify a particular type of community.

Chlorinated hydrocarbons—Synthetic organic compounds containing chlorine, carbon, and hydrogen, many of which are effective as insecticides, such as DDT, endrin, dieldrin, chlordane, heptachlor and toxapane.

Climax—The mature community capable of perpetuation under the prevailing climatic and edaphic conditions.

Commensalism—The living together of two species, usually with benefit to one.

Community—A group of populations of plants and animals living together in a given place, *e.g.*, the community of an oak forest, a marsh, a grassland, a coral reef, a desert, etc.

Compensation point—Depth in a body of water at which the light intensity is such that the amount of oxygen produced by a plant's photosynthesis equals the oxygen it absorbs in respiration; the point at which respiration equals photosynthesis so that net production is zero.

Competition—A interspecific or intraspecific population interaction that occurs when a number of organisms of the same or of different species utilize common resources that are in short supply ("exploitation"); or, if the resources are not in short supply, competition occurs when the organisms seeking that resource, harm one another in the process ("interference").

Competitive exclusion—The ecological principle that if two closely related species compete for a limited resource that is essential for the survival of each, one of the species will tend to be eliminated.

Consumers—Organisms that ingest organic food (usually particulate) or other organisms.

Continental drift—The theory that the continents of the earth were once large land mass that broke into large sections and diverged or drifted apart.

Continuum index—Measure of the position of a community on a gradient defined by the species composition.

Coprophagy—Eating faeces of other animals.

Crude birth rate—The number of births per year for each 1000 individuals in a population.

Crude death rate—The number of deaths per year for each 1000 individuals in a population.

Cryptozoa—Animals living between litter and soil.

Cybernetics—The study of communication and control systems in living organisms and machines.

Deme—An interbreeding group in a population ; also known as local population.

Demography—The mathematical study of populations.

Density—Number of individuals in relation to the space in which they occur.

Detritivore—An organism that lives on detritus.

Detritus—Dead organic material in solid or particulate form.

Diapause—A period of suspended growth or development and reduced metabolism in the life cycle of many insects, in which the organism is more resistant to unfavourable environmental conditions than in other periods.

Dilution rate—A general term to describe the rate of additions to a population from birth and immigrations.

Diversity—Differences, complexity, or variety ; usually in reference to species diversity.

Dominance, ecological—The condition in communities or in vegetational strata in which one or more species, by means of their number, coverage, or size, have considerable influence or control upon the conditions of existence of associated species.

Dynamics—In population ecology the study of reasons for *changes* in population size ; contrast with *statics*.

Ecological longevity—Average length of life of individuals of a population under stated conditions.

Ecologist—An ecological scientist who analyses the environment separately and in correlation to living organisms and tries to solve different human problems such as land use, pollution, energy crisis, flooding, erosion, pesticide contamination, radioisotope accumulation, rural and urban blight, population growth, urbanization, crowding, group conflict, infectious diseases, etc.

Ecology—It is the scientific study of the relationships of living organisms with each other and with their environments. It is the science of biological interactions among individuals, populations and communities, and also of ecosystems—the inter-relations of biotic communities with their non-living environments.

Ecosystem—It is a spatial or organizational unit of biosphere which includes living organisms (biotic community) and non-living substances (abiotic environment) interacting to produce an exchange of materials between the living and non-living parts.

Ecotone—Transition zone between two diverse communities, *e.g.*, the tundra-boreal forest ecotone.

Ecotype—A subspecies or race that is especially adapted to a particular set of environmental conditions.

Edaphic—Constrained by soil conditions.

Elfinwood—The scrubby, stunted growth form of trees, often forming a characteristic zone at the limit of tree growth in mountains.

Elytron/femur ratio (E/F)—Ratio of the length of the elytron (outer wing) of locust to the length of the femur (part of hind leg); used to determine morphological phase status of locusts.

Endemic—Refers to a group of organisms native to a given region.

Environment—All the biotic and abiotic factors that actually affect an individual organism at any point in its life cycle.

Environmental resistance—Those factors that tend to limit the growth and/or distribution of a population.

Eolian deposit—Brought and deposited by wind.

Epidemic—The occurrence in greater numbers than usual of a pathogen, parasite, or pest.

Epidemiology—Branch of medicine dealing with epidemic diseases.

Epifauna—Part of the benthos living on the mud surface.

Epilimnion—Upper layer of warm water in a stratified lake.

Equitability-Evenness of distribution of species abundance patterns; maximum equitability occurs when all species are represented by same number of individuals.

Erosion-The wearing away of the earth's surface by wind and water.

Ethnology—The study of the divisions of mankind.

Ethology—The comparative and biological study of behaviour.

Etiology—The cause or developmental history of a condition.

Eutrophication—The process of enrichment and aging in an aquatic ecosystem.

Evapotranspiration—Sum total of water lost from the land by evaporation and plant transpiration.

Facultative agents—A term used by Howard and Fiske to describe agents of destruction which increase their percentage of destruction as population density rises, synonymous with density-dependent factors.

Fauna—A collective term for all the animals in a given region or geological period.

Fecundity—Potential capability of an organism to produce reproductive units such as eggs, sperms, or asexual structures.

Feedback—Information by which a system returns toward a former condition.

Feral—Domestic plants or animals living as escapes and subject to natural selection.

Fertility-Actual capability of an organism to produce living offspring.

Filter feeder—An organism that acquires food by staining or filtering particles from water.

Fidelity—Degree of regularity or "faithfulness" with which a species occurs in certain plant communities, expressed on a five-part scale : (5) exclusive, (4) selective, (3) preferntial, (2) companion, indifferent, (1) accidental strangers.

Food chain—Figure of speech for the dependence for food of organisms upon others in a series beginning with plants and ending with the largest carnivores.

Forb—Herb other then grass.

Gause's principle—The concept that two species can not occupy exactly the same ecological niche within a community.

Genecology—Study of population genetics in relation to the habitat conditions, the study of species and other taxa by the combined methods, and concepts of ecology and genetics.

Genus—A group of closely related species.

Global stability—Ability to withstand perturbations of a large magnitude and not be affected ; compare with *neighborhood stability*.

Gradocoen—Totality of all factors that impinge on a population, including biotic agents and abiotic factors.

Gross production—Production before respiration losses are subtracted ; photosynthetic production for plants and metabolizable production for animals.

Habitat—The natural abode of an organism including its total environment.

Herbicide—Any chemical that kills or injures plant life.

Herbivore—An animal that feeds on plants.

Hierarchy—A social rank order of animals ; dominance hierarchy.

Homeostasis—The maintenance of equilibria in an organism or biological system.

Host—Organisms that furnishes food, shelter, or other benefits to another organism of a different species.

Hydrophyte—Plant that grows wholly or partly immersed in water.

Hydrosphere—That portion of the earth containing water in liquid or vapour form.

Hypolimnion–A lower layer of cold dense water in a stratified lake.

Indicator organism—An organism that indicates the presence or absence of a certain environmental conditions.

Infauna—Part of the **benthos** which burrows.

Invasion area—In locusts the area colonized by swarms during an outbreak ; the invasion area can not support permanent population of locusts.

Invasive—The tendency to invade or enter a new location or niche.

Irruptive—The tendency to increase suddenly in numbers.

Isotopes—Forms of the same element that differ in atomic weight and organization of the atomic nucleus.

Leaching—The removal or downward movement by water of soluble chemicals from the soil or other materials.

Lentic—Means "still water" and is an obscure way of saying "still water".

Life table—Tabulation presenting complete data on the mortality schedule of a population.

Limiting factor—The ecological influence that limits or controls the abundance and/or distribution of a species.

Littoral—Pertaining to the shore of the sea or of a large lake ; a shallow water zone with light penetration to the bottom and often occupied with rooted aquatic plants.

Loess—Wind-blown silt Many temperate landscapes have thick layers of loess blown there during the last glacial retreat.

Logistic curve—An S-shaped growth curve describing the growth of an organism or a population.

Lotic—Means "running water" and is an obscure way of saying "running water".

Malthusian—Refers to the doctrine of Malthus that living organisms tend to increase faster than their means of support.

Megalopis—A large continuous urban area formed by the confluence of two or more cities.

Mesic—Moderately moist habitat.

Mesophyte—Plant that grows in environmental conditions that are medium in moisture conditions.

Metabolism—The total of chemical processes occurring within an organism or biological system.

Mimicry—A condition by which an organism has selective advantage against predation by appearing dangerous, inedible,or inconspicuous.

Monoculture—A simplified biotic community dominated by one species.

Monogamy—Mating of an animal with only one member of the opposite sex.

Mortality—Refers to death.

Multivolatine—Refers to an organism that has several generations during a single season, contrast with *univolatine.*

Mutagen—A substance that produces changes in genetic material.

Mutualism—A interspecific relationship between two or more species in which both or all benefit from the association and can not live separately.

Mycorrhizae—Roots and fungi living symbiotically or commensally.

Natality—The production of offspring by organisms.

Natural selection—The process of evolution by which some organisms give rise to more descendants than others.

Neighbourhood stability—Ability to withstand perturbations of small magnitude and not be affected; compare with *global stability*.

Nekton—Animals of sea, such as fish, which can control their position by swimming ; sea animals other than plankton.

Neritic—Pertaining to the area of the sea over the continental shelf.

Net primary production—Organic synthesis in a plant in excess of that used in respiration of the plant.

Neuston—Organisms supported on water surface.

Niche—The ecological role or "profession" of an organism in its ecosystem ; its activities and relationships to its community and total environment.

Nutrient—A chemical substance that contributes to the growth of an organism.

Obligate predator (or **parasite**)—Predator that is restricted to eating a single species of prey.

Omnivore—A living organism that consumes both plant and animal food.

Optimum yield—Amount of material that can be removed from a population that will maximize biomass (or number, or profit, or any other type of "optimum") on a sustained basis.

Ordination—Process by which plant or animal communities are ordered along a gradient.

Organism—An individual unit constituted to carry on the activities of life.

Organophosphate—Any one of the several synthetic chemical insecticides, such as parathion and malathion, which are very toxic but non-persistent in the environment.

Outbreak area—In locusts the area that can produce swarms that may then move into the invasion area ; outbreak area may be permanently inhabitable by locusts.

Ozone—A very reactive molecule composed of three oxygen atoms; at the earth's surface ozone is an injurious component of photochemical smog, but in the upper atmosphere it forms a protective layer that filters ultraviolet radiation.

Paleoecology—Ecological studies of the past from the evidence of fossils. To geologists it means reconstructing past environments from assumptions about animals and plants found fossil. To biologists it means reconstructing niches of the past from reconstructions of past environments.

Paleolimnology—Studies of lake ecosystems of the past from evidence preserved in lake mud.

Parasite—An organism that derives its nutrition by feeding upon another living organism (host) ; insect parasites are usually fatal to their host and behave more like vartebrate predators.

Parthenogenesis—Development of the egg of an organism into an embryo without fertilization.

Pathogen—An organism that cause disease.

Pelagic—Refers to the open ocean.

Pesticide—An agent that kills the pests, *e.g.*, insecticide.

Phenology—Study of the periodic (seasonal) phenomena of animal and plant life and their relations to the weather and climate (*e.g.*, the time of flowering in plants).

Phenotype—Expression of the characteristics of an organism as determined by the interaction of its genic constitution and the environment ; contrast with genotype.

Photoperiodism—Response of plants and animals to the relative duration of light and darkness (*e.g.*, a chrysanthemum blooming under short days and long nights).

Photosynthesis—Synthesis of carbohydrates from carbon dioxide and water by chlorophyll using light as energy with oxygen as a by-product

Physiological longevity—Maximum lifespan of individuals in a population under specified conditions ; the organisms die of senescence.

Phytoplankton—Plant plankton, primarily algae; the plant community in marine and freshwater situation which floats free in the water and contains many species of algae and diatoms.

Pioneer—An organism that first appears in a newly exposed or altered environment.

Plankton—Floating or weakly swimming organisms in aquatic environment, primarily microscopic.

Poikilothermic—Of or pertaining to "cold-blooded animals"; organisms having no rapidly operating heat-regulatory mechanism ; contrast with **homeothermic**.

Pollution—The unfavourable alteration of the environment.

Polyandry—Mating of single female with several males.

Polygyne—Mating of single male with several females.

Population—Groups of individuals of single species; a group of interacting individuals in a definable space.

ppm—Parts per million, a measure of concentration, the same as micrograms per milliliter or milligrams per litre.

Predation—The process of one organism killing another for food.

Primary production—The process by which organic compounds are formed in photosynthesis or chemosynthesis.

Promiscuous—Mating system in which males and females are not restricted to one sexual partner.

Proximate factors—In evolutionary terms the mechanisms responsible for an adaptation with reference to its physiological and behavioral operation; the mechanics of how an adaptation operates , opposite of *ultimate factors.*

R A D (Radiation Absorbed Dose)—The amount of radiation absorbed by an object or organism, expressed as the amount of energy (in ergs) per gram of absorbing tissue or material.

Radioactive—The property of some chemical elements to spontaneously emit radiation from unstable atomic nuclei.

Radioisotope—An isotope that is unstable and disintegrates, emitting radiation energy.

Recruitment—Increment to a natural population usually from young animals or plants entering the "adult" population.

Respiration—Complex series of chemical reactions in all organisms by which energy is made available for use ; carbon dioxide, water, and energy are the end products.

Saprophyte—Organism which obtains food from dead or decaying organic matter, *e.g*., many bacteria and fungi. It is also called *Saprophage* or *Saprobe.*

Scavengers—Animals that eat the corpses of other animals and plants.

Secondary production—Production by herbivores, carnivores, or detritus feeders ; contrast with *primary production.*

Self-regulation—Process of population regulation in which population increase is prevented by a deterioration in the quality of individuals that make up the population ; population regulation by internal adjustments in behaviour and physiology within the population, rather than by external forces such as predators.

Senescence—Process of aging.

Sere or **seral stage**—The stage in the successional series ; seral is a series of stages that follow one another in an ecological succession.

Sessile—Animal that is attached to an object or is fixed in place, *e.g* , barnacles.

Smog—Polluted air and water vapour.

Soil—The aggregate of decaying organic materials; living organisms, and weathered substrate.

Stability—Absence of fluctuations in populations ; ability to withstand perturbations without large changes in composition.

Standing crop—The amount of biomass present in a defined area, population, or community at any given time.

Statics—In population ecology the study of the reasons of equilibrial conditions or average values; contrast with dynamics.

Stenoplastic—Having little or no modificational plasticity; steno-means narrow ; opposite of euryplastic.

Steppe—Extensive area of natural, dry grassland ; usually used in reference to grasslands in Southwestern Asia and South eastern Europe ; equivalent to *prairie* in North American usage.

Stress—A condition that forces a deviation from homeostasis.

Stressor—A stimulus or factor producing stress.

Sublittoral—Lower division in the sea from a depth of 40 to 60 m to about 200 m ; below the littoral zone.

Succession—The changes in vegetation and animal life by which one kind of population or community is replaced by others ; it culminates into climax.

Symbiosis—The living together of two or more species.

Synecology—Study of groups of organisms in relation to their environment ; includes populations, community, and ecosystem ecology.

Synergistic—The action of separate agencies that together have a greater effect than the sum of their independent effect.

Systemic—Refers to the entire body of an organism.

Taiga—The northern boreal forest zone, a broad band of coniferous forest south of the arctic tundra.

Teratogen—A chemical substance, such as thalidomide, that can cause developmental malformation in an embryo or foetus.

Territory—An area occupied by an animal or group of animals and defended againt others of the same species.

Thermocline—The subsurface layer of a lake characterized by a significant temperature changes.

Thermoregulation—Maintenance or regulation of a particular temprature of the living body.

Threshold—The level (duration of intensity) of a stimulus required to produce an effect.

Topography—Characteristics of the ground surface in regard to physical features.

Trace element—Chemical elements used by organisms in minute quantities and essential to their physiology.

Transpiration—The loss of water vapour through the surface of leaves and other plant parts.

Trophic level—One stage in a nutritive series including producers or various levels of consumers.

Trophic structure—The organization of a community in trophic levels.

Tundra—Treeless area in arctic and alpine regions, varying from bare area to various types of vegetation consisting of grasses, sedges, forbs, dwarf shrubs, lichens and mosses.

Turbidity—A condition in air or water that reduces the transparency.

Ultimate factor—In evolutionary terms the survival value of the adaptation in question ; the evolutionary reason for the adaptation, opposite of *proximate factors.*

Univolatine—Refers to an organism that has only one generation per year.

Vector—Organism (often insects) that transmits a pathogenic virus, bacteria, protozoans or fungus from one organism to another.

Watershed—The total land area contributing surface or ground water to a lake, river, or **drainage basin.**

Wilting point—Measure of soil water ; the water remaining in the soil when the plants are in a state of permanent wilting from water shortage.

Xeric—Refers to a dry environment, *e.g.*, desert environments.

Xerophile—Drought-loving, a term used for a xerophyte by **De Condolle**, 1874.

Xerophyte—Plant that can grow in dry places, *e.g.*, **cacti**.

Yield—The part of productvity utilized by human or other animals.

Zonation—The state of arrangement into bands or zones.

Zoogeography—The science dealing with the geographic distribution of animals.

Zooplankton—Animal portion of plankton ; the animal community in marine and fresh-water situations which floats free in the water, independent of shore and the bottom, moving passively with the currents.

Bibliography

Ahuja, B.S., 1961. *Studies on Phytosociology, Ecology and Phytogeography of the Vegetation of Humid Tropics*. Ph.D. Thesis, Rajasthan University, Jaipur, India.

Aiyar, R.G. and Panikkar, N.K., 1939. Observations on the breeding of brackish water animals of Madras. *Proc. Ind. Acad. Sci.* **9** (B) : 343—364.

Alcock, John, 1979. *Animal Behaviour : An Evolutionary Approach*. Sinauer Associates, Inc. Publishers, Sunderland, Massachusetts.

Allee, W.C., Emerson, A.E., Park, O., Park, T., and Schmidt, K.P., 1949. *Principles of Animal Ecology*. W.B. Saunders Company, Philadelphia.

Ananthakrishnan, T.N. and Viswanathan, T.R., 1976. *General Animal Ecology*. Macmillan Company of India Ltd., New Delhi, India.

Ananthasubramanian, K.S. and Ananthakrishnan, T.N., 1975. Studies on Indian Membracidae, Part I and II, Homoptera-Insecta. *Rec. Zoo. Surv. India.*

Andrewartha, H.G., **1961**. *Introduction to the Study of Animal Population*. University of Chicago Press, Chicago.

Andrewartha, H.G. and Birch, L.C., 1954. *The Distribution and Abundance of Animals*. University of Chicago Press, Chicago.

Antony Raja, A.B.C., 1959. Studies on distribution and succession of sedentary organisms in Madras Harbour. *J. Mar. Biol. Ass. India*, **1** : 180—197.

Arora, R.K., 1960. The botany of Coorg forests. *Proc. Natn. Acad. Sci. India*, **30** (B) : 289—305.

Ashby, W.R., 1958. *An Introduction to Cybernetics*. John Wiley and Sons., Inc., New York.

Atwal, A.S., 1955. Influence of temperature, photoperiod and food on the speed of development, longevity, fecundity and other qualities of the diamond black moth *Plutella maculipennis* (Curtis) (Tineidae-Lepidoptera). *Aust. J. Zool.*, **3** : 185—221.

Axelrod, D.I., 1950. Evolution of desert vegetation in North America. *Carnegi Inst. Wash. Pub.*, **590** : 215—360.

Axelrod, D.I., 1958. Evolution of Madro-Tertiary geoflora. *Bot. Rev.*, **24** : 433—509.

Ayyar, P.N.K. and Margabandhu, V., 1941. Biology of the cotton stem weevil *Pempharulus affinis* under controlled physical conditions. *Bull. Ent. Res.*, **32** : 61—82.

Baer, J.G., 1951. *Ecology of Animal Parasites.* Univ. Illinois Press, Urbana.

Bains, S.S. and Shukla, K.K., 1976. Effect of temperature on the development and survival of maize borer, *Chilo partellus* (Swinhoe). *Indian J. Ecol.*, **3** (2) : 149—155.

Bapat, S.V., 1970. The Bombay duck, *Harpodon nehereus* (Ham.) *Bull. Cent. Mar. Fish. Res. Inst.*, **21** : 66.

Bazzaz, F.A., 1975. Plant species diversity in old field successional ecosystems in southern Illinois. *Ecology*, **56** : 485—488.

Beeson, C.F.C., 1971. *The Ecology and Control of Forest Insects of India and Neighbouring Countries.* Govt. of India Publication, New Delhi.

Bhalla, O.P., and Venkataraman, T.V., 1963 Ecological studies on *Vipio deisae* (Camp.) (Braconidae-Hymenoptera) a parasite of maize and jowar stalk borer *Chilo zonellus* (Swinh). *Indian J. Ent.*, **25** : 36—47.

Bhargava, G., 1978. Making the cities livable. *The Hindustan Times : Sunday Magzine*, April 9th.

Bhatia, K.K., 1954. *Factors in the Distribution of Teak* (*Tectona grandis* Linn.) *and a Study of Teak Forests in Madhya Pradesh.* Ph. D. Thesis, Saugar University.

Blackman, F.F., 1905. Optima and limiting factors. *Ann. Botany*, **19** : 281—298.

Bode, I.H.R., 1958. *Planta*, **51** : 440—480.

Bolin, B., 1970. The carbon cycle. *Scientific American*, Sept. pp. 125—132.

Bor, N.L., 1942. Common grasses of United Provinces. *Indian For. Rec.* (Botany), **2** (1).

Bormann, F.H. and Likens, G.E., 1967. Nutrient cycling. *Science*, **155** : 424—429.

Bott, T.L. and Brock, T.D., 1969. Bacterial growth rates above **90°C** in Yellow Stone hot springs. *Science*, **164**: 1411—1412.

Boughey, A.S., 1968. *Ecology of Populations.* The Macmillan Co., New York.

Boughey, A.S., 1971. *Fundamental Ecology.* Thomas Y. Crowell Company, New York.

Braun-Blanquet, J., 1932. *Plant Sociology : The Study of Plant Communities.* McGraw-Hill Book Co. Inc., New York.

Brody, S., 1945. *Bioenergetics and Growth.* Reinhold Publishing Corp., New York.

Cain, S.A. and Castro, G.M.de.O., 1959. *Manual of Vegetation Analysis.* Harper and Brothers, New York.

Calhoun, J.B., 1962. Population density and social pathology. *Sci. Amer.*, **206** (2) : 1399—1408.

Champion, H.G., 1936 and 1938. A preliminary survey of the forest types of India and Burma. *Indian For. Rec.*, **1** (1).

Champion, H.G. and Griffith, A.L., 1948. *Manual of General Silviculture for India.* FRI, **94**, Manager of Publications, New Delhi.

Champion, H.G. and Seth, S.K., 1968. *A Revised Survey of the Forest Types of India.* FRI **122**. Manager of Publications, New Delhi.

Chatterji, P.C. and Bhaskar, P.R., 1977. Hydrological surveys. *Proc. of Summer Inst. on Resource Inventory and Land Use Planning*, CAZRI, Jodhpur.

Christian, J.J., 1950. The adreno-pituitary system and population cycles in mammals. *J. Mamm.*, **31** : 247—259.

Christian, J.J., 1963. Endocrine adaptive mechanisms and the physiologic regulation of population growth. In W.V. Mayer and R.G. Van Gelder (eds.), *Physiological Mammalogy, Vol. 1. Mammalian Populations*, pp. 180—353. Academic Press, New York.

Clapham, Jr., W.B., 1973. *Natural Ecosystems.* The Macmillan Company, New York.

Clarke, G.L., 1954. *Elements of Ecology.* John Wiley and Sons, *Inc.*, New York.

Clements, F.E., 1916. *Plant Succession: An Analysis of the Development of Vegetation.* Carnegic Inst. Wash. Publ., Washington.

Clements, F.E., 1936. Nature and structure of the climax. *J. Ecol,*, **24** : 252—284.

Clements, F.E., and Shelford, V.E., 1939. *Bioecology.* John Wiley and Sons, Inc., New York.

Cloud, P.E., 1969. *Resources and Man.* W.H. Freeman and Company, San Francisco.

Cloudsley-Thompson, J.L., 1962. Lethal temperatures of some desert arthropods and the mechanisms of heat death. *Ent. Exp. Appl.*, **5** : 270—280.

Cloudsley-Thompson, J.L., 1975. Adaptations of arthropoda to arid environments. *Ann. Rev. Ent.*, **20** : 261—283.

Cockrum, E.L. and McCauley, M.J,, 1965. *Zoology.* W.B. Saunders Company, Philadelphia.

Coker, R.E., 1939. The problem of cyclomorphosis in *Daphnia. Quart. Rev. Biol.*, **14** : 137—148.

Cole, LaMont C., 1958, The ecosphere. *Sci. Amer.*, **198** (4) : 83—96.

Connell, J.H. and Orias, E., 1964. The ecological regulation of species diversity. *Amer. Nat.*, **98** : 399—414.

Cooke, G.D., 1967. The pattern of autotrophic succession in laboratory microsystems. *Bioscience*, **17** : 717—721.

Cooper, J.G., 1859. On the distribution of the forests and trees of North America, with notes on its physical geography. *Ann. Rep't. Bd. Reg. Smithsonian Inst.*, 246–280.

Cowles, H.C., 1899. The ecological relations of the vegetation of the sand dunes of Lake Michigan. *Bot. Gaz.*, **27** : 95–117; 167–202 ; 281–308 ; 361–391.

Craig, R.A., 1968. *The Edge of Space: Exploring the Upper Atmosphere.* Doubleday and Company, Inc., Garden City, N.Y.

Crombie, A.C., 1947. Interspecific competition. *J. Anim. Ecol.*, **16** : 44–73.

Curtis, Helena, 1975. *Biology.* Worth Publishers, Inc., New York.

Curtis, J.T., 1959. *The Vegetation of Wisconsin.* University of Wisconsin Press, Madison, Wisconsin.

Daniel, A., 1954. Seasonal variation and the succession of fouling communities in the Madras harbour waters. *J. Madras Univ.*, **24** : 189–212.

Dansereau, P., 1957. *Biogeography : An Ecological Perspective.* Ronald Press, New York.

Dansereau, P. and Lems, K., 1957. The grading of dispersal types in plant communities and their ecological significance. *Cont. Inst. Bot.*, Univ. Montreal, No. 17.

Das, T.M., Bhaumik, A., Ghosh, A. and Chakravarty, A., 1981. Trees as dust-filters. *Science Today*, **XV** (2) : 19–22.

Daubenmire, R.F., 1947. *Plants and Environments : A Text-Book of Plant Autecology.* John Wiley and Sons, Inc., New York.

Daubenmire, R.F., 1966. Vegetation : identification of typal community. *Science*, **151** : 291–298.

Daubenmire, R.F., 1968. *Plant Communities.* Harper and Row, New York.

Davenport, C.B., 1903. The animal ecology of the cold spring sand pit, with remarks on the theory of adaptation. *Decen. Pub. Univ.* Chicago, **10** : 157–176.

David, C.C., 1955. *The Marine and Fresh-water Plankton.* University Michigan Press. Ann. Arbor.

David, A., Ray, P., Govind, B.V., Rajgopal, K.V. and Banerjee, R.K., 1969. Limnology and fisheries of the Tungabhadra Reservoir. *C.I.F.R.I., Barrackpore, Bulletin*, No. **13** : 188 p.

Davidar, E.R.C., 1976. Census of the Nilgiri tahr in the Nilgiris, Tamil Nadu. *J. Bombay Nat. Hist. Soc.*, **73** (1) : 142–148.

Davis, J.J. and Foster, R.F., 1958. Bioaccumulation of radioisotopes through aquatic food chains. *Ecology*, **39** : 530–535.

Davis, T.A., De, S.K., and Pal, A.K., 1977. Koel as a brood parasite of crow. *Science Reporter*, **14** (1) : 6–12.

Dawson, W.R. and Bartholomew, G.A., 1968. Temperature regulation and water economy of desert birds, in G.W. Brown (ed.), *Desert Biology*, Academic Press, New York.

Deevey, E.S., 1970. Mineral Cycles. *Scientific American*, **223** : 148–158.

Delwiche, C.C., 1970. The nitrogen cycle. *Sci. Amer.*, **223** (5) : 137–146.

DeWilde, M.A. and Houstan, A.H., 1967. Haematological aspects of the thermo-acclimatory process in the rainbow trout (*Salmo gairdneri*). *J. Fish. Res. Bd. Can.*, **24**: 2267–2281.

Dewsbury, D.A., 1978. *Comparative Animal Behaviour*. McGraw-Hill Book Company, New York.

Dharmakumarsinhji, R.S., 1973 (1975). Migratory and resident waterfowl (Anatidae) of Bhavnagar District, Gujrat, India. *Pavo*, **11** (1/2) : 33–60.

Dice, L.R., 1943. The biotic provinces of North America. *Univ. Michigan Press, Ann. Arbor* : 1–78.

Dokuchaev, V.V., 1889. The zones of nature (in Russian). *Akad. Nauk. Moscow*, Vol. 6.

Dudgeon, W., 1921. *J. Indian Bot. Soc.*, **1** : 296–324.

Ehrenfield, D.W., 1970. *Biological Conservation*. Holt Rinehart & Winston, Inc., New York.

Eisenberg, J.F., 1965. The social organisation of mammals. *Handbuch der Zoologie*, **8** : 1–91.

Elton, Charles, 1927. *Animal Ecology*. Sidwick and Jackson Ltd., London.

Elton, Charles, 1949. Population interspersion : as essay on animal community patterns. *J. Eco.*, **37** : 1–23.

Elton, Charles, 1966. *The Patterns of Animal Communities*. Methuen London.

Eltringham, S.K., 1971. *Life in Mud and Sand*. English Universities Press, London.

Emerson, R., 1968, *The Fungi*, Vol. III. Academic Press, New York.

Errington, P.L., 1946. Predation and vertebrate populations. *Rev. Biol.*, **21** : 144–177; 221–245.

Evans, F.C., 1956. Ecosystem as the basic unit in ecology. *Science*, **123** : 1227–1228.

Farghaly, A.M., Ezzat, A.A. and Sabana, M.B., 1973. Effect of temperature and salinity changes on the blood characteristic of *Tilapia zilli* G. in Egyptian littoral lake. *Comp. Biochem. Physiol.*, **46** (A) : 189–193.

Fischer, A.G., 1960. Latitudinal variations in organic diversity. *Evolution*, **14** : 64–81.

Fogg, G.E., 1955, *Nitrogen fixation*. New Biology, No. 18, pp. 52–71. Penguin Books, London.

Forbes, E., 1844. *Rep. British Association for Adv. Science* (1843): 130–193.

Fosberg, F.R., 1967. A classification of vegetation for general purposes. In G.F. Peterken, (ed.) *Guide to the Checklist for I.B.P. Areas.* I.B.P. Handbook No. 4, Blackwells, **Oxford.**

Foster, M.S., 1969. Synchronized life cycles in the orange-crowned warbler and its mallophagan parasites. *Ecology*, **50** : 315–323.

Fry, F.E.J., 1947. Effect of the environment on animal activity. *Univ. Toronto-Studies, Biol. Ser.* No. **55**: 1–62.

Furon, R. 1957. *The Problem of Water : A World Study.* American Elsevier Publishing Co., Inc., New York.

Ganapathy, P.N. and Rao, R.M., 1954. Studies on planktonic copepods—I. seasonal fluctuations in the distribution with reference to salinity and temperature. *Andhra Univ. Ser.* No. **49** : *Mem. in Oceanogr.*, I : 151–162.

Ganapathy, P.N. and Subba Rao, D.V., 1958. Quantitative study of plankton of Lawson's Bay, Waltair. *Proc. Ind. Acad. Sci.*, **XLVIII** : 184–209.

Ganapathy, P.N. and Lakhmana Rao, M.V., 1962. Studies on the ecology of interstitial sands of Visakhapatnam coast. *Proc. Ist All India Congr. Zool.* 1959, pt **3** : 26–31.

Ganapati, S.V., 1972. Organic production of seven types of aquatic ecosystems in India. In P.M. Golley and F.B. Golley (Eds) *Tropical Ecology with Emphasis on Organic Productivity*, pp. 313–350, Univ. Georgia, Athens.

Garrett, S.D., 1963. *Soil Fungi and Soil Fertility.* Pergamon Press, London.

Garett, S.D., 1973. *Acta. Bot. Indica*, **1** : 1–9.

Gates, D.M., 1962. *Energy Exchange in the Biosphere.* Harper and Row Publishers, Inc., New York.

Gause, G.F., 1934. *The Struggle for Existence.* Williams & Wilkins, Baltimore.

Gause, G.F., 1935. Experimental demonstration of Volterra's periodic oscillations in the numbers of animals. *J. Exp. Biol.*, **12** : 44–48.

George, M.G., 1961. Diurnal variations in two shallow ponds in Delhi, India. *Hydrobiologia*, **18** : 265–273.

George, M.G., 1966. Diurnal variations in physico-chemical factors and zooplankton in the surface layers of three freshwater ponds. *Ind. J. Fisheries*, **13** : 48–82.

Gleason, H.A., 1926. The individualistic concept of the plant association. *Bull. Torrey Bot. Club.* **53** : 7–26.

Goldsby, R.A., 1976. *Biology.* Harper and Row Publishers, New York.

Gopal, B., 1973. A survey of Indian studies on ecology and production of wetland and shallow water communities. *Pol. Arch. Hydrobiol.*. **20** : 21–29.

Gopal, B. and Sharma, K.P., 1978. Light regulated seed germination in *Typha angustata*. *Aquatic Botany* **5**.

Gopal, B. and Bhardwaj, N., 1979. *Elements of Ecology.* Vikas Publishing House Pvt. Ltd., New Delhi.

Gopalakrishnan, K., 1971. The biology of Hooghly–Matla estuarine system (West Bengal) India, with special reference to its fisheries. *J. Mar. Biol. Ass. India,* **13** : 182–194.

Gopinathan, C.P., Ramachandran Nair, P.V. and Kesavan Nair, A.K., 1974. Studies on phytoplankton of the Cochin backwater : a tropical estuary. *Ind. J. Fish.,* **21** (2) : 501–513.

Govindankutty, A.G. and Balakrishnan, Nair, N., 1966. Preliminary observations on the interstitial fauna of the south west coast of India. *Hydrobiologia,* **13** : 182–194.

Graveley, F.H., 1927. The littoral fauna of Krusadai islands in the gulf of Mannar. *Bull. Madras Govt. Mus.,* **1** : 1.

Grewal, G.S. and Atwal, A.S., 1968. Development of *Chrotogonus trachypterus* (Blanch) (Orthoptera-Pyrgomorphidae) in relation to different levels of temperature and humidity. *Indian J. Ent.,* **30** : 1–7.

Grinnell, Joseph, 1917. Field tests of theories concerning distributional control. *Amer. Nat.,* **51** : 115—128.

Grinnell, Joseph, 1924. Geography and evolution. *Ecology,* **5** : 225–229.

Grinnell, Joseph, 1928. Presence and absence of animals. *Univ. California Chron.,* **30** : 429–450.

Gupta, R.K., 1961. Vegetation in the catchment areas of rivers Ganga and Yamuna with reference to their soil and water conservation problems. *J. Indian Bot. Soc.,* **40** : 325–334.

Gupta, S.D., 1976. Macrobenthic fauna of Loni reservoir. *J. Inl. Fish. Soc. India,* **8** : 49–59.

Gupta, R.K. and Prakash, I., 1975. *Wildlife Ecology and Conservation.* English Book Depot., Dehradun.

Haeckel, E., 1866. *Generelle Morphologie der organismen.* 2 vols. Georg Reimer, Berlin.

Hairston, N.G., Allan, J.D., Colwell, R.K., Futuyma, D.J., Howell, J., Lubin, M.D., Mathias, J. and Vandermeer J.H., 1968. The relationship between species diversity and stability : an experimental approach with Protozoa and bacteria. *Ecology,* **49** : 1091–1101.

Hamilton, W.D., 1964. The evolution of social behaviour. *J. Theoretical Biology,* **7** : 1–52.

Hale, W.G., 1967. *Collembola,* in A. Burges and Raw, F. *Soil Biology,* pp. 397–411. Academic Press, London.

Hartung, R., 1967. Energy metabolism in oil-covered ducks. *J. Wildlife Management,* **31** (4) : 798–804.

Herre'd II, C.F., 1977. *Biology.* Macmillan Publishing Co., Inc., New York.

Hoar, S.L. and Cottle, M.K., 1952. Dietary fat and temperature tolerance of gold fish. *Can. J. Zool.,* **30** : 41–48.

Hora, S L., 1930. Ecology, binomics and evolution of the torrential fauna with special reference to the organs of attachment. *Trans. Roy. Soc.*, **218** : 171–282.

Hora, S.L., 1947. Torrential fishes and the significance of their distribution in zoogeographical studies. *Bull. Nat. Geo. Soc. India, Banaras*, **7** : 1–10.

Hornell, J., 1922. The Indian pearl fishery of the gulf of Mannar and Pak Bay. *Madras Fish. Bull.*, **16** : 1–188.

Hornell, J. and Nayudu, M.R., 1924. A contribution to the life history of the Indian oil sardine with notes on the plankton of the Malabar coast. *Madras Fish. Bull.*, **17** : 129–197.

Hornocker, M.G., 1970. An analysis of mountain lion predation upon mule deer and elk in the Idaho primitive area. *Wildlife Monograph*, No. **21**. Wildlife Society, Washington.

Hult, R., 1885. *Medd. Soc. Fenn.*, **12** : 161.

Hutchinson, G.E., 1957. *A Treatise on Limnology, Vol. I : Geography, Physics and Chemistry*. John Wiley and Sons, Inc., New York.

Hutchinson, G.E., 1967. *A Treatise on Limnology, Vol. II : Introduction to Lake Biology and the Limnoplankton*. John Wiley and Sons, Inc., New York.

Hutchinson, G.E., 1969. Eutrophication, past and present. In *Eutrophication : Causes, Consequences, Correctives*. National Academy of Science Washington, pp. 12–26.

Hutchinson, G.E., 1970. The biosphere. *Scientific American*, **223** (3) : 44–53.

Jain, N.K., 1958. Composition of sal (*Shorea robusta*, Gaertn.) forests of Madhya Pradesh. *Proc. Natn. Acad. Sci. India*, **28** : 130–152.

Jenny, H. and Raychoudhury, S.P., 1960. Effects of climate and cultivation on nitrogen and organic matter reserves in Indian soils. *Coun. Sci. Indus. Res.*, India.

Jha, U.S., 1968. *The Pond Ecosystem*. Ph. D. Thesis, Banaras Hindu University.

Johnson, P.L., 1970. Dynamics of carbon dioxide and productivity in an arctic biosphere. *Ecology*, **51** : 73–80.

Jones, S., 1960. Notes on animal associations II. The Scyphomedusa *Acromitus flagellatus* (Steasmy) and young, *Seleroides* ***leptolepis*** (Curer and Valenciennes) with the latter forming a vanguard. *J. Mar. Biol. Ass. India*, **2** : 51–52.

Jones, S. and Mahadevan, 1965. Notes **on** animal associations. The pea crab *Pinnotherus deccanensis* (Chopra) inside the respiratory tree of the sea cucumber *Holothuria scabar* (Jager). *J. Mar. Biol. Ass. India*, **7** : 377–380.

Joshi, S.R., 1958. On the phytosociological studies of *Anogeissus latifolia*, Wall. forests of Madhya Pradesh. *Proc. Natn. Acad. Sci. India*, **28** : 116–129.

Jyoti, M. K., 1976. Seasonal variations in food and feeding habits of some lacustrine fishes of Kashmir. *J. Inl. Fish. Soc. India*, **8**: 24–32.

Kader, H.A., Navaneeth Krishnan, P. and Krishnaswamy, S., 1977. On the occurrence of argillotrophic habitat in Madurai, S. India. *Arch. Hydrobiol.* **79** (3) : 413–415.

Kant, S. and Kachroo, P., 1975. Limnological studies in Kashmir lakes : II **Diurnal** movement of phytoplanktons. *J. Indian Bot. Soc.*, **54** (1/2) : 9–12.

Kaufmann, J.H., 1971. Is territory definable ? In A.H. Esser's (ed.) *Behaviour and Environment : The Use of Space by Animals and Men*, pp. 36–40, Plenum, New York.

Kaul, V., 1971. Production and ecology of some macrophytes of Kashmir lakes. *Hydrobiologia*, **12** : 63–69.

Keast, A., 1959. Australian birds : the zoogeography and adaptation to an arid continent. *Biogeography and Ecology in Australia*, **8** : 89–114.

Kendeigh, S.C., 1941. Length of day and energy requirement for gonad development and egg-laying in birds. *Ecology*, **22** : 237–248.

Kendeigh, S.C., 1945. Community selection by birds on the Helderberg Plateau of New York. *Auk*, **62** : 418–436.

Kendeigh, S.C., 1961. *Animal Ecology*. Prentice Hall, Englewood-Cliffs.

Kendeigh, S.C., 1974. *Ecology with Special Reference to Animals and Man*. Prentice-Hall of India Pvt. Ltd., New Delhi.

Khan, A. and Siddiqui, A.Q., **1974** (1976). Seasonal changes in the limnology of a perennial fish pond at Aligarh. *Indian J. Fish.*, **21** (2) : 463–478.

Khanna, P.K., 1964. *The Succession of Fungi on Some Decaying Grasses*. Ph. D. Thesis. Banaras Hindu University.

Kinsey, A.C., 1930. The gall wasp genus *Cynips*. A study in the origin of species. *Indiana Univ. Stud.*, **16** : 84–86 (1929) : 1–577.

Klein, D.R., 1970. Food selection by North American deer and their response to overutilization of preferred plant species. In Adam Watson's *Animal Population in Relation to Their Food Sources*, pp. 25–46. Blackwell Sci. Publ., Oxford.

Kormondy. E.J., 1976. *Concepts of Ecology*. Prentice-Hall, Engelwood Cliffs.

Krebs, C.J., 1972. *Ecology*. Harper and Row, Publishers, New York.

Krishanamoorthy, B., 1963. On the distribution of six species of polychaets in the Adyar estuary, Madras. *J. Mar. Biol. Ass. India*, **5** : 97–102.

Krivolutskii, D.A., 1972. Current concepts of animal "life forms". *Soviet J. Ecol.* (Trans.), **2** : 202–207.

Kuchler, A.W., 1967. *Vegetation Mapping*. The Ronald Press Co., New York.

Lack, D., 1967. The significance of clutch-size in water fowl. *Wild Fowl Trust* **18**th *An. Rep.* : 125–128.

Lack, D., 1971. *Ecological Isolation in Birds.* Blackwell Sci. Publ., Oxford.

Lafond, E.C., 1952. Marine meterology and its relation to organic production in S.E. Asian waters. *J. Mar. Biol. Ass. India.* **4** : 1–9.

Landsberg, H.E., 1970. Man-made climatic changes. *Science*, **170** : 1265–1274.

Langford, A.N. and Buell, M.F., 1969. Integration, idenity and stability in the plant association. *Adv. in Ecol. Res.*, **6** : 83–135.

Lee, R., 1973. The 'greenhouse' effect. *J. Appl. Meterology*, **12** : 556–557.

Lewis, T. and Taylor, L R., 1967. Introduction to Experimental Ecology, Academic Press, London.

Liebig, Justus, 1840. *Chemistry in its Application to Agriculture and Physiology.* Taylor and Walton, London.

Lieth, H., 1970. Phenology in productivity studies. In D.E. Reichle (ed.) *Analysis of Temperature of Forest Ecosystem.* Springer, Berlin.

Lignon, J.D., 1968. Sexual differences in foraging behaviour in two species of *Dendrocopus* woodpeckers. *Auk*, **85** : 203–215.

Lindauer, M , 1961. *Communication Among Social Bees.* Harvard University Press, Cambridge, Mass.

Lotka, A.J., 1925. *Elements of Physical Biology.* Williams and Wilkins, Baltimore, Md.

Lowe-McConnell, R.H. 1969. Speciation in tropical fresh-water fishes. *Biol. J. Linn. Soc.*, **1** : 52–75.

MacArthur, R.H., 1958. Population ecology of some warblers of northeastern coniferous forests. *Ecology*, **39** : 599–619.

MacArthur, R.H., 1972. *Geographical Ecology : Patterns in the Distribution of Species.* Harper and Row, Publishers, New York.

McCombs, L.W. and Rosa, N., 1978. *What's Ecology* ? Addison-Wesley Publishing Company Menlo Park, California.

McCormick, F., 1968. Succession. *Via*, **1** : 22–35; 131–132.

MacFadyen, A., 1963. *Animal Ecology : Aims and Methods.* Pitman, London.

McIntosh, R.P., 1958. Plant communities. *Science*, **128** : 115–120.

McIntosh, R.P., 1967. An index of diversity and the relation of concepts to diversity, *Ecology*, **48** : 392–403.

McNaughton, S.J. and Wolf, L.L., 1970. Dominance and the niche in ecological systems. *Science*, **167** : 131–139.

McNaughton, S.J. and Wolf, L.L., 1973. *General Ecology.* Holt, Rinehart and Winston, New York.

Maelzer, D.A., 1965. A discussion of components of environment in ecology. *J. Theoret. Biol.*, **8** : 141–162.

Mahadevan, S. and Nayar, K. N., 1968. Underwater ecological observations in the gulf of Mannar off Tuticorin, VII: General topography and ecology of rocky bottom. *J. Mar. Biol.Ass. India*, **9** (1) : 147–163.

Malhotra, S. K., 1971 (1974). Studies on the limestone vegetation of Sahasradhara near Dehradun (U. P.). 7. Phytosociological studies : community co-efficient. *Bull. Bot. Surv. India*, **13** (3/4) : 343–344.

Mall, L. P., 1956. Some aspects of the autecology of *Chrozophora rottleri. A. Juss. Bull. Bot. Soc. Univ. Saugar*, **8** (1) : 13–24.

Mall, L. P., 1957. *J. Univ. Saugar*, **6** (B) : 35–54.

Mall, L. P. and Arzare, K. C., 1956. *Bull. Bot. Soc. Univ. Saugar*, **8** : 69–76.

Mall, L. P. and Raina, D. K., 1957. *Bull. Bot. Soc. Univ. Saugar*, **9** : 21–33.

Mani, M. S., 1957. High altitude insect life on the Northwest Himalaya. **Presidential** address, **26**th Sess. *The Natn. Acad. Sci.*, 41–62.

Mani, M. S., 1968. *Ecology and Biogeography of High Altitude Insects.* Dr. W. Junk, N. V., Publishers, The Hague.

Mani, M. S., 1974. *Ecology and Biogeography of India.* Dr. W. Junk N. V., Publishers, The Hague.

Margalef, R., 1958. Information theory in ecology. *Gen. Syst.*, **3** : 36–71.

Margalef, R., 1968. *Perspectives in Ecological Theory*. University of Chicago Press, Chicago, Ill.

Meher, Homji, V. M., 1962. The bioclimates of India in relation to the vegetational criteria. *Bull. Bot. Surv. India*, **4** (1-4): 105-112.

Merriam, C. H., Vernon, B., Nelson, E. W. and Preble, E. A., 1910. Zone maps of North America. *Biol. Surv., U. S. Dept. Agr. Washington.*

Misra, R., 1938. *J. Ecol.*, **26** : 411–461.

Misra, R., 1944. *J. Indian Bot. Soc.*, **23** : 44–52.

Misra, R., 1946. A study in the ecology of low lying lands. *Indian Ecologist*, **1** : 27–46.

Misra, R., 1955. *Bull. Bot. Soc. Univ. Saugar*, **7** : 90–101.

Misra, R., 1958. *Symp. I. B. S.*, Baroda, India.

Misra, R., 1959. *Proc.* **46**th *Indian Sci. Cong. Assoc.*, 96–107.

Misra, R., 1967. Form, function and factors in ecology. *J. Indian Bot. Soc.*, **46** : 144–153.

Misra, R., 1968. Energy transfer along terrestrial food chains. *Trop. Ecol.*, **9** : 105–118.

Misra, R., 1968. *Ecology Work Book.* Oxford and IBH Publishing Co., New Delhi.

Misra, R., 1969. Studies on the primary productivity of terrestrial communities at Varanasi. *Trop. Eco.*, **10** : 1–15.

Misra, R., 1970. *IUCN Publ. N. S.*, **17** : 230–239.

Misra, R. and Shivarao, B.S., 1948. *J. Indian Bot. Soc.*, **27** : 186—199.

Misra, R. and Shivarao, B. S., 1948. *J. Indian Bot. Soc.*, **29** : 1—13.

Misra, R. and Puri, G. S., 1954. *Indian Manual of Plant Ecology* English Book Depot., Dehradun.

Misra, R. and Singh, K.P., 1971. *J. Ind. Bot. Soc.*, **52** (A): 589–601.

Misra, R., Gopal, R.B., Singh, K.P. and Singh, J. S., 1973. *Progress of Plant Ecology in India*, Vol. 1. Today and Tomorrow Printers and Publishers, New Delhi.

Moen, A. N., 1973. *Wildlife Ecology*. W. H. Freeman and Company, San Francisco.

Monk, C. A., 1967. Tree species diversity in the eastern deciduous forest with particular reference to north-central Florida. *Amer. Naturalist*, **101** : 173–187.

Murray, J. and Hjort, J., 1912. *The Depths of the Ocean*. Macmillan and Co., London.

Nace, R. L., 1969. Human uses of ground water. In R. J. Chorley (ed.) *Water*, *Earth* and *Man*, Methuen, London.

Nair, R. V. and Subrahmanyan, R., 1955. The diatom, *Fragilaria oceanica* Cleve, an indicator of abundance of the Indian oil sardine, *Sardinella longiceps* Cuv. and Val. *Curr. Sci.*, **24**:41–42.

Narayanan, E. S., 1960. Super, multi and hyperparasitism and their effect on the biological control of insect pests. *Proc. Nat. Inst. Sci. India*, **26** : 256–280.

Natrajan, G. M., 1977. They are wild but need our protection. *Science Reporter*, **14** (3) : 144–147.

Nayar, C. K. G., 1970. Studies on rotifer populations of two ponds at Pilani, Rajasthan. *J. Zool. Soc. India*, **22** (1 and 2) : 21–34.

Nayar, K. K. and Prabhu, N. R., 1965. *Studies of the Soil Fauna of the Tea Growing Tracts of Kerala*. Reprort submitted to the Tea Board.

Nayar, N. K. and Mahadevan, S. 1965. Underwater ecological observations in the gulf of Mannar off Tuticorin II. The association between a fish and a sea-snake. *J. Mar. Biol. Ass. India*, **7** : 197–199.

Nayar, N.K. and Mahadeven, S., 1965. *J. Mar. Biol. Ass. India*, **7** : 458–459.

Nicholson, A.J., 1933. The balance of animal populations. *J. An. Ecol.*, **2** : 132—178.

Oden, S. and T. Ahl, 1970. The acidification of scandinavian lakes and rivers. *Ymer Arsbok*, pp. 103–122.

Odum, E.P., 1969. The strategy of ecosystem development. *Science*, **164** : 262--270.

Odum, E.P., 1971. *Fundamentals of Ecology*. W.B. Saunders, Toppan Co. Ltd., Tokyo, Japan.

Odum, H.T. and Pinkerton, R.C., 1955. Time's speed regulator, the optimum efficiency for maximum output in physical and biological systems. *Amer. Sci.*, **43** : 331–343.

Olson, E.C., 1961. The food chain and the origin of mammals. *Internat. Colloq. on the Evolution of Mammals ; pt. I. Brussels : Kon. Vlaamse Acad. Wetensch. Lett. Sch. Kunsten Beigie*, pp. 97–116.

Olson, E.C., 1966. Community evolution and the origin of mammals. *Ecology*, **47** : 291–302.

Olson, J.S., 1958. Rates of succession and soil changes on southern lake Michigan sand dunes. *Bot. Gaz.*, **119**, 125–170.

O'Neill, R.V., 1967. Niche segregation in seven species of diplopods. *Ecology*, **48** : 983.

Oppenheimer, J.R., 1973. Effects of environmental factors on dwelling langurs (Primates) in West Bengal. *Proc. Sixtieth Sess. Sci. Cong.* **IV** : 157.

Osburn, R C., Dublin, L. I., Shimer, H. W. and Lull, R. S., 1903. Adaptation to aquatic, arboreal, fossorial, and cursorial habits in mammals. *Amer.* Nat., **37**: 651–665; 731—736, 819–825; **38**: 1904: 1–11.

Paine, R. T., 1966. Food web complexity and species diversity. *Amer. Nat.*, **100**: 65–75.

Pandey, B. N., 1977. Haematological studies in relation to environmental temperature and different periods of breeding cycle in an air breathing fish *Heteropneustes fossilis*, *Folia Haematol., Leipzig*, **104**: 69—74.

Pandeya, S. C., 1950-51. *Bull. Bot. Soc. Univ. Saugar*, **3** : 4–10.

Pandeya, S. C., 1953. Grasslands of Sagar, Madhya Pradesh. *Indian For.*, **78** : 638–654.

Pandeya, S. C., 1967. *Proc. Indian Acad. Sci.*, **55** : 63–75.

Pandeya, S.C. and Kuruvilla, 1968. *Proc. Symp. Rec. Adv. Trop. Ecol.* : 666–684.

Pandeya, S.C., Singh, J.S. and Puri, G.S., 1968. *Research Methods in Plant Ecology*. Asia Publ. House, Bombay.

Pandian, T.J., 1967. Intake, digestion, absorption and conversion of food in the fishes *Megalops cyprinoides* and *Ophiocephalus striatus*. *Mar. Biol.*, **1** : 16-32.

Panikkar, N.K. and Prasad, R.R., 1952. On an interesting association of ophiuroid fish and crab with the jelly fish *Rhopalima loopidus*. *J. Bombay Nat. Hist. Soc.*, **51** : 95–96.

Patnaik, S., 1971. Seasonal abundance and distribution of bottom fauna of Chilka lake. *J. Mar. Biol. Ass., India*, **13** : 106–125.

Pearl, R., 1925. *The Biology of Population Growth*. Alfred A. Knopf., Inc., New York.

Pearse, A.S., 1939. *Animal Ecology*. McGraw-Hill Book Company, Inc , London.

Perrins, C.M., 1965. Population fluctuations and clutch-size in the great tit, *Parus major*, *L.*, *J. Animal Ecol.*, **34** : 601–647.

Petrides, G.A., 1968. *Proc. Symp. Recent Adv. Trop. Ecol.*, 705–709.

Petersen, C.G.J., 1914, 1915, 1918. The animal associations of the sea-bottom in the North Atlantic. *Kobenhavn Ber. Biol. Sta.*, **22** : 89–98 ; **23** : 1–28 ; **26** : 1–62.

Pianaka, E., 1967. On lizard species diversity : North American flatlands desert. *Ecology*, **48** : 333–351.

Pianaka, E., 1973. The structure of lizard communities. *Ann. Rev. Ecol. Syst.* **4** : 53–74.

Pianaka, E., 1974. *Evolutionary Ecology*. Harper and Row, Publishers, New York.

Pillai, K., 1968 Observations on the plankton off Bombay coast with remarks on the hydrographic conditions and fishery *J. Mar. Biol. Ass India*, **10** : 237–244.

Pillai, V.K. and Sreenivasan, A., 1975 (1976). A note on the carbon and nitrogen status of some lakes, reservoirs and ponds of Tamil Nadu. *J. Inl. Fish. Soc. India*, **7** : 95–97.

Pound, R. and Clements, F.E., 1900. *The Phytogeography of Nebraska*. Univ. Nebraska.

Prakash, Ishwar, 1958. Study of vertebrates in the Indian desert. *Turtox News*, **36** : 266.

Prakash, Ishwar, 1961. Taxonomic and biological observations on bats of Indian deserts. *Rec. Ind. Mus.*, **59** : 149–170.

Prakash, Ishwar, 1963. Zoogeography and evolution of the mammalian fauna of Rajasthan desert (India). *Mammalia*, **27** : 342–351.

Prakash, Ishwar, 1963. Taxonomical and ecological account of the mammals of Rajasthan desert. *Annals of Arid Zone* **I** : 142–162.

Prakash, Ishwar, 1977. Wildlife and its inventory. *Proc. of the Summer Inst. on Resource Inventory and Land Use Planning*, CAZRI, Jodhpur, pp. 269–276.

Prasad, R. R., 1956. Plankton of the inshore waters off Mandapam. *Indian J. Fisheries* **III** : 1–42.

Puri, A. N., 1925. A critical study of the hygroscopic coefficient of soil. *J. Agri. Sci.* **15** : 272–283.

Puri, G. S., 1950. *Indian For.*, **76** : 144–153.

Puri, G. S., 1954. *J. Indian Bot. Soc.*, **33** : 394–416.

Puri, G. S., 1960. *Indian Forest Ecology*, 2 Vols. Oxford Press, New Delhi.

Purohit, K. G., 1968. Ecology of mammals in the Great Indian Desert. In R. Misra and B. Gopal (eds.), *Proc. Symp. Recent Adv. Trop. Ecol.* : 269–287.

Putwain, P. D. and Harper, J. L., 1970. *J. Ecol.*, **58** : 251–264.

Qasim, S. Z., 1957. *Proc. Zool. Soc. London*, **128** : 161–208.

Qasim, S. Z., 1957. *J. Anim. Ecol.*, **26** : 389–401.

Qasim, S. Z., 1966. *Curr. Sci.*, **35** : 140–142.

Qasim, S. Z., 1970. Some characteristics of a *Trichodesmium* bloom in the Laccadives. *Deep-sea Res.*, **30** : 11–16.

Qasim, S. Z., 1971. *J. Sci. Indust. Res.*, **30** : 11–16.

Qasim, S. Z., 1972. *Indian J. Fish.*, **19** : 11–28.

Qasim, S. Z., 1973. Productivity of specialized environments, *Mahasagar*, **6** (2) : 95–100.

Qasim, S. Z., 1973. Some implications of the problem of age and growth in marine fisheries from the Indian waters. *Indian J. Fish.*, **20** (2) : 351–371.

Qasim, S. Z., 1975 (1976). Aquaculture, its potential and scope in India. *Proc. Indian Natl. Sci. Acad., Part B. Bio. Sci.*, **41** (5) : 397–420.

Qasim, S. Z., 1977. Role of aquaculture in rural development. *Invention Intelligence.* **12** (1 and 2) : 107–111.

Qasim, S. Z., Bhattathiri, P. M. A. and Abidi, S.A.H., 1968. Solar radiation and its penetration in a tropical estuary. *J. Exp. Mar. Biol. Ecol.*, **2** : 87–103.

Qasim, S. Z. and Gopinathan, C. K., 1969. Tidal cycle and the environmental features of Cochin Backwater (a tropical estuary). *Proc. Indian Acad. Sci.*, **69** (B) : 336–348.

Qasim, S. Z. and Bhattathiri, P.M.A., 1971. Primary production of a seagrass bed on Kavaratti Atoll (Laccadives). *Hydrobiologia*, **38** : 29–38.

Qasim, S. Z. and Sankaranarayanan, V. N., 1972. Organic detritus of a tropical estuary. *Mar. Biol.*, **15** : 193–199.

Ramakrishnan, P. S., 1958. Ecotypic differentiation in some plants of Varanasi. *Proc. Natn. Acad Sci. India.*, **28** (B) : 68–76.

Ramalingam, K., 1961. *J. Mar. Biol. Ass. India*, **2** : 35–50.

Ramam, S. S., 1961. Soil-root relationships in grassland communities of Varanasi. Ph. D. Thesis, Banaras Hindu University.

Ramam, S. S., 1975. Primary production and nutrient cycling in tropical deciduous forest ecosystem. *Trop. Ecol.*, **16** (2) : 140–146.

Ramamurthy, B. N., 1966. *Proc. Second All India Congr. Zool., Varanasi*, 1962, part II.

Ramamurthy, S., 1965. Studies on the plankton of the North Canara Coast in relation to the pelagic fishery. *J. Mar. Biol. Ass. India*, **VII** : 127—147.

Ramamurthy, S. and Dhawan, S. R., 1963. On the characteristics of the plankton of Kandla in the gulf of Kutch during August 1958 to July 1960. *Ind. J. Fisheries*, **X** : 94—101.

Rao, M. S. V. R., 1962. *Soil Conservation in India.* ICAR, New Delhi.

Rao, B. K. S., Dabral, B. G. and Pande, S. K., 1972. Litter production in forest plantation of chir (*Pinus roxburghii*), teak (*Tectona grandis*) and sal (*Shorea robusta*) at New Forest, Dehradun. In P. M. Golley and F. B. Golley (eds.) *Tropical Ecology with an Emphasis on Organic Production*, pp. 235—243. Univ. Georgia, Athens, USA.

Rao, V. S., 1977. An ecological study of three fresh-water ponds of Hyderabad. India : IV the phytoplankton (Diatoms, Euglenineae and Myxophyceae). *Hydrobiologia*, **53** (1) : 13–32.

Raunkaier, C., 1905. *K. Danske Vidensk. Selsk.*, 347–438.

Raunkaier, C., 1934. *The Life Forms of Plants and Statistical Plant Geography.* Clarendon Press, Oxford.

Ray, H. and Chakravarthy, M., 1934. Lunar periodicity in the conjugation of *Conchophtherius lamellidiens*, Ghosh. *Nature*, **134** : 653–654.

Razi, B. A., 1955. *Proc. Natn. Inst. Sci. India*, **21**B : 79–89.

Razi, B. A., 1955–1956. The phytogeography of the Mysore hill tops. *J. Mysore Univ. Sect. B*, **14** (10) : 87–107 ; **15** (11) : 109–144.

Reddiah, K., 1970. The Appa island and its fringing reef in the gulf of Mannar. *J. Mar. Biol. Ass.* India, **XII** : 57–63.

Ricker, W. E., 1954. Stock and recruitment. *J. Fish Res. Bd. Canada*, **11** : 559–623.

Riemerschmid, G. and Elder, J. S., 1945. The absorptivity for solar radiation of different coloured hairy coats of cattle. *Onderstepoort J. Vet. Sci. Animal Ind.*, **20** (2) : 223–234.

Roberts, M. B. V., 1971. *Biology : A Functional Approach*. The English Language Book Society and Nelson, London.

Root, R. B., 1967. The niche exploitation pattern of the blue-gray gnat-catcher. *Ecol. Monogr.*, **37** : 317–350.

Ryder, J. P., 1970. A possible factor in the evolution of clutch size in Ross goose. Wilson Bull., **82** : 5–13.

Sagreiya, K. P., 1967. *Forest and Forestry*. National Book Trust, India.

Saharia, V. B., 1982. *Wildlife in India*. Natraj Publishers, Dehradun.

Saksena, S. B., 1955. *J. Indian Bot. Soc.*, **34** : 262–297.

Salisbury, E. J., 1946. *Symp. on the Ecology of Closely Allied Species. J. Ecol.*, **33** : 115–116.

Sanders, H. L., 1968. Marine benthic diversity : a comparative Study. *Am. Naturalist*, **102** : 243–282.

Satyanarayana, T., Johri, B. N., and Saksena, S. B., 1977. Seasonal variation in mycoflora of nesting materials of birds with special reference to thermophilic fungi. *Trans. Br. Mycol. Soc.*, **68** (2) : 307–309.

Schmidt-Nielsen, K., 1959. Physiology of the camel. *Sci. Amer.*, **201** : 140–151.

Schmidt-Nielsen, K., 1964. *Desert Animals : Physiological Problems of Heat and Water*. Oxford University Press, London.

Schneider, D., 1974. The sex-attractant receptor of moths. *Sci. Amer.*, **231** : 28–35.

Selleck, G. W., 1960. The climax concept. *Bot. Rev.*, **26** : 534–545.

Sellers, W. D., 1965. *Physical Climatology*. University of Chicago Press, Chicago.

Selye, H., 1955. Stress and disease. *Science*, **122** : 625–631.

Serventy, D. L., 1971. Biology of desert birds, in D. S. Farner and J. R. King (eds.) *Avian Biology*, pp. 287–339. Academic Press, New York.

Seth, S. K., Kaul, O. N. and Gupta, A. C., 1963. *Indian For.*, **89** : 90–98.

Seth, G. K., 1976. Know your environment. *Science Reporter*, **13** (1) : 7–11.

Sharma, B. M., 1961. Studies on the vegetation of arid-zones of India VII–Foliar analysis of certain desert plants. *Ecological Problems in the Tropics, Allahabad* : *Natn. Acad, Sci.* (India).

Sharma, P. D., 1971. *Proc. Natn. Acad. Sci. India*, **41** (B) : 139–142.

Sharma, P. D., 1972. On the fungal succession. *Bull. Meerut College Bot. Soc.*, **1** : 26–30.

Sharma, P. D., 1973. Succession of fungi on decaying *Setaria glauca* Beauv. : A qualitative analysis of the mycoflora. *Ann. Bot.*, **37** : 203–208.

Sharma, P. D., 1974. *Trans. Br. Mycol. Soc.*, **63** (2) : 397–400.

Shelford, V. E., 1913. *Animal Communities in Temperate America.* University of Chicago Press, Chicago.

Shelford, V. E., 1929. *Laboratory and Field Ecology.* Williams and Wilkins Co., Baltimore.

Shelford, V. E., 1932. Life zones, modern ecology and the failure of the temperature summing. *Wilson Bull.*, **44** : 144–157.

Sheshadri, B, 1969. *The Twilight of India's Wildlife.* Oxford University Press, London.

Shidhu, S. S., 1961. Ecology of mangroves. *Symp. Ecological Problems in the Tropics. Natn. Acad. Sci.* (India).

Silliman. R. P. and Gutsell, J. S., 1958. *U. S. Fish. Wildl. Serv., Fish. Bull*, **58** : 215–252.

Simmons, I. G., 1974. *The Ecology of Natural Resources.* Edward Arnold (Publishers) Ltd., London.

Simpson, G. G., 1964. Species density of North American recent mammals. *Syst. Zool.*, **13** : 57–73.

Singh, J., 1962. *Trop. Ecol.*, **3** : 119–132.

Singh, J. and Mukharji, S. P., 1971. *Oriental Insects*, **5** (4) : 487–494.

Singh, J. and Mukharji, S. P., 1973. *Sci. and Cult.*, **39** : 498–500.

Singh, J. and Singh, U. R., 1975. An ecological study of soil microarthropods from soil and litter of tropical deciduous forests of Varanasi (India). *Trop. Ecol.*, **16** (2) : 81–85.

Singh, J. and Pillai, K. S., 1976. *Rev. Ecol. Biol. Sol.*, **13** (2) : 331–336.

Singh, R. N., 1961. *Role of Blue-green Algae in Nitrogen Economy of Indian Agriculture. A monograph.* Indian Council of Agricultural Research, New Delhi.

Singh, K.P., 1962. *Trop. Ecol.*, **3** : 119-132.

Singh, K.P., 1968. Litter production and nutrient turnover in deciduous forests of Varanasi. *Proc. Symp. Recent Adv. Trop. Ecol.*, 655–665.

Sita, A., 1963. On helminth parasites from Madras coast. *J. Mar. Biol. Ass. India*, **5** : 137–139.

Sitaramiah, P., 1966. Studies on the ecology of a fresh-water pond community. *Hydrobiologia*, **27** : 529—547.

Smith, R.L., 1974. *Ecology and Field Biology.* Harper and Row, Publishers, New York.
Smith, R.L., 1977. *Elements of Ecology and Field Biology.* Harper and Row, Publishers, New York.
Smyth, M. and Bartholomew, G.A., 1966. The water economy of the black-throated sparrow and the rock wren. *Condor*, **68** : 447—458.
Sokal, R.R. and Rohlf, F.J., 1973. *Biometry : The Principles and Practice of Statistics in Biological Research.* W.H. Freeman and Company, San Francisco.
Southwick, C.H., 1976. *Ecology and the Quality of Our Environment.* D. Van Nostrand Company, New York.
Sreenivasan, A., 1964. A hydrological study of a tropical impoundment, Bhavanisagar Reservoir, Madras State, India, for the years 1956-1961. *Hydrobiologia*, **XXIV** : 514—539.
Sreenivasan, A., 1964. Limnological studies of tropical impoundment II : Hydrological features and planktons of Bhavanisagar Reservoir (Madras State) for 1961-1962. *Proc. Ind. Acad. Sci.*,) **59** (B) : 53—71.
Sreenivasan, A., 1968. *Arch. Hydrobiol.*, **65** : 205—222.
Sreenivasan, A., 1968. *Phykos*, **7** : 144—160.
Sreenivasan, A., 1977. *Arch. Hydrobiol.*, **80** (1) : 70—84.
Steere, J.B., 1894. *Ibis. 6th Ser.*, **6** : 411—420.
Subrahmanyan, R., 1959. Studies of the phytoplankton of the west coast of India, part 1. *Proc. Indian Acad. Sci.*, **50** (B) : 113—187.
Subramanian, N. and Radhakrishnan, R., 1980. Wind effects on structure. *Seience Reporter*, **17** (2) : 88—95.
Sukumaran, S. and Krishnaswamy, S., 1961. Reactions of *Sellana radiata* (Bom.) (Gastropod) to salinity changes. *Proc. Indian Acad. Sci.* **LIV** (B) : 122—129.
Tansey, M.R., 1973. *Mycologia*, **65** : 594—601.
Tansley, A.G., 1935. *Ecology*, **16** : 284—307.
Tansley, A.G., 1939. *The British Isles and Their Vegetation.* The Cambridge Univ. Press.
Taylor, W.P., 1934. *Ecology*, **15** : 274—379.
Taylor, W.P., 1936. *Ecology*, **17** : 333—336.
Taylor, C.R., 1969. The eland and the *Oryx. Sci. Amer.*, **220** (1) : 88—95.
Thorson, G., 1955. *J. Mar. Res.*, **14** : 387—397.
Thorson, G., 1956. *Trans. N. Y. Acad. Sci.*, **18** : 693—700.
Tilly, L.J., 1968. *Ecol. Monogr.*, **38** : 169—197.
Tramer, E.J., 1975. The regulation of plant species diversity on an early successional old-field. *Ecology.*, **56** : 905—914.
Trivedi, T.S., 1955. *Bull. Bot. Soc., Saugar Univ.*, **7** : 55—60.
Trivers, R.L., 1971. The evolution of reciprocal altruism. *Quart. Review of Biol.*, **46** : 35—37.
Troup, R.S., 1921. *Silviculture of Indian Trees.* 3 vols. Oxford Press.

Turresson, G., 1922. *Heriditas.*, **3** : 211–350.

Udvardy, M.D.F., 1959. Notes on the ecological concepts of habitat, biotope and niche. *Ecology*, **40** : 725–728.

Verma, P.S. and Agarwal, V.K., 1982. *Genetics*. S. Chand and Co. Ltd., New Delhi.

Vernadsky, W.I., 1934. *Studies in Geochemistry*. Leningrad.

Vernberg, F.J. and Vernberg, W.B., 1970. *The Animal and the Environment*. Holt, Rinehart and Winston, New York.

Volterra, V., 1931. Variation and fluctuations of the number of individuals in animal species living together. In R.N. Chapman (ed.) *Animal Ecology*, pp. **409**–448. McGraw-Hill, New York.

Vyas, A.B. and Golley, F.B., 1975. *Ecological Principles*. International Scientific Publications, Jaipur.

Wallwork, J.A., 1970. *Ecology of Soil Animals*. McGraw-Hill Book Company, New York.

Warming, E., 1909. *Oecology of Plants*. Clarendon Press, Oxford.

Wellington, W.G., 1960. Qualitative changes in natural populations during changes in abundance. *Can. J. Zool.*, **38** : 238–314.

Whitford, B.P., 1949. Distribution of woodland plants in relation to succession and clonal growth. *Ecology*, **30** : 199–208.

Whittaker, R.H., 1953. A consideration of the climax theory : the climax as a population and pattern. *Ecol. Monogr.* **23**, : 41–78.

Whittaker, R.H., 1962. Classification of natural communities. *Bot. Rev.*, **28** : 1–239.

Whittaker, R.H., 1970. *Communities and Ecosystems*. Macmillan Co., New York.

Whyte, R.O., Venkatraman, S.V. and Dabadghao, 1954. The grasslands of India. *UNFAO. 1955 Symp. "Vegetation of India"*.

Williamson, P., 1971. Feeding ecology of the red-eyed vireo (*Vireo olivaceus*) and associated foliage-gleaming birds. *Ecol. Mono.*, **41** ; 130–150.

Wilson, E.O., 1975. *Sociobiology : The New Synthesis*. Harvard University Press, Cambridge, Mass.

Wilson, E.O. and Bossert, W.H., 1971. *A Primer of Population Biology*. Sinauer Associates, Inc., Publishers, Stamford, Conn.

Winston, F.W., 1956. The acorn microsere with special reference to arthropods. *Ecology*, **37** : 120–132.

Woodbury, A.M., 1954. *Principles of General Ecology*. McGraw-Hill Book Co., New York.

Wynne-Edwards, V.C., 1962. *Animal Dispersion in Relation to Social Behaviour*. Oliver and Boyd, Edinburgh.

Yadav, A.S., 1966. The ecology of microfungi on decaying stems of *Heracleum sphondylium L. Trans. Brit. Mycol. Soc.*, **49**: 471-485.

Index

A

Abiotic components (substance) of ecosystem, 3, 104
Abiotic environmental factors, 193
Absorption, 231
Abundance, 38
Abyssal plain of ocean, 433
Abyssal zone, 414
Acclimatization, 227, 365
Acid rainfall, 499
Adaptation, 116, 172, 365
Aeolion soil, 128
Aerosols, 498
Aestivation, 215
Age **pyramids**, 320
Aging technique, 44
Aggregation, 330
Agriculture, 10, 526
Agrophyte, 288
Air, 111, 114
Air pollution, 495, 498
Alarm pheromone, 339
Algae as human food, 529
Allelochemic individual, 346
Allelochemicals, 346
Allelomones, 346
Allelopathy, 347
Allen's rule, 223
Allogenic community, 295
Alluvial soil, 128, 304
Alpine biome, 453
Altitudes, 244
Altruism, 333
Altruistic behaviour, 333
Amensalism, 7, 343, 346
Ammonia, 397
Amphiphyte, 289
Amplitude, ecological, 365
Animal community, 256
Animal ecology, 9, 26
Animal population, 40
Annual rhythms, 238
Antibiosis (or antibiotic), 347
Aphotic zone, 229, 433
Applied ecology, 10, 484
Aquaculture, 10, 530, 531
Aquatic ecosystems, 407, 511
fauna, 258
succession, 299
Aquifer, 390
Arboriculture, 532
Asbestosis, 499
Assimilation, 385
Association, plant, 289
Atmophytes, 288
Atmosphere, 6, 58, 94, 111, 390
Atmosphere-reservoir cycles, 387
Atolls, 447
Attribute, 66
Australian region, 478
Autecology, 9, 24, 248
Autogenic community, 295
Automobiles, as source of air pollution, 496
Autotrophic bacteria, 149
Autotrophic communities 256
Autotrophic **succession,** 295
Autotrophs, 374
Auxins, 230
Azimuth, 236
Azonal soil, 129

B

Bacteria, 377
Bar diagram, 71
Barometric pressure, 239
Barriers, 480
Barrier beaches, 427
Basal area, 38
Basal cover, 269
Bathyal zone, 433
Bathymetric distribution, 465
Bathypelagic zone, 433
Beach, sea, 440, 445
Behaviour, 3, 331
Benthic animals (benthos), 30, 106, 417
Benthic zone, 438; biotic communities of, 438
Benthos of lake bottom, 420
Bergman's rule, 223
Bicarbonates, 410
Biocoenosis, 17
Biodegradeable, materials, 380 pollutants, 497
Bioecology, 9
Biogeochemical cycles, 8, 104, 386
Biogeography, 465
Biogeographical realm, 452
Biogeographical regions, 97
Biokinetic zone, 197
Biological clock, 8, 236, 249
Biological control, 12, 515
Biological **indicators,** 368
Biological magnification, 513

Biological oxygen demand (BOD), 505, 508
Biological weathering, 128
Biomass, 6, 307
Biomass density, 313
Biomes, 93, 290, 452, 467
Binomial expression, 81
Bioseres (geoseres), 294
Biosphere, 6, 94, 96, 386
Biostatistics, 61
Biotic components of ecosystem, 7, 104
Biotic communities, 7, 94, 255
Biotic factors, 3, 7, 193
Biotic provinces, 478
Biotic potential, 322
Biotopes, 282
Birds, 117
Birth rate, 7
Black lung disease, coal mining and, 499
Blue green algae, 396
Bogs, 407
Bombykol 338
Boreal forest, 455
Brackish habitat, 6
Brackish lagoons, 427
Brashing, 532
Broad leaved evergreen subtropical forest, 459
Brood parasitism, 352
Brooks, 407
Brown water lakes, 416
Buoyancy, 116, 159, 224
Buoyancy hypothesis, 224

C

Calcareous algae, 447
Calcifiction, 147
Calciphlic species, 241
Calciphytes, 142
Calcium, 142
Calcium carbonate, 128, 411
Calefaction, 510
Calorimeter, bomb, 50
Cannibalism, 341
Canopy, 99, 269
Capillary water, 139
Character displacement, 359
Carbon, 48, 392 ; cycle, 392
Carbon dioxide, 164, 396, 410
Carbon monoxide, 500
Carbonation, 128
Carbonic acid, 128, 410
Carnivores, 8, 105
Carrying capacity, 323
Catchment area, 390
Caves, 241
Celestial orientation, 236
Census, 46
Chaparrel, 457
Chasmophytes, 288
Chemical ecology, 10
Chemical factor, of ecosystem, 6
Chemical weathering, 127
Chernozems, 147
Chersophyte, 169
Chlorophyll, 49, 229
Chresard, 139
Circadian rhythms, 236
Circannual rhythms, 238
Clay, 147
Clear water lakes, 416
Climatic factors, 6
Climax community, 8, 294
Cloud, 112, 120, 247, 388
Coal, 519, 526
Coenocline, 284
Cold desert, 461
Colonization, 330
Colluvial soil, 128
Combined water, 138
Commensalism, 7, 342, 344
Communal territories, 336
Communication, 337
Community (ies), 7, 17, 255
  abundance species, 264
  animal, 257
  biotic, 255
  classification of, 287
  climax, 8, 292, 296
  diversity in, 264
  dominance, 268
  ecology, 10, 248, 254
  evolution, 310
  mass, 269
  plant, 34, 256
  stable, 292
  stability of, 281
  structure of, 256
  stratification of, 260
  trophic structure of, 262
Community dominants, 268
Community retrogression, 309
Compensation level, 414
Compensatory predation, 355
Competition, 7, 95, 296, 343;
  contest, 356
  effects of, on population, 360
  interspecific, 356
  intraspecific, 356
Competitive exclusion 358
Complex, gradient, 284
Condensation nuclei, 498
Conduction, 206
Coniferous forest, 455
Conservation ecology, 10
Conservation, 12, 519, 537
Consumers, 8, 105;
  primary, 106

secondary, 106
Contagious distribution, 262
Continental drift theory, 480
Continental shelf (ves), 433
biotic communities of, 440
Continental slope, 433
Continuous distribution, 466
Convection, 205
Cooperation, 95
Coral reefs, 98, 447; biotic communities of, 448
Core, earth, 122
Cosmic rays, 114
Cover, 38
Crowding, 341
Cropland ecology, 10
Crude density, 313
Crumbs (concretions), 132
Crust, earth, 122
Cryptozoans, 262
Cultivation, sendentary, 527; shifting 527
Cultural eutrophication, 505
Culture, peddy, 528
Current, 57, 116, 420
Cursorial animal, 154
Cushion plants, 121
Cybernetic system, 101
Cycles, biogeochemical, 386
population, 327
Cyclomorphosis, 224
Cyclone, 120

D

Daily rhythms, 236
Dams, 532
Dance, round, 341;
tailwagging, 341
Data, 61, 63, 72
DDT, 513
Deamination, 397
Death rate, 7
Deciduous forest, 456, 471
Decomposers, 105
Decomposition, 129
Decreaser plant, 533
Deforestation, 532
Deltas, 427
Demography, 485
Dendrochronology, 40
Denitrification, 398
Density, 38
Density, population 313
Density dependent factors, 326
Density independent factors, 326
Depletion curve, 523
Depopulation, 325
Depressants, 347
Desert, 23, 98, 148, 461;
trace element, 12
Desert ecology, 10
Dessication, 121
Detergents, 507
Determination, 64
Detritus, 379
Detritus-feeding benthos, 424
Detritus food chain, 379
Detrivores, 379
Deviation, standard, 75
Diapause, 214, 238
Diatoms, 369
Diffusion, 116
Dimictic lake, 415
Dinoflagellate blooms, 507
Directional selection, 363
Discontinuous distribution, 466
Disease, 350
Dispersal, 479
Disphotic zone, 229
Disruptive selection, 364
Division of labour, 330
Diurnal migration, 238
Diurnal variation, 408
Diversity, species, 307
Dominance, community, 268
Dominance hierarchy, 334
Dormancy, 227
Drift, 126; glacial, 128
Drought, 182, 190
Dynamic equilibrium (homeostasis), 292
Duff, 129
Dunes, 304
Dust pollution, 501

E

Earth, 6, 96, 122
Ecad, 366
Echard, 139
Ecogeography, 11
Ecocline, 284, 366
Ecological balance, 518
Ecological density, 313
Ecological efficiency, 384
Ecological energetics, 10
Ecological equivalence, 279
Ecological genetics, 10, 363
Ecological indicators, 368
Ecological niche, 7, 99, 273
Ecological pyramids, 382
Ecological succession, 292
Ecology, 3, 4
Ecophenes, 366
Ecophysiology, 10
Ecospecies, 367
Ecosphere, 6, 96

Ecosystem dynamics, 10
Ecosystems, 6, 94, 100
  artificial, 108
  components of, 104, 105
  definition of, 100
  development of, 292
  dynamics, 10, 371
  energy in, 374
  examples of, 105
  incomplete, 108
  natural, 108
  types of, 107
Ecotone, 271
Ecotypes 366
Ectocrines, 18
Edaphic factor, 122
Edaphology, 123
Eddies, 120
Edge effect, 272
Effective climate, 241
Egg-parasitism, 352
Electro-magnetic spectrum, 200
Elevation gradient, 284
Emigration, 7, 325
Endemics, 466
Endothermic animals, 213
Endothermic reaction, 373
Energy, 51, 373
  fixed, 374
  flow, 377
  kinetic, 373
  potential, 373
  pyramids, 384
  radiant, 374
Entropy, 373
Environment, 3, 5, 110
Environmental biology, 4
Environmental factors, 199
Environmental induction, 16
Environmental hormones, 18
Environmental resistance, 323
Environmental stability, 281
Eolian (loess), 304
Epidemiology, 10
Epilimnion, 208, 412
Epiplankton, 436
Epiphytes, 344
Epiphyton, 39
Epizoite, 344
Eremophyte, 169
Erosion, 524
Essential elements, 194
Estuaries, 427
  biotic communities of, 430
  physico-chemical aspects of, 427
  pollution of, 511
Estuarine ecology, 10
Ethiopian region, 477
Ethology, 11
Eubiosphere, 6
Euhaline, 430,
Euphotic zone, 229
Euryhaline, 161
Eurythermal organisms, 210
Eusocial insects, 332
Eutrophication, 507
Eutrophic lake, 416
Evaporation, 116, 206
Evapotranspiration, 166, 388
Evergreen forests, 457, 470
Evolutionary ecology, 10
Exclusion, competitive, 358
Exosphere, 114
Exploitation, 359
Exothermic reaction, 373
Extrinsic factors, in population, 327

F

Fator, 5
Factor compensation, 365
Factor interaction, 195
Fecundity, 222, 315 ; rate, 314
Feedback, loop, 101;
  negative, 101
  positive, 101
Fertility, 314
Fertilizer, 529
Fick's Law, 116
Field capacity, 139
Fire, 241, 310
Fisheries, 11, 530
Fish farming, 530
Fish mortality, 508
Fission, nuclear, 515
Fitness, inclusive, 333
Flight adaptation, 117
Flight distance, 337
Floods, 532
Flowing waters, 421
Fluctuations, population, 327
Fluorosis, 502
Food, new sources of, 529
Food chains, 8, 34, 107, 377
Food web (s), 8, 107, 382
Forest ecology, 10
Forest, 24, 98, 131, 304, 454;
  conservation of, 532
  deciduous, 130, 456
  vegetation, 470
Forestry, 10, 248, 532
Fossil fuels, 525
Fossorial animal, 154
Founder effect, 364
Frequency, 37, 67
  gene, 362
Fresh water, 407
  ecology, 10
  ecosystems, 407
  habitat, 6
Fringing reefs, 447

Frost, 226
Frost action, 126
Fungi, 150
as human food, 529

G

Gallery forests, 460
Game management, 11
Gaseous cycles, 391
Gause's Principle, 358
Genecology, 366
Genetic drift, 364
Genetic races, 365
Geographical distribution, 465
Geographical ecology, 11
Geomagnetism, 239
Geysers, 6
Gibberellin, 150
Glacial drift, 126
Glaciers, 126
Gliding flight, 117
Gloger rule, 223
Gradient, 284
Gradient analysis, 284
Gradient complex, 284
Grasslands, 25, 98, 147, 473;
biomes, 460
ecology, 10, 24
management, 533
Graviportal, animals, 154
Gravity, 116, 126
Gravitation weathering, 126
Gravitational water, 139
Grazing stress, 533
Grazing food chain, 378, 403
Greenhouse effect, 503
Green Revolution, 528
Gregariousness, 330, 331
Ground water, 139, 390
Group survival value, 331
Growth forms, 257, 285
Growth, population, 323

H

Habitat, 6, 99, 148, 406
Habitat ecology, 9
Hadal zone, 433
Hail, 168
Halocline, 435
Halomorphic soil, 129
Halophyte, 169, 288
Halosere, 295
Heat, 156, 199
Heat budget, 207
Heat capacity of **water**, 156
Heat islands, 246, 502
Heat pollution, 510
Heat resistance, 226
Heavy metal pollution, 508
Heliophytes, 230
Helophyte ,168, 189
Helium, 114
Herbicides, 513
Herbivores, 8, 355
Heterosis, 364
Heterotrophic bacteria, 149
Heterotrophic communities, 256
Heterotrophic organisms, 104
Heterotrophic succession, 295, 305
Hibernation, 18, 215
Histogram, 70
Holard, 139
Holoplanktonic form, 437
Homeostasis, 101
Homeothermic animals, 213
Home range, 45, 336
Horizontal stratification, 262
Host-parasite relationship, 347-351
Host-specificity, 349
Human ecology, 10
Humic acids, 380
Humidity, 58, 166
Humification, 129
Humus, 129
Hurricanes, 120
Hydration, 128
Hydrocoles, 169, 177
Hydroelectric power, 526
Hydrogen, 114
Hydrogen-ion concentration (pH), 142, 165
Hydrologic cycles, 157 ,165, 388
Hydrolysis, 127
Hydromorphic soil, 129
Hydrophyte, 168
Hydrosere, 295, 299
Hydrosphere, 6, 94, 155
Hygroscopic coefficient, 140
Hygroscopic water, 139
Hypolimnion, 208, 412
Hyperparasitoids, 352
Hyperpyrexia, 211
Hyperthermia, 191

I

Igneous rocks, 123
Immigration, 325
Immigration rate, 7
Impoundments, 32, 416
Indices, 47
Induction, environmental, 16
Industrial pollution, 508
Information content, 298
Insects, 117, 513
Insulation, 216
Interference, 359

Interspecific interaction, **329, 342**
Intertidal zone, 433
biotic communities of, **440**
Intraspecific interaction, 329
Intrazonal soil, 129
Intrinsic factors, in **populations, 327**
Invasion, 296
Inversion, temperature, 498
Ionosphere, 114

J

Jet stream, 120
Jordon's rule, 223
J-shaped population growth, 324

K

Kairomone, 346
Kin selection, 333
Klinokinesis, 235
Krill, as human food, 12, 530

L

Lacustrine soil, **129**
Lakes, 32, 409, 412;
biotic communities of, 416
kinds of, 415
origin of, 416
stratification of, 412
succession of, 299
zonation of, 412
Laminar flow of water currents, 421
Land pollution, 512
Landscape, 534
Land use, 536
Latent heat of water, 156
Laterite, 147
Latitude, **433**
Latosols, 147
Law of conservation of matter, 386
Law of minimum, 17, 195
Law of limiting **factors,** 195
Leaching, 147
Lead pollution, 501
Lens weight technique, 45
Lentic ecosystems, 412
Lentic water, 407
Lethal limit, 210
Lichens, 128, 303, 397
Life expectancy, 317
Life-form, 37, 257
Life-tables, 317
Light, 59, 159, 227
Light compass reaction, 236
Limiting factors, 7, 17, 195
Limnetic forms, 416
Limnetic zone 412
biota of, 419
Limnology, 10, 32, **407,**
Limnophytes, **289,**
Lithophyte, 169
Lithosphere, 6, 94, 122
Lithosere, 295, 303
Litter, 41, 129
Littoral zone, 99, 412, 433, 440
biota of, 417
Lotic ecosystem, 420
Lotic water, 407
Lumbering, succession, 310
Lunar, periodicity, 239

M

Macroclimate, 241
Macroconsumers (phagotrops), 105
Macroenvironment, 6
Macrofauna, 149
Macrohabitat, 193
Macronutrients, 195
Magnetic storm, 114
Management of natural resources, 532
Mantle, earth, 122
Mariculture, 531
Marine, communities, 436
ecology, 10
habitat, 6
mineral resources, 522
soil, 126, 129
stratification, 433
Marine ecosystems, 432
Marl, 411
Marshes, 412
Marsh plants, 175
Meadows, 453
Mercury pollution, 508
and minamata disease, **508**
Meroplanktonic forms, 437
Mesocoles, 169, 181
Mesofauna, 149
Mesohaline animals, 430
Mesopause, 114
Mesopelagic zone, 433
Mesosphere, 114
Mesophilous plants, 169
Mesophytes, 169, 180
Metalimnion, 414
Metamorphic rocks, 123
Methane, 410
Microclimate, 241, 273, 533
Microconsumers (saprotrophs), 105
Microenvironment, 6, 273
Microfauna, 149
Microflora, 149
Microhabitats, 273, 305
Micronutrients, 195

Microsuccession, 305
Micro-organisms, 255, 507
Migration, 326, 363
Minamata disease, 510
Mineral cycle, 399
Mineral nutrients, 8
Mineralization, 130
Minerals, 123
Mining, ecological aspects of, 524
Moder, 131
Moisture, 120
Monoculture, 532
Monomictic lake, 416
Monsoon forest, 459
Moon and tides, 241
Mor, humus, 131
Moraines, 128
Mortality, 48, 316
Mosses, 128, 303
Muck, soil, 131
Muddy shores, 441
Mull, humus, 131
Mutualism, 7, 343, 345
Mycorrhizae, 345

N

Natality, 314
National park, 534
Natural history, 14
Natural selection, 363
Neap tide, 241
Nearctic region, 476
Negative feedback, 101
Necktoplankton 417
Nekton, 417
Nematodes, 149
Neotropical region, 477
Neritic sea, 440
Neritic zone, 433
Net ecosystem production, 375
Neuston, 417
Neutralism, 7, 342, 344
Niches, 273
Nitrates, 396
Nitric oxide, 114
Nitrification, 397
Nitrogen, 164, 396
  cycle, 17, 396
  fixation of, 397
  oxides, 3, 500
Nitrogen fixing bacteria, 149, 396
Noise pollution, 516
Nuclear energy, 526
Nucleation particles, 387
Nudation, 295
Nutrients, 123
Nutrient cycle, 31, 131

O

Oceanic region, biotic communities of, 436
Oceanography 10, 26, 27
Oceans, 26, 433
  currents, 435
  pollution of, 511
  salinity of, 435
  stratification of, 433
  zonation of, 433
Oecologie, 3
Oil pollution, 511
Oligohaline animals, 430
Oligotrophic lake, 416
Ombrophytes, 288
Omnivorous, 8
Organism, 3, 5, 92
Oriental region, 477
Osmoregulation, 412
Overgrazing, 533
Overpopulation, 536
Oxidation, 127
Oxygen, 48, 55, 113, 162
Oxygen cycle, 391
Oxylophytes, 142, 169, 288
Ozone, 113, 503
Ozone umbrella, 113, 503
Ozonesphere, 113

P

Pabble, sea shore, 441
Palaearctic region, 476
Palaeoecology, 11
Palaeontology, 11
Palynology, 40
Pantograph, 36
Parabiosphere, 6
Parameter, 77
Parasite chains, 379
Parasite, 7, 347
Parasitism, 26, 343, 347
  social, 352
Parasitoids, 352
Peat, 131
Peck order, 334
Pedogenesis, 123
Pedology, 11, 123
Pedonic forms, 416
Peds, 136
Pelagic zone, 433
  biotic communities of, 436
Periphyton, 38
Permafrost, 145, 453
Pest control, 12
  integrated, 515
Pesticides, 513
Phaenology, 37

Phagotrophs, 105
Pheromones, 337
Phosphates, 402
Phosphorus cycle, 402
Photic zone, 160, 433
Photochemical dissociation or reactions, 113
Photochemical smog, 500
Photodissociation, 392
Photokinesis, 235
Photoperiod, 232, 236
Photoperiodism, 232, 236
Photophobic plants, 230
Photophilous plants, 230
Photosynthesis, 48, 97, 163, 229
Phototaxis, 235
Phototropism, 230
Phylloclade, 183
Physical factors of ecosystem, 6
Physiognomy, 256
Physiological deserts, 167
Physiological ecology, 10
Physiological stress, 197, 351
Physiologically dry soil, 148
Phytoplankton, 30, 38, 48, 106, 299, 418, 436
Phytogeography, 465
Phytosociology, 256
Pioneer communities, 294
Plankton, 31, 38, 165, 417
Plankton bloom, 507
Plant community, 256
Plant ecology, 9, 24
Podzolic soil (podzol), 145
Poikilothermic animals (ectotherms), 212
Polar lakes, 409
Pollution, 12
Polyhaline, 430
Polymictic lake, 416
Polymorphism, 365
Polyps, 330
Ponds, 302, 412
  biotic communities of, 262
  stratification in, 262
  succession in, 299
Population, 7, 19, 34, 64, 255, 313
  age distribution of, 319
  birth rate, 314
  cycles, 327
  definition of, 313
  density, 313
  dispersion, 324
  ecology, 10, 254, 313
  growth form of, 323
  growth rate of, 324
  mendelian, 94
  mortality rate, 316
  natality, 314
  regulation of, 7, 326
  relation among, 329
Population explosion, 324
Positive feed-back, 101
Prairie, 302
Precipitation, 168, 460
Predation, 343, 352;
  compensatory, 355
  components of, 353
  regulating effect of, 354
Predator, 7, 353
Predator chain, 379
Pressure, 161, 239
  partial 116
Prey, 7
Prey-predator relationship, 354
Primary consumers, 379
Primary production, 31, 375
Primary producers, 96, 379
Primary succession, 8, 294
Probability, 77
Producers, 8, 104
Production ecology, 10
Production, net 385
  primary, 25, 48
  secondary, 376
Profundal zone, 414
  biota of, 419
Protocooperation, 7, 343, 345
Protozoa, 78, 148, 161
Psammophyte, 196, 288
Psammosere, 295, 304
psychrophytes, 169, 288
Pyramids, age, 320
  ecological, 383, 384
Pycnocline, 435

Q

Quadrate method, 35

R

Radiant energy, 200
Radiation ecology, 11
Radioactive pollution, 515
Radioactive tracers, 43, 51
Radioisotopes, 515
Rainfall, 59, 168
Rain forest, temperate, 457
  tropical, 457
Random distribution, 262
Range expansion, 479
Range of optimum, 197
Range management, 533
Red data book, 534
Reducers, 8, 105
Reduction, 127
Regolith, 125, 134
Regulation of brood size, 316
Remote sensing, 59

Rensch's rule, 223
Reproductive capacity, 316
Reproductive potential, 7
Reproductive rate, 315
Residual soil, 128
Resistance environmental, 7, 322
Resources, 519
Respiration, 48, 163, 377
Rheokinesis, 235
Rheotaxis, 424
Rhythms, 236
daily, 8, 237
seasonal, 8, 238
Riffle zone, 422
Rivers, 420
Rocks, 123, 422
Rocky, sea shore, 441
biotic communities of, 441

S

S-shaped population growth, 324
Salinity, 115, 161, 428
Salt, 161 ; gland, 192
Salt-water ecosystem, 407
Sample, 63
Sampling units, 35
Sand, 304
Sand dunes, 129, 304
Sandy seashore, 441 ; biotic communities of, 445
Sanitary landfill, 512
Saprobes, 414
Saprophytic fungi, 150
Saprotrophs, 105
Saturation, 139, 162
Savanna, 459
Scatophase, 236
Sciophytes, 230
Sea shore, 440
Secondary consumers, 379
Secondary production, 376
Secondary succession, 8, 294
Sediment, 395
Sediment-reservoir cycle, 387
Sedimentary cycles, 399
Sedimentary rocks, 123
Seral stage (sere), 294
Sewage, 506, 531 ; pollution, 506
Sex attractant, 338
Sex-ratio, 322
Shelford's law of tolerance, 196
Sleet, 168
Signaling, 337
Silt, 135 ; pollution, 511
Silviculture, 532
Smelter, 496
Smog, 500
Snow, 168
Social hierarchy, 335
organization, 331
Soil, 25, 52, 122, 451, 533
conservation, 25, 537
definition of, 123
embryonic, 128
fauna, 149
flora, 149
profile, 132
texture, 135
Solar radiation, 201, 242, 374
Solar energy, 12, 203, 531
Solvent, 158
Space ecology, 11
Specialization, 332
Species, 7 ; exclusive, 271
ubiquitous, 271
Species diversity, 264
Species dominance, 268
Spring tides, 241
Springs, 390
SST, 503
Stabilising, selection, 364
Stability hypothesis, 224
Stenohaline, 161
Stenothermal **organisms,** 210
Steppes, 460
Storms, 496
Stratification, 7, 36, 258
Stratified communities, 260, 264
Stratopause, 113
Stratosphere, 113
Streams, 420
Subclimax, 297
Sublittoral zone, 433
Sucession, 8, 293
primary, 8, 294
secondary, 8, 294
Succulent plants, 182, 280
Sulphur cycle, 400
Sulphur dioxide, 401, 499
Sulphates, 400
Sun, 113, 227, 239 ; light, 200, 227, 239
Sunspots, 239
Surface run off, 390
Surface tension, 158
Survivorship curve, 317
Swamps, 189
Symbiosis, 345
Synecology, 9, 248
Systems ecology, 11

T

Tagging, 43
Taiga, 455
Taxonomic ecology, 10
Telmatophytes, 289
Telotaxis, 236
Temperate deciduous forest, 456
Temperature, 110, 114, 126, 199—227

fluctuation, 207
inversion, 498
Temporal stratification, 263
Terrestrial, ecosystem, 450
habitat, 406
mineral conservation, 524
mineral resources, 520
succession, 303
Terriphytes, 289
Territory, 45, 336
Territoriality, 335
Tertiary consumers, 379
Thermal adaptations of animals and plants, 211
Thermal migrations, 227
Thermal pollution, 510
Thermal stratification, 115, 208
Thermocline, 208, 414
Thermodynamics, law of, 373
Thermographs, 60
Thermosphere, 114
Thigmotaxis, 422
Thining, 532
Threshold, 196
Threshold of security, 355
Tidal energy, 526
Tidal flats, 427
Tides, 241, 428, 435
Till, 128
Timber line, 245
Topography, 244
Tolerance limit of, 197
Tornado, 120
Torrent water, 126
Toxic chemicals, 505
Trace elements, 195
Transect method, 36
Transpiration, 231
Trophallaxis, 339
Trophic ecology, 107,
Trophic levels, 107, 377
Trophogenic zone, 414
Tropholytic zone, 414
Tropical forest, 455
Tropical grassland, 459
Tropical lakes, 409
Tropical rain forests, 457
Tropical savanna, 459
Tropopause, 113
Troposphere, 111, 392
Tundra, 145 ; biome 452; soil, 145
Turbidity, 56
Turbulent flow, 421
Turbulent water current, 421
Typhoon, 120

U

Ultraviolet radiation, 113, 503
energy, 113
rays, 200, 503
Universal solvent, water as, 158
Uranium, 515
Urbanization, 518

V

Vant Hoff's Law, 221
Variable, 63, 65, 86
Variate, 66
Variation, 362
Vegetation, 455, 470
Veldt, 460
Vertical circulation, 207
Vertical stratification, 260
Vietnam, 515
Virus, 350
Valley, as microclimate, 245
Viscosity, 157
Vitality, 37
Volant adaptation, 117
Volcanos, 299, 496

W

Wallace line, 477
Waste disposal, 512, 524
Waste management parks,
Wastes, 511, 531
industrial, 508
recycle, 524
solid, 512
Water, 126, 155
chemical properties of, 162
conservation, 535
current, 435
cycle, 387
distribution of, 387
physical properties of, 156
pollution, 163, 505
resources, 519
vapours, 139
Wave action, 126
Weathering of soil-forming rocks, 125
Weber's line, 477
Weeding, 532
Wetlands, 412
Wild-life, 2, 539
Wild-life management, 534, 538
Wilting point, 140
Wind, 59, 119, 168
ecological effect, 120
Wind, as energy source, 526
Woodlands, 457

X

X-rays, 200

Xerocoles, 169
Xerophilous plants, 169
Xerophytes, 169, 182
Xerosere, 295

Z

Zonal soil, 129, 142
Zonation of marine enviroment, 433
Zones, 99
Zone of capacity adaptation, 197
Zone of compatibility, 197
Zone of resistance, 197
Zone of tolerance, 197
Zone of physiological stress, 197
Zone of upwelling, 436
Zoogeograpy, 11, 465
Zooids, 330
Zooplankton, 31, 46, 106, 425